T5-AGA-024

AFRICAN BIODIVERSITY

Molecules, Organisms, Ecosystems

Bernhard A. Huber
Bradley J. Sinclair
Karl-Heinz Lampe
Alexander Koenig Zoological Research Museum,
Bonn, Germany

ISBN-10 0-387-24315-1 (HB)
ISBN-10 0-387-24320-8 (e-book)
ISBN-13 978-0-387-24315-3 (HB)
ISBN-13 978-0-387-24320-7 (e-book)

Printed in the Netherlands.

9 8 7 6 5 4 3 2 1

springeronline.com

Dedication

This book is dedicated to the memory of Clas M. Naumann

Contents

Contributing Authors *

Alberto Ballerio
Viale Venezia 45, 25123 Brescia, Italy

Wolfgang Böhme
Zoologisches Forschungsinstitut und Museum Alexander Koenig, Adenauerallee 160, 53113 Bonn, Germany

Janice L. Bossart
Rose-Hulman Institute of Technology, ABBE, Terre Haute, IN 47803, USA

Nina Sh. Bulatova
Severtsov Institute of Ecology and Evolution, Russian Academy of Sciences, Leninsky prospect 33, Moscow, 119071 Russia

Adrian J.F.K. Craig
Department of Zoology & Entomology, Rhodes University, Grahamstown, 6140, South Africa

Elmar Csaplovics
Department of Geosciences, University of Dresden, Helmholtzstrasse 10 13, 01062 Dresden, Germany

Christiane Denys
FRE2695: OSEB – Département Systématique et Evolution – MNHN - Mammifères & Oiseaux, Case courrier 051, 55 rue Buffon, 75231 Paris cedex 05, France

* Only first author listed in cases of more than three authors

Jon Fjeldså
Zoological Museum, University of Copenhagen, Universitetsparken 15, DK-2100 Copenhagen, Denmark

Götz Froeschke
Dept. Animal Ecology and Conservation, University of Hamburg, Martin-Luther-King-Platz 3, 20146 Hamburg, Germany

Philippe Gaubert
Unité Origine, Structure et Evolution de la Biodiversité (CNRS FRE 2695), Département Systématique et Evolution, Muséum National d'Histoire Naturelle, CP 51, 57 rue Cuvier, 75231 Paris Cedex 05, France

Birgit Gerkmann
Zoologisches Forschungsinstitut und Museum Alexander Koenig, Adenauerallee 160, 53113 Bonn, Germany

Justin Gerlach
University Museum of Zoology Cambridge, Downing Street, Cambridge CB2 3EJ, U.K.

Peter Giere
Institut für Systematische Zoologie, Museum für Naturkunde, Invalidenstraße 43, 10099 Berlin, Germany

Fabian Haas
Staatliches Museum für Naturkunde Stuttgart, Rosenstein 1, 70191 Stuttgart, Germany

Rainer Harf
Dept. Animal Ecology and Conservation, University of Hamburg, Martin-Luther-King-Platz 3, 20146 Hamburg, Germany

Axel Hochkirch
Universität Osnabrück, FB Biologie/Chemie, Fachgebiet Ökologie, Barbarastraße 11, 49076 Osnabrück, Germany

Joachim Holstein
Staatliches Museum für Naturkunde Stuttgart, Rosenstein 1, 70191 Stuttgart, Germany

Bernhard A. Huber
Zoologisches Forschungsinstitut und Museum Alexander Koenig, Adenauerallee 160, 53113 Bonn, Germany

Rainer Hutterer
Zoologisches Forschungsinstitut und Museum Alexander Koenig, Adenauerallee 160, 53113 Bonn, Germany

Frank Koch
Institut für Systematische Zoologie, Museum für Naturkunde, Humboldt-Universität zu Berlin, Invalidenstrasse 43, 10115 Berlin, Germany

Frank-Thorsten Krell
Soil Biodiversity Programme, Department of Entomology, The Natural History Museum, Cromwell Road, London SW7 5BD, UK

Michael Kuhlmann
Westfälische Wilhelms-University Münster, Institute of Landscape Ecology, Robert-Koch-Str. 26, 48149 Münster, Germany

Karl-Heinz Lampe
Zoologisches Forschungsinstitut und Museum Alexander Koenig, Adenauerallee 160, 53113 Bonn, Germany

Simone Lange
Dept. General Zoology, FB 9, University of Duisburg-Essen, 45117 Essen, Germany

Leonid A. Lavrenchenko
Severtsov Institute of Ecology and Evolution RAS, Leninsky pr., 33, Moscow, 119071 Russia

K. Eduard Linsenmair
Lehrstuhl für Tierökologie und Tropenbiologie, Theodor-Boveri-Institut, Biozentrum, Am Hubland, Universität Würzburg, 97074 Würzburg, Germany

Michel Louette
Royal Museum for Central Africa, B-3080 Tervuren, Belgium

Carolin Mayer
Biocentre Klein Flottbek and Botanical Garden, University of Hamburg, Ohnhorststr. 18, 22609 Hamburg, Germany

Michele Menegon
Museo Tridentino di Scienze Naturali, via Calepina 14,C P.393, I-38100 Trento, Italy

Wolfram Mey
Museum für Naturkunde, Humboldt Universität, Invalidenstr. 43, 10115 Berlin, Germany

Timo Moritz
Lehrstuhl für Tierökologie und Tropenbiologie, Theodor-Boveri-Institut, Biozentrum, Am Hubland, Universität Würzburg, 97074 Würzburg, Germany

Winfred Musila
University of Hohenheim, Institute of Botany (210), 70593 Stuttgart, Germany; and National Museums of Kenya, Nairobi, Kenya

Zoltán Tamás Nagy
Institute of Pharmacy and Molecular Biotechnology, University of Heidelberg, INF 364, 69120 Heidelberg, Germany

Manuel Nieto
Museo Nacional de Ciencias Naturales (CSIC), José Gutiérrez Abascal, 2. Madrid 28006, Spain

Lorenzo Prendini
Division of Invertebrate Zoology, American Museum of Natural History, Central Park West at 79th Street, New York, NY 10024-5192, USA

Marta Puente
Departamento de Bioloxía Animal, Laboratorio de Anatomía Animal, Universidade de Vigo, Spain

Klaus Riede
Zoologisches Forschungsinstitut und Museum Alexander Koenig, Adenauerallee 160, 53113 Bonn, Germany

Beate Röll
Lehrstuhl für Allgemeine Zoologie und Neurobiologie, Ruhr-Universität Bochum, 44780 Bochum, Germany

Francesco Rovero
Vertebrate Zoology Section, Museo Tridentino di Scienze Naturali, Via Calepina 14, 38100, Trento, Italy

Andrew R. Marshall
Centre for Ecology Law and Policy, Environment Department, University of York, Heslington, York YO10 5DD, United Kingdom

Sebastiano Salvidio
DIP.TE.RIS. - Dipartimento per lo Studio del Territorio e delle sue Risorse, Corso Europa, 26, 16132 Genova, Italy

Anja C. Schunke
Zoologisches Forschungsinstitut und Museum Alexander Koenig, Adenauerallee 160, 53113 Bonn, Germany

Ralf Seiler
Department of Geosciences, University of Dresden, Helmholtzstrasse 10 13, 01062 Dresden, Germany

Simone Sommer
Dept. Animal Ecology and Conservation, University of Hamburg, Martin-Luther-King-Platz 3, 20146 Hamburg, Germany

Dirk Striebing
Museum für Naturkunde der Humboldt-Universität zu Berlin, Institut für systematische Zoologie (ZMB), Invalidenstraße 43, 10115 Berlin, Germany

Peter J. Taylor
eThekwini Heritage Department, Natural Science Museum, P.O. Box 4085, Durban 4000, Republic of South Africa

Meike Thomas
Zoologisches Forschungsinstitut und Museum Alexander Koenig, Adenauerallee 160, 53113 Bonn, Germany

Herbert Tushabe
Makerere University Institute of Environment and Natural Resources, P.O. Box 7298, Kampala, Uganda

Miguel Vences
Institute for Biodiversity and Ecosystem Dynamics, Zoological Museum, University of Amsterdam, Mauritskade 61, 1092 AD Amsterdam, The Netherlands

Erik Verheyen
Royal Belgian Institute of Natural Sciences, Vautierstraat 29, 1000 Brussels, Belgium

Geraldine Veron
Unité Origine, Structure et Evolution de la Biodiversité (CNRS FRE 2695), Département Systématique et Evolution, Muséum National d'Histoire Naturelle, CP 51, 57 rue Cuvier, 75231 Paris Cedex 05, France

David R. Vieites
Museum of Vertebrate Zoology and Department of Integrative Biology, 3101 Valley Life Sciences Building, University of California, Berkeley, CA 94720-3160, USA

Elisabeth Vollmer
Department of Geosciences, University of Dresden, Helmholtzstrasse 10 13, 01062 Dresden, Germany

Thomas Wagner
Universität Koblenz-Landau, Institut für Integrierte Naturwissenschaften – Biologie, Universitätsstr. 1, 56070 Koblenz, Germany

James E. M. Watson
Biodiversity Research Group, School of Geography and the Environment, University of Oxford, United Kingdom, OX1 3BW

Alexander Wezel
International Nature Conservation, University of Greifswald; corresponding address: Wilhams 25, 87547 Missen, Germany

Ulrich Zeller
Institut für Systematische Zoologie, Museum für Naturkunde, Invalidenstraße 43, 10099 Berlin, Germany

Karin Zippel
Humboldt-Universität zu Berlin, Fachgebiet Obstbau, Albrecht-Thaer-Weg 3, 14195 Berlin, Germany

Preface

In May 2004, the Alexander Koenig Zoological Research Museum hosted the Fifth International Symposium on Tropical Biology. This series was established at the ZFMK in the early 1980s, and has variably focused on systematics and ecology of tropical organisms, with an emphasis on Africa. Previous volumes are those edited by Schuchmann (1985), Peters and Hutterer (1990), Ulrich (1997), and Rheinwald (2000).

The symposium in 2004 was organized by the Entomology Department under the direction of Michael Schmitt. The intention was to focus on Africa rather than on a particular taxon, and to highlight biodiversity at all levels ranging from molecules to ecosystems. This focus was timely partly because of the currently running BIOTA Africa programmes (BIOdiversity Monitoring Transect Analysis in Africa). BIOTA is an interdisciplinary research project focusing on sustainable use and conservation of biodiversity in Africa (http://www.biote-africa.de).

Session titles were Biogeography and Speciation Processes, Phylogenetic Patterns and Systematics, Diversity Declines and Conservation, and Applied Biodiversity Informatics. Each session was opened by an invited speaker, and all together 77 lectures and 59 posters were presented. There were over 200 participants and it was gratifying to us to meet colleagues from 26 nations, including Russia, Ukraine, Japan, USA, and ten African countries. We thank all participants for their valuable contributions.

The 42 contributions in this volume are opened by four invited speakers: Jon Fjeldså on the application of bird distribution data in conservation, Lorenzo Prendini on patterns and processes of scorpion diversity in southern Africa, Miguel Vences on tempo and pattern of adaptive radiations in Madagascar, and Wolfgang Böhme with an historical contribution on Martin Eisentraut. The other articles are roughly arranged taxonomically, starting with arthropods (mostly insects), fishes, amphibians, reptiles, birds, and mammals. Two articles about biodiversity informatics and landcover change monitoring using satellite technology bridge the gap to the three final contributions on plants. Abstracts of those contributions not included in this volume can be downloaded at http://www.museumkoenig.uni-bonn.de/_pdf/kongresse/dfor_kongress_africanbiodiv_abstracts.pdf.

All contributions were peer-reviewed. We are deeply grateful to the 61 referees who answered our request and provided timely, detailed, and helpful reviews. Further thanks are due to Michael Schmitt, Bernhard Misof and Dieter Stüning for co-organizing the symposium, and to a number of volunteers who made sure things ran smoothly. Birgit Rach was a great help formatting contributions.

We gratefully acknowledge financial support from the DFG (Deutsche Forschungsgemeinschaft) and DAAD (Deutscher Akademischer Austausch-dienst).

Bernhard A. Huber, Bradley J. Sinclair, Karl-Heinz Lampe
Bonn, Germany

REFERENCES

Peters, G., and Hutterer, R., 1990 (eds), Vertebrates in the tropics. Proceedings of the International Symposium on Vertebrate Biogeography and Systematics in the Tropics. Alexander Koenig Research Institute and Zoological Museum, Bonn. 424 pp.

Rheinwald, G., 2000 (ed.), Isolated Vertebrate Communities in the Tropics. Proceedings of the 4th International Symposium of Zoologisches Forschungsinstitut und Museum A. Koenig. *Bonner zool Monogr.* **46**:1-400.

Schuchmann, K.-L., 1985 (ed.), Proceedings of the International Symposium on African Vertebrates: Systematics, Phylogeny and Evolutionary Ecology. Alexander Koenig Research Institute and Zoological Museum, Bonn. 585 pp.

Ulrich, H., 1997 (ed.), Tropical Biodiversity and Systematics. Proceedings of the International Symposium on Biodiversity and Systematics in Tropical Ecosystems. Alexander Koenig Research Institute and Zoological Museum, Bonn. 357 pp.

COMPLEMENTARITY OF SPECIES DISTRIBUTIONS AS A TOOL FOR PRIORITISING CONSERVATION ACTIONS IN AFRICA: TESTING THE EFFICIENCY OF USING COARSE-SCALE DISTRIBUTION DATA

Jon Fjeldså[1] and Herbert Tushabe[2]
[1] *Zoological Museum, University of Copenhagen, Universitetsparken 15, DK-2100 Copenhagen, Denmark*
[2] *Makerere University Institute of Environment and Natural Resources, P.O. Box 7298, Kampala, Uganda*

Abstract: Because of limited resources for conservation, and conflicts with other interests, systematic priority analysis is now a central task in conservation biology. Because of doubts about how efficient conservation schemes based on samples of biomes are for maintaining biodiversity, a major shift has taken place to use species distribution data compiled for large numbers of species. The spatial resolution of distribution data for regional or global analysis must necessarily be rather coarse, and it is therefore unclear whether identified priority areas will hold viable populations of all the species they are assumed to cover. We tested this using more finely resolved distribution data for forest birds of eastern Africa. The broad priority areas identified using coarse-scale data were corroborated using fine-scale data, and they appear to include suitable conservation sites for the majority of species. Exceptions to this were mainly in zones with few strictly forest-dependent species. Procedures for corrections and for moving on to identifying action sites within broader priority areas are discussed.

Key words: Conservation; Africa; complementarity; Aves

B.A. Huber et al. (eds.), African Biodiversity, 1–24.

1. INTRODUCTION

In spite of the daunting dimensions of environmental degradation, conservation initiatives have often been taken *ad hoc*, with a minimum of preceding analysis, by declaring nature reserves in areas, which were hostile to human habitation, but at the same time had grandiose landscapes and a mega-fauna that could attract tourism (MacKinnon and MacKinnon, 1986; Balmford et al., 1992). These are not necessarily the best places for conserving biodiversity in a broad sense, as the biologically richest areas tend to be densely populated (Balmford et al., 2001). Because of the resulting conflicts of interest, and the limited resources available to change the trend of biodiversity loss, we now need to identify conservation gaps (e.g., Fjeldså et al., 2003; Burgess et al., 2004, Rodriguez et al., 2004) and carefully analyze where supplementary initiatives are needed. A large literature has emerged over the last ca. 15 years concerning how to use limited resources strategically and select conservation sites in the most efficient way. This is usually done by examining how to protect as many species as possible on the smallest area (e.g., Margules and Pressey, 2000), but it is also possible to consider other costs than area, for instance development potential (Williams et al., 2003) or prohibitive conflicts (Fjeldså et al., in press).

BirdLife International spearheaded the use of species distribution data for global conservation priority analysis as they compiled the distribution data for all range-restricted bird species (ICBP, 1992). Other conservationists at that time thought that too little data were available for comprehensive mapping of biodiversity, and that it was better to use surrogate measures to identify a representative sample of biomes or land classes for conservation (IUCN, 1993). A number of large-scale studies identified "forest frontiers" (Bryant et al., 1997), "major wilderness areas" and landscapes with exceptional "hotspots" of plant species (WWF and IUCN, 1994, Mittermeier et al., 1998, 2000), and landscape formations with distinct biotic communities (ecoregions, Olsen et al., 2001, and Fa et al., 2004; see also Faith and Walker, 1996).

Most of these priority areas were defined through consensus among experts, but the objectivity of these schemes, or their efficiency in terms of how many species they hold per area unit, was rarely tested. According to the analysis by Burgess et al. (2002), these broad priority areas do not cover more species than randomly chosen areas of similar extension.

Critical studies using species data suggest that landscape classifications, even sophisticated schemes based on remotely sensed data, can hardly predict the occurrence of biological hotspots, and notably not the locally aggregated occurrence of range-restricted and threatened species (Jetz et al., 2004). Methods are therefore now requested that allow objective accounting based on species data. A major shift has taken place to use species ranges as units, either compiling existing records in a grid (e.g., Brooks et al., 2001; Lovett et al., 2000; Burgess et al., 2004), as also done in many national atlas projects, or modeling distribution ranges species by species (e.g., Boitani et al., 1999; Petersen et al., 2000; Tushabe et al., 2001).

Analyzing data in a grid system presents many analytical advantages, compared with presence-absence records in variable-sized landscape units defined in Geographical Information Systems (GIS). Patterns of spatial biotic turnover, biozonation and congruence across larger taxonomic groups can be analyzed more objectively (De Klerk et al., 2002a, b; Williams et al., 1999), and this also applies to testing how well conservation priorities made for well-charted taxonomic groups will capture the known diversity in other groups (Moore et al., 2003). We may also test, in a robust way, to what extent the documented patterns correlate with environmental factors, and thus which factors and processes explain the large-scale variation in biodiversity best (Rahbek and Graves, 2001; Jetz and Rahbek, 2002; de Klerk et al., 2002a). Thus, this kind of data enhances understanding and helps us to move beyond the sands of speculation when discussing what processes we need to maintain.

While the task of compiling distributional data for large taxonomic groups has now proven to be within reachable limits, we still need to assess how useful these data are for conservation planning. Uneven data collecting means that regional and global studies must necessarily have a rather coarse spatial resolution, lacking details about spatial heterogeneity within the species ranges, or about the viability of local populations. It has therefore been suggested that the much-used approach of identifying complementary conservation targets (Margules and Pressey, 2000) is of rather theoretical value (e.g., Leader-Williams and Dublin, 2000). This is because:

(1) the data will suffer from collecting bias towards well studied areas (Maddoc and du Plessis, 1999). This problem is often "solved" by using some degree of interpolation, or even extrapolation, based on expert judgment and available maps, or rule-based GIS modeling of habitat availability. Even the best GIS modeling of species ranges tend to overestimate, as species may be absent from some parts of the potential range because of competition or historical caprices ("errors of omission")

(2) the use of range-maps constructed by filling in collecting gaps gives no guarantee that selected priority areas actually contain viable populations, or any population at all, of the species they are supposed to cover ("errors of commission"; see Loiselle et al., 2003).

(3) the chosen areas could also be marginal populations at biome margins where there are many species replacements (e.g., Branch et al., 1995; Araújo and Williams, 2001).

(4) the result of an analysis may be "grain"-dependent, as the bias towards zones with many species replacements would be most prominent on coarse geographical scales where grid-cells are most likely to span adjacent biomes with complementary biota. Larsen and Rahbek (2003) addressed this question by recompiling distribution data for African mammals in a 1° geographical grid to 2°, 4° and 8° cells: they found that the chosen 1° areas were largely represented within minimum sets at larger scales, at least when flexibility was taken into account. However, we do not know how relevant coarse-scale analysis is for choosing sites at a scale that is relevant for identifying manageable conservation sites.

In this paper we will assess the relevance of these concerns. Out of several databases held at the Zoological Museum, University of Copenhagen, with distribution data for 7800 species of African terrestrial and limnic vertebrates, we used the bird database, which is the most relevant for a critical analysis because of the generous interpolation used. Thus, we use the potentially most problematic dataset to test whether the above-mentioned worries are real.

We will examine whether conservation priorities based on finely resolved data would be nested within priority areas identified from coarse-scale data. We will also use the finely resolved data to examine to what extent the priority areas identified on the coarse scale contain verified records of all species, and interconnected populations, which are most likely to persist in the long term. Finally, we will discuss how to move on from coarse-scale priority areas to identifying sites for practical management.

2. MATERIALS AND METHODS

2.1 Study region

The coarse-scale data set was developed for the whole of Africa south of the Sahara (20°N; see Brooks et al., 2001). This is an area of ca. 23 million km^2, corresponding to the combined area of Europe, China, India and USA. Large parts of the continent are still poorly explored, because of difficult access and lack of security, so it is hardly realistic to try to analyze continent-wide patterns with the spatial resolution finer than the 1x1 geographical degrees. However, we found it feasible to make a more finely resolved dataset for forest birds for eastern Africa from 4°N to 16°S and 28-43°E (Fig. 1). In this sub-region a very large portion of the truly evergreen forests have been visited by ornithologists and some of them are thoroughly studied, especially in the Kenya Highlands (Lewis and Pomeroy, 1989) and the Eastern Arc Mountains and Coastal Forests of Tanzania (Burgess et al.,

1998; Burgess and Clarke, 2000; Baker and Baker, 2002), but also much of the Albertine Rift area (Plumptre et al., 2003).

Much less data exist for the savanna woodlands and bush-lands, although they are dotted with evergreen thickets and small forest patches in places with high groundwater levels, at the base of inselbergs etc. However, surveyed evergreen thickets in these biozones have mainly widespread forest generalists that can also be found here and there in the woodland matrix. It is therefore not very likely that these biozones contain unknown forest patches of critical conservation value.

2.2 Coarse-scale databases

The distributional data have been digitized in the WORLDMAP software (Williams, 1996), which is also used for the analyses. The original databases are in a 1° grid (e.g., Brooks et al., 2001; data sources see http://www.zmuc.dk/commonweb/research/biodata_sources_birds.htm; up-dated version of June 2004).

Because of the collecting gaps that exist, especially in regions of extensive bush, a conservative interpolation was used by assuming continuous presence in grid-cells with appropriate habitat between the confirmed collecting points. This approach seems quite realistic for birds, which are mobile and may occasionally appear in any suitable habitat patch in the matrix of less suitable land within the species' range. The interpolation is a qualified estimate of a species' range based on consensus of many persons with considerable field experience in Africa.

For taxonomy we follow Sibley and Monroe (1990, 1993), but added recently described taxa and taxonomic splits that are well documented by recent molecular studies (e.g., Roy et al., 1998; Bowie, 2002, Bowie et al. in press, Beresford, in press). For the present methodology test we include only the data for forest birds, defined from the principles of Bennun et al. (1996), and using information from regional handbooks. However, it is difficult to apply these in a consistent way as many forest generalists favour forest edges or early-successional forest in some parts of their range (see Imbeau et al., 2003), and even use small patches of evergreen riparian vegetation, or mature woodland habitat, in the savanna regions. As we will see, the inclusion of such borderline species will serve to illustrate how sensitive the results of our analysis can be to the selection of species.

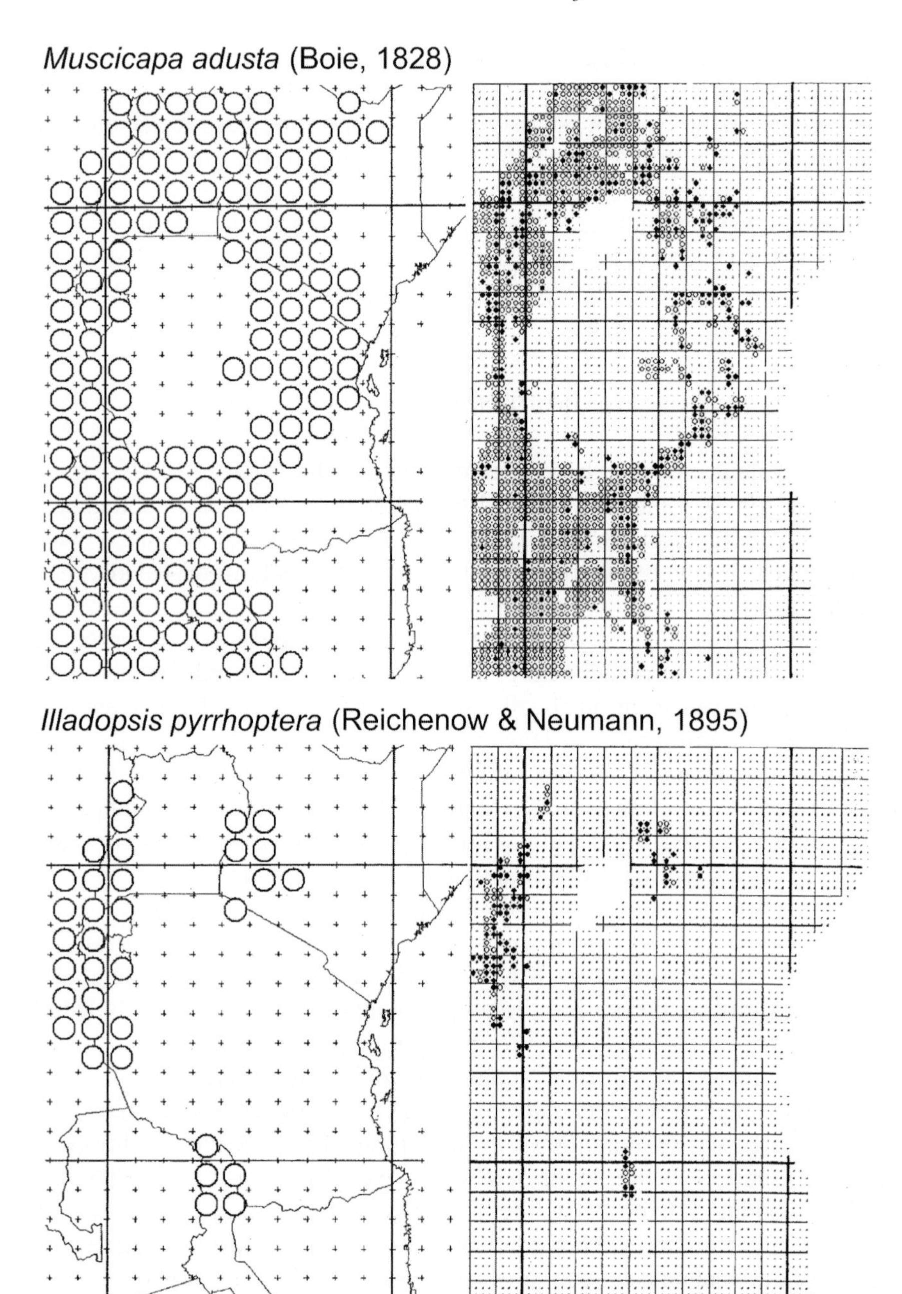

Figure 1. Examples of distribution maps: (a) one forest generalist, Dusky Flycatcher *Muscicapa adusta* (Boie, 1828) and (b) one specialist species, Mountain Illadopsis *Illadopsis pyrrhoptera* (Reichenow and Neumann, 1895) mapped on coarse and fine scales (in 1-degree and 15' grids, with confirmed presence marked black in the fine-scale maps).

2.3 Fine-scale database

The new dataset that was created, in a grid corresponding to 15 x 15 geographical minutes, required a more thorough data compilation, as reviews of specimen data in museums and published sources were supplemented with well targeted fieldwork (since 1991) to many previously unexplored sites, and close contact with many birdwatchers operating in the region (see Acknowledgements). The 15' resolution is fine enough to take into account the exact outline of individual forest tracts. This work was guided by examination of topographic and land cover maps and remotely sensed data available on the internet, such as the Global Land Cover Characterisation ver. 1.2 and Global Topology ver. 3.0 (US Geological Survey, 2000).

Altogether the fine-scale dataset covers 428 species with 114.900 in-grid-cell records, of which 21.897 represent confirmed presence and the rest is interpolation. However, it needs to be said that a large proportion of the interpolation is for forest generalists and edge species, which, according to our field experience, are widely distributed within their range limits (see Fig. 1a). Very little interpolation was done in the case of habitat specialists known from a restricted geographical area, and unrecorded for other well-studied forests (Fig. 1b).

Two important data sources, the bird atlases for Kenya (Lewis and Pomeroy, 1989) and Tanzania (Baker and Baker, in lit.) use half-degree grid-cells, but this provides good guidance for allocating the records to 15' cells when used in conjunction with forest cover maps.

For Uganda we took into account all the point records for forest birds in the National Biodiversity Data Bank (NBDB) at Makerere University. These are georeferenced to actual points of observation, but for estimating the distribution in poorly explored parts of the country (notably the areas of civil unrest in the north) we used a GIS-based overlap analysis inductive model (Brito et al., 1999). The model used by NBDB is a result of testing six environmental parameters (Tushabe et al., 2001; Carswell et al., in press) for their usefulness in predicting suitable habitat for birds. After the testing procedure, rainfall and vegetation type emerged as the best predictors. However, as the approach often overestimates species ranges (Boittani et al., 1999) we subsequently removed those areas of projected occupancy where other evidence (such as negative records in well-studied forests, or presence of competing species) suggests absence. Unfortunately, no GIS model was available for poorly explored areas in the southern woodlands (notably northern Mozambique), where we had to base our judgment on available topographic and habitat maps.

2.4 Conservation priority analysis

For identifying conservation priorities we used the principle of complementarity, which aim to cover all species and at the same time avoid unnecessary duplication (e.g., Margules and Pressey, 2000; Williams, 1998, 2000; Williams et al., 1996). We use two iterative and heuristic search options: (1) a rarity-based algorithm with redundancy back-check identifying complementary sets of areas, and (2) a richness-based ("greedy") algorithm (Williams, 2001). Such approaches have been shown to differ only marginally from optimal solutions using a complete search for the smallest number of grid-cells covering all species (Moore et al., 2003). Optimal solutions require high computing capacity and provide only one solution, and the heuristic approaches have the advantage of being intuitively easier to understand and therefore more transparent, and in providing greater flexibility and possibilities to explore the range of alternative choices (Williams, 2001; Moore et al., cit. op.).

The flexibility can be illustrated in several ways. For the coarse-scale data we simply recorded fully flexible alternative areas in addition to the chosen near-minimum set of areas. For the fine-scale data we used an approach for 16 representations of each species (corresponding to the area of a one-degree cell). This was done by first identifying a minimum set for representing every species in one cell, then a new minimum set after 'blocking' the first set, and so on until all species were covered sixteen times (see Fjeldså, 2001 and Fjeldså et al., in press) (this aim is of course only achieved for species whose total ranges comprise at least 16 cells).

3. RESULTS

3.1 Species richness patterns

The species richness for forest birds in Africa (Fig. 2) peaks on the transition between the Congolian rainforest and the Albertine Rift region, where lowland and highland rainforest is present in the same 1-degree cells. This corresponds to the western part of our eastern African study area (Fig. 3). The highest richness is in the Semliki forest on the Congo/Uganda border, with 214 forest species in a 15' cell, followed by 192 species in Bwindi forest, and 150-180 species per 15' cell also in the Virunga landscape and in the Kivu and Itombwe highlands. A fairly high number of forest bird species, with local peak numbers above 100 forest birds per 15' cell, is also found in the habitat mosaic north of Lake Victoria to Kakamega

Forest in Kenya, and in the larger afromontane and coastal forests of Kenya and Tanzania.

Small and isolated mountains and woodland savannas with interspersed evergreen thickets (especially in Zambia, southern Tanzania, northern Mozambique, and in the northwestern Uelle region) are rather poor. However, the species counts in these habitats would depend strongly on what species we define as forest birds.

The endemism (expressed in Fig. 3b as the most narrowly distributed 25% of the species) shows a quite different and much more locally aggregated pattern (*cf.* Jetz and Rahbek, 2002; Jetz et al., 2004), with very high concentrations in the highlands of the Albertine Rift district and in the Eastern Arc Mountains and some of the adjacent coastal forests. The Kenya Highlands has remarkably few narrow forest endemics, as most Kenyan forest birds are widespread afromontane species, and the species-rich Kakamega forest is "nothing but" an eastern fore-post for widespread Guineo-Congolian birds.

Other priority-setting schemes have ranked the Eastern Arc as the second or third most important area in Africa for the conservation of endemic bird species (ICBP, 1992) and the combined Eastern Arc/Coastal Forests area is

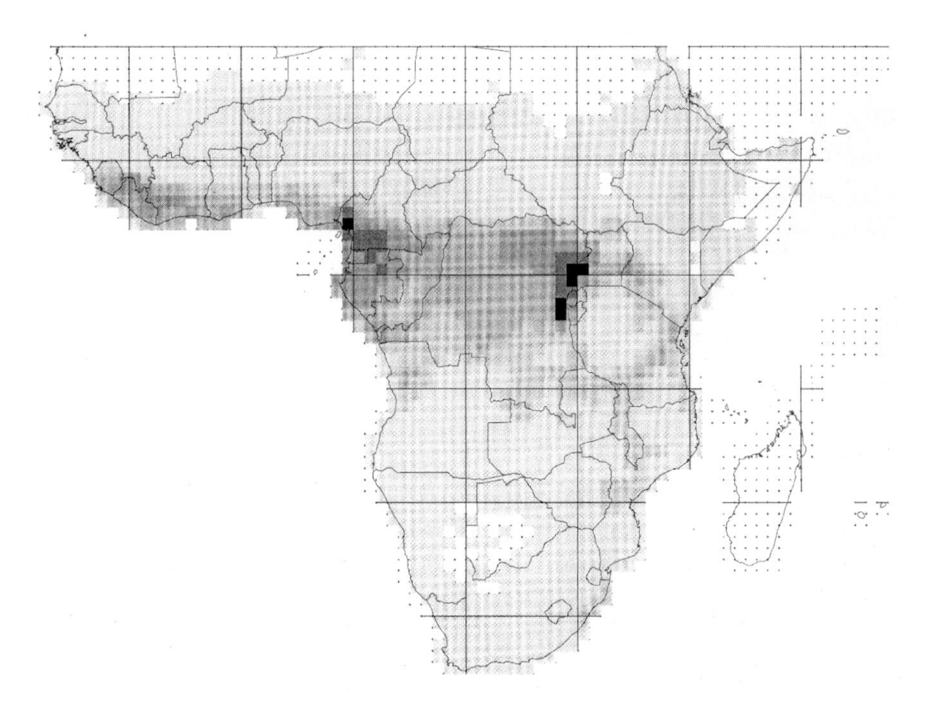

Figure 2. Species richness distribution (absolute scale) for 546 species of African forest birds, the black grid-cells with >259 species.

the second most important 'hotspot' in the Afrotropical Region for total numbers of endemic species (Mittermeier et al., 2000). The Albertine Rift area has been somewhat neglected in the global conservation schemes but has recently been put on the agenda (Plumptre et al., 2003).

3.2 Complementary areas for conservation: the coarse scale

The rarity-based near-minimum set of areas for conserving all forest birds of eastern Africa, on the 1-degree scale, comprises 17 areas, with 12 flexible alternative areas (Fig. 4 a). This is based on a rarity-based algorithm, which first identifies a set of areas needed to cover the most range-restricted species. Thus, eight cells are determined by presence of single-cell endemics, but other cells are compromises for several species, each of which occupy one or more other cells. The 17 cells of the near-minimum set are all essential components of a full conservation plan, but some cells contribute more so-called "goal-essential species" (which are not represented elsewhere in the near-minimum set) than others:

Thus, the Ituri forest cell in northeastern Congo is needed for *Agelastes niger* (Cassin, 1857), *Apalis goslingi* Alexander, 1908, *Apaloderma*

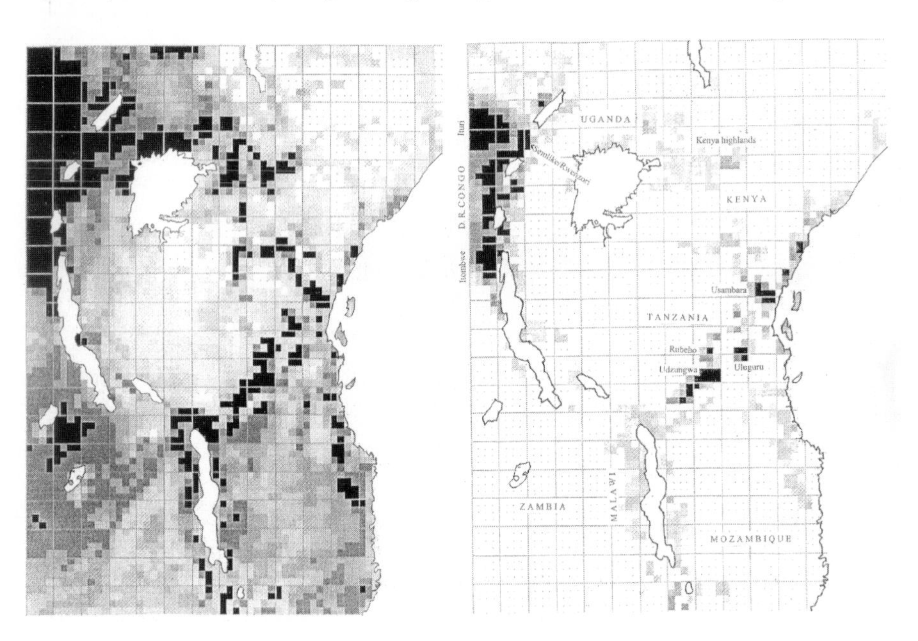

Figure 3. Species richness of forest birds in eastern Africa (proportional scale), in a 15'x15' grid; to the left (a) all 428 species, to the right (b) the "rare quartile" (25%) with most range-restricted species.

aequatoriale Sharpe, 1901, *Dryoscopus sabini* (Gray, 1831), *Muscicapa tessmanni* (Reichenow, 1907) and *M. epulata* (Cassin, 1855) and *Tersiphone rufocinerea* Cabanis, 1875; the adjacent Semliki/Rwenzori cell on the Congo/Uganda border is needed for *Criniger chloronotus* (Cassin, 1860), *Illadopsis puveli* (Salvadori, 1901), *Nectarinia stuhlmanni* Reichenow, 1893, and *Zoothera princei* (Sharpe, 1873); the Itombwe area, SE Congo, is chosen for *Afropavo congensis* Chapin, 1936, *Schoutedenapus schoutedeni* (Prigogine, 1960), *Caprimulgus prigoginei* Louette, 1990, *Hyliota violacea* Verreaux J. & E., 1851, *Malimbus cassini* (Elliot, 1859), *Muscicapa itombwensis* Prigogine, 1951, *Nectarinia rockefelleri* (Chapin, 1932) and *Phodilus prigoginei* Schouteden, 1952. In the Eastern Arc Mountains, a cell in the Udzungwa/Rubeho Mts covers *Xenoperdix udzungwensis* Dinesen et al. 1994, *Nectarinia whytei* Benson, 1948, *N. fuellebornii* Reichenow, 1899 (see Bowie et al. 2004), *N. rufipennis* Jensen, 1983, *Sheppardia lowei* (Grant & Mackworth-Praed, 1941), *S. aurantiithorax* Beresford et al. 2004, *Serinus whytii* Shelley, 1897 and *melanochrous* Reichenow, 1900; and the Taita hills in SW Kenya contribute with *Apalis fuscigularis* Moreau, 1937, *Cinnyricinclus femoralis* (Richmond, 1897), *Turdus helleri* (Mearns, 1913) and *Zosterops silvanus* (Peters & Loveridge, 1935). A cell in the Muchinga highland in the Zambesian woodland biome covers *Phylloscopus laurae* (Boulton, 1931), *Ploceus angolensis* (Bocage, 1878), *Serinus capistratus* (Finsch, 1870) and *Sheppardia bocagei* (Finsch & Hartlaub, 1870). The remaining cells contribute 1-3 more species each to the overall goal.

Also a richness-based ("greedy") algorithm gives 17 areas (Fig. 4 b). This is based on an algorithm where the most species-rich cell is chosen first, then the cell contributing the highest complementary number of species, and so on until all species are covered.

3.3 Complementary areas for conservation: the fine scale

A rarity-based near-minimum set run on the 15' eastern African database (Fig. 4a) comprises 25 areas, 14 of these being inside cells of the coarse-scale near-minimum set, and 4 immediately adjacent to them. For eight representations of each species, 162 cells are needed, 66 inside and 24 adjacent to the coarse-scale priority areas. For 16 representations (corresponding to a one-degree cell in area), 312 cells are needed, 92 inside and 45 adjacent to coarse-scale priority areas, but with many new areas appearing.

A "greedy" area-set based on the 15' database comprises 23 areas (Fig. 4a), 18 inside and 3 immediately adjacent to cells of the coarse-scale priority areas. For eight representations of each species, 162 cells are needed, 77 inside and 28 adjacent to the coarse-scale priority areas. For 16 representations, 310 cells are needed, 103 inside and 46 adjacent to coarse-scale priority areas.

Discrepancies between priorities made on coarse and fine scales are of two kinds:

(1) Mismatch of local distributions of individual species means that species covered in one 1° cell will not necessarily overlap in the same 15' cell. Two or more areas may be needed on the fine scale, and one of these may be outside the 1° area. In the case of the Zambesian woodland biome, the area selection on the fine scale reflects a considerable mismatch in the local occurrence of biome-restricted species that we defined as forest birds.

(2) If we want multiple representations, the area choice may shift considerably as some goal-essential species are recorded in very few 15' cells. For instance, after the two *Apalis lynesi* sites south of Lake Malawi had been picked, the program starts picking areas with other species combinations elsewhere around the Malawi Rift.

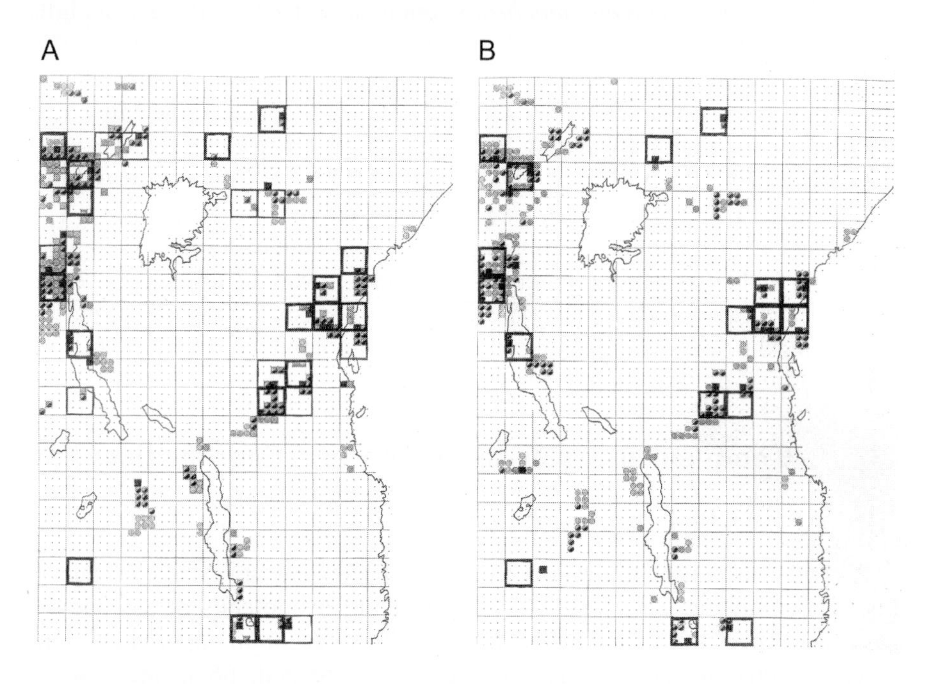

Figure 4. Complementary areas for conserving the forest birds of our study area, based on rarity-based (a) and richness-based ("greedy") algorithms (b). Large squares comprise the near-minimum set of areas (bold frames) and fully flexible alternative areas (thin frames) based on coarse-scale data (1° resolution). Small squares show priority areas based on fine-scale data (15' resolution, filled symbols showing the near-minimum set, half-filled additional areas needed for 8 representations of each species and grey symbols those needed for 16 representations, corresponding to the area of a one-degree cell).

3.4 Are the species actually present in the chosen priority areas?

The fine-scale dataset was carefully reviewed species by species, to check to what extent the coarse-scale near-minimum set of areas (or the fully flexible alternative areas) contained confirmed records. The result is presented in Figure 5.

The far majority of species are well represented. The average for all 428 species is confirmed representations in 3.95 cells of the coarse-scale near-minimum set of areas (rarity-based algorithm; 3.98 cells using the "greedy" algorithm, or 5.87 cells if flexible alternative areas are added). These are of course minimum values, as we do not pretend to know all documented records of forest birds in eastern Africa, and since many species are distributed far beyond our study area, and therefore would be far better represented in a global analysis. Among 69 species with a confirmed representation in one or two cells only, 22% are Guineo-Congolian species with more than 90% of their range outside our East African study area.

Only four species had no confirmed record in the 1° near-minimum set, namely *Apalis argentea* Moreau, 1941 (which is very patchily distributed in the central Albertine Rift area, with confirmed presence in a flexible alternative cell) and three fairly widespread species of the Zambian woodland zone, *Ploceus angolensis*, *Phylloscopus laurae* and *Sheppardia bocagei*, which were picket right at the edge of their respective ranges, near the transition between the humid and drier Zambesian woodland zones (De Klerk et al., 2002b). Over all, the Zambesian forest birds were very poorly represented as a consequence of the small number of coarse-scale priority areas in the southern part of our study region.

This leads us to specifically consider the possible problem of selection of marginal sites (see Introduction). Araújo and Williams (2001), who studied this question for European mammals, birds and reptiles, used a methodology where all cells of a species range were scored using a contagion measure of position on the core-periphery spectrum. This approach is difficult to apply to the forest birds of eastern Africa, as most of them are patchily distributed throughout their range. For 154 species, no single 15' cell is completely surrounded by other occupied cells). Many range-restricted species have dense populations on mountain ridges and scarps immediately adjacent to savanna plains, so isolation at the periphery is not necessarily evidence of low population viability. We therefore instead examined, in the fine-resolution maps, whether the coarse-scale near-minimum set of areas capture marginal isolates of low potential viability (species with >90% of their global range outside our study area omitted from this examination).

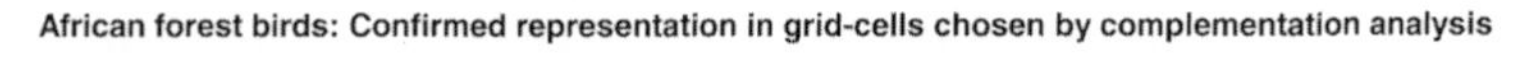

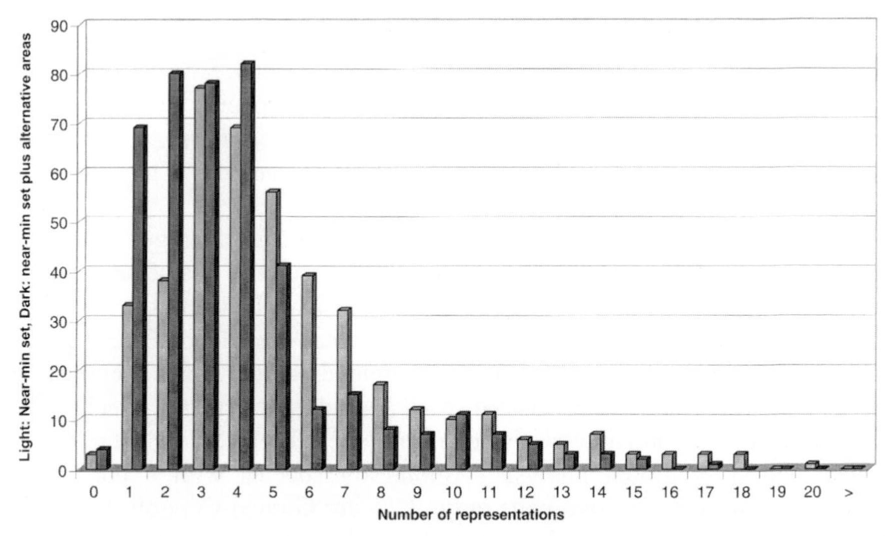

Figure 5. The number of cells in the near-minimum set determined with coarse-scale (one-degree) and interpolated distributional data, in which there are confirmed distribution records of the individual species. Dark columns: near-minimum set calculated for single representation of all species; light columns with flexible alterative also taken into account.

For 83.7% of all species considered, the minimum set included populations interconnected with other populations across at least three one-degree cells. For 15.2% of the species, the minimum set contained only isolated populations (often a tiny cluster of adjacent 15' cells), of which 1/3 are local populations inside a broader range and 2/3 represent species whose entire global range consist of small local populations. Thus, the coarse-scale near-minimum set includes isolated marginal sites only for three species whose ranges include interconnected populations, namely the migratory/nomadic *Ploceus weynsi* (Dubois, 1900) and two Kenyan highland species, *Cinnyricinclus femoralis* and *Zosterops poliogaster* Heuglin, 1861. Of the two latter, *C. femoralis* was picked (along with three single-site endemics) at a peripheral site, Taita hills, with small and vanishing forest patches, and *Z. poliogaster* was picked at its extreme southern margin on Dido hill between the North and South Pare Mts in Tanzania, where there is only 4 km^2 of degraded forest left. In addition to this come the three above-mentioned Zambesian woodland endemics, which were picked at their range margin.

3.5 Inclusion of important bird areas

The coarse-scale priority areas include many of the key sites identified by BirdLife International (Important Bird Areas – or IBAs - for forest birds, Fishpool and Evans, 2001): In D.R. Congo, the Ituri and Itombwe areas, and Kahuzi-Biega and the northern Virunga landscape being flexible alternative areas; in Uganda Semliki and Rwenzori; at the Kenya coast Arabuko Sokoke and several forests in the Shimba Hills area; in Tanzania Udzungwa/Rubeho, Uluguru, W and E Usambaras and S Pare Mts; in Malawi Mt Mulanje, Soche and Thyolo, in Mozambique Mt Namuli. The chosen cell in the Zambesian woodland biome includes the IBA North Swaka, but two of the species for which this area was chosen have not been documented for this IBA.

On the other hand, the important IBAs of the Kenya highlands (viz., Mau and Aberdare forest mosaics) were only recognized as flexible alternatives in our analysis. Other important forest IBAs that are not in the near-minimum set of areas (the Volcans and Nyungwe forests in Rwanda, Kibira in Burundi, Budongo in Uganda, North Ndandi/Kakamega forests in Kenya, Tanzania's volcanic highland and Lindi in the south-east, some highlands in the Eastern Arc and west of Lake Malawi) as they contain no species that are not also present in IBAs inside our priority areas. It is exactly the avoidance of duplicated conservation efforts that is the advantage of the complementarity analysis!

4. DISCUSSION

4.1 Strengths of the complementarity approach

The coarse-scale distribution maps (Fig. 1, left) are range-maps, which tell little about finer distributional details or local distribution gaps. In spite of the potential problems with using such data (p. 3), our analysis reveals that the priority areas chosen from such data contain confirmed records of 99% of all forest birds of the region. They also include some of the most valuable IBAs (Fishpool and Evans, 2001), and sites identified as top priorities by Burgess et al. (1998) and Plumptre et al. (2003).

In addition, the more precise priority areas chosen with fine-resolution data were inside or immediately adjacent to the coarse-scale priority areas. Apart from some problems in the Kenya highlands and Zambesian woodland zones (see below), we may conclude that the problems listed in the Introduction of this paper are of minor importance. This conclusion is on line with that of Larsen and Rahbek (2003), who found that conservation priorities defined on one-degree resolution are nested within areas identified on even coarser scales.

The overall reason for this encouraging conclusion seems to be that the result is driven by the very aggregated distribution of the most range-restricted and patchy (relictual) species, as these aggregates form well-defined centres of endemism, often near the edge of physical domains (Jetz et al., 2004). Because of this nestedness of distributions (Cordeiro, 1998), the complementarity analysis will in most cases prefer peak concentrations rather than the transition zones between broader ecoregions. Over all, most of the interpolation in the dataset is in regions with extensive woodland and bush-land, where there is little endemism.

Ever since the colonial days, ornithologists have had a good intuitive ability to find the unique areas, and much effort has been concentrated in some of these. Although large parts of Africa were under-sampled, we can state today that recent discoveries of new bird species are mainly made in places that have long been known for high endemism (see Raxworthy et al., 2003 for an example of new-discovered reptiles predicted from already-known endemism). Such discoveries, and the taxonomic splitting that takes places, therefore do not alter the overall biogeographic pattern, but instead strengthen and add complexity to the known areas of endemism (Fjeldså, 2003). New taxa that are discovered because of exploration outside these main centers are weakly differentiated (subspecies) in most cases.

The centres of aggregated endemism are valuable conservation targets. Among the goal-essential species that determined the position of the coarse-scale priority areas, 74% are classified as Endangered or Vulnerable (BirdLife, 2000), which *per se* requires conservation actions, and these areas may also play an important role for the regional diversification process. A close spatial congruence between the occurrence of ancient relict species and neo-endemics in these areas suggest ecological stability over millions of years (Fjeldså and Lovett, 1997, and see Fjeldså et al., 1999), in spite of the general instability of the African climate. This could relate to orographic rainfall, or to persistent mist zones caused by weather phenomena such as atmospheric inversions on the interface between warm mountain basins and highlands with cold air ponding.

Thus, the process of speciation may be more than just a consequence of dispersal barriers arising because of rifting and mountain uplift, but may also be linked with intrinsic ecological properties of certain places, allowing populations to persist *in situ* long enough for genetic coalescence, in spite of opportunities for dispersal through surrounding sink habitats. This interpretation is supported by detailed field studies in the Eastern Arc highlands identifying source and sink populations (Fjeldså et al., in press) and comprehensive studies of genetic variation in these forest birds (Bowie, 2002). It is also well on line with recent theoretical studies linking speciation with low climatically driven range dynamics (Jansson and Dynesius, 2003).

Many species are represented only by local and isolated populations in the chosen coarse-scale priority areas, but this reflects to a large extent the patchy nature of the distributions of forest birds in eastern Africa. Exactly for this reason, these species are vulnerable to habitat loss (BirdLife, 2000) and therefore of particular conservation concern.

Considering the projected future changes of the global climate one may fear that such local relict populations are doomed whatever we do. Such worries may be valid for many parts of the world, but the aggregates local occurrence of ancient relict species in certain African highland areas would seem to indicate that these places were little affected by Pleistocene climatic change. For this reason they would be our best guesses of where unique faunas and floras will survive into the future.

The complementarity analysis provides another valuable element for future planning: It helps to tell which of the identified conservation sites (including IBAs) that are redundant and can be dropped if resources are too few to defend all proposed sites.

4.2 Shortcomings

Our optimistic interpretation would probably be less true for savanna biomes and ecologically unstable transition zones, where most organismal groups have life strategies for surviving short- or long-term change. The birds often manage by moving around, so other taxonomic groups with lower dispersal ability could be more interesting for conservation priority analysis, because of higher ability to persist on very local scales (e.g., some plant groups, see Cowling, 2003 and chapters in Friis and Ryding, 2001).

The result of our coarse-scale priority analysis was unconvincing for areas with few zone-restricted birds, such as the Kenyan highland and the southern woodland savannas. Apparently the wide distributions of most species means that no areas stand out here. A cell was therefore chosen where three characteristic species of the humid Zambesian woodland region overlap, at their range margin, with a rather local species of riverine forest. The mismatch between coarse- and fine-scale priority areas in this region (Fig. 4) would disappear if we exclude two species (*Ploceus angolensis* and *Serinus capistratus*), which are only marginally associated with forest. We actually kept these species in the database in order to illustrate how sensitive the priority-setting in such regions can be to contingent decisions. Even without these two "problem species", the precise area choice could easily change by adding another species with a restricted or patchy distribution. A priority analysis based on all birds, which include locally distributed species of savanna and wetland habitats (Broadley and Cotterill, 2004), would have given a more robust area selection here.

Problems can also be seen in the case of the Kenya highlands, as most forest birds are fairly widely distributed all over the highland, or even in adjacent ecoregions. The choice of coarse-scale priority sites is therefore determined by where the zone-restricted species have a marginal range overlap with single-cell endemics in the Taita and Pare Mts, or on Mt Elgon, with a patchily distributed bird (*Apalis pulchra* Sharpe, 1891) that falls outside the main centers of endemism. While the chosen sites are important, *per se*, they are of peripheral importance for other Kenyan highland birds.

4.3 Corrections and an approach towards identifying best conservation sites

Confirmed representations and congruence with fine-scale data suggest that the "greedy" algorithm was the best-performing approach when analysing coarse-scale data. However, it is not perfect. The heuristic analytic approach allows us to explore alternative solutions and reach a meaningful conclusion. However, this is not just done by a computer person, but requires careful judgement by persons with good knowledge about the region.

The corrections require two important steps. First of all, it is necessary to explore why particular cells were chosen. In other words what were the goal essential species? Were they correctly placed? And were these species picked in a suitable place within their range? If the answer to this last question was "no", then the alternative options should be explored, either re-running the analysis after "blocking" the unsatisfactory site(s) to see what area is picked next. In the case of the Kenya highlands, this would suggest a shift of focus to Mau forest complex and Aberdare Mts. In the case of the Zambesian highland, the most meaningful conclusion may be that good management of rather broad areas may be more important than protection of specific sites.

Broad priorities are not about conservation sites, so before we can start practical conservation we need to move on from the broad scale to site selection. This stage requires other data derived from detailed local studies, but the broad priority analysis may nevertheless provide good guidance.

The selection of sites is exemplified in Figure 6, which illustrates the position of prioritised 15' cells (as in Fig. 4) and extent of habitat for goal-essential forest birds in the Udzungwa and Rubeho highlands in the Eastern Arc Mts in Tanzania. Following the rarity-based algorithm the computer first selects two cells marked with "X" which are the upper left cells containing the most range-restricted of the goal-essential species, namely *Xenoperdix udzungwensis obscurata* in Mafwemiro forest and *Nectarinia rufipennis* in the area near Kisinga Rugaro forest. However, based on field studies in all larger forest patches (Fjeldså et al., in press) the core area for both of these

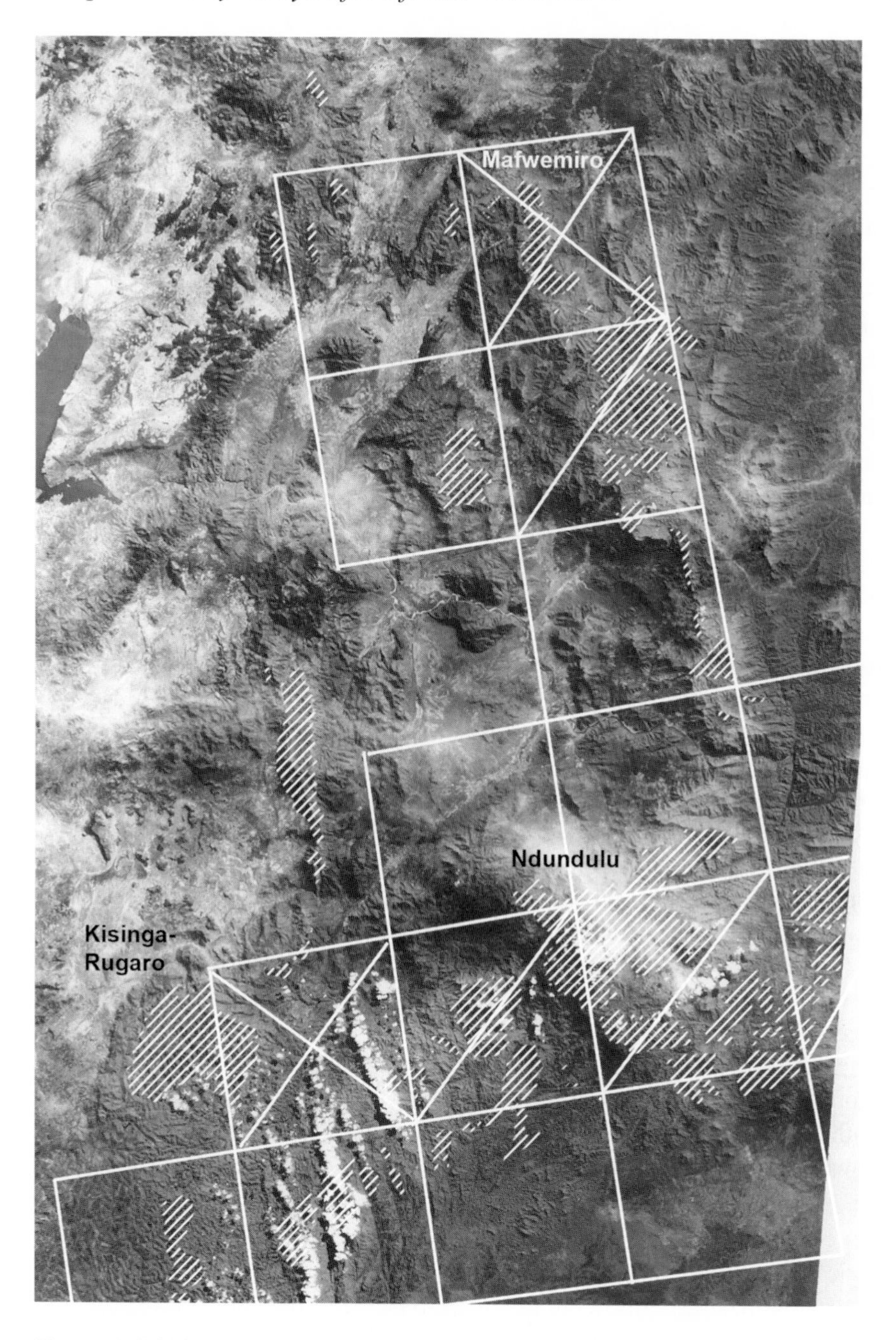

Figure 6. Priority-setting on the local scale: satellite map of part of the Udzungwa and Rubeho highlands (southern and northern parts of the image) in the Eastern Arc Mts of Tanzania, with prioritized 15'-cells marked (cells with x were 1st priority according to the rarity-based analysis, which first selects the upper left cell in the range of the most range-restricted species, but cells marked with / would be better when population sizes and number of species are taken into account). Habitat of goal-essential species (see p. 11) is hatched.

(the nominate subspecies in the case of *X. udzungwensis*) is in the Ndundulu Mts, which also has a broad assortment of other endemic species. Thus, for effective conservation of all target taxa, it would be better to focus on grid-cells in the Udzungwa highland marked with "/". The highland eastwards from central Ndundulu Mts are now in the Udzungwa National Park, and towards the west of these mountains has been the centre for a village-based programme for sustainable land use, funded by the Danish State.

For the planning of action points we now need new kinds of data, which requires local-level surveys of habitat quality as well as threats and land-use methods in order to define where support should be given to more effective food production, for buffer zones, and for gazetting strict protection zones. It is also important that local people learn that intact environments can give income from tourists, and that suitable sites for this activity are identified. We assume that GIS will become a useful tool in this phase.

The declared goal of the 4th Parks Congress in 1992 to conserve 10% of the global land area has been achieved, but very significant portions of the global biodiversity is still unprotected (e.g. Fjeldså et al., 2004; Burgess et al., 2004, Rondinini et al., in press). The main reason for this is that biological hotspots are often situated in regions that are also suitable for people, and therefore populated (Fjeldså et al., 1999; Balmford et al., 2001) and poorly suited as nature parks in the traditional sense. Filling in the network of conservation gaps with traditional reserves will therefore be expensive, and we now need to supplement traditional conservation initiatives with development support to make the land use more sustainable. Fortunately, the montane forest areas that were identified as top priorities in our analysis offer some prospects, because of the positive importance of montane cloud-forest for agriculture in adjacent lowlands and rain-shadow areas (see e.g. Becker, 1999) and important practical experiences have also been gained on how to integrate development and conservation (e.g., Johansson, 2001).

The next big challenges, in our study region, will be to see how a good development can be started in the Kivu area, where existing reserves fail to serve the intended function because of continuous armed conflicts, and in the Itombwe highland, which still has no protection.

ACKNOWLEDGEMENTS

Unpublished data have been provided by the National Biodiversity Data Bank held at Makerere University in Uganda, by BirdLife International in Cambridge, and by many people working in the region, notably E. and N. Baker, N. Burgess, N. Cordeiro, L. A. Hansen, F. P. Jensen, J. Kiure, A. Tøttrup. JF reviewed African birds in museum collections in Chicago,

Copenhagen, Bonn, Lund, Leiden, Milwaukee, Nairobi, Paris, and New York and Tring in part, and hereby thank the curators of these collections for their courtesy. For his helpfulness with data analysis and graphical presentation we finally direct our thanks to L. A. Hansen.

REFERENCES

Araújo, M. B., and Williams, P. H., 2001, The bias of complementarity hotspots toward marginal populations, *Conserv. Biol.* **15:**1710-20.

Baker, N. and E., 2002, Important Bird Areas in Tanzania. WCST, Dar es Salaam.

Balmford, A., Leader-Williams, N., and Green, M. J. B., 1992, Protected areas of Afrotropical forest: history, status and prospects, in: *Tropical Rain Forests: An Atlas for Conservation*, Vol. 2, M. Collins and J.A. Sayer, eds., MacMillan, London, pp. 69-80.

Balmford, A., Moore, J. L., Brooks, T., Burgess, N., Hansen, L. A., Williams, P., and Rahbek, C., 2001, Conservation conflicts across Africa, *Science* **291:**2616-2619.

Becker, C. D., 1999, Protecting a Garúa forest in Ecuador: the role of institutions and ecosystem valuation, *Ambio* **28:**156-161.

Bennun, L., Dranzoa, C., and Pomeroy, D., 1996, The forest birds of Kenya and Uganda, *J. East Afr. Nat. Hist.* **85:**23-48.

Beresford, P. (in press) Toward a phylogenetic biogeography of the Guineo-Congolian biome: molecular systematics of *Bleda* and *Criniger* (Aves: Pycnonotidae), *Mol. Phyl. Evol.*

BirdLife International, 2000, *Threatened birds of the world*, Lynx Edicions and BirdLife, Balcelona and Cambridge, UK.

Boittani, L., Corsi, F., de Biase, A., Carranza, I. D., Ravagli, M., Reggiani, G., Sinibaldi, I., and Trapanese, P., 1999, A Databank for the Conservation and Management of the African Mammals. *Inst. Ecol. Appl., Roma.*

Bowie, R. C. K., 2003, Birds, molecules, and evolutionary patterns among Africa's islands in the sky. Ph.D. thesis, Percy FitzPatrick Institute of African Ornithology, University of Cape Town.

Bowie, R. C. K., Fjeldså, J., Hackett, S. J. & Crowe, T. M., 2004, Systematics and biogeography of the double-collared sunbirds (Nectariniidae) of the Eastern Arc Mountains of Tanzania, *Auk* **121:**660-681.

Branch, W. R., Benn, G. A. & Lombard, A. T. 1995, The tortoises (Testudinidae) and terrapins (Pelomedusidae) of southern Africa: their diversity, distribution and conservation, *S. Afr. J. Zool.* **30:**91-102.

Brito, J. C., Crespo, E. G., and Paulo, O. S., 1999, Modelling wildlife distributions: Logistic multiple regression vs overlap analysis, *Ecography* **22:**251-260.

Broadley, D. G., and Cotterill, F. P. D., 2004, The reptiles of southeast Katanga, an overlooked 'hot spot', *African J. Herpetol.* **53:**35-61.

Brooks, T., Balmford, A., Burgess, N., Fjeldså, J., Hansen, L. A., Moore, J., Rahbek, C., and Williams, P. H., 2001, Toward a blueprint for conservation in Africa, *BioScience* **51:**613-624.

Bryant, D., Nielsen, D., and Tangley, L., 1997, *The last frontier forests: ecosystems and economics on the edge,* World Resources Institute, Washington, D.C.

Burgess, N. D., Nummelin, M, Fjeldså, J., Howell, K. M., Lukumbyzya, K., Mhando, L., Phillipson, P., Vanden Berghe, E., and Depew, L. A. (eds), 1998, Biodiversity and Conservation of the Eastern Arc Mountains of Tanzania and Kenya, *J. East Afr. Nat. Hist.* **87:**1-367.

Burgess, N. D., Rahbek, C., Larsen, F. W., Williams, P., and Balmford, A., 2002, How much of the vertebrate diversity of sub-Saharan Africa is catered for by recent conservation proposals? *Biol. Conserv.* **107:**327-339.

Burgess, N. D., and Clarke, G. P. (eds.) 2000, *The coastal forests of Eastern Africa,* Cambridge University Press, Cambridge, UK.

Burgess, N. D., Rahbek, C., Williams, P., and Balmford, A., 2002, How much of the vertebrate diversity of sub-Saharan Africa is catered for by recent conservation proposals, Biol. Conserv. **107:**327-339.

Burgess, N., Küper, W., Mutke, J., Brown, J., Westaway, S., Turpie, S., Meschack, C., Taplin, J., McClean, C., and Lovett, J. C., 2004, Major gaps in the distribution of protected areas for threatened and narrow range Afrotropical plants, *Biodiver. Conserv.* In press

Carswell, M., Pomeroy, D., Reynolds, J., and Tushabe, H., in press, *The Bird Atlas of Uganda*, BTO.

Chamberlain, D. E., Gouch, S., Vickery, J. A., Firbank, L. G., Petit, S., Pywell, R., and Bradbury, R. B., 2004, Rule-based predictive models are not cost-effective alternatives to bird monitoring in farmland, *Agric. Ecosyst. Environ.* **101:**1-8.

Cordeiro, N. J., 1998, Preliminary analysis of the nestedness patterns of montane forest birds in the Eastern Arc Mountains, *J. East Afr. Nat. Hist.* **87:**101-118.

De Klerk, H. M., Crowe, T. M., Fjeldså, J., and Burgess, N. D., 2002a, Biogeographical patterns of endemic terrestrial Afrotropical birds, *Divers. Distr.* **8:**147-162.

De Klerk, H. M., Crowe, T. M., Fjeldså, J., and Burgess, N. D., 2002b, Patterns of species richness and narrow endemism of terrestrial bird species in the Afrotropical Region, *J. Zool. Lond.* **256:**327-342.

Fa, J. E., Burn, R. W., Stanley Price, M. R., and Underwood, F. M., 2004, Identifying important endemic areas using ecoregions: birds and mammals in the Indo-Pacific, *Oryx* **38:**91-101.

Faith, D. P., and Walker, P. A., 1996, Environmental diversity: on the best-possible use of surrogate data for assessing the relative biodiversity of sets of areas, *Biodiv. Conserv.* **5**:399-415.

Fishpool, L. D. C., and Evans, M. I., 2001, *Important Bird Areas in Africa and Associated Islands. Priority Sites for Conservation*, BirdLife International, Cambridge, UK.

Fjeldså, J., Lambin, E., and Mertens, B., 1999, The relationship of species richness and endemism to ecoclimatic stability - a case study comparing distributions of Andean birds with remotely sensed environmental data, *Ecography* **22:**63-78.

Fjeldså, J., 2001, Cartografiar la avifauna Andina: una base científica para establecer prioridades de conservación, in: *Bosques nublados del neotrópico*, M. Kapelle and A. D. Brown, eds, INBIO, Costa Rica, pp. 125-152.

Fjeldså, J., 2003, Patterns of endemism in African birds: how much does taxonomy matter? *Ostrich* **74:**30-38.

Fjeldså, J., and Lovett, J. C., 1997, Geographical patterns of old and young species in African forest biota: the significance of specific montane areas as evolutionary centers, *Biodiver. Conserv.* **6:**325-347.

Fjeldså, J., Álvarez, M. D., Lazcano, J. M., and Leon, B., in press, Illicit crops and armed conflict as constraints on biodiversity conservation in the Andes Region, *Ambio*.

Foster, P., 2001, The potential negative impacts of global climate change on tropical montane cloud forest, *Earth-Sci. Rev.* **55:**73-106.

Friis, I., and Ryding, O., 2001, *Biodiversity research in the Horn of Africa Region,* Royal Danish Academy of Sciences and Letters, Biol., Vol. 54.

ICBP, 1992, *Putting Biodiversity on the Map: Priority Areas for Global Conservation*, ICBP, Cambridge, UK.

Imbeau, L., Drapeau, P., and Mönkönen, M., 2003, Are forest birds categorized as "edge species" strictly associated with edges? *Ecography* **26:**514-520.

IUCN, 1993, Parks for Life: Report of the IVth World Congress on National Parks and Protected Areas. IUCN, Gland.

Jansson, R., and Dynesius, M., 2002, The fate of clades in a world of recurrent climatic change: Milankovitch oscillations and evolution, *Annu. Rev. Ecol. Syst.* **33:**741-77.

Jetz, W., and Rahbek, C., 2002, Geographic range size and determinants of avian species richness, *Science* **297:**1548-1551.

Jetz, W., Rahbek, C. and Colwell, R. K., 2004, The coincidence of rarity and richness and the potential signature of history in centers of endemism, *Ecology Letters* in press.

Johansson, L., 2001, *Ten Million Trees Later*, GTZ, Eschborn, Germany.

Larsen, F. W., and Rahbek, C., 2003, Influence of scale on conservation priority setting – a test on African mammals, *Biodiver. Conserv.* **12:**599-614.

Leader-Williams, N., and Dublin, H.T., 2000, Charismatic megafauna as 'flagship species', in: Priorities for conservation of mammalian biodiversity: has the panda had its day? A. Entwistle and N. Dunstone, eds., Cambridge Univ. Press, Cambridge, UK, pp. 53-81.

Lewis, A., and Pomeroy, D., 1989, *A Bird Atlas of Kenya*, Balkema, Rotterdam.

Loiselle, B. A., Lowell, C. A., Graham, C. H., Goerck, J. M., Brooks, T. M., Smith, K. G., and Williams, P. H., 2003, Avoiding pitfalls of using species distribution models in conservation planning, *Conserv. Biol.* **14:**1591-1600.

Lovett, J. C., Rudd, S., Taplin, J., and Frimodt-Møller, C., 2000, Patterns of plant diversity in Africa south of the Sahara and their implications for conservation management, *Biodiver. Conserv.* **9:**33-42.

MacKinnon, J. and MacKinnon, K., 1986, Review of the Protected Area System in the Afrotropical Realm, IUCN, Gland.

Maddock, A., and du Plessis, M. A., 1999, Can species data only be appropriately used to conserve biodiversity? *Biodiver. Conserv.* **8:**603-615.

Margules, C. R., and Pressey, R. L., 2000, Systematic conservation planning, *Nature* **405:**243-253.

Moore, J. L., Balmford, A., Brooks, T., Burgess, N. D., Hansen, L. A., Rahbek, C., and Williams, P. H., 2003, Performance of sub-Saharan vertebrates as indicator groups for identifying priority areas for conservation, *Conserv. Biol.* **17:**207-218.

Mittermeier, R. A., Myers, N., Thomsen, J. B., and da Fonseca, G. A., 1998, Biodiversity hotspots and major tropical wilderness areas: approaches to setting conservation priorities, *Conserv. Biol.* **12:**516-520.

Mittermeier, R. A., Myers, N., Gil, P. R., and Mittermeier, C. G., 2000, Hotspot; the Earth's Biologically Richest and Most Endangered Terrestrial Ecoregions. CEMEX and Conservation International, Washington.

Moore, J. L., Folkmann, M., Balmford, A., Brooks, T., Burgess, N., Rahbek, C., Williams, P.H., and Krarup, J., 2003, Heuristic and optimal solutions for set-covering problems in conservation biology, *Ecography* **26:**595-601.

Nicholls, A. O., 1998, Integrating population abundance, dynamics and distribution into broad-scale priority setting, in: *Conservation in a Changing World*, G. M. Mace, A Balmford, J. Ginsberg, eds., Cambridge Univ. Press, Cambridge, UK, pp. 251-272.

Olson, D. M., Dinerstein, E., Wikramanayake, E. D., Burgess, N. D., Powell, G. V. N., Underwood, E. C., D'Amico, J. A., Itoua, I., Strand, H. E., Morrison, J. C., Loucks, C. J., Allnutt, T. F., Ricketts, T. H., Kura, Y., Lamoreux, J. F., Wettangel, W. W., Hedao, P., and Kassem, K. R., 2001, Terrestrial ecoregions of the world: a new map of life on Earth, *BioScience* **51:**933-938.

Plumptre, A. J., Behangana, M., Davenport, T. R. B., Kahindo, C., Kityo, R., Endomba, R., Nkuutu, D., Ssegawa, P., and Eilu, G., 2003, *The Biodiversity of the Albertine Rift,* Wildlife Conservation Society, New York.

Rahbek, C. and Graves, C. R., 2001, Multiscale assessment of patterns of avian species richness, *Proc. Nat. Acad. Sci.* **98:**4534-4539.

Raxworthy, C. J., Martinez-Meyer, E., Horning, N., Nussbaum, R. A., Schneider, G. E., Ortega-Huerta, M. A., and Peterson, A. T., 2003, Predicting distributions of known and unknown reptile species in Madagascar, *Nature* **426:**837-841.

Rodrigues, A. S., Cerdeira, J. O., and Gaston, K. J., 2000, Flexibility, efficiency, and accountability: adapting reserve selection algorithms to more complex conservation problems, *Ecography* **23:**565-574.

Rodrigues, A. S. L., Tratt, R, Wheeler, B. D, and Gaston, K. J., 1999, The performance of existing networks of conservation areas in representing biodiversity, *Proc. Roy. Soc. London Ser. B-Biol.Sci.* **266:**1453-1460.

Rondinini, C., Stuart, S., and Boitani, L. (in press) Habitat suitability models reveal shortfall in conservation planning: the case of African vertebrates, *Conserv. Biol.*

Roy, M. S., Arctander, P., and Fjeldså, J., 1998, Speciation and taxonomy of montane greenbuls of the genus Andropadus (Aves: Pycnonotidae), *Steenstrupia* **24**:51-66.

Spector, S., 2002, Biogeographic crossroads as priority areas for biodiversity conservation, *Conserv. Biol.* **16:**1480-1487.

Tushabe, H., Reynolds, J., and Pomeroy, D., 2001, Innovative aspects of the Bird Atlas of Uganda, *Ostrich* suppl. **15:**183-188.

United States Geological Survey, 2000, Global Land Cover Characterisation. Available at http://edcdaac.usgs.gov/glcc/glcc.html

Williams, P. H,. 1996, Worldmap 4 windows: software and help document 4. London: distributed privately.

Williams, P. H., 1998, Key sites for conservation: area-selection methods for biodiversity, in: *Conservation in a Changing World*, G. M. Mace, A Balmford, J. Ginsberg, eds., Cambridge Univ. Press, Cambridge, UK, pp. 211-249.

Williams, P. H., 2001, Complementarity, in: *Encyclopedia of Biodiversity*, S. A. Levin, ed., Vol. 1, Academic Press, pp. 813-829.

Williams, P. H., Gibbon, D. Margules, C., Rebelo, A., Humphries, C., and Pressey, R., 1996, A comparison of richness hotspots, rarity hotspots and complementarity areas for conserving the diversity of British Birds, *Conserv. Biol.* **10:**155-174.

Williams, P. H., de Klerk, H. M., and Crowe, T. M., 1999, Interpreting biogeographical boundaries among Afrotropical birds: distinguishing spatial patterns in richness gradients and species replacement, *J. Biogeog.* **26:**459-474.

Williams, P. H., Moore, J. L., Kamden Toham, A., Brooks, T. M., Strand, H., A'Amico, J., Wisz, M., Burgess, N. D., Balmford, A., and Rahbek, C., 2003, Integrating biodiversity priorities with conflicting socio-economic values in the Guinean-Congolian forest region, *Biodiver. Conserv.* **12:**1297-1320.

WWF and IUCN, 1994, *Centers of Plant Diversity: a Guide and Strategy for their Conservation. Vol.1. Europe, Africa, SW Asia and the Middle East*, IUCN Publications Unit, Cambridge, UK.

SCORPION DIVERSITY AND DISTRIBUTION IN SOUTHERN AFRICA: PATTERN AND PROCESS

Lorenzo Prendini
Division of Invertebrate Zoology, American Museum of Natural History, Central Park West at 79th Street, New York, NY 10024-5192, USA, E-mail: lorenzo@amnh.org

Abstract: Patterns of scorpion diversity and distribution in southern Africa (south of 15° latitude), and the processes that produced them, are reviewed. A georeferenced presence-only dataset, comprising 6766 point locality records for the 140 scorpion species currently recognised in the subregion, is compiled and analysed with a geographical information system. Hotspots of scorpion species richness and endemism in southern Africa are mapped at the level of a quarter-degree square. The taxonomic composition of the southern African scorpion fauna is assessed and found to comprise distinct western and eastern components. Hotspots of species richness and endemism are concentrated in arid regions with rugged topography, complex geology, or substratal heterogeneity. The distributions of genera and species are discussed in terms of their ecological requirements and modes of speciation within the context of historical events. Historical changes in the geomorphology and climate of southern Africa, coupled with the specific ecological requirements of most southern African scorpions, are proposed as primary causes for their speciation and, ultimately, their high species richness and endemism.

Key words: Chelicerata; Scorpiones; southern Africa; diversity; endemism; distribution; GIS; hotspot; taxonomy; evolution; speciation; biogeography

1. INTRODUCTION

Predictable, non-random patterns characterise the spatial distribution of most taxa (MacArthur, 1972; Myers and Giller, 1988), and scorpions are no exception. The distribution of scorpions has interested arthropod biogeographers for more than a century (Pocock, 1894, Kraepelin, 1905;

B.A. Huber et al. (eds.), African Biodiversity, 25–68.

Birula, 1917a, 1917b, 1925; Hewitt, 1925; Lawrence, 1952; Lamoral, 1980a; Stockwell, 1989; Sissom, 1990; Nenilin and Fet, 1992; Lourenço, 1996, 1998, 2000; Soleglad and Fet, 2003). Scorpions allow opportunities for ecological and historical biogeographic investigation almost unparalleled among Arthropoda. One reason for using scorpions as model taxa in biogeographic studies is the antiquity of the group. Fossil records show that scorpions have existed in the terrestrial environment since the middle Silurian, over 400 MYA, and are almost unchanged today (Jeram, 1990a, 1990b, 1994a, 1994b; Selden, 1993; Sissom, 1990). Scorpion distribution patterns often reveal evidence of continental drift (Koch, 1977, 1981; Lourenço, 1984a, 1991a, 1996, 1998, 2000; Stockwell, 1989; Sissom, 1990) or palaeoclimatic change (Lourenço, 1994a, 1994b, 1996). A similarity in the habits and ecological requirements of fossil and modern scorpions may be inferred from their morphological conservatism (Jeram, 1990b). It is thus reasonable to assume that the ecological factors determining scorpion distributions today are similar to those which determined their distributions in the past.

Scorpions are also useful for biogeographic studies because of their diversity (Lourenço, 1994c). Scorpions occur on all major landmasses except Antarctica, and on many oceanic islands, in all terrestrial habitats – including high-elevation mountaintops, caves, and the intertidal zone – except tundra and high-latitude taiga (Polis, 1990; Gromov, 2001). Diverse communities of scorpions (4-13 species) occur in habitats as different as desert (Williams, 1980) and tropical rainforest (González-Sponga, 1978, 1984, 1996; Lourenço, 1983a). Many ecological factors influence the spatial distribution of scorpions, including temperature, precipitation, soil or rock characteristics, stone or litter cover, topography, vegetation, and environmental physiognomy (Polis, 1990). Some of these factors, e.g. the substratum, may be important determinants of their evolution and may underpin broad patterns of geographical distribution (Lamoral, 1978a; Prendini, 2001a). For example, the sedentary nature of most scorpion species, together with their often narrow physiological and ecological tolerances (Polis, 1990), may have limited their vagility and promoted allopatric speciation by vicariance during periods of palaeoclimatic change (Lamoral, 1978b; Prendini, 2001a), in turn creating high levels of endemism in certain areas, e.g. Baja California (Williams, 1980).

The prevalence of relictual, endemic scorpion species in certain areas may warrant their biogeographic investigation as models for arthropod conservation (Lourenço, 1991b). The importance of knowing the distribution of venomous scorpions, from a medical perspective, is yet another reason to study their biogeography (Newlands, 1978b; Newlands and Martindale, 1980). The ecological abundance and medical importance of scorpions has also ensured their good representation in museum collections. The taxonomic diversity of scorpions is thus better understood, worldwide, than that of many other arthropod taxa. Both factors, in turn, guarantee greater accuracy in biogeographic investigations of the distributions of species or higher taxa.

The rich scorpion fauna of the southern African subregion was recognised at the turn of the previous century. Hewitt (1925) identified a difference between the scorpion species composition of the northeastern and southwestern parts of the subregion, and proposed that the most southerly species were primitive, species further north being more advanced. Hewitt's work was published before the theories of continental drift and plate tectonics were widely accepted, and concentrated on dispersalist rather than vicariance arguments. It also failed to address the ecology of the group. Lawrence (1942, 1952) noted the same distinction between the eastern and western scorpion faunas, but attributed the relative paucity of the eastern fauna entirely to ecological factors, while neglecting the role of history. Later theories about scorpion distribution patterns in southern Africa (Newlands, 1972a, 1972b, 1978a, 1978b, 1980; Lamoral, 1978b, 1979) also failed to accommodate ecological explanations within a historical context.

Such criticisms are not reserved solely for studies of scorpion biogeography, but extend to biogeographic studies of many other taxa. Few authors (see, for example, references in Werger (1978) and Keast (1981)) consider patterns of distribution in terms of ecology and history, despite the undeniable influence of both (Haydon et al., 1994). Myers and Giller (1988: 3) warn that "to progress, biogeography must attempt to ... determine how speciation, adaptation, extinction and ecological processes interact with one another and with geology and climate to produce distributional patterns in the world's biota through time."

In this contribution, a distributional dataset for the southern African scorpion species is compiled and analysed with a geographical information system (GIS). Using evidence from the literature, the observed patterns of diversity and distribution are interpreted as the outcome of two interacting sets of circumstances: (1) ecological factors, i.e. the distribution of environmental gradients, the ecological requirements of the species, and the biotic component of species interactions; (2) historical factors, i.e. past, often chance events that have determined the occurrence of a species in particular localities, e.g. dispersal, vicariance, and speciation (Koch, 1977).

2. METHODS

2.1 Taxonomic considerations

Traditionally, all southern African scorpions were placed into Buthidae C.L. Koch, 1837 or Scorpionidae Latreille, 1802 (Hewitt, 1918, 1925; Lawrence, 1955; Lamoral and Reynders, 1975; Lamoral, 1978b, 1979; Newlands, 1978a, 1978b, 1980; Newlands and Martindale, 1980; Newlands and Cantrell, 1985). Today, they are divided among Bothriuridae Simon, 1880, Buthidae, Liochelidae Fet and Bechly, 2001, and Scorpionidae.

Bothriuridae are represented in southern Africa by two genera: *Brandbergia* Prendini, 2003; *Lisposoma* Lawrence, 1928. Although previously known to share similarities with the bothriurids (Vachon, 1974), *Lisposoma* was retained in Scorpionidae, in a unique subfamily, Lisposominae Lawrence, 1928 (Lawrence, 1955; Lamoral, 1978b, 1979), until Francke (1982) transferred it to Bothriuridae. Lourenço's (2000) creation of a unique family, Lisposomidae Lawrence, 1928, for the genus is unsupported by cladistic analysis (Prendini, 2000a, 2003a, 2003b). *Lisposoma* was revised by Prendini (2003b).

Six genera of Buthidae were recognised in southern Africa (*Buthotus* Vachon, 1949; *Karasbergia* Hewitt, 1913; *Lychas* C.L. Koch, 1845; *Parabuthus* Pocock, 1890; *Pseudolychas* Kraepelin, 1911; *Uroplectes* Peters, 1861) until a seventh, *Afroisometrus* Kovařík, 1997, was created to accommodate a species of *Lychas* from Zimbabwe (FitzPatrick, 1994a). Francke (1985) demonstrated that *Buthotus* is a junior synonym of *Hottentotta* Birula, 1908. Southern African *Parabuthus* have been studied intensively in recent years and 20 species are currently recognised from the subregion (FitzPatrick, 1994b; Prendini 2000b, 2001b, 2003c, 2004a). The species composition of *Karasbergia* and *Pseudolychas* remains unaltered following recent revisions (Prendini 2004b, in press). *Uroplectes* currently contains 19 southern African species (FitzPatrick, 1996, 2001; Fet and Lowe, 2000), but requires extensive revision; the genus is undoubtedly more diverse than *Parabuthus*.

Liochelidae, until recently known as Ischnuridae Simon, 1879, contains three southern African genera formerly assigned to subfamily Ischnurinae Simon, 1879 of Scorpionidae (Lourenço, 1989; Sissom, 1990): *Cheloctonus* Pocock, 1892; *Hadogenes* Kraepelin, 1894; *Opisthacanthus* Peters, 1861. A fourth genus, *Iomachus* Pocock, 1893, with one species, *Iomachus politus* Pocock, 1896, widespread in eastern Africa (Ethiopia, Kenya, Tanzania, Uganda and the Democratic Republic of Congo), and also recorded from northeastern Mozambique (Kraepelin, 1913; Werner, 1936), has not been confirmed as occurring south of 15° latitude; records of *I. politus* from Beira

(Werner, 1936; Aguiar, 1978) are almost certainly erroneous. *Opisthacanthus* was revised by Lourenço (1987a) but the validity of several southern African species remains questionable, as does the validity of *Cheloctonus* and its component species, which remain to be addressed (Prendini, 2000a). *Hadogenes*, presently containing 16 species, is the subject of ongoing revision by the author (Newlands and Prendini, 1997; Prendini, 2001c, in press). Lourenço's (1999, 2000) proposals to transfer this genus to Scorpionidae, or to create a unique family, Hadogenidae Lourenço, 2000, to accommodate it, are unsupported by cladistic analysis (Prendini, 2000a, 2001c).

Opistophthalmus C.L. Koch, 1837 remains the sole representative of Scorpionidae in southern Africa. This diverse genus, also under revision by the author, currently contains 59 valid species (Prendini, 2000c, 2001d; Harington, 2002). The actual number is closer to 80 (Prendini et al., 2003). To date, no records of *Pandinus viatoris* (Pocock, 1890), a scorpionid widespread in central Africa (recorded from the Democratic Republic of Congo, Malawi, Tanzania, Zambia, and northwestern Mozambique), have been confirmed as occurring south of 15° latitude. Alleged records of this species from Zimbabwe (Lamoral and Reynders, 1975; Fet, 2000a) are actually in Zambia (Prendini et al., 2003). A single record from Maputo (Aguiar, 1978) is erroneous.

The status of many southern African scorpion species and subspecies, particularly within *Uroplectes* and the liochelid genera, remains contentious in spite of intensive taxonomic work. However, for the purpose of this study, the taxonomy reflected by the most recent published treatments was employed, according to which 140 species are presently recognised from Africa south of 15° latitude (Appendix 1). Subspecies were not considered, although some will certainly be elevated to species in future revisions.

2.2 Distributional data

Point locality data for each species were collated from available published locality records. Lamoral and Reynders' (1975) catalogue of the scorpions described from the Afrotropical Region up to December 1973 was used for all the early records. Remaining (post-1975) records were obtained from the following works: Eastwood (1977a, 1977b, 1978a, 1978b); Harington (1978, 1984, 2002); Lamoral (1978b, 1979, 1980b); Lourenço (1981, 1987a); Newlands (1980); Newlands and Martindale (1980); Newlands and Cantrell (1985); Fitzpatrick (1994a, 1994b, 1996, 2001); Newlands and Prendini (1997); Prendini (2000b, 2000c, 2001b, 2001c, 2003a, 2003b, 2003c, 2004a, 2004b, in press).

Literature records were supplemented with records of personally identified specimens deposited in the following collections: Austria: Zoologisches Institut und Naturhistorisches Museum, Universität Wien; Belgium: Musée Royal de l'Afrique Centrale, Tervuren; Denmark: Zoological Museum, University of Copenhagen; France: Museum National d'Histoire Naturelle, Paris; Germany: Forschungsinstitut und Natur-Museum Senckenberg, Frankfurt; Zoologisches Forschungsinstitut und Museum Alexander Koenig, Bonn; Zoologisches Institut und Zoologisches Museum, Universität Hamburg; Zoologisches Museum, Universität Humboldt, Berlin; Namibia: Desert Research Foundation of Namibia, Gobabeb; National Museum of Namibia, Windhoek; South Africa: Albany Museum, Grahamstown; Kruger National Park Reference Collection, Skukuza; KwaZulu-Natal Nature Conservation Service, Pietermaritzburg; McGregor Museum, Kimberley; Natal Museum, Pietermaritzburg; National Collection of Insects and Arachnids, Plant Protection Research Institute, Pretoria; National Museum, Bloemfontein; South African Museum, Cape Town; Transvaal Museum, Pretoria; University of Natal, Pietermaritzburg; Sweden: Göteborgs Naturhistoriska Museet, Göteborg; Naturhistoriska Riksmuseet, Stockholm; Zoologiska Institutionen, Lunds Universitet; Switzerland: Musée d'Histoire Naturelle, La-Chaux-de-Fond; Muséum d'Histoire Naturelle, Genève; The Netherlands: Zoologisch Museum, Universiteit van Amsterdam; U.S.A.: American Museum of Natural History, New York; California Academy of Sciences, San Francisco; Field Museum of Natural History, Chicago; Museum of Comparative Zoology, Harvard University, Cambridge, MA; U.S. National Museum of Natural History, Smithsonian Institution, Washington, DC; U.K.: The Natural History Museum, London; Zimbabwe: Natural History Museum of Zimbabwe, Bulawayo.

2.3 Georeferencing

All records of sufficient accuracy were collated to create a point locality geographical dataset for mapping distributional ranges. Only a small proportion of the records were initially accompanied by geographical coordinates or quarter-degree squares (QDS). Localities defined in degrees and minutes latitude and longitude (accurate to within 1.7 km^2) were used preferentially, whenever possible, owing to their high level of spatial resolution. All such localities were, in turn, converted to decimal-degree format for GIS input. Many localities were, however, only available in QDS notation (de Meillon et al., 1961). A decimal-degree approximation for the centroid of the QDS was used for these records, as a centroid would always be situated nearest to the average locality. For example, a locality registered as occurring in QDS SE1927Ac would be recorded as -19.375/27.125. As a QDS covers an area of approximately 26 km^2 in southern Africa (Poynton, 1964), this may result in some loss of resolution. Augmenting accurate locality records with QDS centroids yields distribution maps at least as accurate as maps using the QDS notation system (Newlands, 1980; Newlands and Martindale, 1980).

Available georeferences were checked for accuracy and an attempt was made to trace coordinates for as many remaining records as possible by reference to gazetteers, the official 1:250 000 and 1:500 000 topo-cadastral maps of Namibia, Botswana, Lesotho, South Africa and Swaziland published by the Government Printer of South Africa, and the GEOnet Names Server: http://164.214.2.59/gns/html/cntry_files.html. This database is based on gazetteers published by U.S. National Imagery and Mapping Agency for the U.S. Board on Geographic Names, which contain the standard names approved for official use, unapproved variant names, designations (cities, mountains, rivers, etc.) and coordinates: http://geonames.usgs.gov/bgngaz.html. The following regional gazetteers and lists of place-names were consulted: Gross (1920); Doidge (1950); Webb (1950); National Place Names Committee (1951, 1978, 1988); Ministry of Land and Natural Resources (1959); Gonçalves (1962); Davis & Misonne (1964); Poynton (1964); Copeland (1966); Ministry of Land and Mines (1967); SWALU (1967); Postmaster-General (1970); Surveyor-General (1970, 1988); Coaton and Sheasby (1972); Skead (1973); Haacke (1975); Penrith (1975, 1977, 1979, 1981a,b, 1987); Leistner and Morris (1976); Lamoral (1979); Bamps (1982); Raffle (1984); Poynton and Broadley (1985a,b, 1987); Raper (1987); Irish (1988); Polhill (1988); *Flora Zambesiaca*: *Gazetteer of Localities* (unpublished document, Bolus Herbarium Library, University of Cape Town); Herpetology Department, Transvaal Museum (unpublished records); Entomology Department, South African Museum (unpublished records).

Doubtful localities, discussed in the literature, and single, highly disjunct records that could not be verified, were omitted. After screening for errors, the final species presence-only dataset (*sensu* Mugo *et al.* 1995) was submitted to the GIS for mapping and spatial analysis. A total of 6 766 data points were collated.

2.4 Spatial analysis

Digital distribution maps were produced for each species by superimposing point locality records on a dataset representing the political boundaries of southern Africa (south of 15° latitude), using ArcView GIS Version 3.2 (Environmental Systems Research Institute, Redlands, CA). A spatial join was then conducted by superimposing scorpion distributions on a dataset representing the QDS grid of southern Africa, to determine scorpion hotspots, i.e. areas of high species richness and endemism (Myers et al., 2000), at the scale of a QDS (Lombard, 1995a, 1995b). Hotspots were based on measures of species richness (all 140 species) and endemism (including only southern African endemics), which may reflect centres of endemism, or regions of speciation, given that most scorpion species are sedentary (Polis, 1990; Harington, 1984), and probably experience minimal range-shifting.

Spatial joins were also conducted to determine to what extent scorpion distributions are explained by topography, major sand systems and river drainage in southern Africa. A topographic contour dataset was created from the GTOPO30 raster grid, obtained from the website of the U.S. Government Public Information Exchange Resource: http://edcdaac.usgs.gov/gtopo30/gtopo30.html. The dataset of sand systems was created by clipping and merging relevant polygons extracted from a dataset of African geology from the Department of Marine Geoscience, University of Cape Town, with polygons extracted from a dataset of Namibian landforms from the Namibian National Biodiversity Task Force (Barnard, 1998), downloaded from: http://www.dea.met.gov.na/programmes/biodiversity/countrystudy.htm. A da-taset of major rivers in southern Africa was created by merging polygons extracted from a dataset of Namibian rivers from the website of the Namibian National Biodiversity Task Force (Barnard, 1998) with a dataset of South African rivers from the *Surface Water Resources of South Africa 1990* (WR90) database (Midgley et al., 1994), of the Water Research Commission, South Africa.

3. RESULTS

3.1 Composition of the fauna

Scorpionidae and Buthidae respectively comprise 42 % and 34 % of the southern African scorpion species, whereas Liochelidae and Bothriuridae respectively comprise 22 % and 2 % thereof (Fig. 1; Table 1). Despite a similar proportion of species, buthids comprise 54 % of the southern African scorpion genera, whereas scorpionids comprise only 8 %; liochelids and bothriurids respectively comprise 22 % and 15 % thereof. Eighty-one percent of the southern African scorpion species belong to four genera: *Opistophthalmus*, *Parabuthus*, *Uroplectes*, and *Hadogenes*.

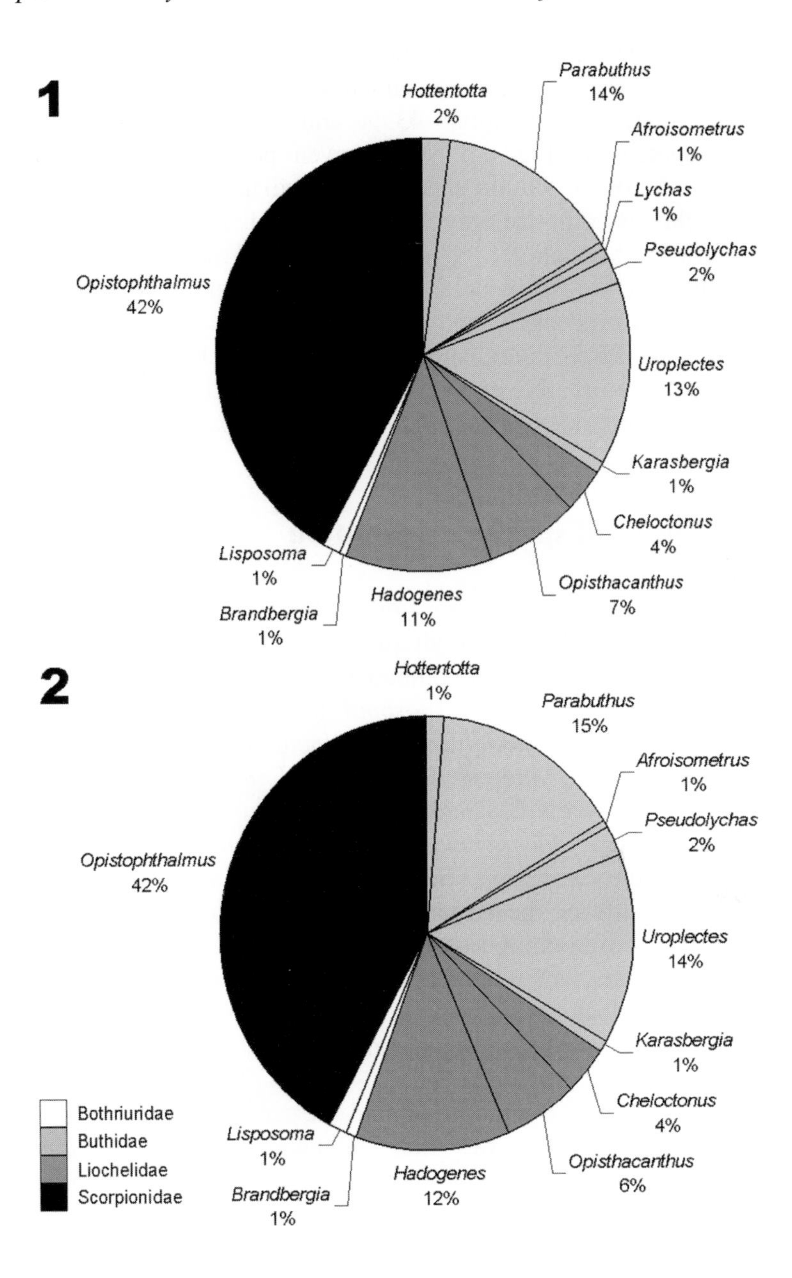

Figures 1,2. 1. Proportion of species in the families and genera of southern African scorpions. 2. Proportion of endemic species in the families and genera of southern African scorpions.

Forty-six percent of the genera and 96 % of the species of scorpions recorded from southern Africa are endemic to the subregion (Table 2). Scorpionids and buthids comprise 43 % and 34 % of southern African endemics, respectively (Fig. 2). Ninety-seven percent of the scorpionid species and 96 % of the buthid species recorded from southern Africa are endemic. Liochelids comprise only 21 % of the endemic scorpion species in southern Africa, although 94 % of the liochelid species recorded are endemic. Three of the six genera endemic to the subregion are buthids (*Afroisometrus*, *Karasbergia* and *Pseudolychas*), two are bothriurids (*Brandbergia* and *Lisposoma*), and the last is a liochelid (*Cheloctonus*). Eighty-four percent of the endemic southern African scorpion species belong to four genera: *Opistophthalmus*, *Parabuthus*, *Uroplectes*, and *Hadogenes* (Fig. 2).

3.2 Patterns of species distribution

Spatial analysis of the distributions of southern African scorpions reveals several patterns (Table 3). The southern African scorpion fauna is divided into distinct western and eastern components. Several genera are restricted to the western (*Brandbergia*, *Lisposoma* and *Karasbergia*) or eastern (*Afroisometrus*, *Lychas*, *Pseudolychas*, *Cheloctonus*, and *Opisthacanthus*) halves of the subregion, whereas others, e.g. *Hottentotta*, *Uroplectes* and *Hadogenes*, are evenly divided into distinct eastern and western elements. The two largest genera, *Opistophthalmus* and *Parabuthus*, dominate the arid western half of the subregion, but each also contains several species endemic to arid parts of the eastern half, such as the Eastern Cape and Limpopo provinces of South Africa.

The distributions of particular species are associated with geomorphological features of the landscape. Certain Buthidae and Scorpionidae are restricted to sand systems such as the Namib, the Kalahari, the Sandveld, the isolated sand systems of southern Namibia and the Northern Cape Province, South Africa, and the sandy areas of the Limpopo Depression. All species of the liochelid genera *Cheloctonus*, *Opisthacanthus*, and *Hadogenes*, with the exception of *C. crassimanus*, *C. jonesii* and *O. asper*, are associated with mountain ranges or regions of rugged topography, as are many species in the scorpionid genus *Opistophthalmus*. *Hadogenes* is absent from the Kalahari and vast areas of the Cape Middleveld, the Free State, and the Karoo, as well as from the Springbok flats of the Limpopo Province, the sandy Mozambique plains and Makatini flats of northern KwaZulu-Natal Province, and the sandy areas of

Table 1. Scorpion diversity in southern African countries, south of 15° latitude. Percentages of the total are provided in parentheses. Questionmarks reflect suspected occurrences that remain to be verified. Abbreviations as follows: A (Angola); B (Botswana); L (Lesotho); Ma (Malawi); Mo (Mozambique); N (Namibia); SA (South Africa); S (Swaziland); Za (Zambia); Zi (Zimbabwe); E (Extralimital).

Family	Genus	Sth. Africa	A	B	L	Ma	Mo	N	SA	S	Za	Zi	E
Bothriuridae	*Brandbergia*	1 (1)						1 (100)					
	Lisposoma	2 (1)						2 (100)					
	Total	3 (2)						3 (100)					
Buthidae	*Afroisometrus*	1 (1)						?	1 (100)			1 (100)	
	Hottentotta	3 (2)	1 (33)	1 (33)		?	1 (33)	2 (67)	2 (67)		1 (33)	1 (33)	1 (33)
	Karasbergia	1 (1)						1 (100)	1 (100)				
	Lychas	1 (1)				1 (100)	1 (100)		1 (100)		1 (100)	1 (100)	1 (100)
	Parabuthus	20 (14)	4 (20)	7 (35)		?	2 (10)	15 (75)	15 (75)		3 (15)	5 (25)	
	Pseudolychas	3 (2)					1 (33)		3 (100)	1 (33)			
	Uroplectes	19 (14)	3 (16)	5 (26)	2 (11)	1 (5)	6 (32)	10 (53)	15 (79)	4 (21)	4 (21)	6 (32)	
	Total	48 (34)	8 (17)	13 (27)	2 (4)	2 (4)	11 (23)	28 (58)	38 (79)	5 (10)	9 (19)	14 (29)	2 (4)
Liochelidae	*Cheloctonus*	5 (4)		?			1 (20)		5 (100)	1 (20)	?	?	
	Hadogenes	16 (11)	1 (6)	2 (13)		?	2 (13)	4 (25)	12 (75)	1 (6)	1 (6)	2 (13)	
	Opisthacanthus	10 (7)		1 (10)	2 (20)	2 (20)	2 (20)		6 (60)	3 (30)	?	2 (20)	2 (20)
	Total	31 (22)	1 (3)	3 (10)	2 (6)	2 (6)	5 (16)	4 (13)	23 (74)	5 (16)	1 (3)	4 (13)	2 (6)
Scorpionidae	*Opistophthalmus*	59 (42)	5 (8)	6 (10)	1 (2)	1 (2)	3 (5)	28 (47)	39 (66)	?	2 (3)	4 (7)	2 (3)
Total	Genera	13	5 (38)	6 (46)	3 (23)	5 (38)	9 (69)	8 (62)	11 (85)	5 (38)	6 (46)	7 (54)	7 (54)
	Species	140	9 (6)	22 (16)	5 (4)	5 (4)	19 (14)	63 (45)	100 (71)	10 (7)	12 (9)	22 (16)	6 (4)

Table 2. Scorpion endemism in southern African countries, south of 15° latitude. Percentages of the total are provided in parentheses. Abbreviations as follows: A (Angola); N (Namibia); SA (South Africa).

Family	Genus	Sth. Africa	A	N	SA
Bothriuridae	Brandbergia	1 (1)		1 (100)	
	Lisposoma	2 (1)		2 (100)	
	Total	3 (2)		3 (100)	
Buthidae	*Afroisometrus*	1 (1)			
	Hottentotta	2 (1)			
	Karasbergia	1 (1)			
	Parabuthus	20 (15)		5 (25)	3 (15)
	Pseudolychas	3 (2)			2 (67)
	Uroplectes	19 (14)		3 (16)	4 (21)
	Total	46 (34)		8 (17)	9 (20)
Liochelidae	*Cheloctonus*	5 (4)			4 (8)
	Hadogenes	16 (12)		1 (6)	8 (5)
	Opisthacanthus	8 (6)			3 (3)
	Total	29 (22)		1 (3)	15 (52)
Scorpionidae	*Opistophthalmus*	57 (43)	1 (2)	19 (32)	27 (46)
Total	Genera	6 (46)		2 (15)	
	Species	134 (96)	1 (1)	31 (22)	51 (36)

the Namib. The scorpion faunas occurring to the north and south of the Orange River are also distinctly different, sharing only 20 species that occur on both sides.

Most southern African scorpion species show discrete restricted distributional ranges (less than 50 QDS), with the exception of a few widespread species, e.g. *P. granulatus*, *O. carinatus* and *O. wahlbergii*. The distributional ranges of closely related species are invariably allopatric or parapatric, especially in the non-buthid genera *Cheloctonus*, *Opisthacanthus*, *Hadogenes*, and *Opistophthalmus*. The distributions of *Opisthacanthus* and *Hadogenes* are almost mutually exclusive, *Opisthacanthus* occupying the length of the eastern escarpment and the Cape Fold Mountains (excluding the Cedarberg) and *Hadogenes* occupying the interior plateau and the Cedarberg.

3.3 Hotspots

Hotspot analysis of species richness indicates that most parts of southern Africa contain at least one scorpion species (Fig. 3). The apparent absence of scorpions from large parts of Mozambique, northern Zimbabwe, and the central Kalahari in Botswana is a sampling artefact. Despite the bias caused by undersampling in these areas, coverage of southern Africa is fairly complete for an arthropod group.

Table 3. Patterns of scorpion distribution in southern Africa, expressed as the number of species with distributions in the west, east or centre of southern Africa; north, south or extending across the Orange River; associated with major sand systems or mountain ranges; and intersecting up to 400 quarter-degree squares (QDS), an index of range size. Percentages of the total are provided in parentheses. Abbreviations as follows: *Brandbergia* (B); *Lisposoma* (Li); *Afroisometrus* (A); *Lychas* (Ly); *Pseudolychas* (Ps); *Hottentotta* (Ho); *Karasbergia* (K); *Parabuthus* (Pa); *Uroplectes* (U); *Cheloctonus* (C); *Opisthacanthus* (Oh); *Hadogenes* (Ha); *Opistophthalmus* (Oo).

		B, Li	A, Ly, Ps	Ho	K	Pa	U	C, Oh	Ha	Oo	Total
Sth. Africa	Western	3 (2)		2 (1)	1 (1)	15 (11)	8 (6)		6 (4)	43 (31)	78 (56)
	Eastern		5 (4)	1 (1)		3 (2)	9 (6)	15 (11)	9 (6)	10 (7)	52 (37)
	Central					3 (2)	2 (1)			6 (4)	11 (8)
Sand Systems	Namib					4 (3)				7 (5)	11 (8)
	Kalahari					5 (4)				4 (3)	9 (6)
	S Namibia-N Cape			1 (1)		4 (3)				4 (3)	9 (6)
	Sandveld									4 (3)	4 (3)
	Limpopo					3 (2)				2 (1)	5 (4)
Mountains	Escarpment		1 (1)			2 (1)		7 (5)	9 (6)	9 (6)	28 (20)
	Cape Fold						3 (2)	4 (3)	1 (1)	7 (5)	15 (11)
	Other	2 (1)		1 (1)		2 (1)	1 (1)	1 (1)	8 (6)	12 (9)	27 (19)
Orange River	North	3 (2)		1 (1)		8 (6)	5 (4)		2 (1)	24 (17)	43 (31)
	South					3 (2)	4 (3)		3 (2)	22 (16)	32 (23)
	Transriverine			1 (1)	1 (1)	8 (6)	2 (1)		2 (1)	6 (4)	20 (14)
Range size (QDS)	1-20	2 (1)	2 (1)	1 (1)		7 (5)	7 (5)	10 (7)	9 (6)	36 (26)	74 (53)
	20-50	1 (1)	3 (2)		1 (1)	2 (1)	3 (2)	2 (1)	2 (1)	16 (11)	30 (21)
	50-100			2 (1)		7 (5)	5 (4)	3 (2)	3 (2)	4 (3)	24 (17)
	100-200					4 (3)	2 (1)		1 (1)		7 (5)
	200-300						2 (1)			2 (1)	4 (3)
	300-400					1 (1)				1 (1)	2 (1)

Twenty-four primary hotspots (11-16 species) occur in the western third of southern Africa. Most are concentrated in northwestern, central, and southern Namibia, and the Richtersveld and Namaqualand regions of the Northern Cape Province, South Africa, with an isolated hotspot in the Breede River Valley, Western Cape Province, South Africa. Secondary hotspots (7-10 species) are concentrated in the same parts of Namibia, as well as the Richtersveld, Namaqualand, and the Western Cape. Additional secondary hotspots occur in Bushmanland and the southern Kalahari, Northern Cape Province, South Africa, the Eastern Cape, Limpopo and Mpumalanga provinces of South Africa, and the Eastern Highlands of Zimbabwe.

Hotspots of endemic species show a similar pattern (Fig. 4). Six primary hotspots (9-12 endemics) occur along the Aus Mountains and the Huib-Hoch Plateau of southern Namibia, the western escarpment, Namaqualand and Breede River Valley in the Northern and Western Cape provinces of South Africa, and the Eastern Cape Province. Secondary hotspots (6-8 endemics) are concentrated in central and southern Namibia, the Richtersveld, Namaqualand and Bushmanland, the Western Cape, Eastern Cape, Limpopo and Mpumalanga provinces.

Most regions of southern Africa contain at least one endemic species, except for Mozambique, northern Zimbabwe, and most of the Kalahari (no scorpion species are endemic to Botswana). The absence of endemics from northern Zimbabwe and the Kalahari cannot be attributed simply to collector bias, however, as some of these areas, particularly eastern Namibia, northern Botswana, and the northeastern border of Zimbabwe have been fairly well sampled. The distributions of several species in these areas extend beyond the boundaries of southern Africa.

Two trends will be noted from the hotspot analyses. Firstly, most hotspots occur in the arid western half of southern Africa, or in arid parts of the eastern half, e.g. the Eastern Cape and the Lowveld of the Limpopo and Mpumalanga provinces, suggesting the influence of climate. Secondly, all hotspots occur in regions of rugged topography, complex geology, or substratal heterogeneity. The Namibian hotspots occur along the Skeleton Coast, the mountainous Kaokoveld and Damaraland, at the interface of the Central Namib Sand Sea and the western escarpment, and in the highlands on the interior plateau (e.g. the Khomas Hochland, the Karasberg, and the Huib-Hoch Plateau). The South African hotspots in the mountainous Richtersveld, Namaqualand and Bushmanland, the sand dunes of the southern Kalahari, and the western escarpment are further evidence of this pattern. Remaining hotspots are associated with the Cape Fold Mountains, the escarpment in Mpumalanga and the Eastern Highlands of Zimbabwe, the Lebombo Mountains, the Magaliesberg, the Waterberg-Soutpansberg

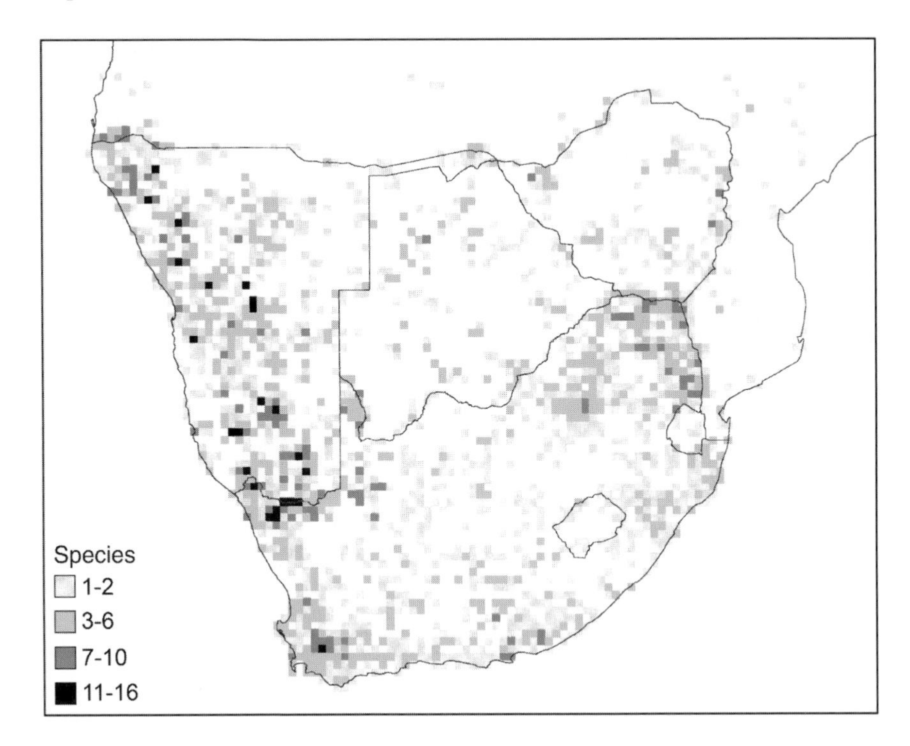

Figure 3. Hotspots of scorpion species richness in Africa, south of 15° latitude, expressed as the total number of species per quarter-degree square.

complex, and the sandy areas of the Limpopo Depression. The hotspot representing the Breede River Valley occurs at the junction of the north-south and east-west axes of the Cape Fold Mountains, and at a transition zone between fynbos and karoo vegetation types, respectively associated with higher and lower rainfall regimes. Notwithstanding the bias caused by undersampling in central Botswana and Mozambique, hotspots are in general poorly represented in areas of uniform topography, geology, and substratum, e.g. the Karoo, the Highveld, the central and northern Kalahari, and the Mozambique plain.

4. DISCUSSION

4.1 Models of speciation

Biogeography should explain the distribution of extant taxa in terms of historical factors together with the use of their contemporary ecology

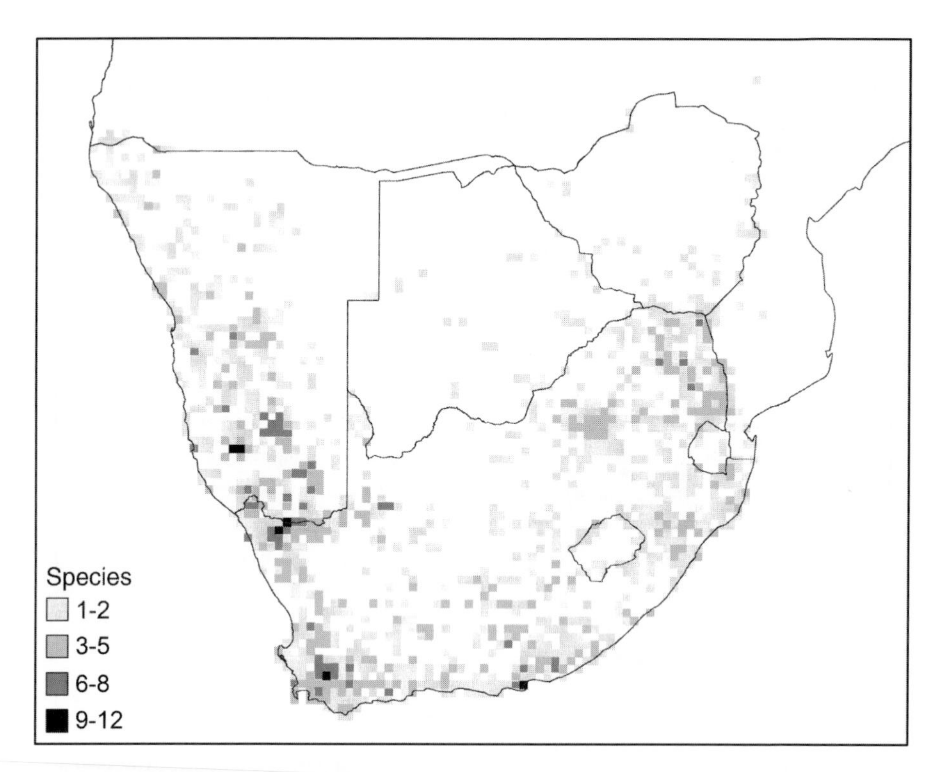

Figure 4. Hotspots of scorpion endemism in Africa, south of 15° latitude, expressed as the number of endemic species per quarter-degree square.

(Myers and Giller, 1988; Haydon et al., 1994). As fossil evidence is absent, biogeographic considerations of southern African scorpions must be derived primarily from ecological sources (Lamoral, 1978b). This approach is considered valid, however, as the habits of scorpions appear to have altered little over time (Jeram, 1990b; Sissom, 1990). The high species richness and endemism of southern African scorpions may be attributed to two main factors: a variable climate and a topographically diverse landscape (associated with complex geology). These factors, together with the specialised ecological requirements of most scorpion species, appear to have promoted speciation, especially during periods of palaeoclimatic change.

Speciation is most often explained by the allopatric model: species evolve from components of larger populations isolated in refugia by a cessation of gene flow between them (Mayr, 1942, 1963). Cessation of gene flow between populations may result from dispersal or vicariance. Although the existence of a physical barrier to gene flow is implied by both, vicariance and dispersal constitute distinct models of disjunction differentiated as follows: vicariance implies that the splitting of an ancestral population was caused by the appearance of a barrier; dispersal implies that the splitting occurred by movement across a pre-existing barrier (Platnick and Nelson, 1978). Neither dispersal nor vicariance explanations should be discarded *a priori* as irrelevant for any particular group of organisms. Biogeographic analysis should allow us to choose objectively between these two types of explanations for particular taxa.

Scorpion dispersal is limited to terrestrial vagility. As aquatic and aerial dispersal are improbable, barring unusual circumstances such as transoceanic rafting and synanthropy (Newlands, 1973, 1978a), the biogeography of scorpion faunas separated by barriers is best explained through a vicariance model, terrestrial dispersal in turn being controlled at any time by various ecological requisites (Lamoral, 1978b). Given the specialised ecological requirements of southern African scorpions, it is feasible to construct a scenario of how repeated speciation might have produced current levels of diversity and endemism in the subregion, in a similar manner to that proposed by Koch (1977, 1981) for the scorpions of Australasia. The remainder of this paper is devoted to a discussion of the ecological and historical factors determining the distribution of southern African scorpions, insofar as these account for their diversity and endemism.

4.2 Ecological factors

4.2.1 Climate

Different genera and species of scorpions have adapted to xeric and mesic habitats in southern Africa, as elsewhere. At the broadest level, the distinct western and eastern elements of the southern African fauna confirm the importance of climate as a determinant of scorpion distribution in the subregion. Two components of climate are thought to affect scorpion distribution, viz. rainfall (relative humidity) and temperature. Rainfall has a noticeable effect on scorpion distribution in Israel (Warburg and Ben-Horin, 1978; Warburg et al., 1980), whereas temperature appears to be the primary factor limiting the southward expansion of tropical scorpion species on the east coast of Australia (Koch, 1977, 1981). The absence of scorpions from higher latitudes may be indicative of the importance of temperature on a global scale (Gromov, 2001); few species occur where snow is a common occurrence (Williams, 1987; Polis, 1990).

Previous authors (Hewitt, 1925; Lawrence, 1942, 1952; Newlands, 1978a, 1978b) observed that several scorpion genera are restricted to, or dominant in, the arid to semi-arid western half of southern Africa, whereas others are restricted to the humid east, a pattern confirmed in the present investigation. The distributions of *Hottentotta* and *Parabuthus* have been related to the 600 mm isohyet, as these genera are generally absent from areas receiving more than 600 mm of annual rainfall (Newlands, 1978b). In contrast, *Afroisometrus*, *Lychas*, *Pseudolychas*, *Cheloctonus* and *Opisthacanthus* require high rainfall, and several species in these genera are endemic to Afromontane forests (Lawrence, 1942, 1953). Their limited distributions may be partly indicative of their relative humidity requirements. *Lisposoma joseehermanorum* is restricted to humid subterranean habitats such as caves, and leaf litter under boulders on south and east-facing slopes in the Otavi Highlands, which receive the highest rainfall in Namibia (Lamoral, 1978b, 1979; Prendini, 2003b). The restricted distributions of certain silvicolous *Uroplectes* (e.g. *U. insignis* on the forested eastern slopes of Table Mountain) may likewise be related to their narrow climatic requirements.

The distinction between the western and eastern components of the southern African scorpion fauna is repeated within more widespread genera such as *Opistophthalmus*, *Parabuthus*, *Hottentotta*, *Hadogenes* and *Uroplectes*, each of which contains particular species or groups of related species that are restricted to either the humid east or the arid west of southern Africa. Newlands (1978a: 688) stated "water is probably the most important single factor affecting the distribution of scorpions in southern Africa in that it is largely responsible for the type of vegetation in any given region and, to an extent, the daily temperature fluctuation of the soil surface."

4.2.2 Substratum

The importance of the substratum in scorpion ecology and distribution is well known (San Martín, 1961; Smith, 1966; Lamoral, 1978a, 1978b, 1979; Bradley, 1986, 1988; Bradley and Brody, 1984; Polis and McCormick, 1986; Fet et al., 1998, 2001; Prendini, 2001a). Lamoral (1978a, 1978b, 1979) established that the distribution of *Opistophthalmus* in southern Africa is determined primarily by soil hardness and, to a lesser degree, by soil texture, each species being restricted to soils within a certain range of hardness, rather than to a particular soil type. Lamoral (1978b: 305) concluded that "the nature of the substratum, taken in its broadest possible definition, is probably the most important single factor that has and still determines the distribution of scorpions ... the nature of the substratum is affected to a greater or lesser extent by vegetation, which is in turn partly the result of prevailing climatic conditions."

Newlands (1978a) and Lamoral (1979) independently classified the southern African scorpions according to their habitat predilections, both recognising a distinction between arboreal, rock-dwelling and burrowing species. Rock-dwelling species were further subdivided into species that inhabit crevices and species that shelter under stones, whereas burrowing species were subdivided according to their occurrence in soft or hard substrata. Bradley (1988) and Polis (1990) redefined these ecomorphotypes for all scorpions (see also Tikader and Bastawade, 1983; Fet et al., 1998). Prendini (2001a) introduced the concept of substratum-specialization and identified stenotopic vs. eurytopic ecomorphotypes, which were then applied to the southern African scorpion species.

Most species of *Parabuthus*, *Opistophthalmus*, and *Cheloctonus* are fossorial. They are divided into psammophilous and pelophilous species (Lawrence, 1969; Newlands, 1972a; Prendini, 2001a). Psammophilous and semi-psammophilous species of *Parabuthus* and *Opistophthalmus* display ecomorphological adaptations to increase locomotor and burrowing efficiency in loose sand (Lawrence, 1969; Newlands, 1972a, 1978a; Polis, 1990; Prendini, 2001a). These adaptations are exaggerated in ultra-psammophilous species (Newlands, 1972a) that occupy shifting sand dune environments (e.g. *O. flavescens* and *O. holmi*). All psammophilous scorpions are specialists, poorly adapted to life outside their sandy habitats. They are unable to burrow in harder or coarser substrata (Polis, 1990), and their distributions are restricted to sand systems. In southern Africa, such habitats occur in the northern and southern Namib, the Kalahari, southern Namibia, the Northern Cape Province, and the Limpopo Province. As species that burrow in soft, sandy soils tend to be restricted to a smaller range of substratum hardness than those burrowing in harder substrata, only a few (e.g. *P. granulatus*) are widely distributed, tracking generalised sandy environments across southern Africa.

Pelophilous species burrow in sandy loam and clay soils, and display ecomorphological adaptations to assist with loosening these compacted substrata (Newlands, 1972a, 1972b, 1978a; Polis, 1990; Prendini, 2001a). Pelophilous species of *Opistophthalmus* are cheliceral burrowers (Newlands, 1972b; Eastwood, 1978b). In contrast, *Cheloctonus jonesii*, a liochelid common in black turf soils (Newlands, 1978a), is a pedipalpal burrower that uses its large, rounded pedipalp chelae for burrowing in this extremely hard clayey substratum (Harington, 1978). Several pelophilous species, e.g. *O. glabrifrons*, are widely distributed, presumably because they are able to burrow in a greater range of soil hardness than is possible for psammophilous species.

Substratum hardness is carried to the extreme in the rock habitats of lithophilous scorpion species, specialists adapted to life in the narrow cracks and crevices of rocks (Newlands, 1972b, 1978a; Polis, 1990; Prendini 2001a, 2001b). The ecomorphological adaptations that characterise lithophilous scorpions facilitate locomotion on rock but hinder locomotion on other substrata. These scorpions are therefore usually restricted to regions of mountainous or rugged topography, their distributions often closely matching the boundaries of mountain ranges and geological formations (Newlands, 1972b, 1978a, 1980; Prendini, 2001b). Lithophilous adaptations are exaggerated in the southern African genus *Hadogenes*, all species of which inhabit the cracks and crevices of weathered outcrops consisting usually of fine-grained quartzite or igneous rocks such as granite, norite, syenite, diorite, and gabbro (Newlands, 1980). *Hadogenes* are absent from vast areas of southern Africa that are either rockless or contain unsuitable geology. Most *Cheloctonus* and *Opisthacanthus* are generalist lithophiles, sheltering under rocks or logs, in addition to occupying crevices (Lawrence, 1953; Newlands, 1972b). Their requirement for higher humidity appears to have restricted their distributions to the eastern escarpment, the Cape Fold Mountains, and the southern and eastern coastal plains. *Hadogenes*, which tolerate lower rainfall, are more widely distributed in the interior (Lawrence, 1942; Newlands, 1980). Males of certain *Opistophthalmus* that excavate shallow scrapes under rocks, or between slabs of rock (e.g. *O. austerus*, *O. karrooensis*, and *O. pallipes*), show similar ecomorphological adaptations to *Hadogenes* (Eastwood, 1978b; Prendini, 2001a). Such semi-lithophilous species are characterised by discrete, allopatric distributions in particular mountain ranges.

Only four obligate corticolous species, that shelter in holes or under the loose bark of trees, often several metres above the ground, occur in southern Africa (Lamoral, 1979; Newlands, 1978a; Newlands and Martindale, 1980; Prendini, 2001b): *Lychas burdoi*; *Uroplectes otjimbinguensis*; *U. vittatus*; *Opisthacanthus asper*. Corticolous species display few ecomorphological adaptations and are widely distributed. All except *U. otjimbinguensis* are restricted to the eastern half of southern Africa.

Lapidicolous scorpions, which shelter under stones or any other available cover, are habitat generalists (Newlands, 1978a; Lamoral, 1978b, 1979; Prendini, 2001a), displaying few ecomorphological adaptations and varied, often widespread distributions (e.g. *U. planimanus* and *U. triangulifer*), governed primarily by climate. In southern Africa, most species of *Hottentotta*, *Pseudolychas*, and *Uroplectes* are lapidicolous. Several normally ground-dwelling species occasionally shelter in arboreal habitats (e.g. *H. conspersus*, *P. villosus*, *U. formosus*, *U. insignis*, *U. lineatus*, and *U. olivaceus*) or are epigeic on vegetation while foraging nocturnally (e.g. *P. villosus* and *U. gracilior*) (Lamoral, 1978b, 1979).

4.2.3 Biotic interactions

Scorpion-scorpion interactions affect the distributions of many scorpion species (Williams, 1970, 1980; Koch, 1977, 1978, 1981; Lamoral, 1978a, 1978b; Shorthouse and Marples, 1980; Bradley and Brody, 1984; Polis and McCormick, 1986, 1987). The exact nature of these interactions is controversial, two main hypotheses with similar predictions having been presented (Polis, 1990): exploitation competition for prey or homesites and interference, manifested as aggression or intraguild predation among potential competitors. Exploitation competition for limited resources selects for ecological divergence of species from one another in resource use, whereas interference competition selects for subordinate species' avoidance of dominant species (Simberloff, 1983). Divergence and avoidance may decrease habitat overlap and even produce competitive exclusion in certain regions, thereby resulting in geographical allopatry or parapatry; e.g. in Australia, *Urodacus* Peters, 1861 species are almost totally absent where *Liocheles waigiensis* (Gervais, 1843) is abundant (Koch, 1977, 1981). Alternatively, the decrease in habitat overlap caused by divergence or avoidance may allow coexistence, resulting in sympatry; e.g. *Hadrurus arizonensis* Ewing, 1928, *Paruroctonus luteolus* (Gertsch and Soleglad, 1966), *Paruroctonus mesaensis* (Stahnke, 1957), and *Vaejovis confusus* Stahnke, 1940 in California (Polis and McCormick, 1986, 1987). In addition, intraguild predation may allow coexistence by preventing exclusion of competitively inferior species through differential exploitation of competitively superior species (Polis and Holt, 1992).

Competition theory predicts that coexistence, and thus sympatry, will be associated with niche differences and "resource partitioning" (Simberloff, 1983): decreased overlap in the use of critical resources, different spatio-temporal patterns, and differences in body size. Evidence suggests that niche differences exist in space and time, but not in the use of food (Polis, 1990). Although there are exceptions (Main, 1956; Koch, 1977, 1981; Lourenço, 1983b), most scorpion species are generalists, eating any prey they are able to capture (Polis, 1979; McCormick and Polis, 1990). Size differences may divide prey eaten by adults of different sizes. For example, in southern Africa, sympatric species of *Opistophthalmus* often differ in size (e.g. *O. flavescens* and *O. holmi* of the Namib, or *O. wahlbergi* and *O. concinnus* of the Kalahari), as do sympatric species of *Hadogenes* (e.g. *H. tityrus* and *H. taeniurus* or *H. zumpti*), implying a difference in ecological niche. However, as adults of large species must grow from a small size at birth, all species overlap in size (and associated size-related prey use) during some part of their lives and such developmental overlap greatly limits the effectiveness of body size differences to divide resources among those species that show a wide range in size during development (Polis, 1990).

Size is nonetheless of paramount importance in interference among scorpions. Intraguild predation is known to occur among at least 30 pairs of scorpion species at six North American sites (Polis et al., 1981; Bradley and Brody, 1984; Polis and McCormick, 1986, 1987; Polis, 1990), and has personally been recorded in southern Africa between *Parabuthus granulatus* and *Opistophthalmus wahlbergi*, *Parabuthus transvaalicus* and *Hadogenes troglodytes*, *H. troglodytes* and *Opistophthalmus boehmi*, *H. troglodytes* and

Opistophthalmus glabrifrons. In California, the impact of intraguild predation by *Paruroctonus mesaensis* and *Hadrurus arizonensis* is reflected in niche shifts among the smaller prey species (*Paruroctonus luteolus* and *Vaejovis confusus*), which are relatively uncommon in the habitats occupied by the intraguild predators (Polis and McCormick, 1986, 1987). The subordinate species occupy microhabitat refuges, avoiding ground predation by foraging primarily from burrow entrances (*P. luteolus*) or on vegetation (*P. luteolus* and *V. confusus*). This is also seen in southern Africa, where smaller species of *Uroplectes* and *Opistophthalmus*, as well as juveniles of larger species (e.g. *Parabuthus*) forage on vegetation, avoiding predation by larger *Parabuthus* and *Opistophthalmus*, including adult conspecifics.

The effect of substratum hardness is an important spatial factor governing the coexistence of different scorpion species. Koch (1977, 1978, 1981) argued that the distribution of species of the fossorial Australian genus *Urodacus* is determined by competition for homesites, citing differential burrow construction among different species as evidence of decreased habitat overlap. Lamoral (1978a, 1978b) argued similarly that specific soil-hardness preferences decrease competition for burrow sites and allow coexistence among southern African *Opistophthalmus*. Species with overlapping hardness preferences tend to be allopatric or parapatric, sympatry usually occurring only when two or more species occupy substrata of very different hardness within the same geographical locality (in which case they are allotopic), as shown in the Kalahari species *O. carinatus* and *O. wahlbergi*, which burrow in calcrete and aeolian sand, respectively (Lamoral, 1978a, 1978b). This may, in part, account for the discrete distributions of most species of *Opistophthalmus*. Sympatric species of *Opistophthalmus* may also coexist by differential burrow location: in northern Namaqualand, *O. granifrons* burrows on sandy-loam flats at the base of inselbergs where *O. peringueyi* occupies a semi-lithophilous niche under stones. Similarly, *O. boehmi* burrows in sandy-loam flats at the base of the northern slopes of the Soutpansberg, where *O. lawrencei* burrows under stones on the lower slopes (Newlands, 1969).

More fossorial buthid species are observed in sympatry than is the case among non-buthids, but this may be due to more subtle size differences and spatio-temporal microhabitat utilisation. In the southern Kalahari, five fossorial *Parabuthus* species (*P. granulatus*, *P. kalaharicus*, *P. kuanyamarum*, *P. laevifrons*, and *P. raudus*) coexist on the same sand dunes by means of differences in size and spatial occurrence on the dune. Sympatric scorpion species may also differ in foraging station: some hunt in vegetation, others forage on the ground, some "doorkeep" at their burrow entrance; some move constantly during foraging, and others are sit-and-wait foragers (Polis, 1990). In the Kalahari, *P. granulatus*, *P. kalaharicus* and *P.*

kuanyamarum are active ground-surface foragers, *P. laevifrons* is epigeic on vegetation, and *P. raudus* is a sit-and-wait forager.

Presumably, exploitation competition for the restricted, two-dimensional rock crevice habitat, or interference through intraguild predation by the larger *Hadogenes* has prevented coexistence with the lithophilous species of *Cheloctonus* and *Opisthacanthus*, the distributions of which are almost entirely allopatric with *Hadogenes*. Exceptions are observed only when there are niche differences. For example, the pelophilous *C. jonesii* is sympatric with the lithophilous *H. newlandsi* and the corticolous *O. asper* in Limpopo Province, South Africa. Generally, species of *Hadogenes* occupy hotter, drier habitats than the smaller lithophilous *Opisthacanthus* and *Cheloctonus*, suggesting that there is also a difference in microclimatic requirements between them. In the Mpumalanga Province, South Africa, *H. bicolor* and *O. validus* occur in sympatry (e.g. at Blyderivierspoort and Haenertsburg), but they are allotopic, *H. bicolor* inhabiting exposed rock outcrops and *O. validus* inhabiting outcrops under forest cover (Newlands, 1980). However, climatic differences fail to explain the replacement, by *Hadogenes minor* in the Cedarberg, of *Opisthacanthus*, which occurs throughout the rest of the Cape Fold Mountains, where geology and climate are largely similar. Exploitation competition or intraguild predation seems a more plausible explanation. Almost all congeneric species of *Hadogenes*, *Cheloctonus* and *Opisthacanthus* occupy discrete, allopatric or parapatric distributional ranges, presumably for the same reason (Newlands, 1980). Exceptions are observed only when there are niche differences. For example, the pelophilous *C. jonesii*, lithophilous *O. laevipes* and corticolous *O. asper* are sympatric in Swaziland and the Mpumalanga and KwaZulu-Natal provinces, South Africa.

4.3 Historical factors

4.3.1 Continental drift

As Gondwanaland fragmented, each of the southern continents carried with it a sample of the ancestral biota. This is well illustrated by several groups (MacArthur, 1972), including scorpions. Bothriurid scorpions, also known from South America, Australia and India, are represented in southern Africa by the genera *Brandbergia* and *Lisposoma* (Francke, 1982; Prendini, 2000a, 2003a, 2003b). Each contains species endemic to northern Namibia that display distributions typical of deserticolous palaeogenic elements (Stuckenberg, 1962). Lamoral (1978b, 1979) suggested that *Lisposoma* is a relict of a formerly tropical forest-dwelling ancestral element that survived the onset of aridification in Miocene times by resorting to a semi-endogean existence, and offered the euedaphic habitat of *L. joseehermanorum* as supporting evidence. The subsequent adoption of a lapidicolous habitat by *L. elegans* may have contributed to its present broader distribution and occurrence in habitats considerably drier than those inhabited by *L. joseehermanorum* (Prendini, 2003b).

A similar evolutionary scenario is proposed for the monotypic buthid genus *Karasbergia*, a fossorial endemic of rocky desert habitats in southern Namibia and the Northern Cape Province, South Africa (Lamoral, 1978b, 1979; Prendini, in press). This species, possibly related to *Charmus* Karsch, 1879 from India and Sri Lanka (Prendini, in press; E.S. Volschenk, pers. comm.), may also be considered a deserticolous palaeogenic element.

Taking possible vicariance and dispersal into account, few affinities exist between the scorpion faunas of the Afrotropical and Neotropical regions. There are, however, three notable exceptions. The first of these is the buthid genus *Ananteris* Thorell, 1891, with 19 species in Central and South America (including the island of Trinidad), and a single species from Guinea and Guinea-Bissau in West Africa, formerly placed in a monotypic genus, *Ananteroides* Borelli, 1911 (Gonzalez-Sponga, 1972, 1980; Lourenço, 1982, 1984b, 1985, 1987b, 1991c, 1993; Lourenço and Flórez, 1989; Fet and Lowe, 2000). The transfer of *Ananteroides feae* Borelli, 1911 to *Ananteris* by Lourenço (1985) was not justified phylogenetically and the monophyly of *Ananteris* remains to be tested.

The second exception is the diplocentrid genus *Heteronebo* Pocock, 1899, currently comprising 15 species. Thirteen species of *Heteronebo* are endemic to the Greater and Lesser Antilles in the Caribbean (Francke, 1978; Francke & Sissom, 1980; Armas, 1981, 1984, 2001; Stockwell, 1985; Armas & Marcano Fondeur, 1987; Prendini, 2000a; Sissom & Fet, 2000). However, two species inhabit Abd-el-Kuri Island off the Horn of Africa, east of Somalia and south of Yemen (Pocock, 1899; Francke, 1979; Sissom, 1990; Prendini, 2000a). Additional records of *Heteronebo*, probably representing undescribed species, were recently obtained from the nearby island of Socotra (W. Wranik & B. Striffler, pers. comm.). Whether the New and Old World species of *Heteronebo* are congeneric remains unclear. Prendini's (2000a) phylogenetic analysis confirmed the monophyly of *Heteronebo* based on the monophyly of a New and an Old World exemplar species of the genus. However, the grouping of New and Old World *Heteronebo* was weakly supported and may not withstand analysis with additional taxa and characters.

The liochelid genus *Opisthacanthus* (Lamoral, 1978b, 1980a) is the third exception. *Opisthacanthus* currently contains seven species in northern South America, southern Central America, the Cocos Islands off the coast of Costa Rica, and Hispaniola Island in the Caribbean, twelve species in Africa, and two species in Madagascar (Lourenço, 1980, 1981, 1987a, 1988,

1991a, 1995, 1997, 2001; Fet, 2000b; Lourenço & Fé, 2003). The origin of the Neotropical species is contentious. Newlands (1973, 1978a) proposed that they dispersed from Africa by trans-Atlantic rafting. However, a more parsimonious conclusion is that they are Gondwana relicts, which reached the Neotropics prior to the African disjunction in the late Cretaceous, and subsequently evolved in isolation (Francke 1974; Lamoral, 1980a; Lourenço, 1984a, 1989; Sissom, 1990). Both hypotheses rest on the assumption that *Opisthacanthus* is monophyletic, an assumption that was questioned on the basis of morphological evidence (Prendini, 2000a). Most of the African species of *Opisthacanthus* appear to be more closely related to other African liochelid genera (e.g., *Cheloctonus*), than to the Neotropical species. The African species of *Opisthacanthus* are not monophyletic either. *Opisthacanthus lecomtei* is more closely related to the South American species, placed in subgenus *Opisthacanthus*, whereas the others, traditionally placed in subgenus *Nepabellus*, are more closely related to *Cheloctonus* (Prendini, 2000a). The relationships of the Malagasy species of *Opisthacanthus*—for which a new subgenus, *Monodopisthacanthus* Lourenço, 2001, was recently erected—to the other species of the genus, also remain unclear. Determining the phylogenetic relationships of the species of *Opisthacanthus* to the other liochelid genera has implications for understanding the biogeography of Liochelidae as a whole.

Opisthacanthus and the related genus *Cheloctonus* conform closely to the criteria proposed for montane palaeogenic elements by Stuckenberg (1962), reiterating a distribution pattern displayed by many other groups of southern African invertebrates. Species of both genera occur in humid habitats at high altitude, on the south- and east-facing slopes of the Drakensberg escarpment (including the Eastern Highlands of Zimbabwe), and the Cape Fold Mountains. *Lychas burdoi* and the related buthids *Afroisometrus* and *Pseudolychas* have similar ecological requirements, and distributional ranges, suggesting that these may also be montane palaeogenic elements.

The genera *Brandbergia*, *Lisposoma*, *Afroisometrus*, *Karasbergia*, *Pseudolychas*, *Cheloctonus*, and *Opisthacanthus* are all relatively basal in their respective clades, and represent an ancient, relictual element in a fauna otherwise characterised by highly derived and speciose genera, e.g., *Parabuthus*, *Hadogenes*, and *Opistophthalmus*, which probably radiated fairly recently in post-Miocene or Pliocene times (Lamoral, 1978b, 1979; Prendini, 2000, 2001b, 2003a, 2003b, 2004b; Prendini et al., 2003).

4.3.2 Geomorphology and climate

Mountain ranges exert a significant influence on rainfall, temperature and vegetation, and may thus have indirectly promoted speciation in taxa with restricted climatic tolerances during periods of palaeoclimatic change. For example, species of *Cheloctonus*, *Opisthacanthus*, *Pseudolychas*, and *Uroplectes* requiring high humidity probably experienced vicariance when formerly continuous patches of Afromontane forest or fynbos became isolated along the eastern escarpment during dry periods from the Pliocene to the Pleistocene (van Zinderen Bakker, 1962, 1976, 1978), and speciated in consequence. The sister species *O. validus* and *O. lamorali*, from the Drakensberg of South Africa and Eastern Highlands of Zimbabwe, respectively, separated by at least 300 km of dry savanna in the Limpopo Depression, are one of several examples. Speciation by vicariance associated with climate-induced expansion and contraction of montane forest or fynbos habitats (e.g. in "Pleistocene refugia") is the favoured explanation for high levels of endemism among montane palaeogenic elements (Stuckenberg, 1962). Most southern African species of *Cheloctonus*, *Opisthacanthus*, and *Pseudolychas*, and several southern African species of *Uroplectes*, may be products of such processes.

Expansion and contraction of the "arid corridor", connecting arid southwestern and northeastern Africa during successive wet and dry phases from the Pliocene to the Pleistocene (Balinsky, 1962; van Zinderen Bakker, 1969) could similarly have induced speciation among arid-adapted species, explaining the existence of *Parabuthus*, *Hadogenes* and *Opistophthalmus* in northeastern Africa (Prendini, 2001b, 2001c; Prendini et al., 2003).

Topography may also be responsible for speciation in scorpions by acting as barriers to dispersal (Koch, 1977). Some of the more important mountain barriers in southern Africa include the Great Escarpment, the Namaqua Highlands, Cape Fold Mountains, Lebombo Mountains, Magaliesberg, Waterberg, and Soutpansberg of South Africa, and the highlands of Namibia, e.g. the Khomas Hochland, Aus Mountains, Huib-Hoch Plateau, Otavi Highlands, and Karasberg. Many of these mountain ranges separate related scorpion species (Lamoral, 1978b).

There is also evidence that sand systems constitute barriers to non-psammophilous scorpions. Lamoral (1978b, 1979) hypothesised that, since the Pliocene, the Kalahari has acted as an agent of vicariance preventing migration of scorpion species along the "arid corridor". Lamoral (1978b) provided an example of how mountains and sand systems may have contributed to the present distribution of *Hottentotta*, a largely Palaearctic genus, in southern Africa. By the end of the Oligocene, the ancestral species of this genus had migrated as far south as their present distributions, facilitated by prevailing tropical and subtropical climates and vegetation. The emergence of the Kalahari sand system during the Pliocene (van Zinderen Bakker, 1975) induced the vicariance of the southwestern and northeastern species groups, resulting in the speciation of *H. trilineatus* in the northeast. Western expansion of the Kalahari forced the ancestral southwestern group to migrate west. Once the western front of the Kalahari sands had reached the Central Highlands up to the 1 500 m contour, presumably in the Pleistocene when climate was generally colder, the ancestral range was effectively bisected and a vicariance established that caused the speciation of *H. conspersus* in the north and *H. arenaceus* in the south. A similar scenario was proposed to account for the distributions of various species of *Parabuthus*, *Uroplectes*, and *Opistophthalmus* north (*P. gracilis*, *P. kraepelini, U. otjimbinguensis*, *U. planimanus*, *O. brevicauda*, *O. cavimanus*, *O. coetzeei*, and *O. ugabensis*) and south (*P. laevifrons*, *P. nanus*, *P. schlechteri*, *U. gracilior*, *U. longimanus*, *U. schlechteri*, *U. tumidimanus*, *O. fitzsimonsi*, *O. gigas*, *O. haackei*, *O. intercedens*, *O. opinatus* and *O. schultzei*) of the Namibian Central Highlands (Lamoral, 1978b). This hypothesis rests on the assumption that the species in question are intolerant of sandy environments and fails to account for the distributions of semi-psammophiles such as *P. gracilis*, *P. laevifrons*, *P. nanus*, *O. coetzeei*, *O. fitzsimonsi*, and *O. intercedens*. Moreover, there is no reason to expect that, for the full duration of the Pleistocene, with its fluctuating warm and cold periods (van Zinderen Bakker, 1962, 1975, 1978), the Central Highlands would have constituted an ecological barrier to the ancestors of pelophilous *Opistophthalmus* which burrow under stones (e.g. *O. cavimanus*, *O. opinatus* and *O. schultzei*), or to species with semi-lithophilous adults (e.g. *O. brevicauda*, *O. ugabensis*, *O. gigas*, and *O. haackei*).

4.3.3 Substratum-specialization

Hotspots of species richness are located primarily in regions of complex topography and geology. The large number of species restricted to particular mountain ranges suggests that speciation often occurs in mountainous areas, rather than adjacent to them. Thus it appears as though intermontane valleys and depressions commonly act as barriers, rather than the other way around.

This may be related to the substratum-specialization of many scorpions, discussed extensively elsewhere (Prendini, 2001a). For example, the erosion of a mountain range, inhabited by a lithophilous scorpion species, would create isolated populations on the resultant inselbergs and ridges. Gene flow would cease once there were no longer rocky outcrops between adjacent inselbergs or ridges, owing to the substratum-specialization of the scorpion species. Observations for *Hadogenes* suggest that very narrow valleys or plains may interrupt gene flow between adjacent populations (Newlands, 1978a, 1980; Newlands and Cantrell, 1985). In the Pretoria area, *H. gunningi* inhabits the Magaliesberg and ridges to the south, whereas *H. gracilis* occurs on a series of ridges running parallel 2-3 km north; the narrow valley between them is free of rock outcrops, and thus acts as a complete barrier to gene flow. The parapatric distribution patterns of *H. trichiurus* in the southern and eastern escarpment and *H. zuluanus* in the Lebombos reveal a similar process: the geomorphology of the hills in the area between differs in that the boulders are composed of basaltic rock, without cracks and crevices, and there are thus no suitable habitats for *Hadogenes* (Newlands, 1980).

Parts of mountain ranges may also have become separated by the invasion of aeolian sand during the Miocene or Pliocene (van Zinderen Bakker, 1975) leading to speciation in the isolated scorpion populations. For example, the Uri-Hauchab Mountains of the southern Namib are separated from the escarpment by a narrow belt of sand which blew in from the coast (Newlands, 1972c, 1978a, 1980). The isolation of these mountains probably took several million years, during which the isolated population of *Hadogenes* evolved into a new species, *H. lawrencei*, morphologically and genetically distinct from its sister species, *H. tityrus*, on the escarpment. As there are no rocks in the area between the Uri-Hauchab Mountains and the escarpment, the 26 km stretch of soft sand provides an insurmountable barrier to their dispersal.

Restricted ecological requirements probably account for the limited distributional ranges of most *Hadogenes*. Presumably, *Hadogenes*, which displays a suite of apomorphic character states, evolved from liochelid ancestors similar to *Opisthacanthus* and *Cheloctonus* (Lamoral, 1978b; Newlands, 1980), most of which occupy a similar, but more generalist lithophilous niche. Specialization in the lithophilous niche and the development of a greater tolerance for high temperatures and low rainfall may have enabled *Hadogenes* to exploit the arid to semi-arid interior of southern Africa, facilitating speciation by vicariance and, ultimately, producing a greater proportion of range-restricted endemics.

Speciation promoted by substratum-specialization may also be extended to fossorial taxa. Psammophilous and semi-psammophilous species, tracking deposits of a particular hardness, may have experienced vicariance when pockets of aeolian sand became separated from larger sand systems – a frequent occurrence during past periods of increased aridity such as the Pliocene, when wind action was believed to have transported vast quantities of sand over the western interior of southern Africa (van Zinderen Bakker, 1975; Lancaster, 1990). The occurrence of isolated populations of psammophilous species in areas of aeolian Kalahari sand support this notion: *P. granulatus*, *P. kuanyamarum* and *O. wahlbergii* occur in sandy areas northwest of the Soutpansberg, several hundred kilometers east of the main Kalahari sand system, but are not found on the intervening calcrete (Newlands, 1969, 1974).

Numerous similar examples of isolated psammophilous and semi-psammophilous *Parabuthus* and *Opistophthalmus* populations occur in

Namibia, especially the sandy regions of the south, and in the Richtersveld and Namaqualand of South Africa, where numerous small dunes or dune-fields, of mixed Namib or Kalahari sand origins, are isolated from the main sand systems against mountains or on sandless plains. Lamoral (1978b) suggested that psammophilous species probably evolved as a result of dispersal into the sand systems, rather than as a result of vicariance. Phylogenetic analyses show that psammophilous and semi-psammophilous species of *Parabuthus*, *Uroplectes* and *Opistophthalmus* are derived and must, therefore, have evolved after the sand systems became well established, speciating after dispersal into an environment that had previously constituted a barrier (Lamoral, 1978b, 1979; Prendini, 2001b). Post-Pliocene adaptation to the sandy substrata was proposed to account for species inhabiting the Kalahari, the Namib, and the sandy areas of southern Namibia and the Northern Cape Province, South Africa. Although it is certainly correct that the psammophilous and semi-psammophilous lineages in these genera evolved after colonising these sandy environments, the actual speciation events that gave rise to current species must have involved vicariance. If cessation of gene flow is an assumed prerequisite for allopatric speciation, the only manner in which populations of a psammophilous or semi-psammophilous scorpion species, unable to burrow in harder soils, could become isolated is through vicariance, i.e. by translocation on small sand dunes or dune-fields that gradually became separated from larger sand systems. Even in historical time, crescent dunes, e.g. barchans, have been found to travel considerable distances over sandless plains, carrying their associated arthropod fauna with them (Penrith, 1979; Endrödy-Younga, 1982). Speciation by vicariance among psammophilous and semi-psammophilous scorpions probably contributed to the high diversity and endemism of *Parabuthus* and *Opistophthalmus*, and may, in part, explain the larger number of hotspots in the arid, sandy western half of southern Africa.

Pelophilous species are capable of utilising a greater range of substratum hardness than other fossorial scorpions (Lamoral, 1978a). Accordingly, they may be less easily isolated and less prone to speciation. The wide distributions of several pelophilous species (e.g. *O. glabrifrons*) contrast markedly with the often highly restricted distributions of most lithophilous and psammophilous species. Nevertheless, many pelophilous species of *Opistophthalmus* also display markedly discrete distributions. Such species tend to be concentrated in the mountainous regions of Namibia and the Northern, Western, and Eastern Cape provinces of South Africa, related species invariably occupying adjacent, but separate mountain ranges, or the intervening valleys. This is particularly clear among semi-lithophilous species of *Opistophthalmus* (Prendini, 2001a). The adults of such species construct shallow scrapes under rocks (Eastwood, 1978b), and may have speciated in a manner similar to that proposed for the lithophilous

Hadogenes, where the ecological requirement for rock cover promoted speciation by vicariance in the various discrete mountain ranges.

4.3.4 Drainage patterns

The most important agents of vicariance in the evolution of southern African scorpions, besides the sand systems and mountain ranges, are represented by major rivers, especially the Orange, the largest river crossing the arid interior and western coastline of southern Africa. The scorpion faunas occurring to the north and south of this river are distinctly different, sharing only 20 species that occur on both sides. All remaining species are endemic to the region north or south of the river and, given their low vagility, are probably autochthonous. The present disjunction is so complete that although several species can be found right up to the northern and southern banks of the river, they do not occur on the opposite sides, a pattern mirrored by flightless tenebrionid and scarabaeid beetles, and lepismatid silverfish (Penrith 1975, 1977; Irish 1990; Harrison 1999).

Other rivers that appear to have acted as agents of vicariance in the evolution of southern African scorpions include the Curoca, Kunene, Hoanib, Huab, Ugab, Swakop, Kuiseb, Koichab, Olifants and Berg rivers of the west coast, the Fish River, a major tributary of the Orange in the southern interior of Namibia, the Breë River of the south coast, and the Limpopo and Zambezi rivers of the east coast. In addition to the Limpopo and the Zambezi, the presence of numerous large, fast-flowing rivers emanating from the Natal Drakensberg section of the eastern escarpment might explain the absence of *Opistophthalmus* species from the region of South Africa between the Natal Drakensberg, the Kei River (Eastern Cape Province) and the Tugela River (KwaZulu-Natal Province). The Kei and Tugela rivers have both incised deep valleys that are inhabited by endemic species of *Opistophthalmus*, but no *Opistophthalmus* species are found in between them. The effects of river drainage on the evolution and biogeography of southern African scorpions are currently under further investigation by the author.

5. CONCLUSIONS

The southern African scorpion fauna contains the following components: ancient elements comprising endemic Gondwanaland relicts (*Brandbergia*, *Lisposoma*, *Karasbergia*, *Opisthacanthus*) and their endemic southern African derivatives (*Cheloctonus*); old elements now widespread in the Afrotropical and Oriental regions (*Hottentotta*, *Lychas* and *Uroplectes*), many of which are also endemic to the southern African subregion, and their endemic derivatives (*Afroisometrus*, *Pseudolychas*); recent elements derived in and largely endemic to the Afrotropical region (*Parabuthus*) or the southern African subregion (*Hadogenes* and *Opistophthalmus*).

Combined effects of geomorphology and palaeoclimatic change have acted as agents of vicariance in determining the evolution of the southern African scorpions, with their restricted climatic and substratal requirements, and are primarily responsible for the high species richness and endemism in the subregion (Prendini, 2001a). The progressive spatial restriction of groups of taxa, such as an entire genus, groups of related species, or single species, is the result of vicariance, in turn facilitated by limited vagility due to narrow ecological tolerances. In many cases, agents of vicariance are still in existence (e.g. mountain ranges, sand systems and palaeodrainage channels), and coincide with disjunct distribution patterns. Specialist lithophilous (*Hadogenes*), semi-psammophilous or psammophilous (*Parabuthus* and *Opistophthalmus*), and pelophilous (*Opistophthalmus*) taxa have speciated extensively, producing a high proportion of range-restricted endemics. *Opistophthalmus*, in particular, with 42 % of the southern African species, has radiated in a similar manner to the burrowing scorpionoid genus *Urodacus* in Australia (Koch, 1977, 1978, 1981). Presumably specialisation into psammophilous, pelophilous, and semi-lithophilous ecomorphotypes promoted rampant speciation by vicariance in both genera (Prendini, 2001a). This pattern contrasts with that observed in the relictual genera (e.g., *Lisposoma*, *Karasbergia*, *Pseudolychas*, *Cheloctonus*, and *Opisthacanthus*), which contain relatively few species.

ACKNOWLEDGEMENTS

This work began at the University of Cape Town and was financially supported by the Foundation for Research Development, Pretoria. Many people and organisations, too numerous to name individually, assisted in various stages of the data acquisition (providing specimens, data or literature), analysis, and development of these ideas. I particularly thank the following: Alec Brown; Amanda Lombard; Daniel Wilson; Elizabeth Scott; H. Peter Linder; Mike Picker; Peter Bradshaw; Steve Thurston; Timothy Crowe; all the curators and managers of the collections from which specimens and locality records were obtained. Bernhard Huber, Wilson Lourenço and Boris Striffler provided helpful comments on the manuscript.

REFERENCES

Aguiar, O. B., 1978, Alguns escorpiões de Moçambique, *Garcia de Orta, Sér. Zool., Jta. Investig. Cien. Ultramar* 7:107-114.

Armas, L. F., de, 1981, Primero hallazgos de la familia Diplocentridae (Arachnida: Scorpionida) en la Española, *Poeyana* **213**:1-12.

Armas, L. F., de, 1984, Escorpiones del Archipiélago Cubano. 7. Adiciones y enmiendas (Scorpiones: Buthidae, Diplocentridae), *Poeyana* **275**:1-37.

Armas, L. F., de, 2001, Scorpions of the Greater Antilles, with the description of a new troglobitic species (Scorpiones: Diplocentridae), in: *Scorpions 2001.* In Memoriam *Gary A. Polis*, V. Fet and P. A. Selden, eds., British Arachnological Society, Burnham Beeches, Bucks, UK, pp. 245-253.

Armas, L. F., de, and Marcano Fondeur, E. de J., 1987, Nuevos escorpiones (Arachnida: Scorpiones) de República Dominicana, *Poeyana* **356**:1-24.

Balinsky, B. I., 1962, Patterns of animal distribution on the African continent, *Ann. Cape Prov. Mus.* **2**:299-309.

Bamps, P., 1982, *Flora d'Afrique Centrale* (*Zaïre-Rwanda-Burundi*). *Répertoire de lieux de récolte*, Jardin botanique national de Belgique, Meise.

Barnard, P., 1998, *Biological Diversity in Namibia*: A *Country Study*, Namibian National Biodiversity Task Force, Windhoek.

Byalynitskii-Birulya, A. A. [Birula, A. A.], 1917a, *Arachnoidea Arthrogastra Caucasica.* Pars I. *Scorpiones*, Zapiski Kavkazskogo Muzeya (Mémoires du Musée du Caucase), Imprimerie de la Chancellerie du Comité pour la Transcaucasie, Tiflis, Russia. Ser A, 5:1-253. [Russian; English translation: Byalynitskii-Birulya, A. A., 1964, *Arthrogastric Arachnids of Caucasia.* 1. *Scorpions*, Israel Program for Scientific Translations, Jerusalem].

Byalynitskii-Birulya, A. A. [Birula, A. A.], 1917b, *Faune de la Russie et des pays limitrophes fondee principalement sur les collections du Musée zoologique de l'Académie des sciences de Russie. Arachnides* (*Arachnoidea*), 1:1-227. [Russian; English translation: Byalynitskii-Birulya, A. A., 1965, *Fauna of Russia and adjacent countries. Arachnoidea.* Vol. I. *Scorpions*, Israel Program for Scientific Translations, Jerusalem].

Birula, A. A., 1925, Skorpiologische Beiträge. 10. Zur geographischen Verbreitung zweier weitverbreiteter Skorpionen-Arten Palaearcticums, *Zool. Anz.* **63**:93-96.

Bradley, R., 1986, The relationship between population density of *Paruroctonus utahensis* (Scorpionida: Vaejovidae) and characteristics of its habitat, *J. Arid Environ.* **11**:165-172.

Bradley, R., 1988, The behavioural ecology of scorpions—a review, in: *Australian Arachnology. Australian Entomological Society Miscellaneous Publication* 5, A. D. Austin and N. W. Heather, eds., Watson, Ferguson and Co., Brisbane, pp. 23-36.

Bradley, R., and Brody, A., 1984, Relative abundance of three vaejovid scorpions across a habitat gradient, *J. Arachnol.* **11**:437-440.

Coaton, W. G. H., and Sheasby, J. L. 1972, Preliminary report on a survey of the termites (Isoptera) of South West Africa, *Cimbebasia Memoir* **2**:1-129.

Copeland, D. J. B., 1966, *Gazetteer of geographical names in the Republic of Zambia*, Government Printer, Lusaka.

Davis, D. H. S., and Misonne, X., 1964, Gazetteer of collecting localities of African rodents, *Documentation Zoologique* (*Tervuren*) **7**:1-100.

De Meillon, B., Davis, D. H. S., and Hardy, F., 1961, *Plague in southern Africa*, Vol 1, Government Printer, Pretoria.

Doidge, E. M., 1950, The South African fungi and lichens to the end of 1945. Locality index, *Bothalia* **5**:967-994.

Eastwood, E. B., 1977a, Notes on the scorpion fauna of the Cape. Part 1. Description of neotype of *Opisthophthalmus capensis* (Herbst) and remarks on the *O. capensis and O. granifrons* Pocock species-groups (Arachnida, Scorpionida, Scorpionidae), *Ann. S. Afr. Mus.* **72**:211-226.

Eastwood, E. B., 1977b, Notes on the scorpion fauna of the Cape. Part 2. The *Parabuthus capensis* (Ehrenberg) species-group; remarks on taxonomy and bionomics (Arachnida, Scorpionida, Buthidae), *Ann. S. Afr. Mus.* **73**:199-214.

Eastwood, E. B., 1978a, Notes on the scorpion fauna of the Cape. Part 3. Some observations on the distribution and biology of scorpions on Table Mountain, *Ann. S. Afr. Mus.* **74**:229-248.

Eastwood, E. B., 1978b, Notes on the scorpion fauna of the Cape. Part 4. The burrowing activities of some scorpionids and buthids (Arachnida, Scorpionida), *Ann. S. Afr. Mus.* **74**:249-255.

Endrödy-Younga, S., 1982, Dispersion and translocation of dune specialist tenebrionids in the Namib area, *Cimbebasia* (A) **5**:257-271.

Fet, V., 2000a, Family Scorpionidae Latreille, 1802, in: *Catalog of the Scorpions of the World* (*1758-1998*), V. Fet, W. D. Sissom, G. Lowe and M. E. Braunwalder, The New York Entomological Society, New York, pp. 427-486.

Fet, V., 2000b, Family Ischnuridae Simon, 1879, in: *Catalog of the Scorpions of the World* (*1758-1998*), V. Fet, W. D. Sissom, G. Lowe and M. E. Braunwalder, The New York Entomological Society, New York, pp. 383-408.

Fet, V., and Lowe, G., 2000, Family Buthidae C. L. Koch, 1837, in: *Catalog of the Scorpions of the World* (*1758-1998*), V. Fet, W. D. Sissom, G. Lowe and M. E. Braunwalder, The New York Entomological Society, New York, pp. 54-286.

Fet, V., Polis, G. A., and Sissom, W. D., 1998, Life in sandy deserts: The scorpion model, *J. Arid Environ.* **39**:609-622.

Fet, V., Capes, M., and Sissom, W. D., 2001, A new genus and species of psammophilic scorpion from eastern Iran (Scorpiones: Buthidae), in: *Scorpions 2001.* In Memoriam *Gary A. Polis*, V. Fet and P. A. Selden, eds., British Arachnological Society, Burnham Beeches, Bucks, UK, pp. 183-189.

FitzPatrick, M. J., 1994a, A new species of *Lychas* C.L. Koch, 1845 from Zimbabwe (Scorpionida: Buthidae), *Arnoldia Zimbabwe* **10**:23-28.

FitzPatrick, M. J., 1994b, A checklist of the *Parabuthus* Pocock species of Zimbabwe with a re-description of *P. mossambicensis* (Peters, 1861) (Arachnida: Scorpionida), *Trans. Zimbabwe Sci. Assoc.* **68**:7-14.

FitzPatrick, M. J., 1996, The genus *Uroplectes* Peters, 1861 in Zimbabwe (Scorpiones: Buthidae), *Arnoldia Zimbabwe* **10**:47-70.

FitzPatrick, M. J., 2001, Synonymy of some *Uroplectes* Peters, 1861 (Scorpiones: Buthidae), in: *Scorpions 2001.* In Memoriam *Gary A. Polis*, V. Fet and P. A. Selden, eds., British Arachnological Society, Burnham Beeches, Bucks, UK, pp. 191-193.

Francke, O. F., 1974, Nota sobre los generos *Opisthacanthus* Peters y *Nepabellus* nom. nov. (Scorpionida, Scorpionidae) e informe sobre el hallazgo de *O. lepturus* en la isla del Coco, Costa Rica, *Brenesia* **4**:31-35.

Francke, O. F., 1978, Systematic revision of diplocentrid scorpions from circum-Caribbean lands, *Spec. Publ. Texas Tech Univ.* **14**:1-92.

Francke, O. F., 1979, Additional record of *Heteronebo* from Abd-el-Kuri Island, P.D.R. Yemen (Scorpiones: Diplocentridae), *J. Arachnol.* **7**:265.

Francke, O. F., 1982, Are there any bothriurids (Arachnida, Scorpiones) in southern Africa?, *J. Arachnol.* **10**:35-39.

Francke, O. F., 1985, Conspectus genericus scorpionum, 1758-1982 (Arachnida, Scorpiones), *Occas. Pap. Mus., Texas Tech Univ.* **98**:1-32.

Francke, O. F., and Sissom, W. D., 1980, Scorpions from the Virgin Islands (Arachnida, Scorpiones), *Occas. Pap. Mus., Texas Tech Univ.* **65**:1-19.

Gonçalves, M. L., 1962, *Índice Toponímico de Moçambique*, Centro de Botânica de Junta de Investigações do Ultramar, Lisboa.

González-Sponga, M. A., 1972, *Ananteris venezuelensis* (Scorpionida, Buthidae) nueva especie de la Guayana de Venezuela, *Mem. Soc. Cienc. Nat. "La Salle"* **32**:205-214.

González-Sponga, M. A., 1978, Escorpiofauna de la Región Oriental del Estado Bolivar, Venezuela. Ed. Roto-Impresos, Caracas.

González-Sponga, M. A., 1980, *Ananteris turumbanensis*, nueva especie de la Guayana de Venezuela (Scorpionida, Buthidae), *Mem. Soc. Cienc. Nat. "La Salle"* **40**:95-107.

González-Sponga, M. A., 1984, *Escorpiones de Venezuela*, Cuadernos Lagoven, Caracas.

González-Sponga, M. A., 1996, *Guía para identificar escorpiones de Venezuela*, Cuadernos Lagoven, Caracas.

Gross, A., 1920, Gazetteer of Central and South Africa, *Geographia* 1-271.

Gromov, A.V., 2001. The northern boundary of scorpions in Central Asia, in: *Scorpions 2001.* In Memoriam *Gary A. Polis*, V. Fet and P. A. Selden, eds., British Arachnological Society, Burnham Beeches, Bucks, UK, pp. 301-306.

Haacke, W. D., 1975, The burrowing geckos of southern Africa, 1 (Reptilia: Gekkonidae), *Ann. Transv. Mus.* **29**:197–241.

Harington, A., 1978, Burrowing biology of the scorpion *Cheloctonus jonesii* Pocock (Arachnida: Scorpionida: Scorpionidae), *J. Arachnol.* **5**:243-249.

Harington, A., 1984, Character variation in the scorpion *Parabuthus villosus* (Peters) (Scorpiones, Buthidae): a case of intermediate zones, *J. Arachnol.* **11**:393-406.

Harington, A., 2001 [2002]. Description of a new species of *Opisthophthalmus* C.L. Koch (Scorpiones, Scorpionidae) from southern Namibia. *Rev. arachnol.* **14**:25-30.

Harrison, J. du G., 1999, Systematics of the endemic south-west African dung beetle genus *Pachysoma* MacLeay (Scarabaeidae: Scarabaeinae), M.Sc. Thesis, University of Pretoria.

Haydon, D. T., Crother, B. I., and Pianka, E. R., 1994, New directions in biogeography?, *Trends Ecol. Evol.* **9**:403-406.

Hewitt, J., 1918, A survey of the scorpion fauna of South Africa, *Trans. R. Soc. S. Afr.* **6**:89-192.

Hewitt, J., 1925, Facts and theories on the distribution of scorpions in South Africa, *Trans. R. Soc. S. Afr.* **12**:249-276.

Irish, J., 1988, Gazetteer of place names on maps of Botswana, *Cimbebasia* **10**:107-146.

Irish, J., 1990, Namib biogeography, as exemplified mainly by the Lepismatidae (Thysanura: Insecta), in: *Namib Ecology: 25 Years of Namib Research*, *Transvaal Museum Monograph* 7, M. K. Seely, ed., Transvaal Museum, Pretoria, pp. 61-66.

Jeram, A., 1990a, Book-lungs in a Lower Carboniferous scorpion, *Nature* **343**:360-361.

Jeram, A., 1990b, When scorpions ruled the world, *New Scientist* **126**:52-55.

Jeram, A., 1993 [1994a], Scorpions from the Viséan of East Kirkton, West Lothian, Scotland, with a revision of the infraorder Mesoscorpionina, *Trans. R. Soc. Edinburgh: Earth Sci.* **84**:283-299.

Jeram, A., 1994b, Carboniferous Orthosterni and their relationship to living scorpions, *Palaeontology* **37**:513-550.

Keast, A., 1981, *Ecological biogeography of Australia*, Junk, The Hague.

Koch, L. E., 1977, The taxonomy, geographic distribution and evolutionary radiation of Australo-Papuan scorpions, *Rec. West. Aust. Mus.* **5**:83-367.

Koch, L. E., 1978, A comparative study of the structure, function and adaptation to different habitats of burrows in the scorpion genus *Urodacus* (Scorpionida, Scorpionidae), *Rec. West. Aust. Mus.* **6**:119-146.

Koch, L. E., 1981, The scorpions of Australia: aspects of their ecology and zoogeography, in: *Ecological biogeography of Australia*, A. Keast, ed., Junk, The Hague, pp. 875-884.

Kraepelin, K., 1905, Die geographische Verbreitung der Skorpione. *Zool. Jahrb., Abt. Syst.* **22**:321-364.

Kraepelin, K. 1912 [1913]. Neue Beiträge zur Systematik der Gliederspinnen. III. A. Bemerkungen zur Skorpionenfauna Indiens. B. Die Skorpione, Pedipalpen und Solifugen Deutsch-Ost-Afrikas, *Jahrb. Hamburg Wiss. Anst.* **30**:123-196.

Lamoral, B. H., 1978a, Soil hardness, an important and limiting factor in burrowing scorpions of the genus *Opisthophthalmus* C.L. Koch, 1837 (Scorpionidae, Scorpionida), *Symp. zool. Soc. Lond.* (London) **42**:171-181.

Lamoral, B. H., 1978b, The scorpions of South West Africa. Ph.D. thesis, University of Natal, Pietermaritzburg.

Lamoral, B. H., 1979, The scorpions of Namibia (Arachnida: Scorpionida), *Ann. Natal Mus.* **23**:497-784.

Lamoral, B. H., 1980a, A reappraisal of the suprageneric classification of Recent scorpions and of their zoogeography, in: *Verhandlungen 8. Internationaler Arachnologen-Kongress Abgehalten an der Universität für Bodenkultur Wien, 7-12 Juli, 1980*, J. Grüber, ed., H. Egermann, Vienna, pp. 439-444.

Lamoral, B. H., 1980b, Two new psammophile species and new records of scorpions from the northern Cape Province of South Africa (Arachnida: Scorpionida), *Ann. Natal Mus.* **24**:201-210.

Lamoral, B. H., and Reynders, S. C., 1975, A catalogue of the scorpions described from the Ethiopian Faunal Region up to December 1973, *Ann. Natal Mus.* **22**:489-576.

Lancaster, N., 1990, Regional aeolian dynamics in the Namib, in: *Namib Ecology: 25 Years of Namib Research*, *Transvaal Museum Monograph* 7, M. K. Seely, ed., Transvaal Museum, Pretoria, pp. 39-46.

Lawrence, R. F., 1942, The scorpions of Natal and Zululand, *Ann. Natal Mus.* **10**:221-235.

Lawrence, R. F., 1952, The unequal distribution of some invertebrate animals in South Africa. *S. Afr. J. Sci.* **48**:308-310.

Lawrence, R. F., 1953, *The Biology of the Cryptic Fauna of Forests, with Special Reference to the Indigenous Forest of South Africa.* A.A. Balkema, Cape Town.

Lawrence, R. F., 1955, Solifugae, scorpions and Pedipalpi, with checklists and keys to South African families, genera and species. Results of the Lund University Expedition in 1950–1951, in: *South African Animal Life*, Vol. 1, B. Hanström, P. Brinck and G. Rudebeck, eds., Almqvist & Wiksells, Uppsala, pp. 152-262.

Lawrence, R. F., 1969, A new genus of psammophile scorpion and new species of *Opisthophthalmus* from the Namib Desert, *Sci. Pap. Namib Desert Res. stn* **48**:105-116.

Leistner, O. A., and Morris, J. W., 1976, Southern African place names, *Ann. Cape Prov. Mus.* **12**:1-565.

Lombard, A. T., 1995a, Introduction to an evaluation of the protection status of South Africa's vertebrates, *S. Afr. J. Zool.* **30**:63-70.

Lombard, A. T., 1995b, The problems with multi-species conservation: Do hotspots, ideal reserves and existing reserves coincide?, *S. Afr. J. Zool.* **30**:145-163.

Lourenço, W. R., 1980, A propósito de duas novas espécies de *Opisthacanthus* para a região Neotropical, *Opisthacanthus valerioi* da "Isla del Coco", Costa Rica e *Opisthacanthus heurtaultae* da Guiana Francesa (Scorpiones, Scorpionidae), *Rev. Nordest. Biol.* **3**:179-194.

Lourenço, W. R., 1981, *Opisthacanthus lamorali*, nouvelle espèce de Scorpionidae pour la Région Afrotropicale (Arachnida: Scorpionidae), *Ann. Natal Mus.* **24**:625-634.

Lourenço, W. R., 1982, Révision du genre *Ananteris* Thorell, 1891 (Scorpiones, Buthidae) et description de six espèces nouvelles, *Bull. Mus. natn. Hist. nat., Paris, 4e sér.* **4**:119-151.

Lourenço, W. R., 1983a, Le faune des scorpiones de Guyane française, *Bull. Mus. natn. Hist. nat., Paris, 4e sér.* **5**:771-808.

Lourenço, W. R., 1983b, Contributions à la connaissance de la faune hypsophile du Malawi (Mission R. Jocque). 2. Scorpions, *Rev. Zool. Afr.* **97**:192-201.

Lourenço, W. R., 1985, Le vériyable statut des genres *Ananteris* Thorell, 1891 et *Ananteroides* Borelli, 1911 (Scorpiones, Buthidae), *Ann. Natal Mus.* **26**:407-416.

Lourenço, W. R., 1984a, La biogéographie des scorpions sud-américains (problèmes et perspectives), *Spixiana* **7**:11-18.

Lourenço, W. R., 1984b, *Ananteris luciae*, nouvelle espèce de scorpion de l'Amazonie brasilienne (Scorpiones, Buthidae), *J. Arachnol.* **12**:279-282.

Lourenço, W. R., 1987a, Révision systématique des scorpions du genre *Opisthacanthus* (Scorpiones, Ischnuridae), *Bull. Mus. natn. Hist. nat., Paris, 4e sér.* **9**:887-931.

Lourenço, W. R., 1987b, Déscription d'une nouvelle espèce d'*Ananteris* collectee dans l'état de Maranhão, Brésil (Scorpiones, Buthidae), *Bol. Mus. Para. Emilio Goeldi, Sér. Zool.* **3**:19-23.

Lourenço, W. R., 1988, Considérations biogéographiques, écologiques et évolutives sur les espèces néotropicales d'*Opisthacanthus* Peters, 1861 (Scorpiones, Ischnuridae), *Stud. Neotrop. Fauna Envir.* **23**:41-53.

Lourenço, W. R., 1989, Rétablissement de la famille des Ischnuridae distincte des Scorpionidae Pocock, 1893, *Rev. Arachnol.* **8**:159-177.

Lourenço, W. R., 1991a, *Opisthacanthus*, genre Gondwanien défini comme groupe naturel. Caractérisation des sous-genres et des groupes d'espèces (Arachnida, Scorpiones, Ischnuridae), *Iheringia, Sèr. Zool.* **71**:5-42.

Lourenço, W. R., 1991b, Scorpion species biodiversity in tropical South America and its application in conservation programmes, *Amer. Zool.* **31**:A85-A85.

Lourenço, W. R., 1990 [1991c], Les scorpions (Chelicerata) de Colombie. II. Les faunes des régions de Santa Marta et de la Cordillère orientale. Approche biogéographique, *Senkenbergiana Biol.* **71**:275-288.

Lourenço, W. R., 1993, A review of the geographical distribution of the genus *Ananteris* Thorell (Scorpiones: Buthidae), with description of a new species, *Rev. Biol. Trop.* **41**:697-701.

Lourenço, W. R., 1994a, Scorpion biogeographic patterns as evidence for a Neblina-São Gabriel endemic center in Brazilian Amazonia, *Rev. Acad. Colomb. Cienc.* **19**:181-185.

Lourenço, W. R., 1994b, Biogeographic patterns of tropical South American scorpions, *Stud. Neotr. Fauna Environ.* **29**:219-231.

Lourenço, W. R., 1994c, Diversity and endemism in tropical versus temperate scorpion communities, *Biogeographica* **70**:155-160.

Lourenço, W. R., 1995, Nouvelles considérations sur la classification et la biogéographie des *Opisthacanthus* néotropicaux (Scorpiones, Ischnuridae), *Biogeographica* **71**:75-82.

Lourenço, W. R., 1996, The biogeography of scorpions, *Rev. Suisse Zool.*, Vol. hors série II:437-448 [Proceedings of the XIIIth International Congress of Arachnology, Geneva, 3-8 September 1995, V. Mahnert, ed.].

Lourenço, W. R., 1997, Synopsis de la faune de scorpions de Colombie, avec des considérations sur la systématique et la biogéographie des espèces, *Rev. Suisse Zool.* **104**:61-94.

Lourenço, W. R., 1998, Panbiogéographie, les distributions disjontes et le concept de familie relictuelle chez les scorpions, *Biogeographica* **74**:133-144.

Lourenço, W. R., 1999, Considérations taxonomiques sur le genre *Hadogenes* Kraepelin, 1894; création de la sous-famille des Hadogeninae n. subfam., et description d'une espèce nouvelle pour l'Angola (Scorpiones, Scorpionidae, Hadogeninae), *Rev. Suisse Zool.* **106**:929-938.

Lourenço, W. R., 2000, Panbiogéographie, les familles des scorpions et leur répartition géographique, *Biogeographica* **76**:21-39.

Lourenço, W. R., 2001, Nouvelles considerations sur la phylogenie et la biogéographie des scorpions Ischnuridae de Madagascar, *Biogeographica* **77**:83-96.

Lourenço, W. R., and Fé, N. F., 2003, Description of a new species of *Opisthacanthus* Peters (Scorpiones, Liochelidae) to Brazilian Amazonia, *Rev. Iber. Aracnol.* **8**:81-88.

Lourenço, W. R., and Flórez, E., 1989, Los escorpiones (Chelicerata) de Colombia. I. La Fauna de la Isla Gorgona. Approximación biogeográfica, *Caldasia* **16**:66-70.

MacArthur, R. H., 1972, *Geographical ecology*, Harper and Row, New York.

Main, B. Y., 1956, Taxonomy and biology of the genus *Isometroides* Keyserling (Scorpionida), *Aust. J. Zool.* **4**:158-164.

Mayr, E., 1942, *Systematics and the Origin of Species*, Columbia University Press, New York.

Mayr, E., 1963, *Animal Species and Evolution*, Harvard University Press, Cambridge, MA.

McCormick, S. J., and Polis, G. A., 1990, Prey, predators, parasites, in: *The Biology of Scorpions*, G. A. Polis, ed., Stanford University Press, Stanford, CA, pp. 294-320.

Midgley, D. C., Pitman, W. V., and Middleton, B. J., 1994, *Surface Water Resources of South Africa 1990* (WR90), CD-ROM, Water Research Commission, Pretoria.

Ministry of Land and Natural Resources (Northern Rhodesia), 1959, *Gazetteer of Geographical Names in the Barotseland Protectorate*, Government Printer, Lusaka.

Ministry of Land and Mines (Zambia), 1967, *Gazetteer of Names in the Republic of Zambia*, Government Printer, Lusaka.

Mugo, D. N., Lombard, A. T., Bronner, G. N., Gelderblom, C. M., and Benn, G. A., 1995, Distribution and protection of endemic or threatened rodents, lagomorphs and macrosceledids in South Africa, *S. Afr. J. Zool.* **30**:115-126.

Myers, A. A., and Giller, P. S., 1988, Process, pattern and scale in biogeography, in: *Analytical Biogeography: An Integrated Approach to the Study of Animal and Plant Distributions*, A. A. Myers and P. S. Giller, eds., Chapman and Hall, London, pp. 3-21.

Myers, N., Mittermeier, R. A., Mittermeier, C. G., da Fonseca, G. A. B., and Kent, J., 2000, Biodiversity hotspots for conservation priorities, *Nature* **403**:853-858.

National Place Names Committee (Department of National Education, Republic of South Africa), 1951, *Official Place Names in the Union and South West Africa* (Approved to end 1948), Government Printer, Pretoria.

National Place Names Committee (Department of National Education, Republic of South Africa), 1978, *Official Place Names in the Republic of South Africa and South West Africa* (Approved to 1 April 1977), Government Printer, Pretoria.

National Place Names Committee (Department of National Education, Republic of South Africa), 1988, *Official Place Names in the Republic of South Africa* (Approved to 1977 to 1988), Government Printer, Pretoria.

Nenilin, A. B., and Fet, V., 1992, [Zoogeographical analysis of the world scorpion fauna (Arachnida: Scorpiones)], *Arthopoda Selecta, Moscow* 1:3-31. [Russian].

Newlands, G., 1969, Two new scorpions from the northern Transvaal, *J. ent. Soc. sth. Afr.* **32**:5-7.

Newlands, G., 1972a, Notes on psammophilous scorpions and a description of a new species (Arachnida: Scorpionida), *Ann. Transv. Mus.* **27**:241-257.

Newlands, G., 1972b, Ecological adaptations of Kruger National Park scorpionids (Arachnida: Scorpiones), *Koedoe* **15**:37-48.

Newlands, G., 1972c, A description of *Hadogenes lawrencei* sp. nov. (Scorpiones) with a checklist and key to the South-West African species of the genus *Hadogenes*, *Madoqua, ser. 2.* **1**:133-140.

Newlands, G., 1973, Zoogeographical factors involved in the trans-Atlantic dispersal pattern of the genus *Opisthacanthus* Peters (Arachnida: Scorpionidae), *Ann. Transv. Mus.* **28**:91-100.

Newlands, G., 1974, Transvaal scorpions, *Fauna and Flora* **25**:3-7.

Newlands, G., 1978a, Arachnida (except Acari), in: *Biogeography and Ecology of southern Africa*, Vol. 2, M. J. A. Werger, ed., Junk, The Hague, pp. 677-684.

Newlands, G., 1978b, Review of southern African scorpions and scorpionism, *S. Afr. Med. J.* **54**:613-615.

Newlands, G., 1980, A revision of the scorpion genus *Hadogenes* Kraepelin 1894 (Arachnida: Scorpionidae) with a checklist and key to the species, M.Sc. thesis, Potchefstroom University for C.H.E.

Newlands, G., and Cantrell, A. C., 1985, A re-appraisal of the rock scorpions (Scorpionidae: *Hadogenes*), *Koedoe* **28**:35-45.

Newlands, G., and Martindale, C. B., 1980, The buthid scorpion fauna of Zimbabwe-Rhodesia with checklists and keys to the genera and species, distributions and medical importance (Arachnida: Scorpiones), *Z. Angew. Zool.* **67**:51-77.

Newlands, G., and Prendini, L., 1997, Redescription of *Hadogenes zumpti* (Scorpiones: Ischnuridae), an unusual rock scorpion from the Richtersveld, South Africa, *S. Afr. J. Zool.* **32**:76-81.

Penrith, M.-L., 1975, The species of *Onymacris* (Coleoptera: Tenebrionidae), *Cimbebasia* (A) **4**:47-97.

Penrith, M.-L., 1977, The Zophosini (Coleoptera: Tenebrionidae) of western southern Africa, *Cimbebasia Memoir* **3**:1-291.

Penrith, M.-L., 1979, Revision of the western southern African Adesmiini (Coleoptera: Tenebrionidae), *Cimbebasia* (A) **5**:1-94.

Penrith, M.-L., 1981a, Revision of the Zophosini (Coleoptera: Tenebrionidae). Part 2. The subgenus *Zophosis* Latreille, and seven related south-western African subgenera, *Cimbebasia* (A) **6**:17-109.

Penrith, M.-L., 1981b, Revision of the Zophosini (Coleoptera: Tenebrionidae). Part 4. Twelve subgenera from arid southern Africa, *Cimbebasia* (A) **6**:125-164.

Penrith, M.-L., 1987, Revision of the genus *Tarsocnodes* Gebien (Coleoptera: Tenebrionidae: Molurini), and a description of a monotypical genus from the Kalahari, *Cimbebasia* (A) **7**:235-270.

Platnick, N. I., and Nelson, G., 1978, A method of analysis for historical biogeography, *Syst. Zool.* **27**:1-16.

Pocock, R. I., 1894, Scorpions and their geographical distribution, *Nat. Sci.* **4**:353-364.

Polhill, D., 1988, *Flora of Tropical East Africa. Index of Collecting Localities*, Royal Botanic Gardens, Kew.

Polis, G. A., 1979, Prey and feeding phenology of the desert sand scorpion *Paruroctonus mesaensis* (Scorpionida: Vaejovidae), *J. Zool.* (Lond.) **188**:333-346.

Polis, G. A., 1990. Ecology, in: *The Biology of Scorpions*, G. A. Polis, ed., Stanford University Press, Stanford, CA, pp. 247-293.

Polis, G. A., and Holt, R. D., 1992, Intraguild predation: the dynamics of complex trophic interactions, *Trends Ecol. Evol.* **7**:151-154.

Polis, G. A., and McCormick, S. J., 1986, Patterns of resource use and age structure among a guild of desert scorpions, *J. Anim. Ecol.* **55**:59-73.

Polis, G. A., and McCormick, S. J., 1987, Intraguild predation and competition among desert scorpions, *Ecology* **68**:332-343.

Polis, G. A., Sissom, W. D., and McCormick, S. J., 1981, Predators of scorpions: field data and a review, *J. Arid Environ.* **4**:309-326.

Postmaster-General (Republic of South Africa), 1970, *List of Post Offices in the Republic of South Africa, in South-West Africa, and other countries of the African Postal Union*, Government Printer, Pretoria.

Poynton, J. C., 1964, The Amphibia of southern Africa: a faunal study, *Ann. Natal Mus.* **17**:1-334.

Poynton, J. C., and Broadley, D. G., 1985a, Amphibia Zambesiaca 1. Scolecomorphidae, Pipidae, Microhylidae, Hemisidae, Arthroleptidae, *Ann. Natal Mus.* **26**:503-553.
Poynton, J. C., and Broadley, D. G., 1985b. Amphibia Zambesiaca 2. Ranidae, *Ann. Natal Mus.* **27**:115-181.
Poynton, J. C., and Broadley, D. G., 1987, Amphibia Zambesiaca 3. Rhacophoridae and Hyperoliidae, *Ann. Natal Mus.* **28**:161-229.
Prendini, L., 2000a, Phylogeny and classification of the superfamily Scorpionoidea Latreille 1802 (Chelicerata, Scorpiones): An exemplar approach, *Cladistics* **16**:1-78.
Prendini, L., 2000b, A new species of *Parabuthus* Pocock (Scorpiones, Buthidae), and new records of *Parabuthus capensis* (Ehrenberg), from Namibia and South Africa, *Cimbebasia* **16**:201-214.
Prendini, L., 2000c, Chelicerata (Scorpiones), in: *Dâures – Biodiversity of the Brandberg Massif, Namibia*, A.H. Kirk-Spriggs and E. Marais, eds., *Cimbebasia Memoir* **9**:109-120.
Prendini, L., 2001a, Substratum specialization and speciation in southern African scorpions: the Effect Hypothesis revisited, in: *Scorpions 2001.* In Memoriam *Gary A. Polis*, V. Fet and P. A. Selden, eds., British Arachnological Society, Burnham Beeches, Bucks, UK, pp. 113-138.
Prendini, L., 2001b, Phylogeny of *Parabuthus* (Scorpiones, Buthidae), *Zool. Scr.* **30**:13-35.
Prendini, L., 2001c, Two new species of *Hadogenes* (Scorpiones, Ischnuridae) from South Africa, with a redescription of *Hadogenes bicolor* and a discussion on the phylogenetic position of *Hadogenes*, *J. Arachnol.* **29**:146-172.
Prendini, L., 2001d, A review of synonyms and subspecies in the genus *Opistophthalmus* C.L. Koch (Scorpiones: Scorpionidae), *Afr. Entomol.* **9**:17-48.
Prendini, L., 2003a, A new genus and species of bothriurid scorpion from the Brandberg Massif, Namibia, with a reanalysis of bothriurid phylogeny and a discussion of the phylogenetic position of *Lisposoma* Lawrence, *Syst. Ent.* **28**:149-172.
Prendini, L., 2003b, Revision of the genus *Lisposoma* Lawrence, 1928 (Scorpiones: Bothriuridae), *Insect Syst. Evol.* **34**:241-264.
Prendini, L., 2003c, Discovery of the male of *Parabuthus muelleri*, and implications for the phylogeny of *Parabuthus* (Scorpiones: Buthidae), *Amer. Mus. Novit.* **3408**:1-24.
Prendini, L., 2004a, The systematics of southern African *Parabuthus* Pocock (Scorpiones, Buthidae): Revisions to the taxonomy and key to the species, *J. Arachnol.* **32**:109-186.
Prendini, L., 2004b, Systematics of the genus *Pseudolychas* Kraepelin (Scorpiones: Buthidae), *Ann. Entomol. Soc. Amer.* **97**:37-63.
Prendini, L., in press, On *Hadogenes angolensis* Lourenço, 1999 syn. n. (Scorpiones, Liochelidae), with a redescription of *H. taeniurus* (Thorell, 1876). *Rev. Suisse Zool.*
Prendini, L., in press, Revision of *Karasbergia* Hewitt (Scorpiones: Buthidae), a monotypic genus endemic to southern Africa, *J. Afrotrop. Zool.*
Prendini, L., Crowe, T. M., and Wheeler, W. C., 2003, Systematics and biogeography of the family Scorpionidae Latreille, with a discussion of phylogenetic methods, *Invert. Syst.* **17**:185-259.
Raffle, J. A., 1984, *Third report of the Place Names Commission*, Government Printer, Gaborone.
Raper, P. E., 1987, *Dictionary of Southern African Place Names*, Lowry Publishers, Johannesburg.
San Martín, P. R., 1961, Observaciones sobre la ecología y distribución geográfica de tres especes de escorpiones en el Uruguay. *Rev. Fac. Human. Cien., Univ. Repúb.* (Montevideo) **19**:1-42.
Selden, P. A., 1993, Fossil arachnids—recent advances and future prospects, *Mem. Queensl. Mus.* **33**:389-400.

Shorthouse, D. J., and Marples, T., 1980, Observations on the burrow and associated behaviour of the arid-zone scorpion *Urodacus yaschenkoi* (Birula), *Aust. J. Zool.* **28**:581-590.

Simberloff, D., 1983, Sizes of coexisting species, in: *Coevolution*, D.J. Futuyma and M. Slatkin, eds., Sinauer Associates, Sunderland, MA, pp. 404-430.

Sissom, W. D., 1990, Systematics, biogeography and palaeontology, in: *The Biology of Scorpions*, G.A. Polis, ed., Stanford University Press, Stanford, CA, pp. 64-160.

Sissom, W. D., and Fet, V., 2000, Family Diplocentridae Karsch, 1880, in: *Catalog of the Scorpions of the World (1758-1998)*, V. Fet, W. D. Sissom, G. Lowe and M. E. Braunwalder, The New York Entomological Society, New York, pp. 329-354.

Skead, C. J., 1973, Zoo-historical gazetteer, *Ann. Cape Prov. Mus.* **10**:1-259.

Smith, G. T., 1966, Observations on the life history of the scorpion *Urodacus abruptus* Pocock (Scorpionida), and an analysis of its home sites, *Aust. J. Zool.* **14**:383-398.

Soleglad, M. E., and Fet, V. 2003, High-level systematics and phylogeny of the extant scorpions (Scorpiones: Orthosterni), *Euscorpius* **11**:1-175.

Stockwell, S. A., 1985, A new species of *Heteronebo* from Jamaica (Scorpiones, Diplocentridae), *J. Arachnol.* **13**:355-361.

Stockwell, S. A., 1989, Revision of the phylogeny and higher classification of scorpions (Chelicerata), Ph.D. Thesis, University of California, Berkeley.

Stuckenberg, B. R., 1962, The distribution of the montane palaeogenic element in the South African invertebrate fauna, *Ann. Cape Prov. Mus.* **2**:190-205.

Surveyor-General (Republic of South Africa), 1970, *Index of Orange Free State Farms*, Department of Public Works and Land Affairs, Pretoria.

Surveyor-General (Republic of South Africa), 1988, *Alphabetical list of farms in the Province of Transvaal*, Department of Public Works and Land Affairs, Pretoria.

SWALU, 1967, *Swalurama Farms Directory*, SWALU, Windhoek.

Tikader, B. K., and Bastawade, D. B., 1983, *Fauna of India*, Vol 3., *Scorpions*, Zoological Survey of India, Calcutta.

Vachon, M., 1973 [1974], Étude des caractères utilisés pour classer les familles et les genres de scorpions (Arachnides). 1. La trichobothriotaxie en arachnologie. Sigles trichobothriaux et types de trichobothriotaxie chez les scorpions, *Bull. Mus. Natn. Hist. Nat., Paris* (3) **140**:857-958.

van Zinderen Bakker, E. M., 1962, Botanical evidence for Quaternary Climates in Africa, *Ann. Cape Prov. Mus.* **2**:16-31.

van Zinderen Bakker, E. M., 1969, The "arid corridor" between south-western Africa and the Horn of Africa, *Palaeoecol. Afr.* **4**:139-140.

van Zinderen Bakker, E. M., 1975, The origin and palaeo-environment of the Namib Desert biome, *J. Biogeography* **2**:65-73.

van Zinderen Bakker, E. M., 1976, The evolution of late-Cenozoic palaeo-climates of southern Africa, *Palaeoecol. Afr.* **9**:160-202.

van Zinderen Bakker, E. M., 1978, Quaternary vegetation changes in southern Africa, in: *Biogeography and Ecology of southern Africa*, M. J. A. Werger, ed., Junk, The Hague, pp. 131-143.

Warburg, M. R., and Ben-Horin, A., 1978, Temperature and humidity effects on scorpion distribution in northern Israel, *Symp. zool. Soc. Lond.* (London) **42**:161-169.

Warburg, M. R., Goldenberg, S., and Ben-Horin, A., 1980, Scorpion species diversity and distribution within the Mediterranean and arid regions of northern Israel, *J. Arid Environ.* **3**:205-213.

Webb, R. S., 1950, *Gazetteer for Basutoland.* The first draft of a list of names, with special reference to the 1:250 000 map G. S. G. S. No. 2567 of June 1911: "Basutoland, from a reconnaissance survey made in 1904-9 by Captain M. C. Dobson, R. F. A.", A. H. Fischer and Sons, Paarl.

Werger, M. J. A., 1978, *Biogeography and Ecology of southern Africa*, Junk, The Hague.
Werner, F., 1936, Neu-Eingänge von Skorpionen im Zoologischen Museum in Hamburg, *Festschr. 60. Geburtstage Prof. Dr. Embrik Strand* **2**:171-193.
Williams, S. C., 1970, Coexistence of desert scorpions by differential habitat preference, *Pan-Pacific Entomol.* **46**:254-267.
Williams, S. C., 1980, Scorpions of Baja California, Mexico, and adjacent islands, *Occas. Pap. California Acad. Sci.* **135**:1-127.
Williams, S. C., 1987, Scorpion bionomics, *Ann. Rev. Entomol.* **32**:275-295.

Appendix 1. Scorpion genera and species recorded from southern African countries, south of 15° latitude. Questionmarks reflect suspected occurrences that remain to be verified. Abbreviations as follows: A (Angola); B (Botswana); L (Lesotho); Ma (Malawi); Mo (Mozambique); N (Namibia); SA (South Africa); S (Swaziland); Za (Zambia); Zi (Zimbabwe); E (Extralimital).

Species	A	B	L	Ma	Mo	N	SA	S	Za	Zi	E
Bothriuridae											
Brandbergia haringtoni Prendini, 2003						+					
Lisposoma elegans Lawrence, 1928						+					
Lisposoma joseehermanorum Lamoral, 1979						+					
Buthidae											
Afroisometrus minshullae (FitzPatrick, 1994)							+			+	
Hottentotta arenaceus (Purcell, 1901)						+	+				
Hottentotta conspersus (Thorell, 1876)	+					+					
Hottentotta trilineatus (Peters, 1861)		+		?	+		+		+	+	+
Karasbergia methueni Hewitt, 1913						+	+				
Lychas burdoi (Simon, 1882)				+	+		+		+	+	+
Parabuthus brevimanus (Thorell, 1876)	+					+	+				
Parabuthus calvus Purcell, 1898							+				
Parabuthus capensis (Ehrenberg, 1831)						+	+				
Parabuthus distridor Lamoral, 1980							+				
Parabuthus gracilis Lamoral, 1979						+					
Parabuthus granulatus (Ehrenberg, 1831)	+	+				+	+			+	
Parabuthus kalaharicus Lamoral, 1977		+				+	+				
Parabuthus kraepelini Werner, 1902	?					+					
Parabuthus kuanyamarum Monard, 1937	+	+				+	+		+	+	
Parabuthus laevifrons (Simon, 1888)		+				+	+				
Parabuthus mossambicensis (Peters, 1861)		+		?	+		+		+	+	
Parabuthus muelleri Prendini, 2000						+					
Parabuthus namibensis Lamoral, 1979						+					
Parabuthus nanus Lamoral, 1979						+	+				
Parabuthus planicauda (Pocock, 1889)							+				
Parabuthus raudus (Simon, 1888)	+	+				+	+		+	+	
Parabuthus schlechteri Purcell, 1899						+	+				
Parabuthus stridulus Hewitt, 1913						+					
Parabuthus transvaalicus Purcell, 1899		+			+		+			+	
Parabuthus villosus (Peters, 1862)	?					+	+				
Pseudolychas ochraceus (Hirst, 1911)		?	?				+			?	
Pseudolychas pegleri (Purcell, 1901)					+		+	+			
Pseudolychas transvaalicus Lawrence, 1961							+				
Uroplectes carinatus (Pocock, 1890)	+	+				+	+			+	
Uroplectes chubbi Hirst, 1911		+			+		+		+	+	
Uroplectes flavoviridis Peters, 1861				+	+		+		+	+	?
Uroplectes formosus Pocock, 1890			+		+		+	+			
Uroplectes gracilior Hewitt, 1914						+	+				
Uroplectes insignis Pocock, 1890							+				
Uroplectes lineatus (C. L. Koch, 1844)							+				
Uroplectes longimanus Werner, 1936						+					
Uroplectes marlothi Purcell, 1901							+				
Uroplectes olivaceus Pocock, 1896					+		+	+		+	
Uroplectes otjimbinguensis (Karsch, 1879)	+					+					

Appendix 1. (continued).

Species	A	B	L	Ma	Mo	N	SA	S	Za	Zi	E
Buthidae (continued)											
Uroplectes planimanus (Karsch, 1879)	+	+			+	+	+		+	+	?
Uroplectes schlechteri Purcell, 1901						+	+				
Uroplectes teretipes Lawrence, 1966						+					
Uroplectes triangulifer (Thorell, 1876)			+				+	+			
Uroplectes tumidimanus Lamoral, 1979		+				+	+				
Uroplectes variegatus (C. L. Koch, 1844)							+				
Uroplectes vittatus (Thorell, 1876)		+			+	+	+	+	+	+	
Liochelidae											
Cheloctonus anthracinus Pocock, 1899							+				
Cheloctonus crassimanus (Pocock, 1896)							+				
Cheloctonus glaber Kraepelin, 1896							+				
Cheloctonus intermedius Hewitt, 1912							+				
Cheloctonus jonesii Pocock, 1892		?			+		+	+	?	?	
Hadogenes bicolor Purcell, 1899							+				
Hadogenes gracilis Hewitt, 1909							+				
Hadogenes granulatus Purcell, 1901		+			+				+	+	
Hadogenes gunningi Purcell, 1899							+				
Hadogenes lawrencei Newlands, 1972						+					
Hadogenes longimanus Prendini, 2001							+				
Hadogenes minor Purcell, 1899							+				
Hadogenes newlandsi Prendini, 2001							+				
Hadogenes phyllodes Thorell, 1876						+	+				
Hadogenes taeniurus (Thorell, 1876)	+					+					
Hadogenes tityrus (Simon, 1888)						+	+				
Hadogenes trichiurus (Gervais, 1843)							+				
Hadogenes troglodytes (Peters, 1861)		+			+		+			+	
Hadogenes zuluanus Lawrence, 1937							+	+			
Hadogenes zumpti Newlands, 1985							+				
Opisthacanthus asper (Peters, 1861)		+			+		+	+	?	+	+
Opisthacanthus basutus Lawrence, 1955			+								
Opisthacanthus capensis Thorell, 1876							+				
Opisthacanthus diremptus (Karsch, 1879)							+				
Opisthacanthus laevipes (Pocock, 1893)					+		+	+			
Opisthacanthus lamorali Lourenço, 1981										+	
Opisthacanthus piscatorius Lawrence, 1955							+				
Opisthacanthus rugiceps Pocock, 1897				+							+
Opisthacanthus rugulosus Pocock, 1896				+							
Opisthacanthus validus Thorell, 1876			+				+	+			
Scorpionidae											
Opistophthalmus adustus Kraepelin, 1908						+					
Opistophthalmus ammopus Lamoral, 1980							+				
Opistophthalmus ater Purcell, 1898							+				
Opistophthalmus austerus Karsch, 1879							+				
Opistophthalmus boehmi (Kraepelin, 1896)		+			+		+			+	+
Opistophthalmus brevicauda Lawrence, 1928						+					
Opistophthalmus capensis (Herbst, 1800)							+				
Opistophthalmus carinatus (Peters, 1861)	+	+			+	+	+		+	+	
Opistophthalmus cavimanus Lawrence, 1928						+					

Appendix 1. (continued).

Species	A	B	L	Ma	Mo	N	SA	S	Za	Zi	E
Scorpionidae (continued)											
Opistophthalmus chrysites Lawrence, 1967						+					
Opistophthalmus coetzeei Lamoral, 1979						+					
Opistophthalmus concinnus Newlands, 1972		+				+	+				
Opistophthalmus crassimanus Purcell, 1899							+				
Opistophthalmus fitzsimonsi Hewitt, 1935		+				+	+				
Opistophthalmus flavescens Purcell, 1898						+					
Opistophthalmus fossor Purcell, 1898							+				
Opistophthalmus fuscipes Purcell, 1898							+				
Opistophthalmus gibbericauda Lamoral, 1979	+					+					
Opistophthalmus gigas Purcell, 1898						+	+				
Opistophthalmus glabrifrons Peters, 1861		+		+	+		+	?		+	+
Opistophthalmus granicauda Purcell, 1898							+				
Opistophthalmus granifrons Pocock, 1896							+				
Opistophthalmus haackei Lawrence, 1966						+	+				
Opistophthalmus harpei Harington, 2001						+					
Opistophthalmus holmi (Lawrence, 1969)						+	+				
Opistophthalmus intercedens Kraepelin, 1908						+					
Opistophthalmus intermedius Kraepelin, 1894							+				
Opistophthalmus jenseni (Lamoral, 1972)						+					
Opistophthalmus karrooensis Purcell, 1898							+				
Opistophthalmus keilandsi Hewitt, 1914							+				
Opistophthalmus lamorali Prendini, 2000						+					
Opistophthalmus laticauda Purcell, 1898							+				
Opistophthalmus latimanus C. L. Koch, 1841							+				
Opistophthalmus latro Thorell, 1876							+				
Opistophthalmus lawrencei Newlands, 1969							+				
Opistophthalmus leipoldti Purcell, 1898							+				
Opistophthalmus litoralis Lawrence, 1955	+					+					
Opistophthalmus longicauda Purcell, 1899							+				
Opistophthalmus lornae Lamoral, 1979						+	+				
Opistophthalmus luciranus Lawrence, 1959	+										
Opistophthalmus macer Thorell, 1876							+				
Opistophthalmus nitidiceps Pocock, 1896							+				
Opistophthalmus opinatus (Simon, 1888)						+					
Opistophthalmus pallipes C. L. Koch, 1842						+	+				
Opistophthalmus pattisoni Purcell, 1899							+				
Opistophthalmus penrithorum Lamoral, 1979						+					
Opistophthalmus peringueyi Purcell, 1898							+				
Opistophthalmus pictus Kraepelin, 1894							+				
Opistophthalmus pluridens Hewitt, 1918							+				
Opistophthalmus praedo Thorell, 1876							+				
Opistophthalmus pugnax Thorell, 1876			+				+				
Opistophthalmus pygmaeus Lamoral, 1979						+					
Opistophthalmus scabrifrons Hewitt, 1918						+					
Opistophthalmus schlechteri Purcell, 1898							+				
Opistophthalmus schultzei Kraepelin, 1908						+					
Opistophthalmus setifrons Lawrence, 1961						+					
Opistophthalmus ugabensis Hewitt, 1934						+					
Opistophthalmus wahlbergii (Thorell, 1876)	+	+				+	+		+	+	

MADAGASCAR AS A MODEL REGION FOR THE STUDY OF TEMPO AND PATTERN IN ADAPTIVE RADIATIONS

Miguel Vences
Institute for Biodiversity and Ecosystem Dynamics, Zoological Museum, University of Amsterdam, 1092 AD Amsterdam, The Netherlands; E-mail: vences@science.uva.nl

Abstract: The comparative study of adaptive radiations is a fruitful field that allows to test hypotheses and predictions of evolutionary theory. Madagascar is ideal for such studies because a large number of such radiations evolved on this island in isolation. In vertebrates, the historical biogeography of many Malagasy groups has recently been elucidated by molecular phylogenetic analyses, and by molecular time estimates. Such molecular clocks are a suitable method to estimate colonization times. However, as reviewed herein, they need to be applied cautiously due to the recent discovery of various potential artefacts and pitfalls. As an example of possible comparative studies across radiations, divergences of the Malagasy vertebrate clades to their closest non-Malagasy relatives in the mitochondrial 16S rRNA gene were found to be correlated to species diversity of the Malagasy clades. This suggests that the more diverged clades are richer in species. Pseudoxyrhophiine snakes and mantellid frogs contain distinctly more species than would be expected by this correlation alone. Further such comparative analyses appear to be promising but require a yet better state of knowledge on phylogeny and divergence times of the Malagasy taxa.

Key words: Adaptive radiation; molecular clock; biogeography; vertebrate origins; Madagascar

1. INTRODUCTION

Radiations of organisms long have attracted the attention of evolutionary biologists. Definitions of the term 'radiation' are remarkably vague but usually start from the assumption that this process is related to speciation,

B.A. Huber et al. (eds.), African Biodiversity, 69–84.

and differs from 'normal' speciation by its higher rate of rapid splitting events (Schmitt, 2004). Whereas some organisms, often called living fossils, retain a conserved morphology and low diversity over millions of years, others are able to spawn hundreds of new species in very short time spans. These radiations can be adaptive i.e., speciation is triggered by differential adaptation to a variety of niches, or non-adaptive, in which speciation occurs in the absence of clear niche differentiation (Gittenberger, 1991, 2004).

Studying radiations may permit insights into mechanisms and factors influencing the process of speciation and thereby help us to understand the origins of biodiversity. Are intrinsic or extrinsic factors and innovations most relevant for diversification? Is fast molecular and karyological evolution a consequence of or a prerequisite for cladogenesis? Is the genetic diversity within species-rich crown groups (many young species each with few private alleles) necessarily larger than in their species-poor sister clades (with few old species that possibly contain many different genetic lineages)? How common is adaptive vs. non-adaptive diversification? Answering such general questions calls for an approach beyond single-case studies.

The most famous radiations are doubtless those that took place on islands (e.g., Gillespie and Roderick, 2002). Examples include Darwin's finches of Galapagos (Grant and Grant, 2002), Caribbean anoline lizards (Losos, 1994; Losos et al., 1998; Losos and Schluter, 2000) or Hawaiian *Drosophila* and other insects (e.g., Wagner and Funk, 1995; Shaw, 2002) and invertebrates (Schubart et al., 1998). The Canary Islands have been much used as a model region for the study of radiations (Juan et al., 2000). Ancient lakes, famous for their radiations of cichlid fishes and invertebrates (Martens et al., 1994; Schön and Martens, 2004; Verheyen et al., 2003), are nothing but islands of water from the perspective of aquatic organisms.

Such islands and archipelagos offer spectacular examples of radiations but the number of these examples is usually small. Some of the model regions are remote oceanic landmasses that have been reached by only a limited number of successful colonizers. In other cases simple area effects provide constraints on the number and extent of radiations: it would be difficult to imagine many species-rich radiations of large-sized taxa, e.g. reptiles or mammals, on small volcanic islands. As a consequence, many of the famous radiations are unique - they are exciting to study, they provide interesting conclusions and predictions, but the possibility of rigorously testing these predictions is limited. In contrast, on large continental entities the number of taxa and monophyletic groups is large enough for comparative studies, but any generalization is made more difficult by long and often extremely complex geographical, geological and biological histories. For instance, in Europe, the repeated immigrations after each glacial cycle contributed to the current mosaic of taxa with partly very different phylogeographic origins (e.g., Taberlet et al., 1998).

Experimental biologists can repeat their experiments under identical conditions to statistically verify their hypotheses. Evolutionary biologists who wish to study general patterns of radiations under natural conditions require systems in which evolution itself has repeated its experiments. To reach a deeper understanding of a radiation process, one case should be compared with parallel radiations of other related taxa (Sudhaus, 2004) - and with those of unrelated taxa as well.

This paper is a plea to introduce Madagascar as one geographical region that fulfills the requirements for comparative studies on evolutionary diversification. Being separated from other landmasses for over 80 million years, Madagascar is similar to islands in providing, once reached by new colonizers, a uniform and isolated playground for evolution. And simultaneously its large surface of almost 590,000 square kilometres qualifies it as veritable microcontinent with a high number of endemic organismal groups. This fourth largest island in the world harbours not only an extremely high faunal and floral endemism, larger than 90% in many taxa, but also a striking diversity in species and ecological adaptations (Goodman and Benstead, 2003).

Many if not most of Madagascar's endemic clades of organisms qualify as adaptive radiations. Among birds, the vangas (Vangidae) and Malagasy songbirds rival Darwin's finches from the Galapagos Islands in ecological diversity (Cibois et al., 2001; Yamagishi et al., 2001). Mantellid frogs show a remarkable convergence in morphology and reproductive biology to Asian and African lineages (Bossuyt and Milinkovitch, 2000), and pseudoxyrhophiine snakes are morphologically so diverse that they were classified in different non-endemic subfamilies until recently (Nagy et al., 2003).

In the following I will first review how recent methodological progress in the field of molecular clocks, and a growing amount of molecular phylogenetic results, have contributed to elucidate the biogeographic origins of the endemic vertebrate clades of Madagascar. I will then provide a first example of cross-radiation analysis by testing for correlation between genetic divergence of clades and their species diversity, and give an outlook on the future potentials and perspectives of such studies.

2. DATING THE MALAGASY RADIATIONS: POTENTIALS AND LIMITS OF THE MOLECULAR CLOCK

As Krause et al. (1997) have pointed out, less than 10 years ago the biogeographic origins of Madagascar's vertebrate fauna were still among the greatest unsolved mysteries of natural histories. To reconstruct historical biogeography and understand macroecology, it is crucial to integrate both topological and temporal information (Donoghue and Moore, 2003), plus data on the phylogenetic relationships among taxa (Wiens and Donoghue, 2004). Because Tertiary terrestrial fossils from Madagascar are lacking (Krause et al., 1997), researchers who attempt to elucidate the phylogenetic and temporal patterns of these radiations have to rely almost entirely on molecular data from recent taxa. While the value and reliability of molecular phylogenies is in general beyond doubt, the use of DNA sequences to infer divergence times is much more controversial. Such "molecular clock" approaches consist of a variety of statistical methods that make use of the fact that mutations in any DNA fragment accumulate over evolutionary time, except where selective pressure against such a process is very strong. Once the gene flow between two clades ceases (speciation under an evolutionary species concept; Wiens, 2004), the pairwise divergence among them in a homologous DNA fragment will increase over time. Hence, pairwise sequence divergence is correlated with time since divergence, although this relation is not necessarily linear.

The most straightforward of these methods, described at length in Hillis et al. (1996), assumes a *global molecular clock* and relies on the assumption that the relation between sequence divergence and time is linear. For calibration, taxa are needed for which the divergence time is known very precisely (e.g., through a sequence of fossils) and the sequence divergence can be assessed (i.e., usually extant taxa) (Fig. 1a). A regression line is drawn through one or several of these calibration points and prediction confidence intervals calculated (Fig. 1b). Any pair of taxa of unknown age but of known sequence divergence are plotted on the regression line and the divergence time can be estimated with 95% confidence (Fig. 1c). The more calibration points are available, the more reliable the estimates are. The sequences in each taxon pair (those used for calibration and those tested), and all sequences among each other, need to have similar molecular substitution rates. This can be verified with relative rate tests (e.g., Takezaki et al., 1995). Hedges and Kumar (2003, 2004) argue that the precision of such regression-based global clock calculations can be increased by performing parallel estimates based on independent genes, lowering the standard errors of the eventual multiple estimate. However, this method has been heavily criticized (e.g., Graur and Martin, 2004). Rodriguez-Trelles et al. (2002) further provided evidence for a statistical bias in molecular clocks that may result consistently in overestimates of divergence times. A major problem, beside the uncertain assumption of rate constancy and the disputes about calculation of standard errors, is the general imprecision of many fossil data that provide time intervals rather than precise points of divergence for calibration (Graur and Martin, 2004; Reisz and Müller, 2004).

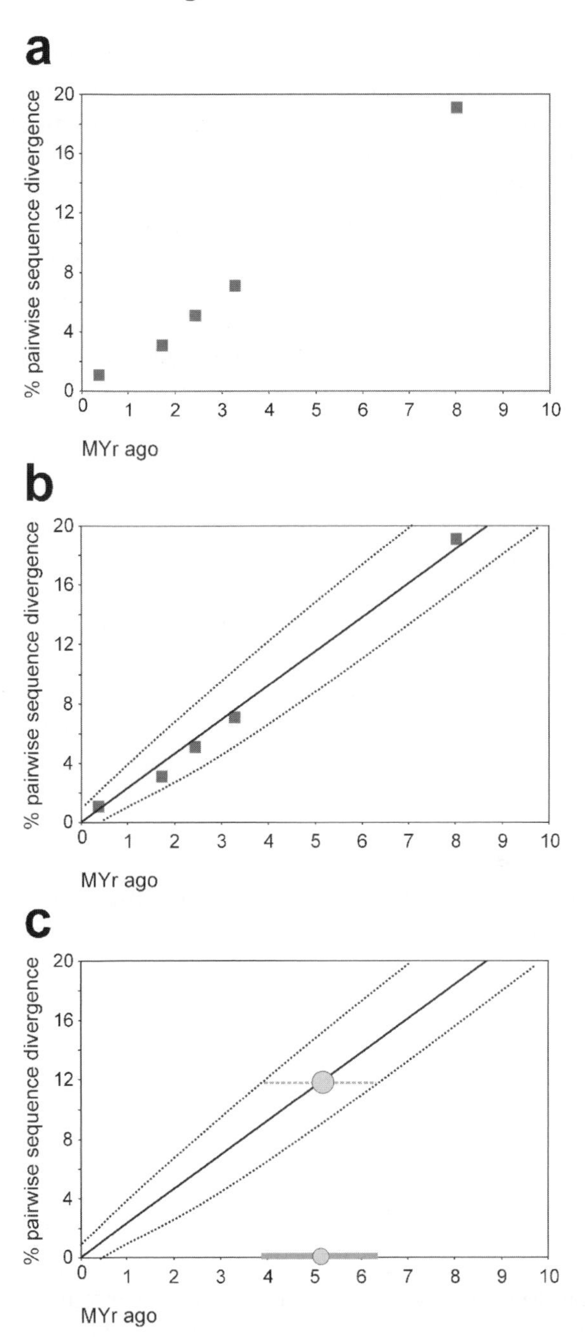

Figure 1. Schematic representation of regression-based molecular clock methods (global molecular clock). (a) Pairwise sequence divergence is plotted against age since divergence for taxa pairs with a well known fossil record or well known vicariance time. (b) A regression line with prediction confidence intervals is drawn. (c) The age of a species pair of known sequence divergence can be assessed using this regression.

Direct inclusion of such calibration intervals (as minimum or maximum time constraints) is possible in the various applications of *local molecular clocks*, which use different approaches and algorithms to take local variations and autocorrelation of substitution rates into account. The most commonly used programs, at present, are (1) Multidivtime, written by J. Thorne, which makes use of a Bayesian approach described by Thorne et al. (1998) and Thorne and Kishino (2002); and (2) r8s, written by M. Sanderson, which uses non-parametric rate smoothing or penalized likelihood (Sanderson, 1997; 2002). Both are tree-based, i.e., an important part of the input to be provided is a tree with branch lengths (Fig. 2a). After applying a calibration (Fig. 2b) and using various statistics this additive tree is then transformed into an ultrametric tree (Fig. 2c) in which branch length is directly equivalent to time. These methods are more widely applicable than global clocks because they allow the use of genes and taxa of unequal molecular substitution rates, and take uncertainty of calibrations into account. In addition, phylogeneticists more often will have a set of sequences from few genes in many taxa than many genes from few taxa, leaving local clock methods as best suited for their data. However, local molecular clocks, too, may not be free of artifacts. Hoegg et al. (2004) found that likelihood tree reconstruction algorithms tend to assign shorter branch lengths to paraphyletic basal taxa when these are sister to a species-rich crown group. This might be due to an artifact paralleling the node density effect described by Sanderson (1990) for parsimony methods, although maximum likelihood should be more robust to node density effects (Bromham et al., 2002). This might influence tree-based molecular clocks, and the finding of Yang and Yoder (2003) of divergence time estimates varying with the number of taxa included in crown groups may be related to this problem.

Despite these unsolved uncertainties there is little doubt that molecular clocks do keep time as convincingly demonstrated by the general correlation of fossil and molecular times of divergence (Hedges and Kumar, 2003; Douzery et al., 2004). As stressed by Graur and Martin (2004), the most important aspect is to report on divergence time estimates together with appropriate information about statistical uncertainty. C. Poux and colleagues (unpubl. manuscript) argue that for a reliable dating of the time of colonization of Madagascar by mammals, two separate dates need to be combined: the split of the Malagasy clade from its closest non-Malagasy sister group, and the first split within the Malagasy group (the onset of the Malagasy radiation). Taking only the split among the Malagasy clade and its closest extant non-Malagasy sister clade does not suffice, because the actual closest sister clade may have gone extinct. Similarly, using only the onset of the Malagasy radiation provides only an upper limit for the time of colonization, because the radiation may not immediately follow the

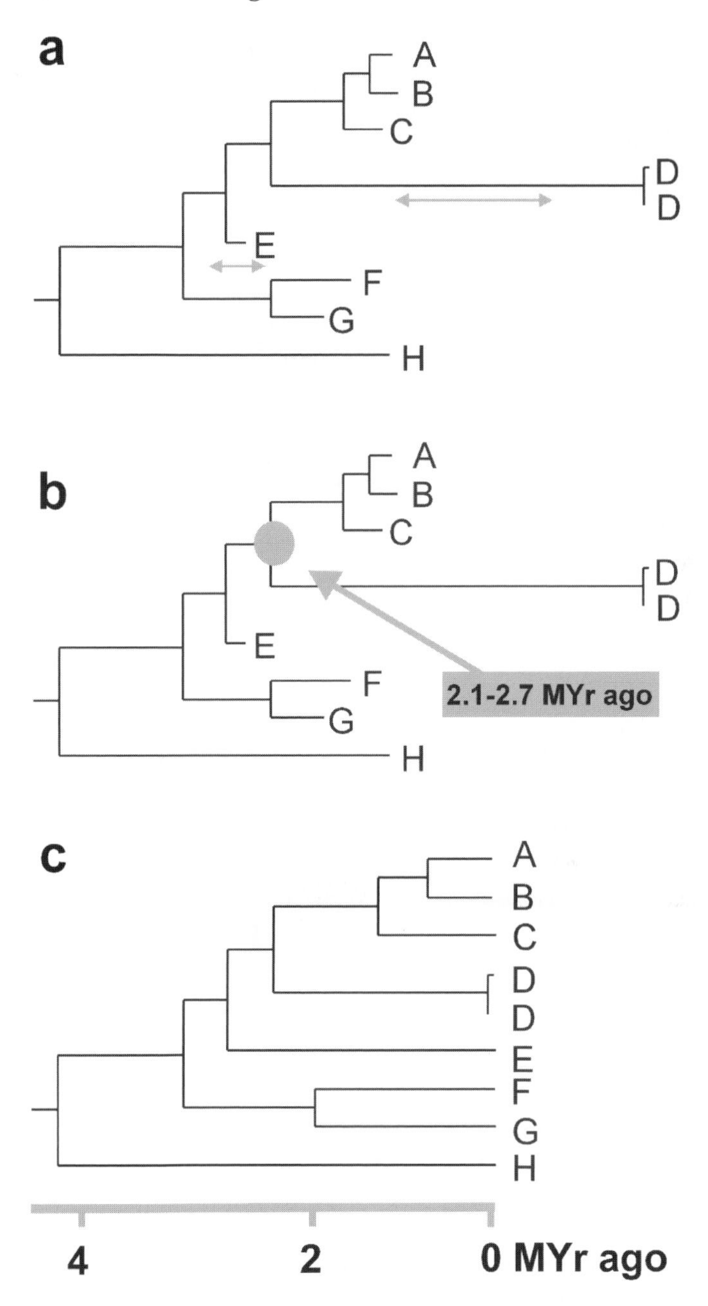

Figure 2. Schematic representation of tree-based local molecular clock methods. (a) A phylogram (additive tree) is constructed; branch lengths and therefore substitution rates often vary among lineages. (b) A calibration point or calibration interval is fixed among the ingroup taxa. (c) Taking the branch length differences into account, the tree is converted into an ultrametric tree in which branch lengths are strictly equivalent to time, e.g. through non-parametric rate-smoothing (Sanderson, 1997) or a Bayesian method of calculating autocorrelation among estimated local substitution rates (Thorne et al., 1998).

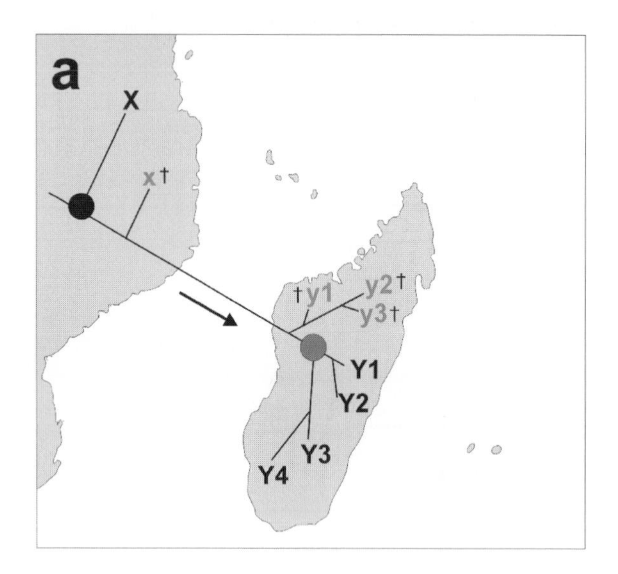

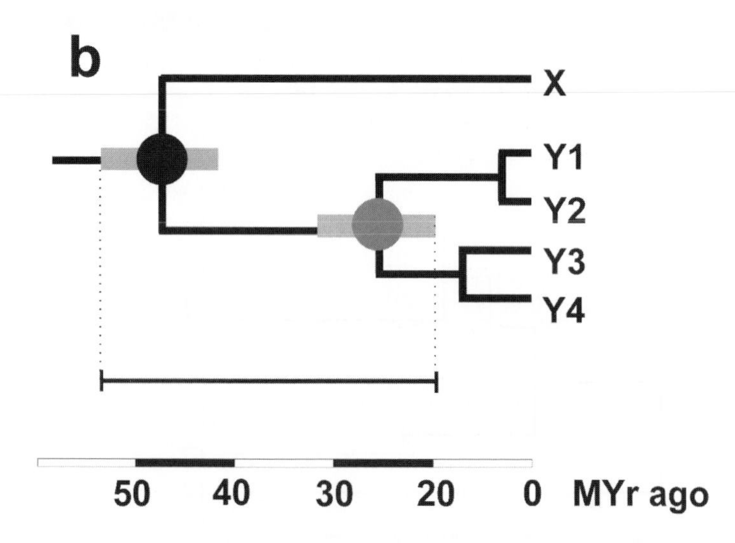

Figure 3. Estimating colonization times of islands (here Madagascar) using molecular clock data from extant taxa following the rationale of Poux et al. (submitted). (a) Estimates are available for the time of divergence of the extant insular taxa Y from their nearest extant non-insular sister taxon X (black circle), and of the earliest divergence among the extant insular taxa Y1-Y4 (grey circle). Any of these two estimates may be very close to the actual time of colonization, but extinct species from the mainland, x, may have been closer to the colonizing ancestor, early radiations in Madagascar, y1-y3, may have gone extinct, and the time from colonization to the onset of the radiation may be long. (b) Time estimates and confidence intervals from a local clock approach. The window of possible colonization times is given with 95% significance by the upper confidence interval of the first estimate (black circle) and the lower confidence interval of the second estimate (grey circle).

Table 1. Species richness of clades of non-flying (terrestrial and freshwater) vertebrates in Madagascar, and their uncorrected pairwise sequence divergence (in percent) in the 16S or 12S rRNA genes to their non-Malagasy sister-groups or close relatives. In some cases, phylogenetic knowledge is still incomplete, (see footnotes; these groups are only tentatively included here). Number of species refers exclusively to species occurring on the Malagasy mainland; nevertheless, a radiation is considered as endemic and monophyletic even if it includes species or populations on the Comoros and small islands such as Europa, Glorioso or Cosmoledo. The 16S divergences are largely based on data presented in Vences (in press). The list is not complete; clades were excluded in cases of incomplete knowledge on their phylogeny or molecular differentiation. All data are from Vences (in press) except those for tenrecs (Douady et al., 2002), carnivorans (Yoder et al., 2003), *Lygodactylus* (Puente et al., 2005), *Hemidactylus* (Vences et al., 2004), *Phelsuma* (Sound, Kosuch, Veith, Vences, unpublished)

Taxonomic group	Monophyletic	N species in Madagascar	Sister group or close relative	p_{16S} to sister group (%)
Fishes				
Malagasy Cichlidae	No[1]	33	African cichlids	12
[Cichlidae: *Paretroplus*]	Yes	15	Indian *Etroplus*	8
Malagasy aplocheiloids	Yes[2]	6	Indian aplocheiloids	13
Amphibians				
Mantellidae	Yes	143	Rhacophoridae	19
Hyperoliidae: *Heterixalus*	Yes	10	*Afrixalus* (Africa)	13
Ranidae: *Ptychadena mascareniensis*[3]	Yes	1	*Ptychadena* species (Africa)	6
Reptiles				
Lygodactylus	Uncertain[4]	21	African *Lygodactylus*	18
Phelsuma	No[5]	24	*Phelsuma ocellata* (Africa)	18
Hemidactylus mercatorius[3]		1	African *Hemidactylus mercatorius*	7
Malagasy gherrosaurids	Yes	18	African gherrosaurids	18
Colubridae: Malagasy pseudoxyrhophiines	Yes	73	*Ditypophis* (Sokotra)	7
Colubridae: *Mimophis*[3]	Yes	1	African psammophiines	6
Malagasy Boidae	Yes	4	South American boids	8
Chamaeleonidae: *Calumma* clade 1	No?	15	*Bradypodion*?[6]	13
Chamaeleonidae: *Furcifer*	Yes	18	*Chamaeleo*?[6]	14
Testudines: land tortoises[7]	Yes	6	African *Geochelone*	8
Testudines: *Erymnochelys*[3]	Yes	1	South American podocnemids	13
Scincidae: *Euprepis*	Yes	13	African *Euprepis*	7
Scincidae: *Cryptoblepharus*[3]	Yes	1	Other *Cryptoblepharus*	3

Table 1. continued

Taxonomic group	Monophyletic	N species in Madagascar	Sister group or close relative	p_{16S} to sister group (%)
Ratite birds				
Mullerornis[7]	Yes?	8	Australian/New Zealand ratites	8
Mammals				
Malagasy tenrecs	Yes	27	Otter shrews (Africa)	21
Malagasy carnivores	Yes	9	Herpestids (Africa)	8
Lemurs	Yes	43	Lorisiformes	18
Nesomyine rodents	Yes	22	African murids (*Cricetomys* and others)	12

[1]The African and Indian cichlids together make up the sister group of the African cichlid clade.
[2]Includes a Seychellean species.
[3]Group included here although consisting of only one species in Madagascar.
[4]Malagasy *Lygodactylus*, including *Microscalabotes*, may be paraphyletic with respect to African species of the genus; here included only tentatively.
[5]The non-Malagasy *Phelsuma*, i.e., the species from Comoros, Mascarenes, Seychelles, Aldabra, etc., are likely to have colonized these islands from Madagascar.
[6]Chameleon phylogeny is not yet clarified. Following the phylogeny of Raxworthy et al. (2002), the Malagasy chameleons as a whole are paraphyletic. One clade of *Calumma* splits off the tree after *Bradypodion*, and therefore these two taxa are compared here. *Furcifer*, with another clade of *Calumma*, was placed sister to *Chamaeleo*.
[7]Includes extinct taxa.

colonization, or because taxa from early radiations may have gone extinct as well. Hence, it is the lower 95% confidence interval of the first of these estimates and the upper 95% confidence interval of the second, which provide a reliable time window in which the colonization of Madagascar has taken place. For mammals, a group with a rich fossil history, even such a conservative approach is able to provide sufficiently accurate estimates of colonization times if the data set is large and appropriate (Poux et al., unpublished data).

In Malagasy vertebrates, molecular phylogenetic studies have led to an enormous progress in the understanding of their origins. If the majority of Madagascar's fauna and flora was a relict of Gondwanan times and originated by vicariance during these plate tectonic events, then most sister groups of Malagasy taxa would be expected to occur in India, the landmass that was last connected to Madagascar (Briggs, 2003). However, as reviewed by Vences (in press), the sister groups of most Malagasy clades have been identified from Africa. Only aplocheiloid and cichlid fishes, as well as mantellid frogs, have their probable closest relatives in India. Boas, iguanas and podocnemine turtles are related to South American forms. And the extinct elephant birds appear to have been related to ratites from Australia and New Zealand.

Molecular clock analyses have added the second crucial bit of information. Madagascar separated from Africa at 160-158 mya and from India at 96-84 mya (Briggs, 2003). The molecular clock evidence in most cases points to Cenozoic divergences of the Malagasy taxa from their non-Malagasy sister groups, hence during periods when Madagascar was fully isolated from other land masses (e.g., Yoder et al., 2003, Vences et al., 2003; vertebrates reviewed in Vences, in press). This applies even to such groups where the area cladograms are in agreement with the pattern of Gondwana fragmentation (Vences et al., 2001).

As a result of these studies, a picture is now emerging in which many of the Malagasy vertebrate clades originated by Cenozoic overseas dispersal, agreeing with the fossil record from the Latest Cretaceous which contains primitive groups with little resemblance to extant taxa (Krause et al., 1997, 2003). Nevertheless, vicariance cannot be excluded as origin for many groups, and is still a likely explanation in some cases. At least 20-30 independent colonizations of, or vicariance origins in, Madagascar have been identified. Almost certainly the remaining phylogenetic uncertainties regarding these groups will be clarified in the next decade, and times of colonization should soon be assessed with sufficient significance as well. This provides a unique opportunity to start with the search for general patterns among these radiations.

3. COMPARING PATTERNS OF RADIATIONS: FIRST EXERCISES FROM THE PERSPECTIVE OF MALAGASY VERTEBRATES

The species diversity of the Malagasy vertebrate clades in relation to time is one fundamental question that can only be answered across taxa, not in single-case studies. It can be approached from two perspectives. First, we can test whether the number of species in a certain clade of Malagasy vertebrates is related to the time elapsed since colonization. Whether species diversity is correlated with the age of the clade, that is, whether more ancient clades are more speciose than younger ones, is not trivial because some radiations, such as the cichlid fishes of Lake Victoria, have spawned enormous species diversities in extremely short time spans (Verheyen et al., 2003). Second, we can assess whether species richness shows important differences among radiations that originated before or after a certain key event, such as for example the origin of rainforests on Madagascar (probably in the Eocene/Oligocene; Wells, 2004). Such analyses are not yet possible

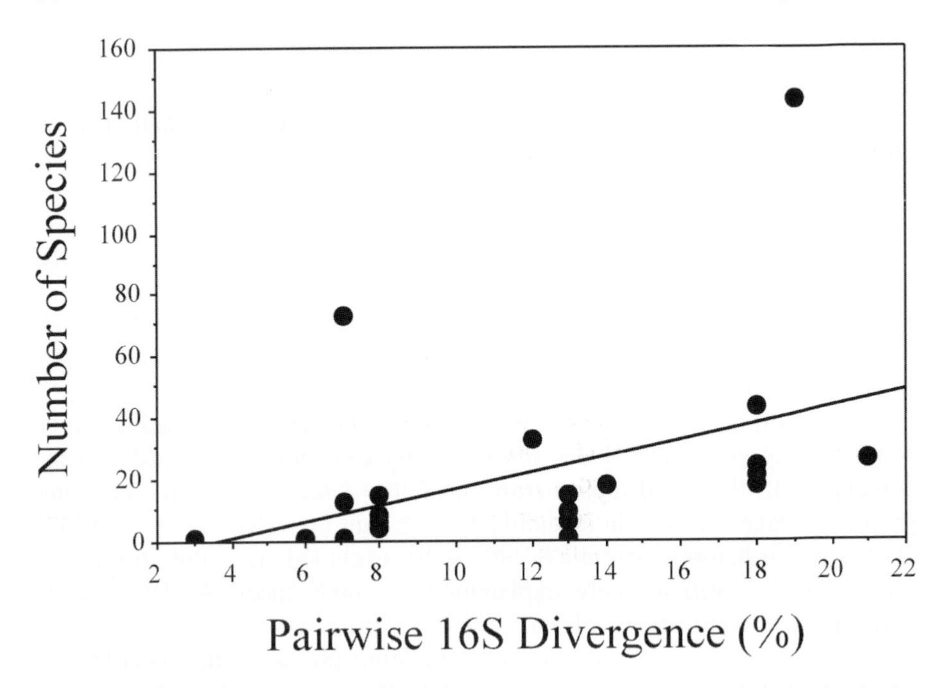

Figure 4. Relation between of species diversity of Malagasy vertebrate clades and pairwise 16S rRNA divergences between the Malagasy taxa and their non-Malagasy sister groups or close relatives. Data from Table 1 (see sources and references there). The correlation is significant (see text).

because reliable molecular clock estimates have not yet been published for a sufficient number of taxa. However, it is possible to elaborate a first prediction by comparing genetic divergences in a fragment of the mitochondrial 16S rRNA gene.

The correlation among 16S divergence of the Malagasy taxa from their sister group with their species diversity (Table 1 and Fig. 4) was significant in a parametric analysis (Pearson correlation coefficient = 0.432; P = 0.039) and a non-parametric analysis (Spearman correlation coefficient = 0.657; P = 0.001). The significance of the non-parametric correlation remained after deleting all groups that contained only a single Malagasy species (Spearman correlation coefficient = 0.5; P = 0.0.035), whereas the parametric analysis was not significant under these conditions (P = 0.163). Although the taxa included differ in their substitution rates, the genetic divergence data should be roughly equivalent to divergence time. We can therefore predict that there might be a correlation also between species diversity of Malagasy clades and their age. This hypothesis is to be tested once that reliable age estimates for more clades become available.

Two of the clades included here deviate strongly from the general pattern by being more diversified than expected (the two upper dots in Fig. 3): mantellid frogs and pseudoxyrhophiine snakes. If a radiation is to be defined by an extraordinary rate of species formation, these two clades therefore comply with the definition. However, such reasoning needs to keep in mind that species numbers in different clades are not fully comparable; for instance top predators such as carnivorans will diversify less than other taxa, and therefore it is not a surprise that this clade contains few species.

4. OUTLOOK: MADAGASCAR AS EVOLUTIONARY MODEL SYSTEM

Extending the above example to correlate species diversity with the time since colonization is only one perspective for comparative studies on the Malagasy radiations. Exploring how molecular substitution rates are correlated with species diversity (e.g., Barraclough and Savolainen), and with the rate of speciation, is another exciting field. Such studies need not be restricted to species diversity but can be extended to morphological, ecological or genetic diversity of clades. By reconstructing character states of the ancestors of radiations it might be possible to test the hypothesis that generalists are more likely to spawn radiations than specialists (e.g., Nosil, 2002). However, although such studies are discernible as a perspective, they still suffer from the lack of sufficiently reliable phylogenies and datings for many groups. Molecular phylogenetic work on the Malagasy vertebrates is therefore still a necessity, and should be expanded to more invertebrate and plant groups as well.

Madagascar has been described by many superlatives - it is one of the hottest biodiversity hotspots for conservation priorities (Myers et al., 2000), it is a "Sanctuaire de la Nature" (Oberlé, 1981), the origin of its extant biota was one of the "greatest unsolved mysteries of natural history" (Krause et al., 1997). In addition, this microcontinent is on its way to become one of the most promising model regions for evolutionary studies.

ACKNOWLEDGEMENTS

This paper has benefited from discussions with many colleagues, although not in all aspects it necessarily reflects their opinions. In particular I would like to thank Franco Andreone, Wolfgang Böhme, Wilfried W. de Jong, Frank Glaw, Ole Madsen, Axel Meyer, Celine Poux, David R. Vieites and David B. Wake. My work in Madagascar would not have been possible without cooperating with the Département de Biologie Animale of the Université d'Antananarivo, in particular with Daniel Rakotondravony, Olga Ramilijaona and Noromalala Raminosoa. The Malagasy authorities are acknowledged for providing research and export permits. Financial assistance was provided by the Deutsche Forschungsgemeinschaft, the Deutscher Akademischer Austauschdienst, the Netherlands Research council (NWO/WOTRO), and the BIOPAT and Volkswagen foundations.

REFERENCES

Barraclough, T. G., and Savolainen, V., 2001, Evolutionary rates and species diversity in flowering plants, *Evolution* **55:**677–683.

Bossuyt, F., and Milinkovitch, M. C., 2000, Convergent adaptive radiations in Madagascan and Asian ranid frogs reveal covariation between larval and adult traits, *Proc. Natl. Acad. Sci. USA* **97:**6585-6590.

Briggs, J. C., 2003, The biogeographic and tectonic history of India, *J. Biogeogr.* **30:**81–388.

Cibois, A., Slikas, B., Schulenberg, T. S., and Pasquet, E., 2001, An endemic radiation of Malagasy songbirds is revealed by mitochondrial DNA sequence data, *Evolution* **55:**1198-1206.

Donoghue, M. J., and Moore, B. R., 2003, Toward an integrative historical biogeography, *Integr. Comp. Biol.* **43:**261-270

Douady C. J., Catzeflis F., Kao D. J., Springer M. S., and Stanhope M. J., 2002, Molecular evidence for the monophyly of Tenrecidae (Mammalia) and the timing of the colonization of Madagascar by Malagasy tenrecs, *Mol. Phylogenet. Evol.* **22:**357-363.

Douzery, E. J. P., Snell, E. A., Bapteste, E., Delsuc, F., and Philippe, H., 2004, The timing of eukaryotic evolution: does a relaxed molecular clock reconcile proteins and fossils?, Proc. Natl. Acad. Sci. USA **101:**15386–15391.

Gillespie, R. G., and Roderick, G. K., 2002, Arthropods on islands: colonization, speciation and conservation, *Annu. Rev. Entomol.* **47:**595–632

Gittenberger, E., 1991, What about non-adaptive radiation?, *Biol. J. Linn. Soc.* **43:**263-272.

Gittenberger, E., 2004, Radiation and adaptation, evolutionary biology and semantics, *Org. Divers. Evol.* **4:**135-136.

Goodman, S. M., and Benstead, J. P., eds, 2003, *The Natural History of Madagascar,* The University of Chicago Press, Chicago and London.

Graur, D., and Martin, W., 2004, Reading the entrails of chickens: molecular timescales of evolution and the illusion of precision, *Trends Genet.* **20:**80-86.

Grant, P. R., and Grant, B.R., 2002, Adaptive radiation of Darwin's Finches, *American Scientist* **90:**130-139.

Hedges, S. B., and Kumar, S., 2003, Genomic clocks and evolutionary timescales, *Trends Genet.* **19:**200-206.

Hedges, S. B., and Kumar, S., 2004, Precision of molecular time estimates, *Trends Genet.* **20:**242-247.

Hillis, D. M., Mable, B. K., and Moritz, C., 1996, Applications of molecular systematics, in: *Molecular systematics*, 2nd ed, D. M. Hillis, C. Moritz, and B. K. Mable, eds, Sinauer Associates, Sunderland, MA, pp. 515-543.

Hoegg, S., Vences, M., Brinkmann, H., and Meyer, A., 2004, Phylogeny and comparative substitution rates of frogs inferred from three nuclear genes, *Mol. Biol. Evol.* **21:**1188-1200.

Jansa, S. A., Goodman, S. M., and Tucker, P. K., 1999, Molecular phylogeny and biogeography of the native rodents of Madagascar (Muridae: Nesomyinae): a test of the single-origin hypothesis, *Cladistics* **15:**253-270.

Juan, C., Emerson, B. C., Oromí, P., and Hewitt, G. M., 2000, Colonization and diversification: towards a phylogeographic synthesis for the Canary Islands, *Trends Ecol. Evol.* **15:**104-109.

Krause D. W., 2003, Late Cretaceous vertebrates of Madagascar: a window into Gondwanan biogeography at the end of the age of dinosaurs, in: *The Natural History of Madagascar*, S. M. Goodman, and J. P. Benstead, eds, The University of Chicago Press, Chicago and London, pp. 40-47.

Bromham, L., Woolfit, M., Lee, M. S. Y., and Rambaut, A., 2002, Testing the relationship between morphological and molecular rates of change along phylogenies, *Evolution* **56:**1921-1930.

Krause, D. W., Hartman, J. H., and Wells, N. A., 1997, Late Cretaceous vertebrates from Madagascar. Implications for biotic changes in deep time, in: *Natural change and human impact in Madagascar,* S. M. Goodman, and B. D. Patterson, eds, Smithsonian Institution Press, Washington, pp. 3-43.

Losos, J. B., 1994, Integrative approaches to evolutionary ecology: *Anolis* lizards as model systems, *Annu. Rev. Ecol. Syst.* **25:**467-493.

Losos, J. R., Jackman, T. R., Larson, A., de Queiroz, K., and Rodriguez-Schettino, L., 1998, Contingency and determinism in replicated adaptive radiations of island lizards, *Science* **279:**2115-2118.

Losos, J. B., and Schluter, D., 2000, Analysis of an evolutionary species-area relationship, *Nature* **408:**847-850.

Martens, K., Goddeeris, B., and Coulter, G., eds., 1994, *Speciation in ancient lakes,* Archiv für Hydrobiologie - Advances in Limnology, Volume 44.

Myers, N., Mittermeier, R. A., Mittermeier, C. G., Fonseca, G. A. B., and Kent, J., 2000, Biodiversity hotspots for conservation priorities, *Nature* **403:**853-858.

Nagy, Z. T., Joger, U., Wink, M., Glaw, F., and Vences, M., 2003, Multiple colonizations of Madagascar and Socotra by colubrid snakes: evidence from nuclear and mitochondrial gene phylogenies, *Proc. Roy. Soc. London B* **270:**2613-2621.

Nosil. P., 2002, Transition rates between specialization and generalization in phytophagous insects, *Evolution* **56:**1701-1706.

Oberlé, P., ed., 1981, *Madagascar: Un Sanctuaire de la Nature*, Lechevallier, Paris.

Poux, C., Chevret, P., Huchon, D., de Jong, W. W., and Douzery, E. J. P., unpublished manuscript, Arrival and diversification of caviomorph rodents and platyrrhine primates in South America.

Puente, M., Thomas, M., and Vences, M., 2005, Phylogeny and biogeography of Malagasy dwarf geckos, *Lygodactylus* Gray, 1864: preliminary data from mitochondrial DNA sequences (Squamata: Gekkonidae). this volume.

Raxworthy C. J., Forstner M. R. J., Nussbaum R. A., 2002, Chameleon radiation by oceanic dispersal, *Nature* **415:**784-786.

Reisz, R. R. and Müller, J., 2004, Molecular timescales and the fossil record: a paleontological perspective, *Trends Genet.* **20:**37-241.

Rodriguez-Trelles, F., Tarrio, R., and Ayala, F. J., 2002, A methodological bias toward overestimation of molecular evolutionary time scales, *Proc. Natl. Acad. Sci. USA* **99:**8112-8115.

Sanderson, M. J., 1990, Estimating rates of speciation and evolution: a bias due to homoplasy, Cladistics **6:**387-391.

Sanderson, M. J., 1997, A nonparametric approach to estimating divergence times in the absence of rate constancy, *Mol. Biol. Evol.* **14:**1218-1231.

Sanderson, M. J., 2002, Estimating absolute rates of molecular evolution and divergence times: a penalized likelihood approach, *Mol. Biol. Evol.* **19:**101-109.

Schmitt, M., 2004, The 44th Phylogenetisches Symposium - Bonn, 22-24 November 2002 - on 'adaptive radiation', *Org. Divers. Evol.* **4:**125.

Schön, I., and Martens, K., 2004, Adaptive, pre-adaptive and non-adaptive components of radiations in ancient lakes: a review, *Org. Divers. Evol.* **4:**137-156.

Schubart, C. D., Diesel, R., and Hedges, S. B., 1998, Rapid evolution to terrestrial life in Jamaican crabs, *Nature* **393:**363-365.

Shaw, K. L., 2002, Conflict between nuclear and mitochondrial DNA phylogenies of a recent species radiation: What mtDNA reveals and conceals about modes of speciation in Hawaiian crickets, *Proc. Natl. Acad. Sci. USA* **99:**16122-16127.

Springer, M. S., Debry, R. W., Douady, C., Amrine, H. M., Madsen, O., de Jong, W. W., and Stanhope, M. J., 2001, Mitochondrial versus nuclear gene sequences in deep-level mammalian phylogeny reconstruction, *Mol. Biol. Evol.* **18:**132-143.

Sudhaus, W., 2004, Radiation within the framework of evolutionary ecology, *Org. Divers. Evol.* **4:**127-134.

Taberlet, P., Fumagalli, L., Wust-Saucy, A.-G., and Cosson, J.-F., 1998, Comparative phylogeography and postglacial colonization routes in Europe, *Mol. Ecol.* **7:**453-464.

Takezaki, N., Rzhetsky, A. and Nei, M., 1995, Phylogenetic test of the molecular clock and linearized trees, *Mol. Biol. Evol.* **12:**823-833.

Thorne, J. L., and Kishino, H., 2002, Divergence time and evolutionary rate estimation with multilocus data, *Syst. Biol.* **51:**689-702.

Thorne, J. L., Kishino, H., and Painter, I. S., 1998, Estimating the rate of evolution of the rate of molecular evolution, *Mol. Biol. Evol.* **15:**1647-1657.

Vences, M., in press, Origin of Madagascar's extant fauna: a perspective from amphibians, reptiles and other non-flying vertebrates, *Ital. J. Zool.*

Vences, M., Freyhof, J., Sonnenberg, R., Kosuch, J., and Veith, M., 2001, Reconciling fossils and molecules: Cenozoic divergence of cichlid fishes and the biogeography of Madagascar, *J. Biogeogr.* **28:**1091-1099.

Vences, M., Vieites, D. R., Glaw, F., Brinkmann, H., Kosuch, J., Veith, M., and Meyer, A., 2003, Multiple overseas dispersal in amphibians, *Proc. Roy. Soc. London B* **270:**2435-2442.

Vences, M., Wanke, S., Vieites, D. R., Branch, W. R., Glaw, F., and Meyer, A., 2004, Natural colonization or introduction? Phylogeographical relationships and morphological differentiation of house geckos (*Hemidactylus*) from Madagascar, *Biol. J. Linn. Soc.* **83:**115-130.

Verheyen, E., Salzburger, W., Snoeks, J., and Meyer, A., 2003, The origin of superflock of cichlid fishes from Lake Victoria, East Africa, *Science* **300:**325-329.

Wagner, W. L. and V. A. Funk, eds., 1995, *Hawaiian Biogeography: Evolution on a Hot Spot Archipelago*, Smithsonian Institution Press, pp. 467.

Wells N. A., 2003, Some hypotheses on the Mesozoic and Cenozoic paleoenvironmental history of Madagascar, in: *The Natural History of Madagascar*, S. M. Goodman, and J. P. Benstead, eds, The University of Chicago Press, Chicago and London, pp. 16-34.

Wiens, J. J., and Donoghue, M., 2004, Historical biogeography, ecology and species richness, *Trends Ecol. Evol.* in press.

Yamagishi, S., Honda, M., Eguchi, K., and Thorstrom, R., 2001, Extreme endemic radiation of the Malagasy vangas (Aves: Passeriformes), *J. Mol. Evol.* **53:**39-46.

Yang, Z., and Yoder, A. D., 2003, Comparison of likelihood and bayesian methods for estimating divergence times using multiple gene loci and calibration points, with application to a radiation of cute-looking mouse lemur species, *Syst. Biol.* **52:**705–716.

Yoder, A. D., Burns, M. M., Zehr, S., Delefosse, T., Veron, G., Goodman, S. M., and Flynn, J., 2003, Single origin of Malagasy Carnivora from an African ancestor, *Nature* **421:**734-737.

COMMEMORATING MARTIN EISENTRAUT (1902-1994) - IMPORTANT EXPLORER OF TROPICAL AFRICAN VERTEBRATES

Wolfgang Böhme
Zoologisches Forschungsinstitut und Museum Alexander Koenig, Adenauerallee 160, 53113 Bonn, Germany, E-mail: w.boehme.zfmk@uni-bonn.de

Abstract: The present paper was read as a plenary lecture during the 5th ZFMK International Symposium on Tropical Biology ("African Biodiversity") to commemorate the scientific life and legacy of Professor Martin Eisentraut, who was director of ZFMK from 1957 - with two years interruption - to 1977. The paper is aimed to show appreciation to his many scientific achievements in vertebrate zoology by giving an outline of his career, which focussed on the exploration of Central African, particularly Cameroonian vertebrates.

Key words: Martin Eisentraut; research on bats; Balearic lizards; Chacoan vertebrates; tropical vertebrates of Cameroon

Martin Eisentraut with two viverrids. Photo: Steinbach.

B.A. Huber et al. (eds.), African Biodiversity, 85–98.

1. INTRODUCTION

Martin Eisentraut was born on October 21, 1902 at Groß-Töpfer, Thuringia, as the 2nd son of the priest Jakob Eisentraut (1867-1947) and his wife Anna, born Bischoff (1866-1946). It was from his mother that Martin received his first support for developing an interest in nature. He finished high school at a gymnasium with a particularly long and famous tradition at Halle/Saale ("Latina der Francke'schen Stiftungen") in 1921 and entered the University of Halle the same year. He studied zoology, botany and geology and passed his doctoral examination only four years later in summer 1925, under the supervision of the famous geneticist Professor Valentin Haecker (1864-1927). His thesis dealt with orthopteran chromosomes.

2. THE FIRST YEARS IN BERLIN: HIBERNATION AND BAT RESEARCH

Through the intermediation of one of his fellow students in Professor Haecker's group, viz. Bernhard Rensch (1900-1990), Martin Eisentraut received a volunteer position (first without payment) at the Natural History Museum of Berlin. The director at that time, Professor Carl Zimmer (1873-1950) entrusted him with drafting and preparing a new exhibition concept for a so-called biological hall, where visitors were to be confronted with biological and ecological problems rather than with traditional taxonomic arrangements of stuffed specimens. The topic to be dealt with was hibernation. Eisentraut started with the European hamster (*Cricetus cricetus*), went to the field and excavated these rodents from their burrows. Dealing with hibernation and hibernation physiology in mammals resulted in several important publications (e.g. Eisentraut 1933, 1955) and led him also to develop a particular interest in bats. There was, however, a second reason for him to concentrate on bats rather than on birds, which had originally been his primary interest. The reason was the personality of Professor Erwin Stresemann (1889-1972) of the Berlin Museum. Although Stresemann proved to be most influential on Eisentraut's fellow student Ernst Mayr (born 1904), he led Martin Eisentraut **not** to become an ornithologist, obviously due to their incompatible characters (as Eisentraut told me later several times). For him, bats became an excellent replacement group for birds, the more as they fitted also in the hibernation problem.

Eisentraut soon became most impressed to see the large numbers of overwintering bats in caves and mine tunnels around Berlin and immediately began asking the question where these animals might go in spring. The idea

Figure 1. Collecting hibernating bats in a mine tunnel at Rüdersdorf/Berlin (Eisentraut on the left). Photo: ZFMK archives.

Figure 2. Preparation for bat banding. Same site as Fig. 1 (Eisentraut on the left). Photo: ZFMK archives.

of bat-banding was born (Figs. 1, 2) and Eisentraut thus became the "father of bat-banding" in Europe. His intensive research on various aspects of bat biology and ecology resulted in his early but nonetheless classical monograph: "Die deutschen Fledermäuse, eine biologische Studie" (= The German bats, a biological study, Eisentraut 1937).

Bats continued to play an important role in his further research interests, particularly also in tropical species.

Approximately one fifth of his publications is exclusively devoted to bats, many of his other papers (for a complete list see Böhme and Hutterer 1999), however, deal with physiological, faunistic, systematic and zoogeographical questions concerning bats; some of them highlight their economic importance, the experiences with bat-banding, entire bat faunas (e.g. Fernando Poo = Bioko Island), or contribute to encyclopedias (e.g. "Grzimek's animal life"). Of his ten papers that collectively appeared in book form, a second one (apart from the above-mentioned 1937 monograph) is devoted exclusively to bats: "Aus dem Leben der Fledermäuse und Flughunde" (= Life of Micro- and Megachiroptera, 1957). American chiropterologists honoured him by publishing a laudatio for his 80th birthday in "Bat Research News", which I was allowed to write (Böhme 1984).

3. THE LIZARD INTERMEZZO: RESEARCH ON BALEARIC LIZARDS

Eisentraut's interest in lizards and their variation on and between islands arose quite spontaneously but generated very remarkable results. It was provoked by two travels (1928 and 1930) to the Balearic Islands. Eisentraut started first to compete with some German herpetologists - first of all Lorenz Müller (1868-1953) - in describing and naming new microinsular taxa ("Rassen" = races, see Fig. 3) but subsequently extended his research interests to more general problems of island lizard biology and ecology (particularly island melanism) and he again summarized all his experience and knowledge in yet another classical book: "Die Eidechsen der spanischen Mittelmeerinseln und ihre Rassenaufspaltung im Lichte der Evolution" (= The lizards of the Spanish Mediterranean islands and their raciation in the light of evolution, Eisentraut 1950) (see also Böhme 1994, 2004).

As a part of his studies, Eisentraut (1930) made also field experiments by translocating lizards of different phenotypes to so far uncolonized offshore islets (Dado Grande, today Dau gran, off Ibiza/Eivissa), viz. normally patterned, green lizards together with melanistic ones, in order to study the inheritance of melanism under field conditions. The results obtained found a long-term interest and have taken effect up to the present (Böhme and Eisentraut 1981, Mayol 2004).

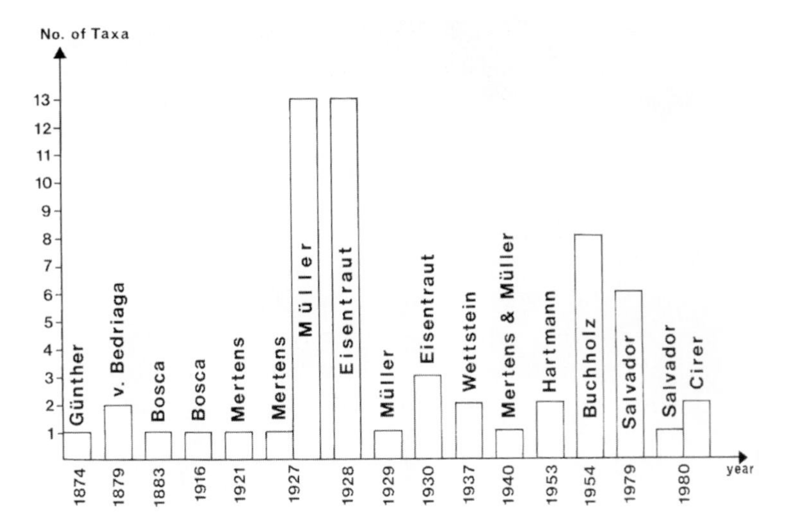

Figure 3. Diagram showing Eisentraut's portion of infraspecific Balearic lizard taxa described. From Böhme (2004).

4. THE BOLIVIA CHACO EXPEDITION 1930

When in Ibiza, Eisentraut received a telegram inviting him to participate in a zoological expedition to the Gran Chaco of Bolivia, and he accepted at once. Remarkably, his wife followed him some months later, travelling alone by boat from Hamburg to Buenos Aires, then by train to the Bolivian border where her husband expected her with two horses (Fig. 4). Among the numerous results of this expedition (see the bibliography compiled by Böhme and Hutterer 1997), two aspects were most interesting anticipations of discoveries made by other zoologists some four decades later:

1. Eisentraut collected (respectively shot) a broad-snouted caiman (*Caiman latirostris*) in a pond near Villa Montes, and when he dragged the dead specimen, a female, out of the water with a lasso, a group of squeaking baby caimans (now deposited in the Berlin Museum) followed their dead mother (now in ZFMK). At that time, nobody knew anything about the highly developed maternal care in crocodiles which first became known to science through Pooley's (1974, and subsequent) observations and descriptions on the Nile crocodile (*Crocodylus niloticus*).

2. Also already in 1930, Eisentraut saw a group of peccaries differing from the two known species, viz. the collared (*Tayassu tayacu*) and the white-lipped peccary (*Tayassu pecari*) mainly by its bigger size, but was unable to collect one of them. In 1937, he received a letter by a German settler named Carl Berkhan who lived in the Bolivian Chaco and reported on

Figure 4. Martin and Johanna Eisentraut in front of their house in the Bolivian Chaco (1930). Photo: ZFMK archives.

this third, unknown species, including its preference for opuntia fruits wherefore the locals called it "quimilero". Only in 1975, this third species was "officially" discovered in the Paraguayan Chaco (where it is called "tagua") and identified as a form which was known already as a late Pleistocene/ early Holocene fossil! It bore already the paleospecies name *Catagonus wagneri*. Its discovery as a member of the recent fauna was a sensation (Wetzel et al. 1975), but Eisentraut received late and double satisfaction for his failure to collect a specimen in 1930: by having been proven correct with his assumption that he had met with an unknown species in 1930, and by providing at least the first proof of its occurrence in Bolivia (Eisentraut 1986); the corresponding voucher specimen was for many years on public display in Museum Koenig (Fig. 5).

Figure 5. Mounted specimen of *Catagonus wagneri* in the former exhibit of Museum Koenig. First voucher of the occurrence of this "living fossil" from Bolivia. Photo by W. Böhme.

However, the German settler Berkhan cited above had written to Eisentraut even about the existence of a fourth peccary species, even bigger in size and - in contrast to the former - of red-brown coloration, called "rosado" by the local hunters. It is unknown to science until today, but as the German planter Berkhan was right with anticipating a third peccary, why should he be wrong with the fourth? The very recent discovery of a new "giant" peccary in Amazonian Brazil highlights the potential of South America to further add new species to the list of big mammals today - a truly cryptozoological but nonetheless serious issue.

5. CAMEROON (1938 AND 1954-1973)

Despite all these fascinating research activities of Martin Eisentraut, his main field - Africa, and in particular Cameroon - has still to be started with. He got first acquainted with this country in 1938, arriving on a banana steamboat from Hamburg. His aim was to study the physiological problems he had worked on in Central European bats, also in tropical species. On this occasion he became fascinated by Mt. Cameroon and by the altitudinal zonation of the fauna. But his plan to return had to be postponed for years because of World War II. However in 1954, Eisentraut became the first German citizen after the war to receive a special permit by the British mandatory authorities to again visit West Cameroon. In the meantime, he had left Berlin and its Natural History Museum because of the political difficulties in the divided city. The museum was in the eastern sector, his house, however, in West Berlin, and it turned out to be impossible to cover life expenses in West Berlin with an East German salary. He preferred to go to West Germany, viz. to the Stuttgart Museum where he became curator of mammalogy. From here he was able to equip his next expedition to Cameroon, which proved very successful. It concentrated on the study of vertebrates in the different altitudinal zones of Mt. Cameroon. His plan, to return again to Cameroon in 1957, was delayed by a remarkable event: During the preparations of this trip, he received the offer to become the director of Museum Koenig in Bonn. Eisentraut accepted and started the planned expedition with taxidermists of both the Stuttgart and the Bonn Museum as companions.

The next decade was characterised by further expeditions to Cameroon, which were now extended to the important great island Fernando Poo (today Bioko) (Fig. 6), which is a geological continuation of the Cameroonian volcanic ridge towards the south-west ("Dorsale camerounaise"). He also extended his trips to the hinterland of Mt. Cameroon, the so-called Bamenda Banso highlands. Eisentraut visited Mt. Kupe, the Manenguba Mts., the

Figure 6. Martin Eisentraut on Fernando Poo (today Bioko, Equatorial Guinea) with natives of the Bube tribe. Photo by H. Dischner.

Rumpi Hills (Fig. 7) and also Mt. Oku, where he discovered many endemic mammals and birds. Lower vertebrates were also collected by him, and he entrusted his colleagues Ethel Trewavas (Natural History Museum London) and Robert Mertens (Senckenberg Museum, Frankfurt) with the study of the fishes and the amphibians and reptiles respectively. A most remarkable chameleon from the Rumpi Hills was named after him (Mertens 1968, see Fig. 8).

The manifold scientific results of these travels led to numerous papers which were summarized by Eisentraut in two comprehensive, again classical monographs: "Die Wirbeltiere des Kamerungebirges" (= The vertebrates of Mt. Cameroon, Eisentraut 1963) and "Die Wirbeltierfauna von Westkamerun und Fernando Poo" (= The vertebrate fauna of western Cameroon and of Bioko, Eisentraut 1973).

In 1969, at the age of 65, Martin Eisentraut had to retire from the directorship of Museum Koenig after 12 years. However, his retirement proved to be only a break rather than the beginning of a phase of his life without administrative duties. Since his successor was suspended from directorship in 1971, Eisentraut was asked to return again for a couple of years to guide the museum, and it was this year that I entered Museum Koenig myself. Two years later, Eisentraut planned another research trip to Cameroon where he wanted to extend his activity for the first time to the arid

Figure 7. One of Eisentraut's true (in the literal sense!) expeditions, here to Rumpi Hills (1967). Photo by W. Hartwig.

north of this country; and it was me whom he asked to accompany him. This trip led us to the Sudanian and Sahelian savanna belts of Cameroon near Lake Tchad (Figs. 9, 10), to the southern rainforests near Kribi and to Mt. Cameroon. For me, it was probably the most impressive event in my life to see the tropics for the first time, in company of such an experienced and knowledgeable man. It was the starting point of my own research interest and activity in tropical Africa, and it certainly helped to maintain the African focus in ZFMK's research profile until today.

Figure 8. The endemic Rumpi Hills' chameleon *(Chamaeleo eisentrauti).* Photo by C. Wild.

Figure 9. Martin Eisentraut at Figuig (North Cameroon) waiting for bush taxi transport to Mora and Waza, near Lake Tchad (1973). Photo by W. Böhme.

Figure 10. Taking photos of savanna elephants at Waza National Park, North Cameroon (1973). Photo by W. Böhme.

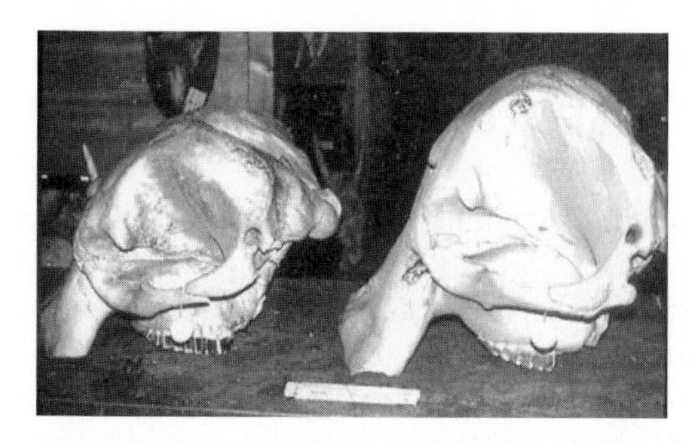

Figure 11. Skulls of an adult pygmy (left) and an adult forest elephant in the Tervuren Museum collection. Note the different shape of the orbital region. Photo by W. Böhme.

Eventually, this trip led us, at Kribi, to the acquaintance with a German amateur zoologist, Ulrich Roeder, who had spent many years in the study of the pygmy elephant problem (Roeder 1970, 1975). Lower jaws and tusks of these elephants seen in settlements of the pygmies near Kribi seemed convincing to us that actually two distinct elephant forms were occurring in sympatry and even in syntopy in the southern rainforest. Subsequently, we began to study the elephant skull collection of the Congo Museum at Tervuren and found marked differences in skull shape and proportions between the two forms (Eisentraut and Böhme 1989, see Fig. 11). Moreover, field observations seemed to corroborate the existence of these two forms (Böhme and Eisentraut 1990, see Fig. 12). Although a recent study of mtDNA (Debruyne et al. 2003) did not demonstrate the existence of two distinct species in the forest, the overall evidence is still conflicting between morphology and genetics, and further studies including broader DNA sampling are required to solve this problem that, in the positive case, would also have highest conservation priority as the world's largest land animals are concerned.

Figure 12. The famous photo of a pygmy elephant herd with their own young. Scale is provided by a white egret *(Casmerodius albus)* partly hidden behind the head of the guiding female. Photo by H. Nestroy.

6. FRUITFUL YEARS AFTER THE FINAL RETIREMENT IN 1979

Next to the pygmy elephant problem, with which Eisentraut occupied himself intensively, there were several other projects which he followed up after his second and final retirement in 1979. Several papers (see Böhme and Hutterer 1999 for details) added new data to his big monograph "Das Gaumenfaltenmuster der Säugetiere und seine Bedeutung für stammesgeschichtliche und taxonomische Untersuchungen" (= The pattern of palatal ridges in mammals and its importance for phylogenetic and taxonomic studies: Eisentraut 1976) (a reference unfortunately lacking in the bibliography by Böhme and Hutterer l.c.). But his most important works in these years were two books where he revisited his great expeditions in a popular, narrative form. One of them, entitled "Im Land der Chaco-Indianer" (= In the land of the Chaco Indians, Eisentraut 1983) dealt with the Bolivian Chaco which he had visited again in 1977, 75 years old, accompanied by his daughter Hannelore Vaassen, and he devoted much space to the miserable fate the Indian tribes had undergone since his first expedition in 1930.

The other, more voluminous book appeared one year earlier. It was concerned with his many expeditions to Cameroon and to Fernando Poo: "Im Schatten des Mongo-ma-loba" (= In the shadow of Mongo-ma-loba; the latter being the native name of Mt. Cameroon, Eisentraut 1982). As also with the Chaco book, it is illustrated with colour plates and numerous line drawings made by Wolfgang Hartwig, chief taxidermist of Museum Koenig, renowned animal painter, and several times faithful companion of Eisentraut's expeditions to Africa, also of two of myself. In this book, Eisentraut did not only describe his zoological experiences, but also the philosophical attitudes that make the life and the work of a field zoologist so fascinating.

In these years, Martin Eisentraut published also some books that were not zoological. Two small volumes dealt with contemporary history, one of them within the framework of his familial history. One booklet, at last, was a collection of poems that he published under a pseudonym, throwing new light on his facet-rich personality. The very last publication was a booklet with a tape cassette, again devoted to West Africa and to his primarily most beloved animal group, the birds: "Die Biotope westafrikanischer Vogelarten und deren Rufe und Gesänge" (= The biotopes of West African birds and their calls and songs; Eisentraut 1993).

One year before, on October 21, 1992, Martin Eisentraut was able to celebrate his 90th birthday, and many colleagues and the current staff of Museum Koenig congratulated him (Fig. 13). He could look back to a long, for the most part happy life, and to a rich scientific career that included

Figure 13. Reception and celebration of Professor Eisentraut's 90th birthday in the ornithological library of Museum Alexander Koenig: Congratulations by the late director Prof. Clas Naumann (1992). Photo by K. Busse.

remarkable public honours such as the honorary membership of the German Mammalogical Society (1972) or the Great Commander's Cross of the Order of Merit of the Federal Republic of Germany, awarded to him the same year. But more important for him was the certainty that the next generation of zoologists in his museum felt and still feels the obligation to maintain his scientific legacy, including one important focus on the exploration of tropical African vertebrates.

ACKNOWLEGEMENTS

I thank Klaus Busse, Wolfgang Hartwig, ZFMK, and Chris Wild, Nyasoso/Cameroon, for the permission to use their photographs, and Ursula Bott and Philipp Wagner, ZFMK, for technical assistance during the preparation of this paper.

REFERENCES

Böhme, W., 1984, Martin Eisentraut - Philosopher, Scientist, and Explorer: a tribute, *Bat Res. News* **25(2):**13-14.

Böhme, W., 1994, In memoriam Prof. Dr. Martin Eisentraut (1902-1994) – lacertidenkundliche Aspekte seines zoologischen Werkes, *Die Eidechse* **5(13):**1-3.

Böhme, W., 2004, The German contributions to Mediterranean herpetology with special reference to the Balearic Islands and their lacertid lizards, in: V. Perez-Mellado, N. Riera, and A. Perera eds., *The biology of lacertid lizards - Evolutionary and ecological perspectives, Recerca (Inst. Menorquí d'Estudis*), **8:**63-82.

Böhme, W., and Eisentraut, M., 1981, Vorläufiges Ergebnis eines unter natürlichen Bedingungen angesetzten Kreuzungsversuches bei Pityusen-Eidechsen, *Podarcis pityusensis* (Boscà, 1883) (Reptilia: Lacertidae), *Bonn. zool. Beitr.* **32(1/2):**145-155.

Böhme, W., and Eisentraut, M., 1990, Zur weiteren Dokumentation des Zwergelefanten (*Loxodonta pumilio* Noack, 1906), *Z. Kölner Zoo* **33(4):**153-158.

Böhme, W., and Hutterer, R., 1999, Leben und Werk von Martin Eisentraut (1902-1994), *Bonn. zool. Beitr.* **48(3/4):**367-382.

Debruyne, R., Van Holt, A., Barriel, V. and Tassy, P., 2003, Status of the so-called African pygmy elephant (*Loxodonta pumilio* (Noack 1906)): phylogeny of cytochrome b and mitochondrial control region sequences, *C.R- Biologies* **326:**687-697.

Eisentraut, M., 1930, Beitrag zur Eidechsenfauna der Pityusen und Columbreten, *Mitt. zool. Mus. Berlin* **16:**397-410.

Eisentraut, M., 1933, Winterstarre, Winterschlaf und Winterruhe. Eine kurze biologisch-physiologische Studie, *Mitt. zool. Mus. Berlin* **19:**48-63.

Eisentraut, M., 1937, *Die deutschen Fledermäuse - eine biologische Studie,* Leipzig (Schöps), 184 pp.

Eisentraut, M., 1950, *Die Eidechsen der spanischen Mittelmeerinseln und ihre Rassen-aufspaltung im Lichte der Evolution,* Akademie-Verl., Berlin, 225 pp.

Eisentraut, M., 1955, *Überwinterung im Tierreich*, Kosmos-Verl., Stuttgart, 80 pp.

Eisentraut, M., 1957, *Aus dem Leben der Fledermäuse und Flughunde*, VEB G. Fischer, Jena, 175 pp.

Eisentraut, M., 1963, *Die Wirbeltiere des Kamerungebirges, unter besonderer Berück-sichtigung des Faunenwechsels in den verschiedenen Höhenstufen*, Parey Verl., Hamburg, 353 pp.

Eisentraut, M., 1973, Die Wirbeltierfauna von Fernando Poo und Westkamerun unter besonderer Berücksichtigung der Bedeutung der pleistozänen Klimaschwankungen für die heutige Faunenverbreitung, *Bonn. zool. Monogr.* **3:**1-428.

Eisentraut, M., 1982, *Im Schatten des Mongo-ma-loba.* Busse-Verl., Bonn, 241 pp.

Eisentraut, M., 1983, *Im Land der Chaco-Indianer. Begegnungen mit Menschen und Tieren in Südost-Bolivien,* Biotropic-Verl., Baden-Baden, 108 pp.

Eisentraut, M., 1986, Über das Vorkommen des Chaco-Pekari, *Catagonus wagneri*, in Bolivien, *Bonn. zool. Beitr.* **37(1):** 43-47.

Eisentraut, M., 1993, *Die Biotope westafrikanischer Vogelarten und deren Rufe und Gesänge,* Brigg-Verl., Augsburg, 48 pp. + tape.

Eisentraut, M., and Böhme, W., 1989, Gibt es zwei Elefantenarten in Afrika? *Z. Kölner Zoo* **32(2):**61-68.

Mayol, J., 2004, Survival of an artificially hybridized population of *Podarcis pityusensis* at Dau gran: evolutionary implications, in: V. Perez-Mellado, N. Riera, and A. Perera eds., *The biology of lacertid lizards - Evolutionary and ecological perspectives. Recerca (Inst. Menorquí d'Estudis)* **8:**239-244.

Mertens, R., 1968, Beitrag zur Kenntnis der Herpetofauna von Kamerun und Fernando Poo, *Bonn. zool. Beitr.* **19:**69-84.

Pooley, A. C., 1974, How does a baby crocodile get to water? *Afr. Wildlife* **28(4):**8-11.

Roeder, U., 1970, Beitrag zur Kenntnis des afrikanischen Zwergelefanten, *Loxodonta pumilio* Noack 1906, *Säugetierk. Mitt.* **18:**197-215.

Roeder, U., 1975, Über das Zwergelefanten-Vorkommen im Südwesten des Kameruner Waldgebiete,. *Säugetierk. Mitt.* **23:**73-77.

Wetzel, R. M., Dubos, R. E., Martin, R. L., and Myers, P., 1975, *Catagonus*, an “extinct” peccary, alive in Paraguay, *Science* **189:**379-38.

EARWIGS (DERMAPTERA: INSECTS) OF KENYA - CHECKLIST AND SPECIES NEW TO KENYA

Fabian Haas[1], Joachim Holstein[1], Anja Zahm[1], Christoph L. Häuser[1] and Wanja Kinuthia[2]

[1] *Staatliches Museum für Naturkunde Stuttgart, Rosenstein 1, D-70191 Stuttgart, Germany; E-mail: haas.smns@naturkundemuseum-bw.de, holstein.smns@naturkundemuseum-bw.de, azahm@t-online.de, chaeuser@gmx.de*

[2] *National Museum of Kenya, PO Box 40658, Nairobi, Kenya; E-mail: eafrinet@africaonline.co.ke*

Abstract: A checklist of Kenyan Dermaptera was composed from the literature, material of the National Museum Kenya, a database by one of us, and recently collected material. In Kenya 46 species in 23 genera from 10 families occur, with 5 endemic and 6 cosmopolitan species. Two new records were based on our personal collecting. Two species were previously not recorded: *Diplatys ugandanus* Hincks, 1955 and *Haplodiplatys kivuensis* (Hincks, 1951) from Kakamega Forest Reserve. The species numbers for the neighbouring countries was compiled from an online database (see http://www.earwigs-online.de). We believe that the relief of Kenya, with large areas above 1500 to 2000 m and the moderate temperatures and large dry areas in Northern Kenya provide unfavourable conditions for Dermaptera.

Key words: Dermaptera; Orthoptera; checklist; East Africa; Kenya; regional fauna

1. INTRODUCTION

A recent travel to the Kakamega Forest Reserve (western Kenya, approx. 50 km north of Kisumu, Lake Victoria, for coordinates see Table 2) was taken as an occasion to survey the Kenyan Dermaptera fauna. It was last treated in two short contributions by Kevan (1952, 1954) and later in Brindle's (1973, 1978) milestone on 'The Dermaptera of Africa'. After

B.A. Huber et al. (eds.), African Biodiversity, 99–107.

about 30 years, a re-examination of Kenya's Dermaptera fauna seemed worthwhile, especially as a checklist for Tanzania has become available recently (Haas and Klass, 2003). These data were supplemented by a database on dermapteran distribution (http://www.earwigs-online.de), compiled from the literature on the region and on selected taxa, and original specimen data from several museums (MNHN, NHM and ZMUC) which one of us [FH] could visit. On this field trip, data were recorded in the NMK and specimens were collected at different locations in Kenya.

2. MATERIAL AND METHODS

The Dermaptera were collected with several standard techniques, such as light traps, sweep netting and manual examination of live and rotting vegetation. The leaf sheaths of bamboo proved to be a preferred hiding place for some species. We also searched under the bark of logs and trees. No canopy fogging was done. All specimens were preserved in alcohol and stored in the SMNS. Voucher specimens are transferred to the NMK.

Museum abbreviations are as follows: MNHN: Museum National d'Histoire Naturelle, Paris, France; NHM: The Natural History Museum London, UK; NMK: National Museum of Kenya, Nairobi; SMNS: State Museum for Natural History, Stuttgart, Germany; ZMUC: Zoological Museum of the University of Copenhagen, Denmark.

3. RESULTS

A checklist with 46 species of Dermaptera occurring in Kenya was compiled (Table 1). Kakamega Forest Reserve was of particular interest. Amongst the seven species found there (Table 2), are the first Kenyan records of *Diplatys ugandanus* Hincks, 1955 and *Haplodiplatys kivuensis* (Hincks, 1955) (Fig. 1).

In addition, two larvae of other dermapteran taxa were found in Kakamega Forest, which cannot be identified to species (Table 2).

4. DISCUSSION

Comparing the Kenyan species list with those of the adjacent countries (alphabetic order), DR Congo has 122 species, Eritrea 10, Ethiopia 27,

Figure 1. Three species of Dermaptera from Kakamega Forest Reserve. A) *Haplodiplatys kivuensis* (Hincks, 1951). B) and D) *Spongovostox assiniensis* (Bormans, 1893). C) *Diplatys ugandanus* Hincks, 1955. Copyrights for all photos: J. Holstein.

Rwanda 21, Somalia 4, Sudan 13, Tanzania 102 (31 endemics, Haas and Klass, 2003), Uganda 55, the number of 46 species (5 endemic) is mid level low. Tanzania has almost twice the number of species as Kenya. Uganda, which is much smaller than Kenya, still has a few more species. This could be due to different collecting intensity over the last century, and the low number of publications dedicated to Kenyan Dermaptera supports this assumption to some extent.

However, we also assume that the relief and climate of Kenya are important factors for the comparatively low diversity in Kenyan Dermaptera. A greater part of Kenya lies at an altitude between 1500 and 2000 m asl or above, keeping the temperatures, especially at night, lower than expected for an equatorial area. The altitude may also have negative effects on the amount of rainfall. In countries with lower elevation and higher rainfall, such as Uganda, Rwanda (high species number relative to surface area) and the DR Congo, the diversity of Dermaptera is markedly increased, whereas neighbour countries with pronounced dry conditions, such as Somalia, Ethiopia and Sudan, have an even poorer Dermaptera fauna.

The effect of climate is also evident within Kenya itself. The region with the highest temperatures and the greatest amount of rain of our travel, Kakamega Forest Reserve, holds the largest number of species (7) we could find in one place (Table 2). In any other place we could only find one or two species.

Table 1. Checklist for the Dermaptera of Kenya 'LT': Locus typicus, indicates that a species was first described from Kenya; COS: cosmopolitan species; END: endemic to Kenya. For museum acronyms see Material and methods. The occurrence of species in countries neighbouring Kenya (DR Congo: ZR, Eritrea: ER, Ethiopia: ET, Rwanda: RW, Somalia: SO, Sudan: SD, Tanzania: TZ, Uganda: UG) is given to show faunal relations. Eritrea is mentioned because it was part of Ethiopia, and older 'Ethiopian' records might refer to this area.

Systematic position and Species	Status	Source	Occurrence in countries neighbouring Kenya
Hemimeridae			
1. *Hemimerus deceptus* Rehn & Rehn, 1936		NMK	TZ, UG
2. *Hemimerus hanseni* Sharp, 1895		NMK, Nakata & Maa (1974)	ZR, UG
3. *Hemimerus sessor* Rehn & Rehn, 1936	LT	Kevan (1954), Nakata & Maa (1974)	UG
4. *Hemimerus vosseleri* Rehn & Rehn, 1936		NMK, Nakata & Maa (1974)	ZR, TZ
Karschiellidae			
5. *Bormansia africana* Verhoeff, 1902		NHM, ZMUC, Hincks (1959), Kevan (1952), Menozzi (1938)	ZR, RW, TZ, UG
6. *Bormansia impressicollis* Verhoeff, 1902	LT	MNHN, Borelli (1915), Hincks (1959), Rehn (1924)	TZ, UG
Diplatyidae			
7. *Diplatys fella* Burr, 1911		MNHN	ZR, SD, TZ, UG
8. *Diplatys macrocephalus* Palisot de Beauvois, 1805		MNHN, Brindle (1973), Hincks (1959)	ZR, TZ, UG
9. *Diplatys pictus* (Zacher, 1910)	LT, END	MNHN, Burr (1911a), Hincks (1959)	none
10. *Diplatys ugandanus* Hincks, 1955		this contribution	UG
11. *Haplodiplatys kivuensis* (Hincks, 1951)		this contribution	ZR
12. *Lobodiplatys coriaceus* (Kirby, 1891)		Brindle (1973), Hincks (1959)	ZR, UG
Pygidicranidae			
13. *Echinosoma afrum* (Palisot de Beauvois, 1805)		NMK, this contribution	ZR, UG
14. *Echinosoma wahlbergi* Dohrn, 1863		NMK, Brindle (1971), Hincks (1943, 1951, 1959)	ZR, TZ, UG
15. *Dacnodes caffra* (Dohrn, 1867)		NHM, NMK, ZMUC, Hincks (1959)	SO, TZ, UG
16. *Dacnodes separata* (Burr, 1908)		Burr (1912)	ZR, RW, TZ, UG

Table 1. (continued).

17. *Dacnodes wigginsi* (Burr, 1914)		Burr (1911b), Hincks (1951)	
Apachyidae			
18. *Apachyus depressus* (Palisot de Beauvois, 1805)		ZMUC, Kevan (1952)	ZR, TZ
Labiduridae			
19. *Labidura riparia* (Pallas, 1773)	COS	NMK, Brindle (1973), Kevan (1954)	ZR, ET, SO, SD, TZ, UG
20. *Nala lividipes* (Dufour, 1828)	COS	NMK, Brindle (1973)	ZR, ET, SO, TZ, UG
Anisolabididae			
21. *Anisolabis felix* (Burr, 1907)		NMK, ZMUC, Brindle (1978), Hincks (1943), Kevan (1954)	ET, TZ, UG
22. *Anisolabis jeanneli* (Menozzi, 1938)	LT, END	MNHN, Brindle (1978), Menozzi (1938)	none
23. *Euborellia annulipes* (Lucas, 1847)	COS	NMK	ZR, ER, ET, TZ, UG
24. *Euborellia cincticollis* (Gerstaecker, 1883)		Brindle (1978), this contribution	ZR, TZ, UG
25. *Euborellia tellinii* (Borelli,1908)		MNHN	ER, UG
26. *Gonolabis hincksi* (Brindle, 1964)	LT, END	NMK, Brindle (1964, 1978)	none
Spongiphoridae			
27. *Pseudovostox truncatus* Brindle, 1970	LT, END	Brindle (1973)	none
28. *Labia minor* (Linnaeus, 1758)	COS	NMK, Brindle (1973)	ZR, ET, SD, TZ, UG
29. *Paralabella borellii* (Burr, 1908)	LT	Brindle (1968, 1973)	ZR, TZ
30. *Chaetospania pauliani* Hincks, 1948		Brindle (1969a, 1971)	ZR, UG
31. *Chaetospania rodens* Burr, 1907		Brindle (1973), Hincks (1943, 1950)	ZR, TZ
32. *Spongovostox assiniensis* (Bormans, 1893)		NMK, Brindle (1973), this contribution	ZR, RW, TZ, UG
33. *Spongovostox marginalis* (Thunberg, 1827)	LT	NMK, Brindle (1973)	ZR, TZ, UG
34. *Spongovostox quadrimaculatus* (Stal, 1855)		Burr (1911c)	ZR, TZ
Forficulidae			
35. *Cosmiella adolfi* (Burr, 1909)		MNHN	ZR, RW, TZ, UG

Table 1. (continued).

36. *Cosmiella bilobata* (Brindle, 1973)	LT	NMK, Brindle (1973), this contribution	TZ
37. *Cosmiella pygidiata* (Brindle, 1973)	LT, END	Brindle (1973)	none
38. *Diaperasticus bonchampsi* (Burr, 1904)		Brindle (1967)	ZR, ET, SD, TZ
39. *Diaperasticus erythrocephalus* (Olivier, 1791)		NMK, Brindle (1973), Burr (1907), this contribution	ZR, ER, ET, RW, SO, SD, TZ, UG
40. *Diaperasticus sansibaricus* (Karsch, 1886)		MNHN, NMK, Brindle (1973), Kevan (1952)	ZR, RW, TZ, UG
41. *Forficula auricularia* Linnaeus, 1758	COS	Brindle (1973)	ZR, TZ
42. *Forficula beelzebub* (Burr, 1900)		Brindle (1973)	none (but Nepal, India)
43. *Forficula senegalensis* Audinet-Serville, 1839		NMK, Brindle (1973), Burr (1952), Kevan (1954), this contribution	ER, ET, RW, SD, TZ, UG
44. *Guanchia rehni* (Burr, 1953)	LT	MNHN, NMK, Brindle (1969b, 1973)	TZ
45. *Guanchia sjoestedti* (Burr, 1907)		Brindle (1966), Hincks (1950, 1955a, b), this contribution	ZR, RW, TZ
Chelisochidae			
46. *Chelisoches morio* (Fabricius, 1775)	COS	Brindle (1973)	ER, TZ

Furthermore, a recent survey of the Tanzanian earwig fauna by Haas and Klass (2003) showed that its diversity is not evenly distributed over the country but significantly increased in the "Eastern Arc", a patchwork of mountain forests in eastern Tanzania. This area has high levels of biodiversity and endemism for many taxa (references in Haas and Klass, 2003), and 24 out of 31 endemic Tanzanian Dermaptera occur there (Haas and Klass, 2003). A comparable area, or biodiversity hotspot, has not been reported for Kenya so far.

In the Kakamega Forest Reserve no canopy fogging was done, however, in Tanzania this method proved of limited usefulness for collecting Dermaptera. Only three species of earwigs were found, none of them restricted to the canopy (Haas and Klass, 2003). Canopy fogging, however, might be useful for establishing natural history data of Dermaptera.

Table 2. List of specimens collected for this contribution

Systematic position and Species	location / altitude	date / legit	notes
Karschiellidae			
Bormansia sp. Nymph	S 00° 20,719' E 034° 51,851' 1600 m	19 I 2004 F Haas	Kakamega Forest Reserve, Colobus Trail near Udo's Camp, secondary forest
Diplatyidae			
Diplatys ugandanus	S 00° 20,719' E 034° 51,851' 1667 m	27 I 2004, 19.30 – 22.00 hrs J Holstein	Kakamega Forest Reserve, at light trap on top of view point at Buyango Hill
Diplatys ugandanus	S 00° 20,719' E 034° 51,851' 1667 m	18 I 2004, 18.30 – 23.30 hrs F Haas	Kakamega Forest Reserve, at light trap on top of view point at Buyango Hill
Haplodiplatys kivuensis	S 00° 20,719' E 034° 51,851' 1667 m	30 i 2004 J Holstein	Kakamega Forest Reserve, at light trap on top of view point at Buyango Hill
Haplodiplatys kivuensis	S 00° 20,719' E 034° 51,851' 1667 m	27 I 2004, 19.30 – 22.00 hrs J Holstein	Kakamega Forest Reserve, at light trap on top of view point at Buyango Hill
Haplodiplatys kivuensis	N 00°19' E 34°52' 1580 m	26 I 2004, 20.00 – 23.00 hrs J Holstein	Kakamega Forest Reserve, at light trap, Zalazar Circuit near Buyango Hill, primary forest
Pygidicranidae			
Dacnodes sp. nymph	S 00° 20,719' E 034° 51,851' 1667 m	18 I 2004, 18.30 – 23.30 hrs F Haas	Kakamega Forest Reserve, at light trap on top of view point at Buyango Hill
Echinosoma afrum	S 00° 20,719' E 034° 51,851' 1667 m	18 I 2004, 18.30 – 23.30 hrs F Haas	Kakamega Forest Reserve, at light trap on top of view point at Buyango Hill
Anisolabididae			
Euborellia cincticollis	S 00° 23,256' E 036° 49,043' 2290 m	15/16 I 2004 F Haas	Aberdare Range Camp III, near Tusk Camp site, close to Ruhuruini Gate, bamboo forest, found breeding under bark of dead tree
Spongiphoridae			
Spongovostox assiniensis	S 00° 20,719' E 034° 51,851' 1667 m	18 i 2004, 18.30 – 23.30 hrs F Haas	Kakamega Forest Reserve, at light trap on top of view point at Buyango Hill
Spongovostox sp.	S 00° 20,719' E 034° 51,851' 1667 m	30 i 2004 J Holstein	Kakamega Forest Reserve, at light trap on top of view point at Buyango Hill
Forficulidae			
Cosmiella bilobata	S 00° 10,213' E 037° 12,83' 3048 m	12 i 2004 F Haas	Mt. Kenya Research & Meterological Station, Naro Moru, bamboo forest
Guanchia sjoestedti	S 00° 10,535' E 037° 08,800' 2442 to 2450 m	10 i 2004 F Haas	At light trap, Mt. Kenya, Naro Meru, Camp site near main gate, close to ranger house, extensive afro-montane forest
Guanchia sjoestedti	S 00° 23,256' E 036° 49,043' 2290 m	15/16 i 2004 F Haas	Aberdare Range Camp III, near Tusk Camp site, close to Ruhuruini Gate, bamboo forest

ACKNOWLEDGEMENTS

We would like to thank for the NMK's permission to collect specimens and examine them in Germany. This project was supported by the BMBF through BIOLOG / BIOTA E06 and we are grateful for this support.

Data recording in the MNHN, NHM and ZMUC was funded by the EU/IHP programme and we are grateful to the host of Dr. N. M. Andersen (COBICE, Copenhagen), Mrs. J. Marshall (SYS-RESOURCE, London) and Dr. C. Amadegnato (COLPARSYST, Paris) for their help and patience.

We would also like to thank the anonymous referee for his suggestions, which improved the manuscript.

REFERENCES

Borelli, A., 1915, Insectes Orthopteres 1. Dermaptera, *Résult. scient. Voy. Ch. Allaud et R. Jeannel en Afrique orientale* (1911-1912). **1:**3-22.

Brindle, A., 1966, A new species of *Forficula* L. (Dermaptera, Forficulidae) from Africa, *Ent. Mon. Mag.* **102**:147-149.

Brindle, A., 1967, A key to the Ethiopian genus *Diaperasticus* Burr (Dermaptera: Forficulidae), *Proc. R. Ent. Soc. Lond.* (B) **36:**147-152.

Brindle, A., 1968, Contribution a la faune du Congo (Brazzaville) Mission A Villiers and A Descarpentrie. LXXIII Dermaptera, *Bull. l'IFAN, Ser. A*, **30:**772-779.

Brindle, A., 1969a, A key to the African species of the genus *Chaetospania* Karsch (Dermaptera, Labiiade), *Proc. R. Ent. Soc. Lond.* (B), **38:**95-100.

Brindle, A., 1969b, Recoltes de Ph. Bruneau de Miré au Cameroun, *Bull. de l'IFAN, Ser. A, 31:*58-72.

Brindle, A., 1971, Le massif des monts Loma (Sierra Leone) Fascicule I. XI. Dermaptera, *Mém.Inst. fondamental Afr. noire*, **86:**265-281.

Brindle, A., 1973, The Dermaptera of Africa. Pt I, Ann. Musée R. Afr. Cent. Ser. 8, Sci. Zool. **205:**1-335.

Brindle, A., 1978, The Dermaptera of Africa. Pt II, Ann. Musée R. Afr. Cent. Ser. 8, Sci. Zool. **225:**1-204.

Burr, M., 1907, Dermapteren von Madagaskar, den Comoren und Britisch-Ostafrika, *Voeltzkow Reise in Ostafrika in den Jahren 1903-1905* **2:**55-58.

Burr, M., 1911a, Dermaptera, *Genera Insect.* **122:**1-112.

Burr, M., 1911b, Dermaptera, Wissenschaftliche Ergebnisse der Deutschen Zentral-Afrika-Expedition 1907-1908 unter Führung Adolf A. Friedrichs Herzog zu Mecklenburg Part 3: 455-460.

Burr, M., 1911c, Notes on the Forficularia. XVIII, More new species, *Ann. Mag. Nat. Hist.* (8) **8:**39-50.

Burr, M., 1912, Die Dermaptera des k.k. naturhistorischen Hofmuseums in Wien, *Annl. Naturh. Hofmus. Wien* **26:**63-108.

Burr, M., 1952, What is *Forficula rodziankoi* Semenov? *Proc. R. Ent. Soc. Lond.* (B) **21:**164-166.

Haas, F., and Klass, K. D., 2003, The Dermaptera of Tanzania: checklist, faunal aspects and fogging canopies (Insecta: Dermaptera), *Entomol. Abh.* **60:**45-67.

Hincks, W. D., 1943, Coryndon Memorial Museum Expedition to the Chyulu Hills - IX. Dermaptera, *J. East Afr. Nat. Hist. Soc.* **17:**33-38.

Hincks, W. D., 1950, Some Dermaptera from Kilimanjaro, Tanganyika, *Ent. Mon. Mag.* **86:**179-181.
Hincks, W. D., 1951, The Dermaptera of the Belgian Congo. *Ann. Musée Congo Belge Sci. Zool. Ser. 8* **8:**1-50.
Hincks, W. D., 1955a, A systematic monograph of the Dermaptera of the world based on material in the British Museum (Natural History). 1. Pygidicranidae Subfamily Diplatyinae, *Brit. Mus. (Nat. Hist.),* 132 pp.
Hincks, W. D., 1955b, Contributions a l'etude de la faune entomologique du Ruanda-Urundi (Mission P. Basilewsky 1953), *Ann. Musée Congo Belge Sci. Zool. Ser. 8* **40:**73-81.
Hincks, W. D., 1959, A systematic monograph of the Dermaptera of the world based on material in the British Museum (Natural History). 2. Pygidicranidae excluding Diplatyinae, *Brit. Mus. (Nat. Hist.) Ent.,* 218 pp.
Kevan, D. K. McE., 1952, Dermaptera, Manteodea, Phasmatodea and Saltatoria collected by Dr. B. Benzon and Dr. F.W. Bræstrup in Southern and Central Kenya Colony, *Ent. Meddel.* **26:**222-230.
Kevan, D. K. McE., 1954, Dermaptera from Northern Kenya. *Entomologist* **87:**75-76.
Menozzi, C., 1938, Mission scientifique de l'Omo, 4, fasc. 34 (Orthoptera), 11 (Dermaptera), **4:**135-143.
Nakata. S. and Maa, T. C. 1974, A review of the parasitic earwigs (Dermaptera: Arixeniina; Hemimerina), *Pac. Insects* **16:**307-374.
Rehn, J. A. G., 1924, The Dermaptera of the American Museum Congo Expedition, with a catalogue of the Belgian Congo Species, *Bull. Am.Mus. Nat. Hist.* **49:**349-413.

MORPHOMETRIC DIFFERENTIATION IN THE EAST AFRICAN GRASSHOPPER GENUS *AFROPHLAEOBA* JAGO, 1983 (ORTHOPTERA: ACRIDIDAE)

Axel Hochkirch
Universität Osnabrück, FB Biologie/Chemie, Fachgebiet Ökologie, Barbarastraße 11, 49076 Osnabrück, Germany, E-mail: hochkirch@biologie.uni-osnabrueck.de

Abstract: Research on speciation processes in tropical evergreen forests is of high importance for the understanding of the genesis of biodiversity. The Eastern Arc Mountains in Tanzania are a biodiversity "hotspot", including many single mountain endemics. One endemic genus of this area is the flightless grasshopper genus *Afrophlaeoba*. The morphologically very similar species of this genus were studied by means of multivariate morphometrics. Although significant differences were found between all species in each character studied as well as in multidimensional space, there was a slight degree of overlap between some of the species. This was particularly true for the species *A. nguru* and *A. longicornis*, while the sister genus *Parodonomelus* differed in many regards from the *Afrophlaeoba* species. The phenetic distance between the species correlates well with the genetic distances based upon mtDNA sequences. The genus seems to represent a neoendemic radiation with uncertain species status. Although genetic, ecological and morphometric differences are rather low, the differences in courtship behaviour are pronounced. Based upon the phenetic and phylogenetic relationships, a typo-phylogeographic scenario can be proposed, in which riverine and coastal forests acted as habitat corridors in humid periods. The rainshadow of Zanzibar island might be the most important barrier for southern and northern sister taxa endemic to the coastal forests and the Eastern Arc.

Key words: Morphometrics; Eastern Arc; Orthoptera; African biodiversity; rainforest; phylogeography

B.A. Huber et al. (eds.), African Biodiversity, 109–118.

1. INTRODUCTION

Tropical evergreen forests are excellent regions for studies of speciation processes, since biodiversity and the number of small range endemics is exceptionally high in these ecosystems (Erwin, 1982). One of the worldwide "hotspots" of biodiversity are the Eastern Arc Mountains in Tanzania (Myers et al., 2000). This area consists of several separated mountain blocks, in each of which many genera have endemic species (Burgess et al., 1998). The evergreen forests of these mountains are separated by savannah and dry woodland (Lovett and Wasser, 1993). Hence, they are comparable to forested islands. According to the refuge concept (Haffer, 1974), tropical forest species have evolved by isolation in such persistent areas. The distribution pattern of the grasshopper genus *Afrophlaeoba* Jago, 1983 is typical for flightless insect genera of the Eastern Arc Mountains, with endemic species in single mountain blocks. The species are genetically, morphologically and ecologically rather similar (Hochkirch, 2001). Multivariate morphometrics can be a useful tool at the taxonomical level of species and below to find dissimilarities in organisms, which differ in only a few qualitative characters (Blackith and Reyment, 1971). The purpose of this study was to investigate the degree of geographical variation in morphometrics, to analyse the phenetic relationships and compare them with phylogenetic reconstructions, based upon mtDNA sequences (Hochkirch, 2001).

2. METHODS

The genus *Afrophlaeoba* consists of four small, flightless species, each of which is endemic to a single mountain block of the Eastern Arc Mountains in Tanzania. *Afrophlaeoba usambarica* (Ramme, 1929) occurs in the East Usambara Mts., *A. euthynota* Jago, 1983 in the Uluguru Mts., *A. nguru* Jago, 1983 in the Nguru Mts. and *A. longicornis* Jago, 1983 in the Rubeho and Ukaguru Mts. (Jago, 1983). Additionally, a sister genus was included in the analysis, *Parodontomelus arachniformis* Jago, 1983. This species occurs in the coastal forests of Tanzania, including the East Usambaras (Jago 1983). Thirty males of each species were collected and 26 body dimensions were measured. The characters were selected according to the descriptions of Jago (1983). The two first flagellar segments had to be combined, since they were fused in some cases. All measurements were performed with a stereo-dissecting microscope controlled by a step-motor.

The data were analysed by means of a discriminant analysis, performed with SPSS 11.5.1. This multivariate method is an ordination technique for displaying and describing the differences between group centroids by extracting the eigen vectors from the pooled variance-covariance matrix of the within-group. Although the method is robust to considerable deviations from the theoretical requirement of homogeneity (Blackith and Reyment, 1971), the data were log-transformed to achieve an approximate normal distribution. No correction for size was performed, since body length is thought to be the most important character for identifying *A. euthynota*. Hence, it is likely that one of the discriminant functions will represent size. The most frequently used criterion for evaluating the discriminance is Wilks' Lambda, which is an inverse measure. In contrast, F statistics evaluate the difference between the groups, but not the discriminating power. The contribution of each character to the discriminant functions is usually given by standardized canonical coefficients (Backhaus et al., 2000). In a stepwise discriminant analysis only those characters, which minimize Wilks' Lambda, are included in each step. This algorithm automatically sorts the variables according to their importance by including the most important characters first (Backhaus et al., 2000). For the stepwise computation the maximum significance of the F value for the inclusion of a variable was set to 0.01; the minimal significance for the exclusion was set to 0.15. Specimens with missing flagellar segments were excluded. The sample, therefore, consisted of 29 *P. arachniformis*, 27 *A. euthynota*, 31 *A. longicornis*, 30 *A. nguru* and 29 *A. usambarica*. A first analysis was made including *P.arachniformis*, the second analysis excluded the outgroup.

Due to the lack of dissimilarities in qualitative characters (Hochkirch, 2001), phylogenetic analyses based on morphological data were not yet possible. The metric scale of morphometric data does not conform to a conventional phylogenetic analysis. The discriminant analysis, however, reveals pairwise squared Mahalanobis distances between the groups (Blackith and Reyment, 1971). The roots of these distances were used to infer phenograms by nighbor joining. The calculation was performed by the computer program MEGA 2.1 (Kumar et al., 2001). It should be noted that the phenogram represents relationships based on phenetic similarity rather than on common descent (Blackith and Reyment, 1971).

3. RESULTS

The first stepwise discriminant analysis, which included the outgroup, has a high discriminating power, which is illustrated by an exceptionally low

Table 1. Included characters in the stepwise discriminant analysis in order of the steps in which they were included (data included *P. arachniformis*)

Step	Character	Lambda	F	df1	Df2	P
1	Tegminal width	0.150	199.5	4	141.0	< 0.001
2	Vertertex length	0.051	119.7	8	280.0	< 0.001
3	Posterior distance between lateral pronotal carinae	0.029	86.1	12	368.1	< 0.001
4	Vertex width	0.014	79.8	16	422.2	< 0.001
5	Head width	0.009	72.4	20	455.3	< 0.001
6	Anterior distance between lateral pronotal carinae	0.006	65.1	24	475.7	< 0.001
7	Length of flagellar segment 3	0.005	59.6	28	488.2	< 0.001
8	Length of flagellar segment 4	0.004	54.4	32	495.8	< 0.001
9	Width of flagellar segment 9	0.003	49.9	36	500.2	< 0.001
10	Tegminal length	0.003	46.1	40	502.4	< 0.001

Wilks' Lambda (Λ = 0.003; chi^2= 803.2; df = 44; P < 0.001). The classification phase assigned 89.7% of the cases correctly. The univariate F-statistics revealed significant differences between the species for all 26 characters (P < 0.001). Ten characters were included by the stepwise procedure (Table 1). The multivariate analysis revealed significant pairwise differences between all species for all four functions (F statistics; P < 0.001). A high morphometric distance between the two genera *Parodontomelus* and *Afrophlaeoba* is illustrated by an extremely high eigen value of the canonical function 1 (CAN1: 86.5%). The highest standardized coefficients of CAN1 are vertex width (-1.18), head width (0.95) and the posterior distance between the lateral pronotal carinae (0.76). It is obvious that all these characters represent body widths, which characterize the difference between the two genera. The functions CAN2, CAN3 and CAN4 discriminate the four *Afrophlaeoba* species. However, the differences become more obvious if the discriminant analysis is calculated without *P. arachniformis*.

Table 2. Included characters in the stepwise discriminant analysis in order of the steps in which they were included (data excluded *P. arachniformis*)

Step	Character	Lambda	F	df1	df2	p
1	Vertex length	0.264	104.9	3	113.0	< 0.001
2	Length of flagellar segment 9	0.165	54.6	6	224.0	< 0.001
3	Length of flagellar segment 4	0.137	38.0	9	270.3	< 0.001
4	Length of flagellar segment 3	0.108	32.0	12	291.3	< 0.001
5	Length of flagellar segment 1+2	0.090	28.0	15	301.3	< 0.001
6	Posterior distance between lateral pronotal carinae	0.078	24.9	18	306.0	< 0.001
7	Tegminal width	0.067	23.0	21	307.8	< 0.001

In the second discriminant analysis (excluding *P. arachniformis*), seven characters were included (Table 2). The univariate F-statistics revealed significant differences between all 26 characters ($P < 0.001$). All three functions show significant differences between the four species in the multivariate analysis (F statistics; $P < 0.001$). Wilks' Lambda is rather low ($\Lambda = 0.067$; $chi^2 = 299.5$; $df = 21$; $P < 0.001$). For CAN2 to CAN4 it is 0.299 ($chi^2 = 133.5$; $df = 12$; $P < 0.001$), for CAN3 to CAN4 0.689 ($chi^2 = 41.1$; $df = 5$; $P < 0.001$). The classification phase of the discriminant analysis assigned 89.1% of the cases correctly. CAN1 discriminates the *A. euthynota* data from the other three species. It explains 66.5% of the variance. This comparatively high value is mainly based on the smaller body size of *A. euthynota*. However, a discriminant analysis with size correction revealed similar results. CAN2 explains 24.9% of the variance and CAN3 8.6%. The highest standardized coefficient of CAN1 is the vertex length (0.68). This character is also the most effective discriminating character, as is shown by its low Wilks' Lamda. CAN2 discriminates *A. usambarica* from all other species (Fig. 1). The highest standardized coefficients on this function are

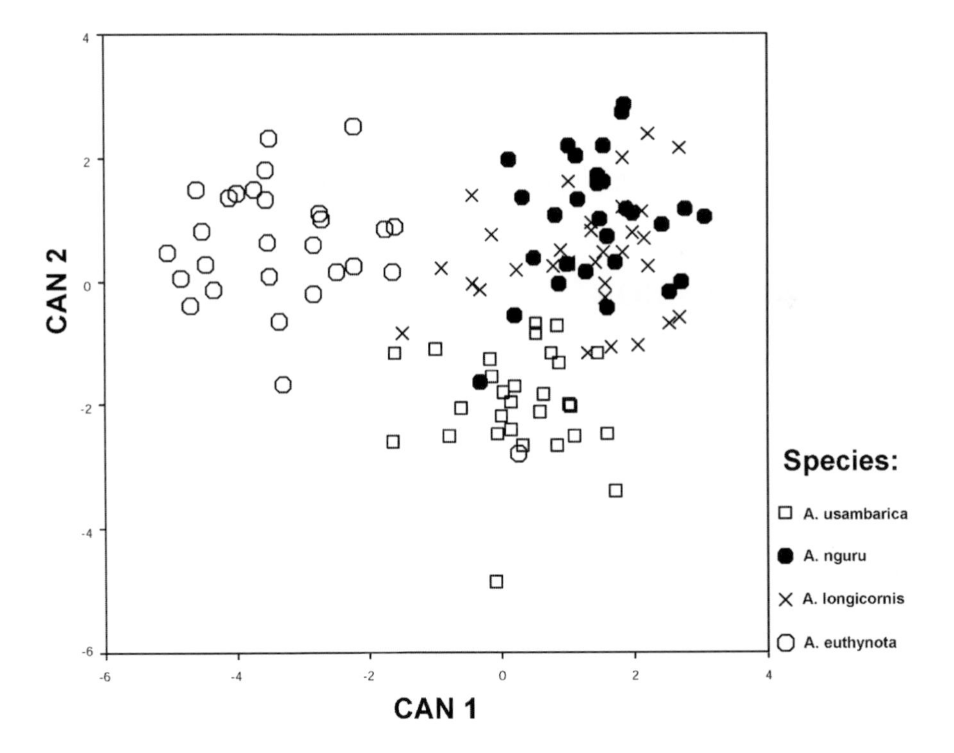

Figure 1. Plot of canonical variates 1 and 2 of the discriminatory topology for the *Afrophlaeoba* analysis.

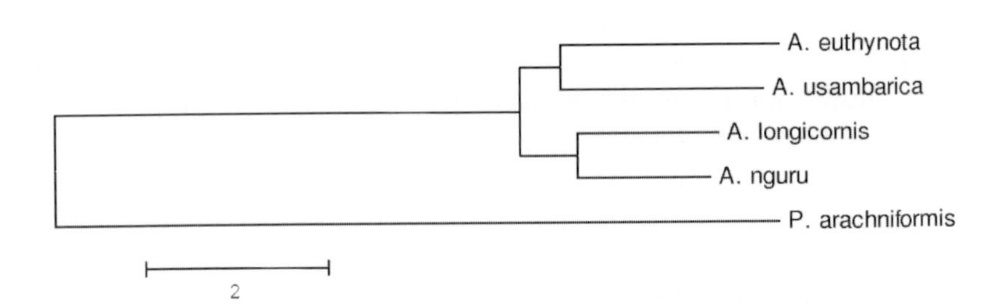

Figure 2. Phenogram inferred from the rooted pairwise Mahalanobis distances.

the lengths of the flagellar segments 3, 4 and 6 (0.81; -0.87; 0.99). On CAN3 *A. nguru* and *A. longicornis* are separated. These two species are still plotted close together and have a rather high overlap, with 13% of the *A. longicornis* classified as *A. nguru* and 7% vice versa. CAN3 has a proportional eigen value of only 8.6%. The highest coefficient of this function is the length of the first two flagellar segments (0.78).

The phenogram inferred by neighbor joining clearly illustrates the high morphometric distance between *P. arachniformis* and the *Afrophlaeoba* species (Fig. 2). Within *Afrophlaeoba* two groups can be distinguished, *euthynota-usambarica* and *nguru-longicornis*. The latter group is branched together due to the lowest morphometric distance of all species pairs.

4. DISCUSSION

There is a high structural similarity of the neighbor-joining phenogram and the gene trees based upon mtDNA (Hochkirch, 2001), which suggest a high correlation of morphological and genetic distances. The two genera *Afrophlaeoba* and *Parodontomelus* can be clearly separated by the different shape of the lateral caerinae (Hochkirch, 1999). The results show that the keys and descriptions of the *Afrophlaeoba* species (Jago, 1983) are not reliable. *Afrophlaeoba longicornis* and *A. nguru* are morphometrically very similar, with a high degree of overlap. However, a typical characteristic of *A. euthynota*, the small body size, can be confirmed. At present, DNA sequencing represents a more powerful tool for identification than morphology (Hochkirch, 2001). Since the four species occur allopatrically, identification should not be problem.

The question arises, whether all taxa represent "good" species in sense of the biological species concept (Mayr, 1942). Genetic drift will lead to morphometric differences even between closely located populations (Gries et al., 1973). However, differences in the courtship behaviour indicate prezygotic isolation (Hochkirch, 2001). A final conclusion regarding the species status is only possible by interbreeding experiments. The behavioural differences are particularly high between the two most closely related species *A. nguru* and *A. longicornis* (Hochkirch, 2001), indicating secondary contact. The close morphometric relationship between the two species supports the hypothesis that these two species have been separated rather recently.

The Eastern Arc Mountains are known for high rates of endemism, including palaeoendemic and neoendemic taxa (Fjeldså and Lovett, 1997). Based upon the low morphometric and genetic distances, the *Afrophlaeoba* species have to be regarded as neoendemics. The small ranges of all species suggests that they are restricted to evergreen rainforests (Jago, 1983). Although this is true on a geographic scale, they occur rather at forest edges and clearings (Hochkirch, 1996), which allows them to survive even under single, evergreen trees with dense canopy (Hochkirch, 1998). This ecology has also consequences for phylogeographic conclusions, since no closed forest connection has to be proposed to connect populations.

Since the Eastern Arc forests are separated by dry savannah and woodland, two elements might have connected the ancestral range of the flightless insects: coastal forests and riverine forests. If a scenario of the evolutionary divergence of the genus has to be set, the ancestral distribution would cover the formerly widespread Coastal Forests, adjacent riverine forests and parts of the Eastern Arc (Fig. 3). The Wami River basin might have acted as a corridor for the mountain and coastal populations. The phenetic and phylogenetic relationships suggest an initial fragmentation of the "Wami River Connection" close to the coast, probably in the Zanzibar rainshadow, which is the driest area at the coast (Hawthorne, 1993). The breakdown of this habitat connection was followed by a fragmentation of the Coastal Forests shortly after that. The "Wami River Connection" between the Nguru Mountains and the Rubeho Mountains must have remained stable for a longer time and might even still exist. Molecular data suggests a Pleistocene breakdown of the habitat connections (Hochkirch, 2001), as has been suggested for other recent radiations within the Eastern Arc (Fjeldså and Lovett, 1997). The present north-south disjunctions in the distribution patterns of other taxa confirm this hypothesis (Hawthorne, 1993; Robbrecht, 1996; Tattersfield, 1998; Lindqvist and Albert, 1999). There is unequivocal evidence that Coastal Forests were once more widespread in areas that are now woodland (Clarke and Karoma, 2000). These landscape changes were possibly affected by natural desiccation as well as by human disturbance. The importance of riverine forests as corridor for Eastern Arc elements

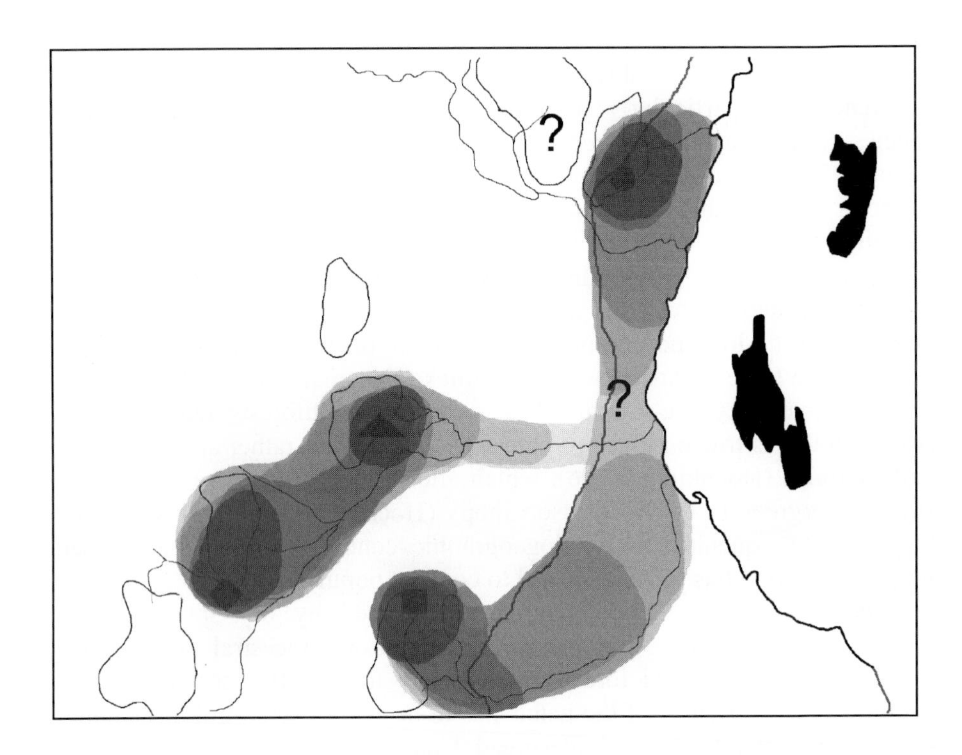

Figure 3. Synopsis of a vicariance scenario for the separation of populations of *Afrophlaeoba* (question marks represent unconfirmed recent records, lighter shadings represent a higher ages; the off-shore islands of Zanzibar and Pemba are black).

is supported by other taxa as well (Hamilton, 1982; Hoffmann, 1993; Kellman et al., 1994; Brandl et al., 1996). It seems appropriate to include the gallery forests of Tanzania into future research, since they might represent important links and provide answers to the history of interconnections.

ACKNOWLEDGEMENTS

I would like to thank my supervisor, Prof. Dr. Dietrich Mossakowski (Bremen, Germany) and my local contact, Prof. Kim Howell (Dar es Salaam, Tanzania). Technical advice was provided by Dieter Wienrich (Bremen, Germany). This work was financially supported by the "Stifterverband für die Deutsche Wissenschaft" and by a "DAAD Doktorandenstipendium im Rahmen des gemeinsamen Hochschulsonderprogramms III von Bund und Ländern." The Commission for Science and Technology (COSTECH, Dar es Salaam, Tanzania) provided the research clearance for this study.

REFERENCES

Backhaus, K., Erichson, B., Plinke, W., and Weiber, R., 2000, *Multivariate Analysemethoden – Eine anwendungsorientierte Einführung*, Springer, Berlin.

Blackith, R. E., and Reyment, R. A., 1971, *Multivariate morphometrics*, Academic Press, London.

Brandl, R., Bagine, R. N. K., and Kaib, M., 1996, The distribution of *Schedorhinotermes lamanianus* (Isoptera: Rhinotermitidae) and its termitophile Paraclystis (Lepidoptera: Tineidae) in Kenya: Its importance for understanding East African biogeography, *Global Ecol. Biogeogr. Letters* **5**(3):143-148.

Burgess, N. D., Fjeldså, J., and Botterweg, R., 1998, Faunal importance of the Eastern Arc Mountains of Kenya and Tanzania, *J. East Afr. Nat. Hist.* **87:**37-58.

Clarke, G. P., and Karoma, N. J., 2000, History of anthropic disturbance, in: *Coastal Forests of Eastern Africa*, N. D. Burgess and G. P. Clarke, eds., IUCN, Gland, pp. 251-261.

Erwin, T. L., 1982, Tropical forests: their richness in Coleoptera and other arthropod species, *Coleopts. Bull.* **36:**74-75

Fjeldså, J., and Lovett, J. C., 1997, Geographical patterns of old and young species in African forest biota: the significance of specific montane areas as evolutionary centres, *Biodivers. Conserv.* **6:**325-346.

Gries,B., Mossakowski, D., and Weber, F., 1973, Coleoptera Westfalica: Familia Carabidae, Genera *Cychrus*, *Carabus* und *Calosoma*, *Abh. Landesmus. Münster* **35**(4):1-80.

Haffer, J., 1974, Avian speciation in tropical South America, *Publs. Nuttal. Ornithol. Club* **14:**1-390.

Hamilton, A. C., 1982, *Environmental history of East Africa: a study of the Quaternary*, Academic Press, London.

Hawthorne, W. D., 1993, East African coastal forest botany, in: *Biogeography and Ecology of the Rain Forests of Eastern Africa*, J. C. Lovett and S. K. Wasser, eds., Cambridge University Press, pp. 57-99.

Hoffman, R. L., 1993, Biogeography of East African montane forest millipedes, in: *Biogeography and Ecology of the Rain Forests of Eastern Africa*, J. C. Lovett and S. K. Wasser, eds., Cambridge University Press, pp. 103-114

Hochkirch, A., 1996, Habitat preferences of grasshoppers (Orthoptera: Acridoidea, Eumastacoidea) in the East Usambara Mts., NE Tanzania, and their use for bioindication, *Ecotropica* **2:**195-217.

Hochkirch, A., 1998, A comparison of the grasshopper fauna (Orthoptera: Acridoidea) of the Uluguru Mts. and the East Usambara Mts. (Tanzania*), J. East Afr. Nat. Hist.* **87:**221-232.

Hochkirch, A., 1999, *Parodontomelus luci* n. sp. (Orthoptera: Acrididae, Acridinae), a new grasshopper species from the Udzungwa Mountains, Tanzania, *J. Orthoptera Res.* **8:**39-44.

Hochkirch, A., 2001, *A phylogenetic analysis of the East African Grasshopper Genus Afrophlaeoba Jago, 1983*, Cuvillier Verlag, Göttingen.

Jago, N. D., 1983, Flightless members of the *Phlaeoba* Genus group in eastern and north-eastern Africa and their evolutionary convergence with the Genus *Odontomelus* and its allies (Orthoptera, Acridoidea, Acrididae, Acridinae), *Trans. Am. Entomol. Soc.* **109:**77-126.

Kellman, M., Tackaberry, R., Brokaw, N., and Meave, J., 1994, Tropical gallery forests, *Nat. Geol. Res. Expl.* **10:**92-103.

Kumar, S., Tamura, K., Jakobsen, I. B., and Nei, M., 2001, MEGA2: Molecular Evolutionary Genetics Analysis software, *Bioinformatics* **17**(12):1244-1245.

Lindqvist, C., and Albert, V. A., 1999, Phylogeny and conservation of African violets (*Saintpaulia*: Gesneriaceae): new findings based on nuclear ribosomal 5S non-transcribed spacer sequences, *Kew Bull.* **54:**363-377.

Lovett, J. C., and Wasser, S. K., 1993, *Biogeography and Ecology of the Rain Forests of Eastern Africa*, University Press, Cambridge.
Mayr, E., 1942, *Systematics and the origin of species*, Columbia University Press, New York.
Myers, N., Mittermeier, R. A., Mittermeier, C. G., da Fonseca, G. A. B., and Kent, J., 2000, Biodiversity hotspots for conservation priorities, *Nature* **403:**853-858
Robbrecht, E., 1996, Generic distribution patterns in subsaharan African Rubiaceae (Angiospermae), *J. Biogeogr.* **23:**311-328.
Tattersfield, P., 1998, Patterns of Diversity and Endemism in East African Land Snails, and the implication for conservation, J. Conchol., *Special Publication* **2:**77-86.

KATYDIDS AND CRICKETS (ORTHOPTERA: ENSIFERA) OF THE KAKAMEGA FOREST RESERVE, WESTERN KENYA

Joachim Holstein[1], Fabian Haas[1], Anja Zahm[1], Christoph L. Häuser[1] and Wanja Kinuthia[2]

[1] *Staatliches Museum für Naturkunde Stuttgart, Rosenstein 1, 70191 Stuttgart, Germany, E-mail:* holstein.smns@naturkundemuseum-bw.de, haas.smns@naturkundemuseum-bw.de, azahm@t-online.de, chaeuser@gmx.de

[2] *National Museums of Kenya, Invertebrate Zoology Department, P.O. Box 40658, Nairobi, Kenya; E-mail: eafrinet@africaonline.co.ke*

Abstract: In the BIOTA East Project E06 (BMBF project ID 01LC0025), three field trips to the Kakamega Forest Reserve were undertaken between September 2002 and January 2004. Among other Orthoptera representatives 122 specimens of Ensifera were recorded. Most of the specimens were collected manually, using a net, or recorded at light traps. The number of species occurring in Kenya is about 200, according to the NMK collection, literature data (Kevan, 1950; Kevan and Knipper, 1961; Otte et al., 1988), and OSF (Otte and Naskrecki, 1997). According to the Orthopteran Species File OSF (http://osf2x.orthoptera.
org/osf2.2/ OSF2X2Frameset.htm) the type locality of 94 species is located in Kenya. Both figures will most probably increase significantly in the future. From the Kakamega Forest region, 33 species in seven families are reported to date, based on our own field collecting (122 specimens) and NMK data (about 60 specimens). The estimated minimal number of occurring species is 60.

Key words: Orthoptera; Ensifera; katydids; crickets; check-list; Kenya; regional fauna; Kakamega Forest

1. INTRODUCTION

The main focus of the BIOTA East Sub-Project E06 (Häuser et al., 2003) is to assess the butterfly and moth fauna present in the Kakamega Forest Reserve which is situated in western Kenya, approx. 50 km north of Kisumu,

B.A. Huber et al. (eds.), African Biodiversity, 119–124.

Lake Victoria (see Kokwaro, 1988). During field work on Lepidoptera for this project, numerous Orthoptera specimens, mainly Ensifera and Caelifera, were also collected in this area. The National Museums of Kenya, Nairobi (NMK) holds a remarkable collection of Kenyan Orthoptera, which we could study during our visits to Kenya. This paper presents a preliminary survey of the Orthopteran fauna for the Kakamega Forest Reserve based on faunistic data from these two sources. A main goal of this contribution is to encourage further research and studies of these interesting and ecologically important insects.

2. MATERIALS AND METHODS

Field recorded specimens were either collected manually using a sweep net or they were attracted to the light traps regularly used for recording nocturnal Lepidoptera (Häuser et al., 2003). Orthopteran specimens collected were pinned, set, and dried in the field. For lack of time we could not remove intestines, and in a few cases this caused some serious changes to body colour. The removal and examination of genitalia is intended during future taxonomic work.

The specimens collected are temporarily stored in the collection of the State Museum of Natural History Stuttgart, Germany (SMNS), and after further taxonomic examination, vouchers will be transferred to the collections of the Department of Invertebrate Zoology, National Museums of Kenya, Nairobi (NMK) for permanent storage. Identification of specimens is based on determined reference material in NMK and SMNS, and standard literature references (Beier, 1962; Chopard, 1943; Kevan and Knipper, 1961). Additional information has been obtained from the Orthopteran Species File (OSF; http://osf2x.orthoptera.org/osf2.2/ OSF2X2Frameset.htm and http://www.tettigonia.com/).

As the Caelifera collection of NMK has been recurated and is now in very good condition, the Ensifera collection should also be improved. The specimens are stored in older drawers, some rather desolate. There is some material collected in the early 20th century, the bulk from the mid-20th century up to the latest research projects like BIOTA East in 2003. Many specimens of Ensifera were determined by Ragge in the 1980s, Townsend labelled the Gryllotalpidae.

In the following, we restrict our analysis to Ensifera. Caelifera and Mantodea will be considered in future publications.

3. RESULTS

During the authors' field work, a total of 122 Ensifera specimens were recorded from the study area. The preliminary list of records identified to genus from the Kakamega Forest Reserve is shown in Table 1. So far, a total of 33 Ensifera species representing 15 tribes in seven families have been identified from within the Kakamega Forest Reserve area. Although the 33 species are clearly distinguishable "morphospecies", which could be determined down to the genus level, full identification to species level is still pending for 14 species. A few specimens (less than 10) are only determined to a level higher than genus and are not listed in Table 1.

The number of Kakamega records in the NMK collection is relatively low (less than 60 specimens), most of them were collected after 1960, several after 2000 (BIOTA material), but many species we recorded are represented in NMK from other Kenyan localities.

Both in the NMK collection and in our field recorded material at SMNS, there still are several specimens which have not yet been determined at all, and a more complete species list can be expected at a later stage of investigation. From the information provided through the Orthopteran Species File, a total of 94 Ensifera species have been described from Kenya, and about 200 species can be expected to occur in the entire country (Kevan and Knipper, 1961; Otte and Naskrecki, 1997, Otte et al., 1988). The Tettigonioidea website of Otte and Naskrecki (1997) lists 49 species occurring in Kenya (IX/2004). Based on these figures and other published studies on Orthoptera from East African locations (Hemp and Hemp, 2003; Kevan, 1950), the actual total number of species present in the Kakamega Forest area is estimated to reach at least 60.

4. DISCUSSION

As the 33 species listed in Table 1 are still to be complemented by a number of unidentified specimens (not listed in Table 1) from NMK and possibly other collections taken in the Kakamega Forest region, the data for comparing this figure with other inventory lists or to comparatively assess the Orthopteran diversity for the area are insufficient yet. Up to now we have had no opportunity to investigate material which was collected by fogging by another BIOTA subproject dealing with beetles, and is stored at the Museum Alexander Koenig in Bonn, Germany. In addition, we have little doubt that some of the currently unidentified species existing in collections will reveal additional, new species through future taxonomic revisions. Thus the estimated number of 60 species occurring in the Kakamega Forest

Table 1. Identified Ensifera species from Kakamega Forest Reserve as recorded during BIOTA field work or present in the NMK collections. SMNS indicates records based on specimens collected by the authors, and temporarily stored in the State Museum of Natural History Stuttgart. Nomenclature follows OSF, listed synonyms are taxon names which are still used in the NMK collection.

	Taxon	Collection
	Gryllidae	
	Gryllinae, Gryllini	
1.	*Grylloderes maurus* (Afzelius & Brannius, 1804) (= *Gryllulus morio* F.)	NMK, SMNS
2.	*Acheta* spec. (= *Gryllulus* spec.)	NMK, SMNS
3.	*Scapsipedus marginatus* (Afzelius & Brannius, 1804)	NMK
4.	*Teleogryllus gracilipes* (Saussure, 1877) (= *Gryllulus gracilipes*)	NMK
5.	*Teleogryllus xanthoneurus* (Gerstaecker, 1869) (= *Gryllulus xanthoneurus*)	NMK
	Oecanthidae	
	Oecanthinae, Oecanthini	
6.	*Oecanthus* spec.1	NMK
7.	*Oecanthus* spec.2	SMNS
	Phalangopsidae	
	Homoeogryllinae, Endacustini	
8.	*Phaeophilacris* spec.	SMNS
	Podoscirtidae	
	Euscyrtinae	
9.	*Euscyrtus bivittatus* Guerin-Meneville, 1884	SMNS
	Trigonidiidae	
	Nemobiinae, Pteronemobiini	
10.	*Pteronemobius* spec	SMNS
	Trigonidiinae, Trigonidiini	
11.	*Trigonidium* spec.	SMNS
12.	*Trigonidium* (Trigonidomorpha) spec.	NMK, SMNS
	Gryllotalpidae	
	Gryllotalpinae, Gryllotalpini	
13.	*Gryllotalpa africana* Beauvois, 1805	SMNS
14.	*Gryllotalpa microptera* Chopard, 1939	NMK
	Tettigoniidae	
	Conocephalinae, Conocephalini	
15.	*Megalotheca longiceps* (Peringuey, 1918) (= *Xiphidion longiceps)*	NMK, SMNS
16.	*Conocephalus* (*Anisoptera*) *maculatus* (Le Guillou, 1841) (= *Anisoptera continua* Walker)	NMK
17.	*Conocephalus* (*Anisoptera*) spec.	SMNS
18.	*Conocephalus* (*Conocephalus*) spec.	SMNS
	Conocephalinae, Copiphorini	
19.	*Ruspolia* spec.	NMK, SMNS
	Hetrodinae, Enyaliopsini	
20.	*Enyaliopsis carolinus* Sjœstedt, 1913	NMK
21.	*Enyaliopsis durandi* (Lucas, 1884)	NMK

Table 1. continued

	Meconematinae	
22.	*Amytta* spec.	SMNS
	Mecopodinae, Mecopodini	
23.	*Anoedopoda erosa* Karsch, 1891	NMK
	Phaneropterinae, Acrometopini	
24.	*Horatosphaga leggei* (Kirby, 1909)	NMK
25.	*Horatosphaga meruensis* (Sjœstedt, 1909) (= *Conocephalus meruensis*)	NMK, SMNS
	Phaneropterinae, Phaneropterini	
26.	*Phaneroptera nana* Fieber, 1853	NMK, SMNS
	Phaneropterinae, unassigned	
27.	*Arantia rectifolia* Brunner von Wattenwyl, 1878	NMK, SMNS
28.	*Eurycorypha prasinata* Stål, 1874	NMK
29.	*Eurycorypha* spec. 1	SMNS
30.	*Eurycorypha* spec. 2	SMNS
31.	*Gelatopoia bicolor* Brunner von Wattenwyl, 1891	SMNS
32.	*Tetraconcha banzyvilliana* Griffini, 1909	NMK
	Pseudophyllinae, Pleminiini	
33.	*Lichenochrus* spec.	SMNS

Reserve is most likely still conservative. This figure should probably be regarded as a minimum number and the true species diversity may turn out to be much higher after more thorough investigations, which we hope will be stimulated by these preliminary findings.

ACKNOWLEDGEMENTS

For granting permission to conduct field work and collect specimens in the Kakamega Forest Reserve and for much logistic support received on the ground, we would like to thank the authorities of the Kenya Wildlife Service (KWS), and the National Museums of Kenya, Nairobi. Financial support for this project from the German Federal Ministry for Education and Research (BMBF) through its BIOLOG program (BIOTA-East, project ID: 01LC0025) is gratefully acknowledged.

REFERENCES

Beier, M., 1962, Orthoptera Tettigoniidae (Pseudophyllinae I), Lieferung 73, in: *Das Tierreich*, R. Mertens, W. Hennig and H. Wermuth, eds., De Gruyter, Berlin, pp 468.

Chopard, L., 1943, *Faune de L'Empire Français I: Orthopteroides de L'Afrique du Nord*, Librairie Larose, Paris, pp. 450.

Häuser, C., Kühne, L., Njeri, F. and Holstein, J., 2003, Lepidoptera as indicators of human impact of tropical rainforest ecosystems in Eastern Africa, in: *Sustainable use and conservation of biological diversity – A challenge for society*, Federal Ministry of Education and Research (BMBF) & German Aerospace Center, Project Management (DLR - PT) eds., International Symposium, 1-4 December 2003, Berlin. Symposium Report, Part A. Deutsches Zentrum für Luft- und Raumfahrt, Projektträger des BMBF (DLR-PT), Bonn, pp. 110-111.

Hemp, C. and Hemp, A., 2003, Saltatoria cenoses of high altitude grasslands on Mt. Kilimanjaro, Tanzania (Orthoptera: Saltatoria), *Ecotropica* **9**(1/2)**:**71-97.

Kevan, D. K. McE., 1950, Orthoptera from the hills of Southeast Kenya, *J. East Afr. Nat. Hist. Soc.* **19:**192-224.

Kevan, D. K. McE. and Knipper, H., 1961, Geradflügler aus Ostafrika, *Beitr.Ent.* **11**(3/4)**:**356–413.

Kokwaro, J. O., 1988, Conservation status of the Kakamega Forest in Kenya: The easternmost relic of the equatorial rain forests of Africa, *Monogr. Syst. Bot. Missouri Bot. Garden* **25**:471-489.

Otte, D. and Naskrecki, P., 1997, Orthoptera Species Online. http://osf2x.orthoptera.org/osf2.2/OSF2X2Frameset.htm (12/VII/2004) and http://www.tettigonia.com/ (27/IX/2004) for Tettigonioidea.

Otte, D., Toms, R. B., and Cade, W., 1988, New species and records of East and Southern African crickets (Orthoptera: Gryllidae: Gryllinae), *Ann. Transvaal Mus.* **34**(19)**:**405–418.

ECOLOGY AND DIVERSITY OF CANOPY ASSOCIATED CERATOCANTHIDAE (INSECTA: COLEOPTERA, SCARABAEOIDEA) IN AN AFROTROPICAL RAINFOREST

Alberto Ballerio[1] and Thomas Wagner[2]
[1] *Viale Venezia 45, I-25123 Brescia, Italy, E-mail: alberto.ballerio.bs@numerica.it*
[2] *Universität Koblenz-Landau, Institut für Integrierte Naturwissenschaften – Biologie, Universitätsstr. 1, D-56070 Koblenz, Germany, E-mail: thwagner@uni-koblenz.de*

Abstract: Within a survey of canopy arthropods made by insecticidal fogging in the Budongo Forest, a semideciduous rainforest in Western Uganda, during the rainy season and the dry season, an unusually large number of Ceratocanthidae (Coleoptera: Scarabaeoidea) was collected. Five species were found: *Callophilharmostes fleutiauxii* (Paulian, 1943), *Petrovitzostes guineensis* (Petrovitz, 1968), *Philharmostes (Holophilharmostes) badius* Petrovitz, 1967, and *Philharmostes (Philharmostes) adami* Paulian, 1968, all recorded for the first time from Uganda, and *Carinophilharmostes vadoni* (Paulian, 1937). This is the best documented evidence of an association of Ceratocanthidae with the canopy of understory trees and the first time that this phenomenon is discussed. Furthermore, observations about seasonality, reproductive biology, sex ratio and abundance are presented.

Key words: Forest canopy; Uganda

1. INTRODUCTION

The Ceratocanthidae (formerly known as Acanthoceridae) comprise about 340 described species in 40 genera. They are small beetles (2–12 mm), usually very convex, capable of rolling their body up, with all body parts perfectly fitting together to form a compact ball. This group exhibits a pantropical distribution, with the majority of species occurring in tropical rainforests. The 15 genera and 73 species described from the Afrotropical

B.A. Huber et al. (eds.), African Biodiversity, 125–132.

region are distributed into four main centres of diversity, the Guineo-Congolian rainforest block, the Eastern Arc rainforests of Kenya and Tanzania, Madagascar and the afrotemperate forests of South Africa (Ballerio, unpubl. data). Most of the biology of these beetles is unknown. Most species are found in dead logs and leaf litter. They are supposed to feed on soft food, probably fungi (Scholtz, 1990). In recent literature (Choate, 1987; Hammond, 1990; Davies et al., 1997; Ballerio, 2000a) and in museum material (Ballerio, unpubl. data) there is scattered evidence of the existence of canopy dwelling species, collected by beating leaves at forest edges or glades, by canopy fogging or by netting, but this evidence has never been discussed or adequately highlighted.

Extensive collecting by canopy fogging in a semideciduous rainforest in Uganda provided sound evidence of an association of some ceratocanthid species with the forest canopy. It was possible to collect about 660 ceratocanthids belonging to five species. The aim of this paper is to discuss the results and to infer some data on the biology of canopy associated Ceratocanthidae.

2. METHODS

Research was carried out by Th. W. in the Budongo Forest Reserve in Western Uganda, around the field station of the Budongo Forest Project (1°45'N, 31°35'E), 1250 m a.s.l. This is a semideciduous rainforest consisting of logged compartments next to old primary forests and swamp forests (Wagner, 2000). The mean annual rainfall is about 1600 mm, with a relatively constant precipitation from March to November and a strong dry season from December to February.

Four tree species (*Rinorea beniensis* (Welwitsch ex Olivier), *Trichilia rubescens* Olivier, *Cynometra alexandri* C. H. Wright and *Teclea nobilis* Delile) were fogged in three different stands, as primary, secondary and swamp forest. Per site and season, eight conspecific trees were fogged, which is subsequently named "collecting unit". Trees were neither flowering nor fruiting and belong mainly to the understory with fogged tree heights between 7-20 m. Insecticidal fogging was made during the rainy season (June/July 1995) and during the dry season (January 1997) in primary, secondary and swamp forest, where 64 and 48 trees respectively were fogged. Further details can be found in Wagner (2000, 2001). Specimens were then prepared and labelled, and identified, dissected and sexed by A. B.

3. RESULTS

3.1 Taxonomic composition, diversity and abundance of the ceratocanthid community

The following five Ceratocanthidae species were collected by canopy fogging: *Callophilharmostes fleutiauxii* (Paulian, 1943) (Fig. 3b), first record for Uganda, *Carinophilharmostes vadoni* (Paulian, 1937) (Fig. 3a), *Petrovitzostes guineensis* (Petrovitz, 1968) (Fig. 3c), first record for Uganda, *Philharmostes (Holophilharmostes) badius* Petrovitz, 1967 (Fig. 3d), first record for Uganda, and *Philharmostes (Philharmostes) adami* Paulian, 1968, first record for Uganda. A total of 665 individuals were collected, which equals 1.45% of all beetles collected. Cerathocanthids are the most important scarab group to be collected, and this is the largest sample of Ceratocanthidae by canopy fogging ever found at one site. Apart from *Philharmostes (Philharmostes) adami* of which only two individuals were collected, of all other species gained by canopy fogging we found unusually high numbers (Fig. 1). Among mycophagous beetles, Corylophidae, Phalacridae and Latridiidae were common in the samples. The latter were the most abundant of all beetles collected (Wagner, 2000).

3.2 Specificity to tree species, forest type and seasonality

There is no evidence of association of any of the collected ceratocanthid species with any particular tree species. All ceratocanthid species have been collected in all three forest types and from all tree species examined (Fig. 1). *Rinorea* yielded the largest number of individuals, but most trees fogged belong to this tree species. In both wet and dry seasons, the secondary forest yielded the largest numbers of individuals.

The species composition of the ceratocanthid community does not seem to be significantly affected by seasonality, although the total number of individuals strongly decreased in the dry season compared to the wet season (dry season: 225 individuals, wet season: 488 individuals (53 individuals from *Teclea* and *Trichilia* trees, which were not collected during the dry season were eliminated from this comparison). We observed about the same species composition in the dry and wet seasons. However, the correspondence analysis highlights a stronger seasonal change in the primary forest (Fig. 2), while data from the secondary and swamp forest show no clear seasonal pattern. The secondary forest proved to be much richer in individuals and species during the dry season than the primary forest (182 individuals, four species, secondary forest; 21 individuals, two species,

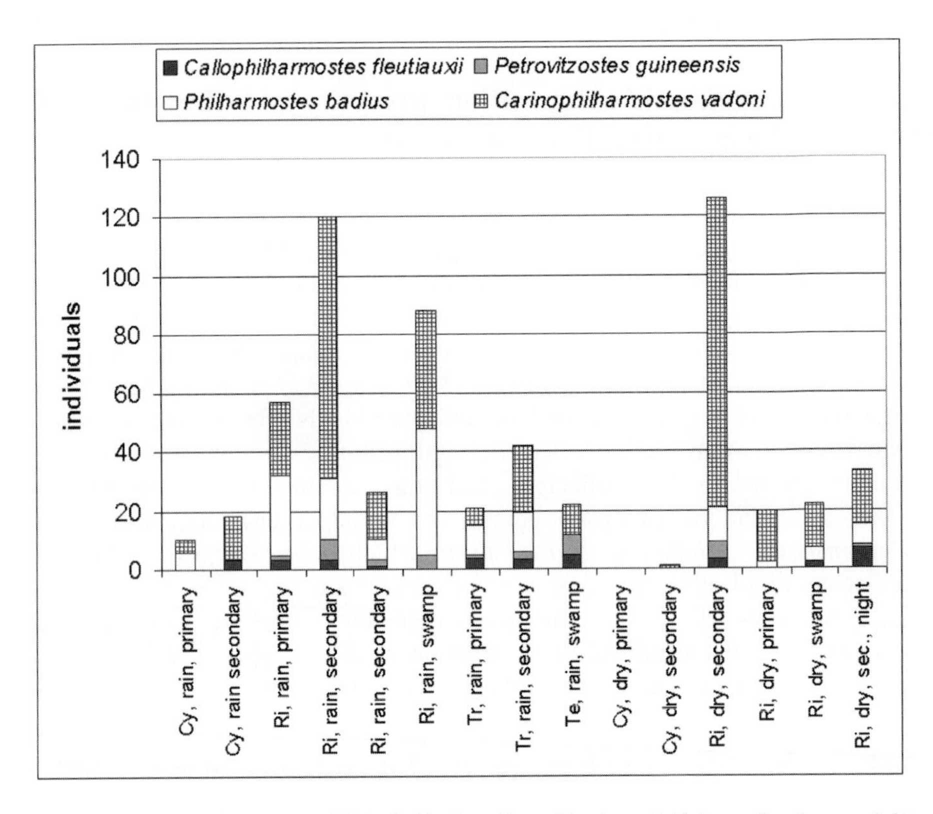

Figure 1. Number of ceratocanthid individuals collected by insecticidal tree fogging on eight con-specific trees (collecting unit: Cy: *Cynometra*, Ri: *Rinorea*, Tr: *Trichilia*, Te: *Teclea*) per season (rain, dry) and forest type (secondary, primary swamp).

primary forest). Another remarkable result is that the Ceratocanthidae during the dry season have been collected almost exclusively from *Rinorea beniensis* trees which are relatively small trees, offering a dense shady crown, which favours maintenance of humidity (Wagner, 2001). *Rinorea* trees growing in the secondary forest have also on average a smaller stature and denser leaf cover than in the other forest types. All these factors create a comparatively wetter microhabitat that can explain accumulation of Ceratocanthidae on these trees.

3.3 Effect of ant dominance on Ceratocanthidae abundance

In Budongo Forest ants are the dominant arthropod group, with 29.0% of all arthropods collected (Schulz and Wagner, 2002). This abundance of ants

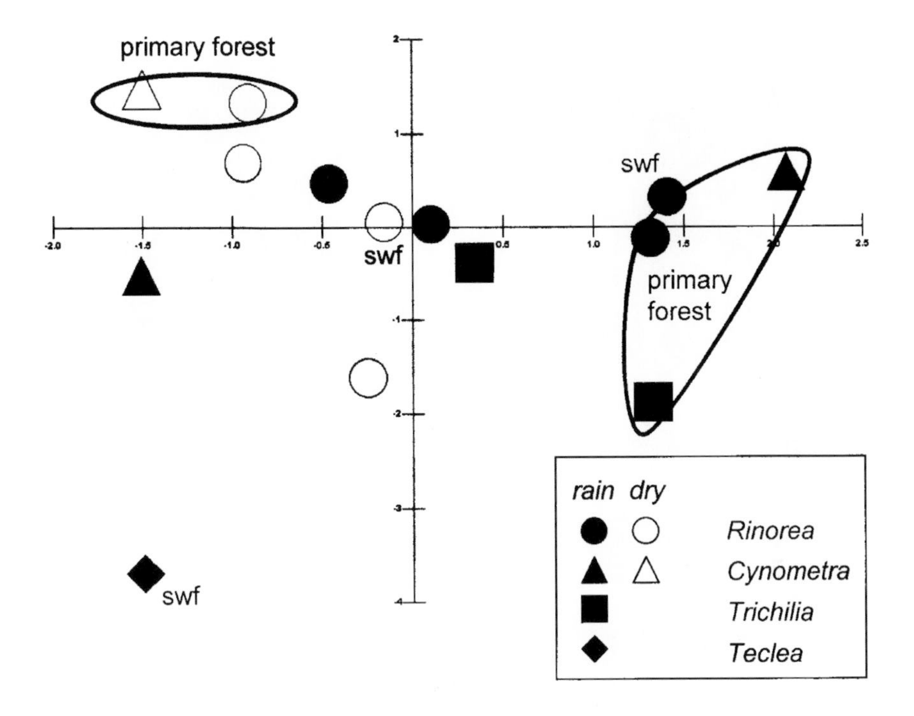

Figure 2. Correspondence analysis for five species and 665 individuals of Ceratocanthidae. Primary forest collecting units outlined, three collecting units from swamp forests marked (swf), all other markings from secondary forest.

seems to have a strong effect on beetle assemblages. The high number of individuals of Ceratocanthidae in Budongo Forest could possibly be explained by their capability of rolling up and avoiding ant attacks.

4. DISCUSSION

4.1 Species diversity and biogeographic patterns

The five species of Ceratocanthidae collected in Budongo Forest belong to the *Philharmostes* genus-group, which is endemic to tropical Africa and Madagascar (Ballerio, 2000b, 2001). The species of this group represent about one half of the Afrotropical ceratocanthid fauna and are usually found in leaf litter, in termite nests or by beating leaves in forests. The species found in Budongo show a typical Guineo-Congolian distribution, occurring from the Guinea Gulf countries eastwards to Uganda. They all are volant and

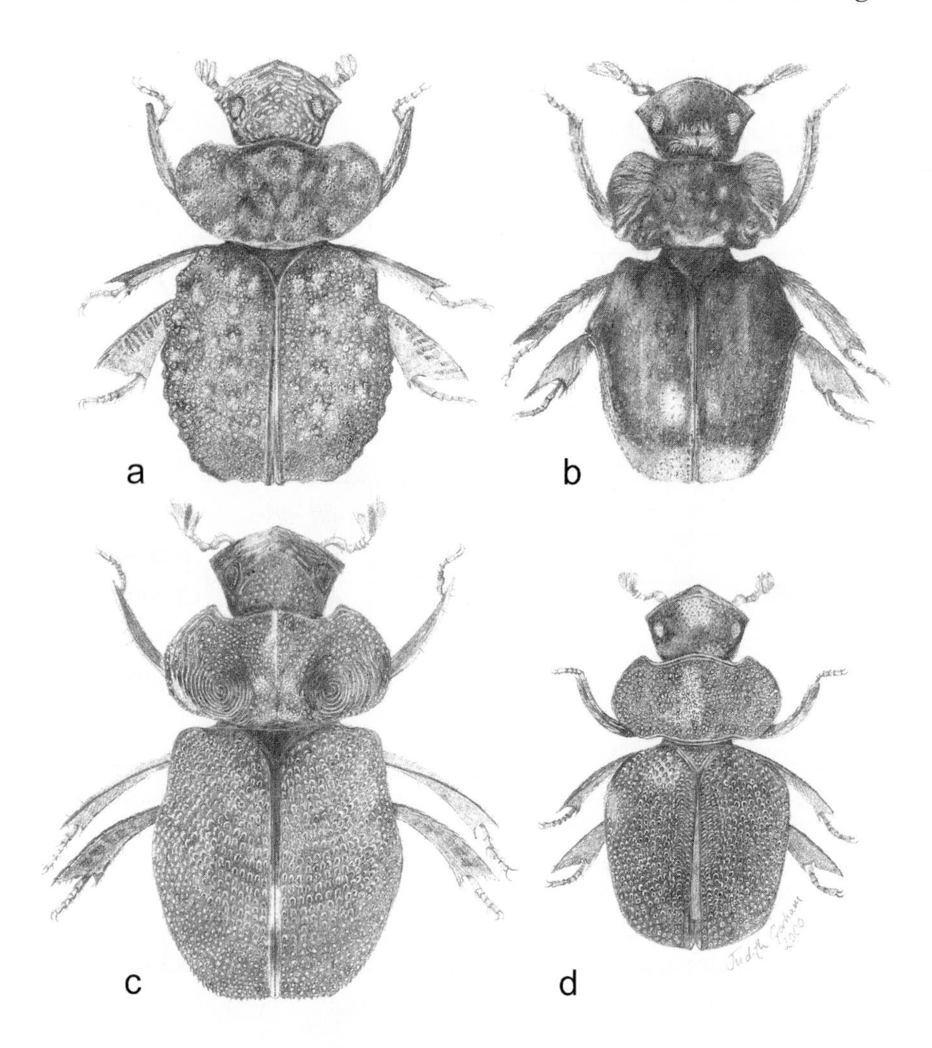

Figure 3. Habitus of *Carinophilharmostes vadoni* (a), *Callophilharmostes fleutiauxii* (b), *Petrovitzostes guineensis* (c) and *Philharmostes badius* (d).

seem to occur often syntopically all over their range. We are aware of at least four localities where all or at least three of these species occur together, i.e. the Budongo Forest (present paper), the Nimba Mountains (Paulian, 1993), Nkolentangan in Ecuatorial Guinea (Paulian, 1977) and the Kindamba region in Democratic Republic of Congo (Paulian, 1968, 1977).

4.2 Abundance and seasonality

There are no available data on the phenology of Ceratocanthidae, and this is probably the first time that quantitative data are recorded from both dry and wet season for these beetles. Since there is no clear seasonal pattern, with exception of the primary forest, it may be that a dry season of two months in Budongo Forest is too short to cause a real change in beetle diversity and life cycle. The large number of individuals seems to be a distinctive trait of Budongo Forest, since canopy fogging with the same techniques in various other sites belonging to the same biogeographic and ecological region, such as gallery forests along the Akagera River (East Rwanda), Irangi (Congo) and Semliki Forest (Uganda), did not yield any Ceratocanthidae but one single specimen of *P. guineensis* from Irangi.

4.3 Sex ratio and reproduction cycle

We did not notice any significant difference in sex ratio between rainy and dry seasons. Particularly interesting is the observation that the majority of dissected females from both seasons had a very large spermatophore inside the bursa copulatrix. This indicates that mating had just occurred and this means that the reproductive season could be very extended or that there could to be two distinct reproductive seasons.

5. CONCLUDING REMARKS

Up to now, there are about ten genera of canopy associated Ceratocanthidae known, occurring in different parts of the tropics. Some of them show morphological and biological adaptations to life in the forest canopy, like *Eusphaeropeltis* Gestro, 1899 (Oriental Region) and *Ceratocanthus* White, 1842 (Neotropical Region), which represent two lineages with the most evident adaptations to arboreal life. The Guineo-Congolian genera found in Budongo Forest do not show such strong adaptations. Probably they live in rotten leaves and soil portion on larger branches and possibly feed on fungi growing on leaves and trunks, but this will be the topic of further research.

ACKNOWLEDGEMENTS

We thank Judith Gorham (Brescia) for executing the pencil habitus drawings.

REFERENCES

Ballerio, A., 2000a, A new genus of Ceratocanthidae from the Oriental Region (Coleoptera: Scarabaeoidea), *Elytron* **13:**149–164.

Ballerio, A., 2000b, A new genus and species of Ceratocanthidae from Tanzania (Coleoptera: Scarabaeoidea), *Afr. Zool.* **35**(1):131–137.

Ballerio, A., 2001, Description of *Philharmostes werneri* n. sp. from Tanzania with notes on the "Philharmostes" generic group (Coleoptera, Ceratocanthidae), *Fragm. Entomol.* **33**(2):147–157.

Choate, P. M., 1987, Biology of *Ceratocanthus aeneus* (Coleoptera: Scarabaeidae: Ceratocanthinae), *Fla Entomol.* **70**(3):301–305.

Davies, J. G., Stork, N. E., Brendell, M. J. D. and Hine, S. J., 1997, Beetle species diversity and faunal similarity in Venezuelan rainforest tree canopies, in: *Canopy Arthropods*, N.E. Stork, J. Adis and R. K. Didham, eds., Chapman & Hall, London, pp. 85–103.

Hammond, P. M., 1990, Insect abundance and diversity in the Dumoga-Bone National Park, N. Sulawesi, with special reference to the beetle fauna of lowland rain forest in the Toraut region, in: *Insects and the Rain Forests of South East Asia (Wallacea)*, W. J. Knight and J. D. Holloway, eds., The Royal Entomological Society of London, London, pp. 197–254.

Paulian, R., 1968, The Scientific Results of the Hungarian Soil Zoological Expedition to the Brazzaville-Congo. 33. Espèces de la famille Acanthoceridae (Coleoptera: Scarabaeoidea), *Opusc. Zool.* **8:**87–98.

Paulian, R., 1977, Révision des Ceratocanthidae (Coleoptera, Scarabaeidae). I. – Les formes africaines, *Rev. Zool. Afr.* **91:**253–316.

Paulian, R., 1993, Quelques Cératocanthides des Monts Nimba en Guinée, Afrique Occidentale [Coleoptera, Scarabaeoidea], *Rev. Fr. Entomol. (nouv. ser.)* **15:**145–147.

Schulz, A., and Wagner, T., 2002, Influence of forest type and tree species on canopy ants (Hymenoptera: Formicidae) in Budongo Forest, Uganda, *Oecologia* **133:**224–232.

Scholtz, C. H., 1990, Phylogenetic tends in the Scarabaeoidea (Coleoptera), *J. Nat. Hist.* **24:**1027-1066.

Wagner, Th., 2000, Influence of forest type and tree species on canopy-dwelling beetles in Budongo Forest, Uganda, *Biotropica* **32**(3)**:**502–514.

Wagner, Th., 2001, Seasonal changes in the canopy arthropod fauna in *Rinorea beniensis* in Budongo Forest, Uganda, *Plant Ecol.* **153:**169–178.

HUMAN INFLUENCE ON THE DUNG FAUNA IN AFROTROPICAL GRASSLANDS (INSECTA: COLEOPTERA)

Frank-Thorsten Krell[1], Vincent S. Mahiva[2], Célestin Kouakou[3,4], Paul N'goran[3,4], Sylvia Krell-Westerwalbesloh[1,5], Dorothy H. Newman[1,5,6], Ingo Weiß[7], Mamadou Doumbia[4]

[1] *Soil Biodiversity Programme, Department of Entomology, The Natural History Museum, Cromwell Road, London SW7 5BD, UK, E-mail: f.krell@nhm.ac.uk*
[2] *Biodiversity and Conservation Programme, International Centre of Insect Physiology and Ecology (ICIPE), Duduville, Kasarani, P.O. Box 30772-00100, Nairobi, Kenya*
[3] *Centre Suisse de Recherches Scientifiques, 01 B.P. 1303, Abidjan 01, Côte d'Ivoire*
[4] *Université Abobo-Adjamé UFR des Sciences de la Nature, Filière Gestion et Valorisation des Ressources Naturelles (Biodiversité et Gestion Durable des Ecosystèmes), 02 B.P. 801, Abidjan 02, Côte d'Ivoire*
[5] *Zoologisches Forschungsinstitut und Museum Alexander Koenig, Adenauerallee 160, D-53113 Bonn, Germany*
[6] *Department of Agriculture, The University of Reading, Earley Gate, P.O. Box 236, Reading RG6 6AT, U.K.*
[7] *Häusernstraße 26, D-83671 Benediktbeuern, Germany*

Abstract: Dung beetles are the main dung recyclers in most Afrotropical environments. We compare dung beetle abundance in herbivore dung in West and East African grasslands that are subject to various anthropogenic disturbances. Dung beetles are similarly abundant in grasslands with indigenous wild herbivores and with extensive cattle farming. However, if herbivore dung is regularly removed (i.e. collected by local people for domestic and agricultural use) or not present locally at all, the abundance of dung beetles decreases. The lowest numbers were found in an urban settlement.

Key words: Dung beetles; West Africa; East Africa; savanna; disturbance; land use

B.A. Huber et al. (eds.), African Biodiversity, 133–139.

1. INTRODUCTION

Afrotropical grasslands form the basis of many African economies, because they are used for the majority of agricultural, horticultural and stock farming activities. Thus, with a growing population, their indigenous fauna and flora are under enormous threat. Coprophagous scarab beetles play a decisive role in tropical grasslands in maintaining soil fertility by returning nitrogen, phosphorus and other nutrients from faeces on the soil surface into the soil and by increasing soil aeration and water capacity (Hanski and Cambefort, 1991). With this study, we present a preliminary view of the overall abundance of dung beetles and other coprophilous beetles of coprocenoses in several Afrotropical grasslands affected by various levels of anthropogenic disturbance.

2. MATERIAL AND METHODS

To obtain comparable quantitative data on dung beetle assemblages we deposited standardized portions (ca. 900 ml, i.e. 1 kg) of fresh cattle or buffalo dung on the ground and flooded them after half a day in a bucket of water together with the populated soil below. Because of fundamental diel differences (Krell et al., 2003), day and night assemblages were sampled separately. In each site, ten night and ten day samples were taken, randomly distributed over the experimental period. The method is described and discussed in detail by Krell et al. (2003). All samples were collected in the first half of the main rainy season to reduce seasonal effects.

The following sites were sampled. They can be ordered along a disturbance gradient with the most natural condition boasting a constant supply of faeces of native large mammals ('Comoé'), intermediate conditions showing a replacement of the native mammal faeces by livestock faeces ('Bringakro P', 'Shiveye') and the most disturbed condition without constant herbivore dung supply (though human faeces are regularly present) ('Bringakro I', 'Buyangu', 'Youpogon').

a) 'Comoé': Northeastern Côte d'Ivoire, southern Parc National de la Comoé, 8°45'02"N, 3°48'58"W (09.vii.-25.vii.1997), see figures 1 and 4 in Krell et al. (2003); savanna parkland ('savane arbustive') with indigenous large herbivore fauna, not used for agriculture or farming, but annually burned and regularly poached;

b) 'Bringakro P': Central Côte d'Ivoire, V-Baoulé, Bringakro, 6°25'07"N, 5°04'01"W (27.vi.-2.vii.2003); savanna parkland used for extensive cattle farming, annually burned, native large mammals almost extinct;

c) 'Shiveye': Western Kenya, outside Kakamega Forest, Shiveye village, 0°09'52" N, 34°48'04"E (6-23.iv.2003); grassland of a small farm, used as cattle pasture, dung not used by humans, native large mammals extinct;

d) 'Bringakro I': Central Côte d'Ivoire, V-Baoulé, Bringakro, 6°25'00"N, 5°05'31"W (27.vi.-2.vii.2003); grassland near village (mainly *Imperata* sp.), not used for cattle farming, annually burned, native large mammals extinct;

e) 'Buyangu': Western Kenya, Kakamega Forest, Buyangu, 0°21'13"N, 34°51'49"E (9-19.iv.2003); grassland not used for cattle farming, but surrounded by farms, in which 80% of the fresh cattle dung is regularly and all year round collected by farmers for various purposes (survey with 40 farmers in the area), native large mammals extinct;

f) 'Youpogon': Southern Côte d'Ivoire, Abidjan-Youpogon, 5°19'26"N, 4°04'58"W (16-29.vii.2002); strip of grass and herbaceous vegetation in urbanized residential area between tarmac street and housing, no cattle or other herbivores present.

The material is deposited in The Natural History Museum London except for the samples from Bringakro, which were destroyed by airport authorities in Brussels despite valid export papers. A part of the Kenyan material will be deposited in the National Museums of Kenya after species identification.

We present the data as pooled absolute abundance of individuals at family level. For this brief overview, we refrain from presenting means with standard deviation because the generally high standard deviations we find with dung beetle assemblages mainly reflect the trivial weather fluctuations during the sampling period since dung beetle abundance is strongly influenced by weather conditions (Davis, 1995). We also refrain from presenting medians because outlying activity peaks of abundant species are regular and significant events in dung beetle assemblages and would be neglected if considering medians and quartiles only.

We present the most abundant taxa Scarabaeoidea (proper dung beetles, coprophagous), Histeridae (clown beetles, predators), Hydrophilidae (water beetles, predators and coprophagous), and Staphylinidae (rove beetles, predators and coprophagous) and pool the remaining taxa ('rest', mainly Carabidae, predators).

3. RESULTS AND DISCUSSION

Although disturbed by poaching and affected by bush fires, the site in the Parc National de la Comoé is the most natural reference site. Anthropogenic bush fires have been an integral component of the West African savanna ecosystems for several hundred to 50,000 years (Meurer et al., 1994). Thus,

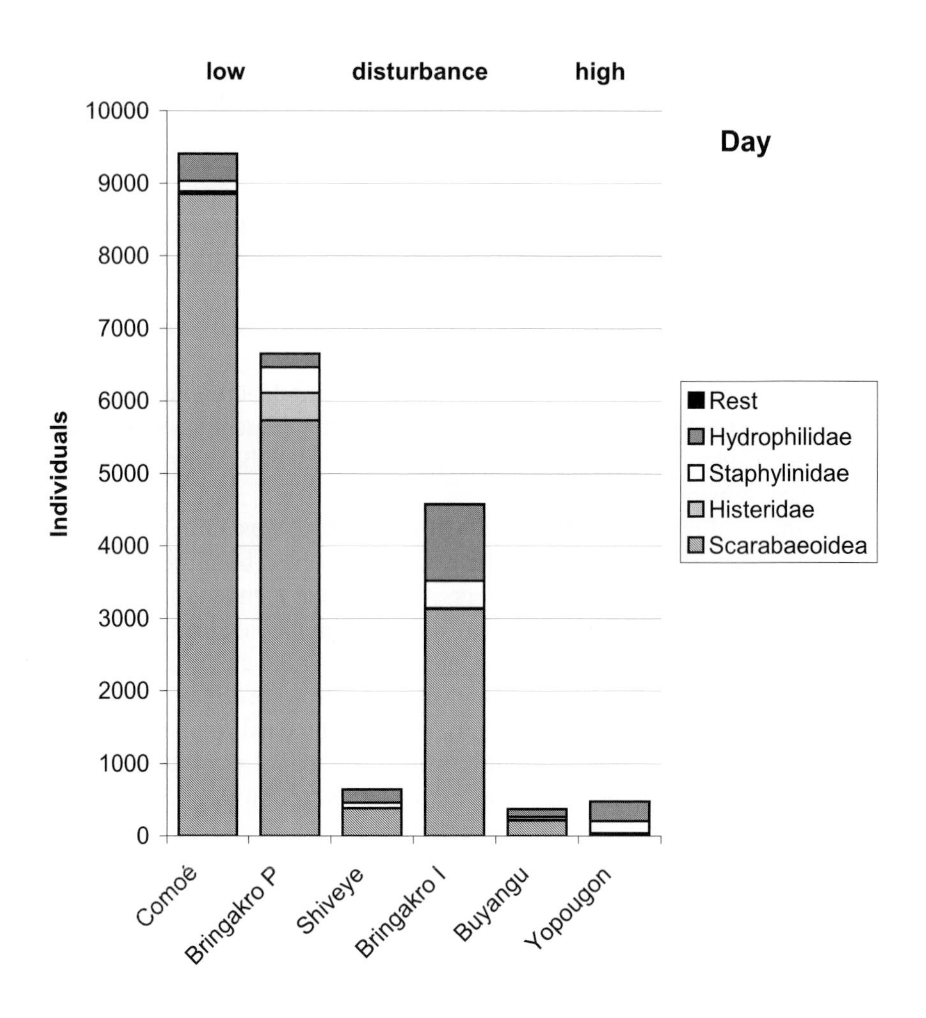

Figure 1. Abundance of beetles inhabiting ten 1 kg portions of bovid dung per site, exposed from 6:00 h to 16:00 h (diurnal beetles).

we do not consider fire as a disturbance for the purpose of this study. Savannas with constant dung supply ('Comoé' with native herbivores, 'Bringakro P' with livestock) show the highest abundance of coprophagous beetles both during the day and the night (Figs 1-2). Although the dung in the savanna of Bringakro is supplied by livestock and not by the original fauna, dung beetle abundance during the day does not differ significantly from the Comoé site ($p=0.169$; exact permutation test, SsS 1.1g, Rubisoft Software GmbH; Crowley, 1992) and is even significantly higher at night ($p=0.009$).

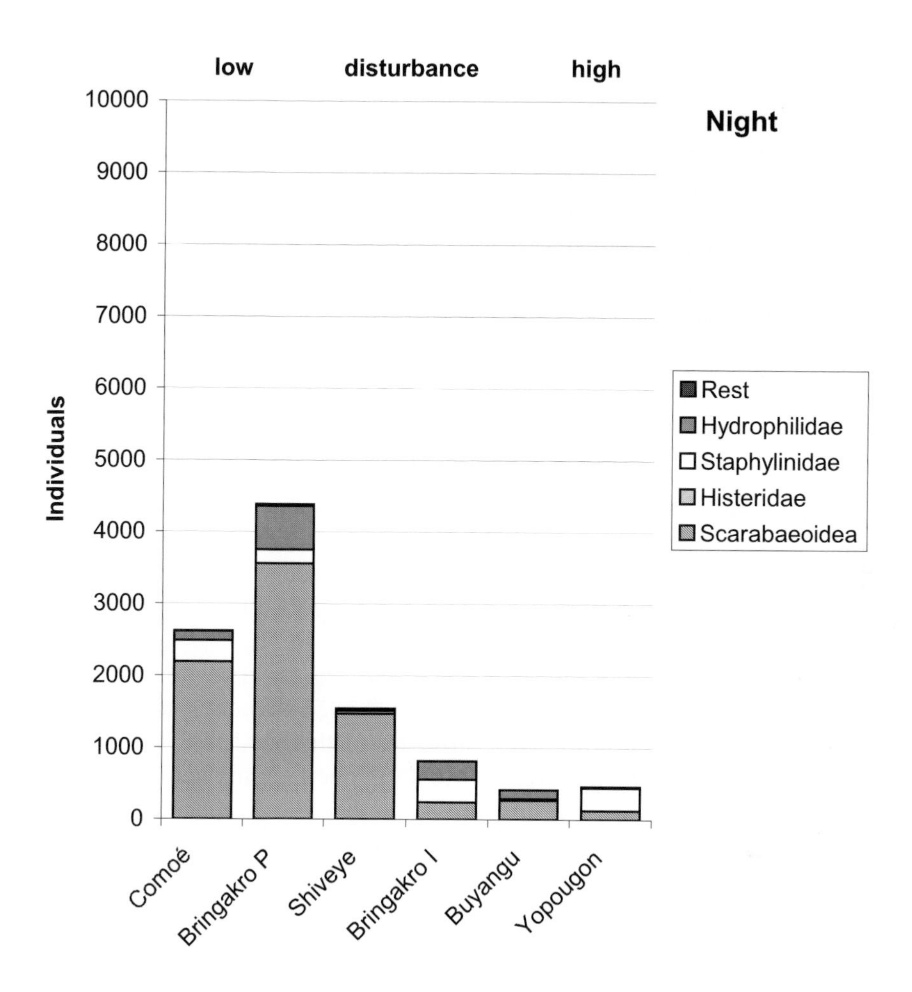

Figure 2. Abundance of beetles inhabiting ten 1 kg portions of bovid dung per site, exposed from 18:00 h to 6:00 h (nocturnal beetles).

Dung beetle abundance on the farm with constant dung supply (Shiveye) is much higher at night than during the day. This meadow surrounded by fields and gardens might be either too derived from natural grasslands (causing a different microclimate particularly during the day) or simply too small to support large dung beetle populations. However, compared to an area without constant dung supply 17 km away (Buyangu), the proper dung beetles are significantly more abundant (p_{day}=0.020; p_{night}=0.005).

'Buyangu' and 'Yopougon' are the most disturbed sites since in the whole area herbivore dung is not regularly available, either because most of the cattle dung is removed by farmers or large herbivores are lacking. The urban site at Youpogon, which additionally suffers from pedestrian and nearby vehicle traffic, shows the lowest dung beetle abundance. Although the pooled number of beetles is higher than in rural Buyangu, this difference is not significant (p_{day}=0.504; p_{night}=0.859).

The other rural site without herbivore dung supply is an *Imperata* grassland 'Bringakro I'. *Imperata* grass is not a palatable food-source for the local cattle, but pasturelands are within a distance of 0.9 km (well within flight range of dung beetles). Compared with the grazed savanna 'Bringakro P' (which is at a distance of 2.8 km), the dung beetle abundance is, however, significantly lower (p_{day}=0.012; p_{night}=0.002).

4. CONCLUSION

Despite the high mobility and dispersal power of dung beetles, local dung supply seems to influence their abundance. Although omnipresent human faeces are attractive to most dung beetle species (Walter, 1978), for maintaining high abundance of herbivore-dung beetles the presence of herbivore dung is needed.

For sheer abundance, livestock faeces are suitable to replace the droppings of indigenous wild herbivores, however, this does not necessarily extend to diversity and dominance structure of the beetles. Removing dung from farmland apparently decreases the dung beetle fauna, and although the usage of dung as fertilizer or building material means an economic advantage for the local farmers, possible consequences of a diminished dung recycler fauna for local hygiene and soil fertility, albeit unknown so far, might be worth investigation.

ACKNOWLEDGEMENTS

The project was funded by a grant of the Deutsche Forschungsgemeinschaft (DFG) to K.E. Linsenmair, Universität Würzburg (Li 150/18-1,2,3) and by the BIOLOG programme of the German Federal Ministry of Education and Research and was a part of section E09 in the BIOTA research initiative. We are deeply indebted to late Prof. Clas Naumann, former director of Zoologisches Forschungsinstitut und Museum Alexander Koenig, Bonn, who co-initiated BIOTA and was responsible for the realization of our project. Fieldwork was permitted by the Ministère de l'Agriculture et des Resources Animales de Côte d'Ivoire, Abidjan, the

Ministry of Education, Science and Technology, Nairobi, Kenya, and the Kenyan Wildlife Service.

REFERENCES

Crowley, P. H., 1992, Resampling methods for computation-intensive data analysis in ecology and evolution, *Ann. Rev. Ecol. Syst.* **23:**405-447.

Davis, A. L. V., 1995, Daily weather variation and temporal dynamics in an Afrotropical dung beetle community (Coleoptera: Scarabaeidae), *Acta Oecol.* **16:**641-656.

Hanski, I., and Cambefort, Y., 1991, *Dung Beetle Ecology*, Princeton University Press, Princeton, xiii + 481 pp.

Krell, F.-T., Krell-Westerwalbesloh, S., Weiß, I., Eggleton, P., and Linsenmair, K. E., 2003, Spatial separation of Afrotropical dung beetle guilds: a trade-off between competitive superiority and energetic constraints (Coleoptera: Scarabaeidae), *Ecography* **26:**210-222.

Meurer, M., Reiff, K., Sturm, H.-J., and Will, H., 1994, Savannenbrände in Tropisch-Westafrika, *Petermanns Geogr. Mitt.* **138:**35-50.

Walter, P., 1978, Recherches écologiques et biologiques sur les scarabéides coprophages d'une savane du Zaïre. Thèse Docteur d'État, Université des Sciences et Techniques du Languedoc, Montpellier, 366 pp., annex, pls.

GENUINE AFRICANS OR TERTIARY IMMIGRANTS? – THE GENUS *HYDROPSYCHE* IN THE AFROTROPICAL REGION (INSECTA, TRICHOPTERA: HYDROPSYCHIDAE)

Wolfram Mey
Museum für Naturkunde, Humboldt Universität, Invalidenstr. 43, D – 10115 Berlin, Germany, E-mail: wolfram.mey@museum.hu-berlin.de

Abstract: The distinctive character of the Afrotropical fauna has developed during a long period of isolation. About 30 million years ago the African and Eurasian plates collided and numerous groups migrated in and out of Africa. Successful colonizers adapted and became part of the endemic communities. Based on a biogeographic and phylogenetic analysis, the African species of the genus *Hydropsyche* were identified as belonging to this group of early immigrants. The genus now contains 17 afrotropical species all belonging to the *propinqua*-group which is endemic to Africa. The ranges of the related groups are in the Holarctic Region (*angustipennis*-group), and India to South East Asia (*newae*-group). This area contains the diversity centre of the genus. The African species are rather homogenous in their morphology, and their great similarity is indicative of a single immigration event. The species are unevenly distributed over the continent, with a concentration in the central African Rift system. A continental species colonised Madagascar, and subsequently Mauritius and Reunion. In an attempt to reconstruct the expansion routes, the phylogeography of the group is examined and briefly discussed.

Key words: Trichoptera; *Hydropsyche* Pictet; *Stenopsyche* McLachlan; immigration; rivers; radiation; taxonomic diversity centre; biogeography; phylogeny; Asia; Africa; Afromontane Biome

B.A. Huber et al. (eds.), African Biodiversity, 141–150.

1. INTRODUCTION

Each continent has its own characteristic entomofauna, which encompasses species, genera, tribes, subfamilies, families and even orders that are endemic for the continent. They are used for distinguishing between faunal regions on a world-wide scale. In Asia, the caddisfly genus *Stenopsyche* McLachlan, 1868 of the family Stenopsychidae is a typical faunal element. The range of the genus is shown in Figure 1. The majority of the some 80 described species occur in North India, Myanmar and South China. It was therefore a considerable surprise when a species of the genus was found in the Kivu Region of Eastern Congo and in Rwanda (Marlier, 1950). The species was named *S. ulmeriana* Marlier, 1950 and was associated with the *pubescens*-group which is an assemblage of about 15 derived species (Schmid, 1969). *Stenopsyche ulmeriana* has to be regarded as a neoendemic in Africa. Its ancestor was undoubtedly of Asian stock and was able to immigrate into the Afrotropical Region.

Interestingly, there is a fossil record of the genus from Baltic Amber (Ulmer, 1912). In the Tertiary, the range of the genus must have been much wider than today and must certainly have come closer to the African continent, which facilitated a successful colonisation. The case of *Stenopsyche ulmeriana* provides clear evidence of immigration by an Asian species into the Afrotropics, very probably in Tertiary times. It is to be expected that further species in addition to *Stenopsyche* were able to invade in the same way.

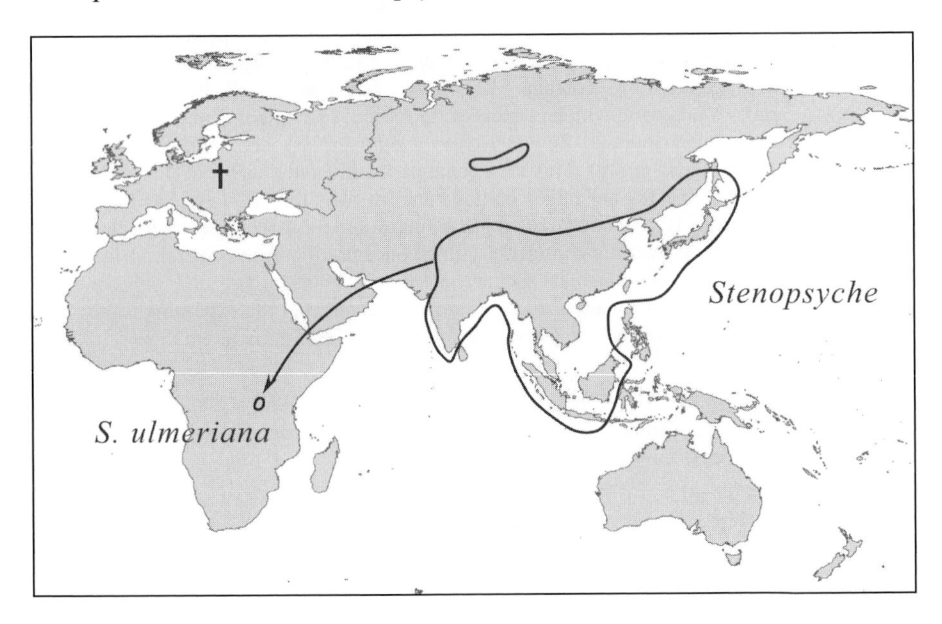

Figure 1. Geographic range of the genus *Stenopsyche* McLachlan (Stenopsychidae).

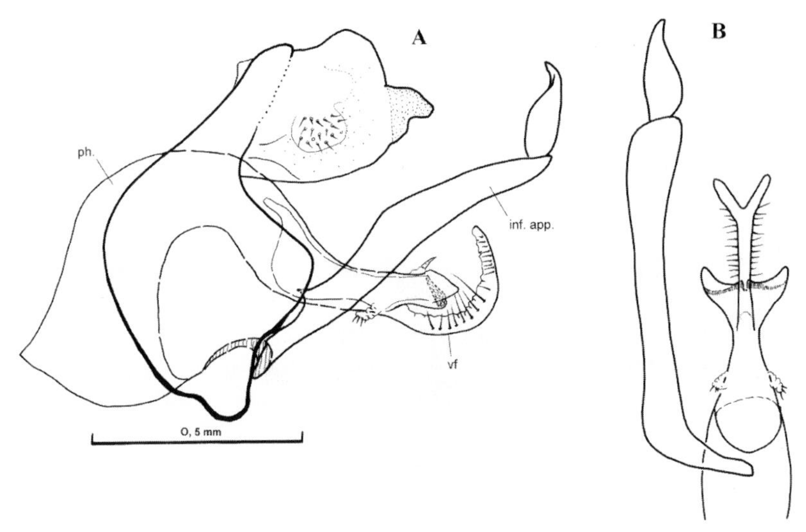

Figure 2. Male genitalia of *Hydropsyche marlieri* Jaquemart & Statzner, 1983, A - lateral view, B – ventral view (inf. app. – inferior appendages, ph – phallotheca, vf – ventral fork).

I have chosen this case to demonstrate the principal pattern. However, the biogeographic distribution and phylogeny of other taxa are not so clear and straightforward for recognising an extra-African origin. The genus *Hydropsyche* Pictet, 1834 is an example of this.

2. THE GENUS *HYDROPSYCHE* PICTET IN AFRICA

With about 400 species this is one of the largest genera of the Trichoptera. Nearly 30 species have been described from Africa south of the Sahara, of which only 17 have proven to be valid (Mey, 2005). They are placed in the *propinqua* group, which is clearly distinct from the remainder of *Hydropsyche*. The main synapomorphy is the presence of a ventral fork (vf) in the phallic apparatus of the male (Fig. 2). The morphological features of the group appeared to be important enough that some authors established a separate genus (Navas, 1931; Ulmer, 1907).

The adults are medium-sized insects and have a rather uniform appearance (Fig. 3). The larvae live in streams and rivers and are omnivorous. The final instars construct capture nets to catch other invertebrates or to strain food from the current. The larvae are often very abundant and dominate the benthic macroinvertebrate community. With their cases and nets they can cover large stretches of the river bottom,

Figure 3. Adult of *Hydropsyche jeanneli* Mosely (Hydropsychidae) from Kakamega Forest, Kenya.

especially in stony areas such as rapids or riffles. They make up a significant portion of the food for fish species in running waters.

In Africa, the typical habitats for *Hydropsyche* species are confined to or close to the Afromontane Biome (Fig. 6). The species do not occur in the middle and lower courses of large rivers like Niger, Nile, Kongo, Sambezi, Okavango.

3. BIOGEOGRAPHY AND PHYLOGENY OF THE GENUS

The range of *Hydropsyche* covers the Northern Hemisphere, Africa and extends southward through the West Pacific to New Zealand (Fig. 4). A closer inspection of the distribution map reveals three important facts:

- Africa is the only southern continent that possesses *Hydropsyche* species. The genus is absent from South America and Australia.
- The African species belong only to a single species group, whereas on other continents two or more species groups occur sympatrically.
- The taxonomic diversity centre comprising the highest number of species groups is located in Asia. It extends from Middle Asia along mid-elevations of the Himalaya Mountain Chain, to Yunnan and North Vietnam.

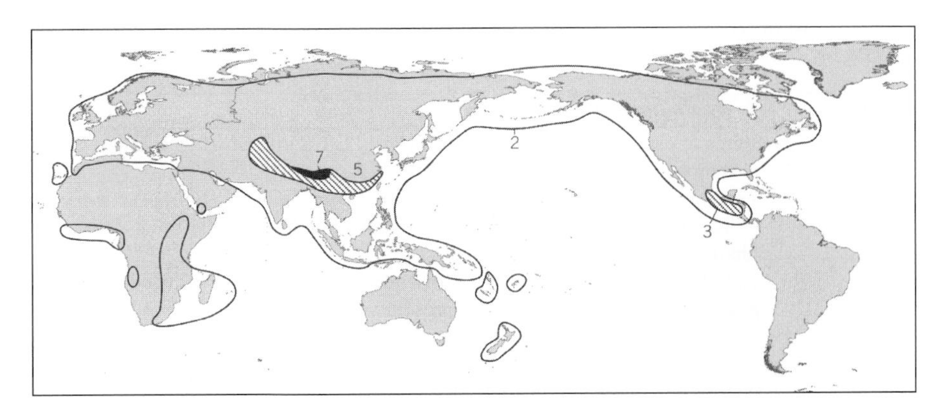

Figure 4. Geographic range of the genus *Hydropsyche* Pictet. The black and striped area is the taxonomic diversity centre of the genus with the highest number of sympatrically occurring species groups.

These three points alone make the African range of the genus a candidate for being a secondary extension from the main range in Asia. Substantial support for this hypothesis might be provided by analysing the phylogeny of the genus. If the most primitive or ancestral species occur in, or close to, the taxonomic diversity centre, then the origin of the genus should probably be there too. On the other hand, if the African group proves to comprise the most primitive species, then we would have to consider the group as a genuine Afrotropical element with a dispersal out of Africa.

A taxonomic revision of the genus which would provide the basis for a phylogenetic analysis is not available. However, a number of species-groups have been revised in recent years (Malicky, 1977; McFarlane, 1976; Mey, 1996, 1998, 1999, 2003, 2005; Nimmo, 1987; Schefter & Wiggins, 1986; Schefter et al., 1986). A subdivision or taxonomic structure for the genus was proposed by Ross & Unzicker (1977) and Malicky & Chantaramongkol (2000). The approach of Ross & Unzicker (1977) was based on an evolutionary scenario which was derived from morphological features of the male genitalia. Although they considered only a part of the genus and concluded with brief division of *Hydropsyche* into several genera, this approach was nonetheless fruitful and of great heuristic value. The same approach was used to analyse certain Asian species groups (Mey, 1999, 2003) but without introducing new generic names. A preliminary application of the method to the complete spectrum of *Hydropsyche* resulted in a phylogenetic hypothesis which is shown in Figure 5. The 400 species are assigned to 13 major species groups, which are defined by morphological characters of the male phallic apparatus. Only unique synapomorphies have been used. They are summarised in Table 1.

Table 1. List of recognised synapomorphies of *Hydropsyche* which were used to construct the cladogram of the genus in Fig. 5

Number	Synapomorphies
1	stems of Cu 1 and M, and bases of M3 and Cu1a in hindwings parallel, very close together
2	- presence of at least five pairs of endothecal spines - endotheca with a singular process between phallotrema and tip of phallotheca
3	endotheca with paired processes (or endotheca secondarily reduced)
4	phallotremal sclerite or sclerites present
5	dorsolateral pair of endothecal process sclerotised, enlarged and without spines
6	phallotrema with an unpaired sclerite bearing a ventral tongue
7	phallotremal sclerites paired, concave internally

It is beyond the scope of the present paper to discuss here the details and implications of the cladogram. This will be the subject of a separate article. For present purposes it is used as an orientation and for recognising the basal and terminal groups.

The exciting questions were: what is the position of the *propinqua*-group in the topography of the phylogenetic tree and what are the related groups? The cladogram in Figure 5 gives the answers: The most ancestral groups of the genus do indeed occur in the taxonomic diversity centre in Asia. The area contains, in addition, species of the related, ancestral genera *Hydromanicus* Brauer, 1865, *Hydatopsyche* Ulmer, 1926 and *Hydatomanicus* Ulmer, 1951. The adelphotaxon of the *propinqua*-group is the *angustipennis*-group.

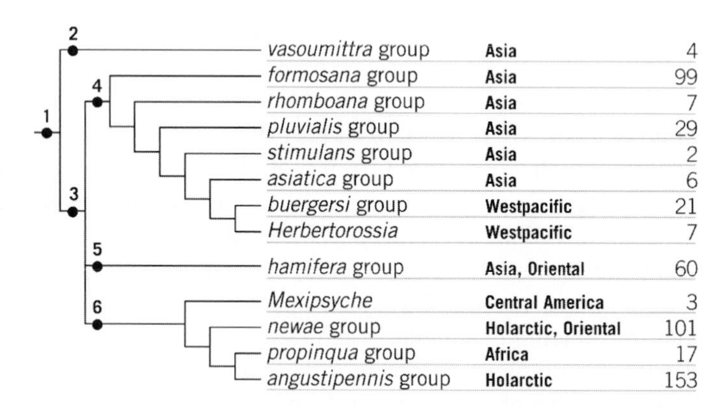

Figure 5. Species group – area cladogram of *Hydropsyche* Pictet, with figures of currently known species on the right-hand side. Numbers on the cladogram denote synapomorphies (the genera *Aoteapsyche* McFarlane, *Caledopsyche* Kimmins, and *Orthopsyche* McFarlane are included in the *H. buergersi*-group).

Both are terminal clades that evolved from descendants of the *newae* – group, as already suggested by Ross & Unzicker (1977). The fusion of the phallotremal sclerites with the sclerotised phallotheca is the synapomorphy of the two groups. The *angustipennis*- and *newae*-groups are the most species-rich groups in the genus. They both inhabit a wide range in the Holarctic Region. Although some species live in North Africa and in the Jordan Valley in the Near East, no member of the *angustipennis* -group has found its way to the Nile valley and further south into the Afrotropical region.
In conclusion, the genus *Hydropsyche* in Africa is a relatively young offshot of an Asian ancestor. It is by no means an ancient element of Gondwanan

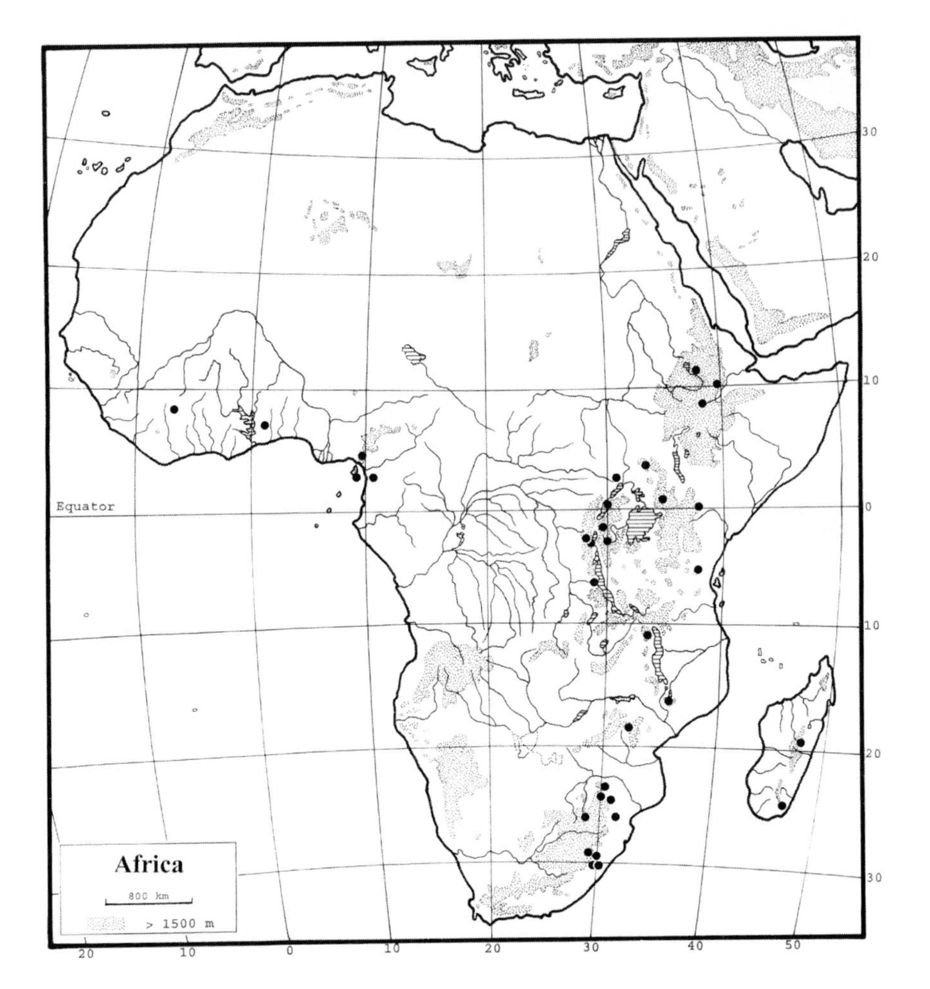

Figure 6. Distribution of the *propinqua*-group in Africa. Species occur also on Mauritius and Réunion, but are not shown here.

origin, but is an addition to the African fauna which probably emerged in the Tertiary. The currently known distribution in Africa shows a close association with montane areas, following the East African Rift System from Ethiopia to Malawi, and further south to the Drakensberg Mountains in South Africa (Fig. 6).

Its occurrence in Cameroon, West Africa and Madagascar conforms with the general distribution pattern of Afromontane elements (cf. Kingdon, 1989; Werger, 1978). The Asian origin supports the hypothesis that the immigration of the ancestor of the group took place from the Northeast of the African –Arabian Plate. Perhaps the north-south alignement of the mountains dictated the direction of the dispersal towards the south? If the dispersal was gradual and accompanied by speciation processes, the resulting north-south sequence of species could produce a comb-like cladogram, a so-called "Hennigian Comb". This congruence of area and phylogeny facilitates the recognition of the dispersal route. The pattern is known as Hennig`s Progression Rule.

An analysis of the phylogenetic relationships among the African species showed that the speciation pattern predicted by this model did not take place. Although the species are very similar to each other and most of the

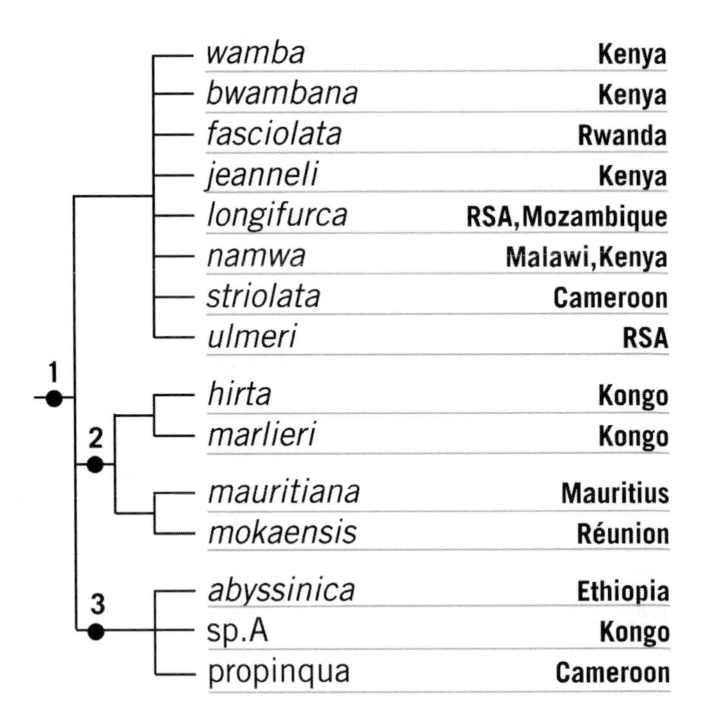

Figure 7. Cladogram of the *H. propinqua*-group. The numbers refer to synapomorphies explained in the text. (Species known in the female sex only are not considered).

differences merely enable the species to be separated, they can be grouped by three derived characters:

1 – phallotheca with a subapical, ventral process, upwardly bent and with a bifid tip (= ventral fork), 2 – ventral fork with membraneous extensions bearing small spines or teeth, 3 – dorsal endothecal processes absent.

Three separate clades were identified (Fig. 7). They do not show a north-south arrangement of species. The northernmost species, *H. abyssinica* is a derived species with some new characters, whilst the southernmost species, *H. longifurca*, belongs to the polytomous clade containing most of the East African species. This pattern points instead to a colonisation of the Afromontane Biome by a single species. Later, it was followed by subsequent radiation and speciation in different mountain ranges during times of reduction and isolation of the Afromontane Biome (Moreau, 1963). A continental species colonised Madagascar, and subsequently Mauritius and Reunion. The strong similarity between the species is indicative of a relatively young age for this radiation. The Pleistocene climatic fluctuations are probably the driving force which triggered the speciation (Moreau, 1963). The distinctiveness of the *propinqua*-group as a separate entity within *Hydropsyche*, however, is suggestive of a greater age which certainly dates back to some period in the Mid-Tertiary. The collision of the African with the Eurasian plate in the Miocene provided a land connection which enabled the onset of faunal exchange processes to take place. Together with other dispersing species the ancestor of the *Hydropsyche propinqua*-group could have taken this route and have eventually established successfully in Africa.

ACKNOWLEDGEMENTS

I gratefully acknowledge the support of Niels Hoff from the graphics department of the Museum for producing the distribution maps, and the help of Adrian Pont (Reading) for correcting the English text. Karl Kjer reviewed the manuscript. Financial support for the research reported here was provided in part by the Deutsche Forschungsgemeinschaft (Mey 1085/3, Me 1085/5).

REFERENCES

Kingdon, J., 1989, Island Africa. *The evolution of Africa`s rare animals and plants*. Princeton University Press, Princeton, pp. 287.

Malicky, H., 1977, Ein Beitrag zur Kenntnis der *Hydropsyche guttata*-Gruppe (Trichoptera, Hydropsychidae), *Z. Arbeitsgem. Österr. Entomologen* **29:**1-28.

Malicky, H., and Chantaramongkol, P., 2000, Ein Beitrag zur Kenntnis asiatischer *Hydropsyche*-Arten (Trichoptera, Hydropsychidae), *Linzer biol. Beitr.* **32**(2):791-860.

Marlier, G., 1950, Une famille de Trichoptères nouvelle pour l'Afrique: les Stenopsychidae, *Bull. Ann. Soc. Entomol. Belg.* **86:**207-218.

McFarlane, A. G., 1976, A generic revision of New Zealand Hydropsychinae (Trichoptera), *J. Royal Soc. New Zealand* **6:**23-35.

Mey, W., 1996, Zur Kenntnis der *Hydropsyche pluvialis* - Gruppe in Südostasien (Trichoptera: Hydropsychidae), *Entomol. Z.* **106:**121-164.

Mey, W., 1998, The genus *Hydropsyche* Pictet, 1834 on islands in the West Pacific Region and description of new species (Trichoptera, Hydropsychidae), *Tijdschr. Entomol.* **140:**191-205.

Mey, W., 1999, The *Hydropsyche formosana* group in the Oriental Region: taxonomy, distribution and phylogeny (Insecta, Trichoptera: Hydropsychidae), in: *Proceedings of the 9th International Symposium on Trichoptera, Chiang Mai 1998*, H. Malicky, and P. Chantaramongkol, eds., Faculty of Science, University of Chiang Mai, pp. XIII + 479.

Mey, W., 2003, Insular radiation of the genus *Hydropsyche* Pictet, 1834 in the Philippines and its implications for the biogeography of SE Asia (Insecta, Trichoptera: Hydropsychidae), *J. Biogeogr.* **30:**227-236.

Mey, W., 2005, Revision of the *propinqua*-group of the genus *Hydropsyche* Pictet, 1834 in Africa (Insecta, Trichoptera), *Aquatic Insects* (in press).

Moreau, R. E., 1952, Africa since the Mesozoic: with particular reference to certain biological problems, *Proc. Zool. Soc. London* **121**:869-913.

Moreau, R. E., 1963, Vicissitudes of the African biomes in the Late Pleistocene, *Proc. Zool. Soc. London* **141**:395-421.

Nimmo, A. P., 1987, The adult Arctopsychidae and Hydropsychidae (Trichoptera) of Canada and adjacent United States, *Quaestiones Entomol.* **23:**1-189.

Navás, L., 1931, Insectos del Museo de Paris, *Broteria, Ser. Zoologia* **27:**101-135.

Ross, H. H., and Unzicker, J. D., 1977, The relationships of the genera of American Hydropsychinae as indicated by phallic structures (Trichoptera, Hydropsychidae), *J.Georgia Entomol. Soc.* **12:**298-312.

Schefter, P. W., and Wiggins, G. B., 1986, A systematic study of the Nearctic larvae of the *Hydropsyche morosa* group (Trichoptera: Hydropsychidae), *Life Sciences Miscellaneous Publications*, Royal Ontario Museum, pp. 94.

Schefter, P. W., Wiggins, G. B., and Unzicker, J. D, 1977, A proposal for assignement of *Ceratopsyche* as a subgenus of *Hydropsyche*, with new synonyms and a new species (Trichoptera: Hydropsychidae), *J. N. Am. Benthol. Soc.* **5:**67-84

Schmid, F., 1969, La famille de Sténopsychides (Trichoptera), *Can. Ent.* **101**(2):187-224.

Ulmer, G., 1907, Neue Trichopteren, *Notes Leyden Mus.* **29:**1-53.

Ulmer, G., 1912, Die Trichopteren des Baltischen Bernstein, *Beiträge zur Naturkunde Preussens*, Königsberg, **10:**1-380.

Werger, M. J. A., (ed.), 1978, *Biogeography and ecology of southern Africa.* Junk Publishers, The Hague, 2 volumes, pp. 1-1439.

FRUIT-FEEDING BUTTERFLY COMMUNITIES OF FOREST 'ISLANDS' IN GHANA: SURVEY COMPLETENESS AND CORRELATES OF SPATIAL DIVERSITY

Janice L. Bossart[1], Emmanuel Opuni-Frimpong[2], Sylvestor Kuudaar[3] and Elvis Nkrumah[3]

[1] *Rose-Hulman Institute of Technology, ABBE, Terre Haute, IN 47803, USA; E-mail: bossart@rose-hulman.edu*

[2] *School of Forest Resources and Environmental Science, Michigan Technological University, Houghton, MI 49931, USA*

[3] *Forestry Research Institute of Ghana, Kumasi, Ghana*

Abstract: We conducted a year long survey of the fruit-feeding butterfly fauna of sacred forest groves and forest reserves of Ghana. About one-third of all fruit-feeding species recorded for Ghana were trapped, but our total sample had not yet reached the point of species saturation. Rarefied species richness was higher in the larger forest reserves than in the small sacred groves, as predicted by species-area relationship theory. *Bebearia* and *Euriphene* showed the greatest decline in species richness across sites, demonstrating that common, wide-spread species can be vulnerable to fragmentation.

Key words: Lepidoptera; indigenous reserves; species' vulnerability

1. INTRODUCTION

Estimates are that anywhere from 70 to 90% of Ghana's original high canopy climax forest has been eliminated (Hawthorne and Abu-Jaum, 1995; Weber et al., 2001). Very little original forest-cover exists outside of demarcated forest reserves and much of this relict habitat is sacred lands that were set aside by indigenous peoples hundreds of years ago and strictly protected via religious sanctions and taboos (Lebbie and Freudenberger,

B.A. Huber et al. (eds.), African Biodiversity, 151–158.

1996; Ntiamoa-Baidu, 2001). Ghana's long-protected sacred forest groves were originally embedded within continuous forest cover, but now exist as isolated habitat 'islands'. These indigenous reserves likely stand as biodiversity repositories of forest communities, and stepping stones that help link discrete forest blocks. In highly degraded regions, sacred forest groves constitute the only examples of old growth forest that remain in the country and some harbor the only surviving specimens of plant species associated with disappearing forest subtypes (Hall and Swain, 1981).

The long term goal of this project is to identify key predictors of community composition and species persistence and extinction in highly modified landscapes. Butterflies are excellent model systems for this task. They show a wide diversity of relative sensitivities to environmental change, are tightly linked to ecological systems as both primary consumers (herbivores) and food items, are easily collected and identified, and additionally elicit the emotional concern necessary to bring about conservation action in the face of the conflicting socioeconomic priorities that take precedence over nature conservation in developing countries.

2. METHODS

The fruit-feeding butterfly species (which account for approximately 1/3 of the ~730 forest-endemic butterfly species in Ghana; (Larsen, 2001)) were inventoried at two forest reserves and four sacred grove sites located in the moist semi-deciduous forest zone using typical fruit-baited traps. These remnant forests differ greatly in size. The forest reserves, Bobiri and Owabi, are 5000 and 1200 ha., respectively. The sacred groves, Asantemanso, Gyakye, Bonwire, and Kajease, are 259, 11.5, 8, and 6 ha., respectively.

Fruit-bait traps were hung in the forest understory, approximately 10 cm above the ground. Traps were baited with mashed, fermented banana, and butterflies were collected approximately 24h later. A total of 61 traps were installed across all sites. Conscious effort was taken to install all traps in similar microhabitats within areas of closed canopy forest. Five traps were hung in each of three discrete areas at Bobiri and Owabi, and in two at Asantemanso, in order to gain a more representative sample from these larger forests. Four traps were hung in each of two separate areas at Kajease and Bonwire, but the distance between sub-areas was small and restricted by the size and irregular shape of these forests. Five traps total were hung at Gyakye. Individual traps within sub-areas were separated from each other by at least 50 m and by no more than 200 m. Trap sampling occurred at regular intervals for one year. Sites were sampled from 18-22 times during the course of the study.

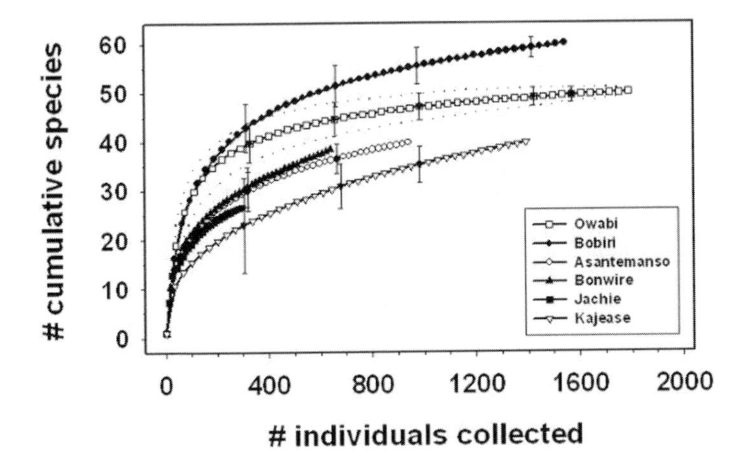

Figure 1. Rarified species richness curves showing numbers of species accumulated at each site as a function of increasing numbers of individuals trapped through time. Sub-samples (and 95% confidence intervals) from larger total samples were calculated relative to the total sample at each site with a smaller total sample. For example, five sub-samples are shown for Owabi, the site with the largest total sample, each of which corresponds to the sample size at a site with a smaller total sample.

Trap data were pooled for each site-date combination to minimize variance associated with different traps, e.g., due to differential trap attractiveness, annihilation of trap collections by driver ants, etc. Collections from each site were rarefied using EcoSim700 (Gotelli and Entsminger, 2003) to allow for comparison of richness among sites that had total samples of different sizes (Gotelli and Colwell, 2001). Rarefaction is a robust statistical technique that calculates estimates of species richness for sub-samples of a specified size drawn at random from the total sample. Comparable sized sub-samples can then be compared across sites. Rarefaction curves are similar to species accumulation curves but cannot be used to extrapolate estimates of total species richness.

3. RESULTS

Nearly 7000 specimens were collected from our baseline surveys (Fig. 1), representing 78 species and 34% of the total fruit-feeding butterfly fauna known for Ghana (Table 1). Traps were most successful at collecting species with an affinity for the moist forest subtype and species found in all forest subtypes (34 and 51% of those species possible, respectively). Only five species were collected that are viewed as non-specialized on forest habitats. Traps were also more successful at collecting Satyrinae (54% of fruit-

Table1. Sampling completeness. Species are categorized based on habitat affinities and scarcity rankings[1] and grouped by subfamily (from Larsen, 2001). The first number in each cell is the number of species trapped during this survey; the second number is the actual number of species in that category. Dashes indicate non-existent habitat-scarcity combinations. Rows were not included in cases where all cells were empty. Numbers in parentheses are species trapped/actual species of each habitat subtype and percentage represented. VC = very common, C = common, NR = not rare, R = rare, and VR = very rare.

	Argynninae	Limenitinae	Charaxinae	Nymphalinae	Satyrinae
Wet Forests (11/54 = 20%)					
C	--	0/2	--	--	--
NR	--	5/11	1/2	0/2	3/9
R	--	2/16	0/7	--	0/3
VR	--	0/2	--	--	--
Moist Forests (22/64 = 34%)					
C	0/2	7/12	0/1	--	4/4
NR	--	3/16	3/6	0/2	3/5
R	--	2/11	0/4	--	--
VR	--	0/1	--	--	--
Dry Forests (5/18 = 26%)					
C	0/1	--	--	--	1/1
NR	--	4/6	0/3	0/3	0/1
R	--	0/1	0/2	--	--
All Forests (35/68 = 51%)					
VC	--	1/3	--	2/2	3/3
C	--	16/22	6/10	3/7	1/1
NR	--	1/7	1/5	0/1	1/3
R	--	0/2	0/2	--	--
Ubiquitous & Savannah (5/25 = 20%)					
VC	--	--	--	0/1	--
C	--	0/3	1/4	0/4	1/4
NR	--	--	0/1	0/3	3/3
R	--	--	0/2	--	--
Total (78/229)	0/3	41/115	12/49	5/25	20/37

feeding species) relative to other subfamilies (36, 24 and 20% of Limenitinae, Charaxinae, and Nymphalinae fruit-feeders, respectively).

Total samples collected from each site had not reached species saturation when trapping was terminated. New species continued to accumulate as additional specimens were added (Fig. 1). Only the Owabi sample began to approach an asymptote, indicating that continued trapping at this site would

Table 2. Numbers and percent of total Satryinae and select Limenitinae trapped at each site; N=total species

	Satyrinae (N=19)	**Limenitinae**	
		Euphaedra (N=15)	*Bebearia + Euriphene* (N=19)
Bobiri	19 (100%)	12 (80%)	18 (95%)
Owabi	18 (95%)	11 (73%)	12 (63%)
Asantemanso	13 (68%)	9 (60%)	10 (53%)
Gyakye	9 (47%)	6 (40%)	4 (21%)
Bonwire	15 (80%)	8 (53%)	10 (53%)
Kajease	12 (63%)	10 (67%)	5 (26%)

have yielded few additional species. Rarefied species richness was highest in trap collections from forest reserves compared to sacred groves at all sub-sample sizes. The largest forest reserve, Bobiri Forest, had the highest species richness of any site at any given sample size.

Relative vulnerability to forest fragmentation appears non-random across species' groups, especially when considered relative to numbers trapped from Bobiri, the most speciose community (Table 2). Of the three species groups well represented in the overall sample, the *Bebearia* and *Euriphene* showed the greatest decline in numbers of species trapped relative to Satyrinae and *Euphaedra*. Owabi, for example, had 95% as many Satyrinae and 92% as many *Euphaedra* as Bobiri, but only 67% as many *Bebearia* and *Euriphene*. The relative discrepancy was even greater at certain other sites, e.g., Kajease.

4. DISCUSSION

That only slightly more than one-third of the possible fruit-feeding species were collected seems low at first glance since each forest site was surveyed 18-22 times. But species restricted to wet or dry forest and non-forest specialists should have been less commonly trapped given our surveys targeted forest remnants of the moist semi-deciduous subtype. The fact that some of these unexpected species were collected in our traps indicates that sites surveyed have elements characteristic of wet or dry forests, habitat affinities of these particular species are less defined than originally thought, and/or species were attracted from outside the forest proper. Rare species should also have been less frequent in trap samples because of their slower rate of accumulation relative to more common species. Species accumulation curves were still slowly climbing, which suggests additional rare species would have been collected had trapping continued. Puzzling absences from our trap collections concern those species viewed as frequently encountered or generally common inhabitants of Ghana's moist forests, such as certain Limenitinae or Charaxinae. Trapping protocol might explain why such species were under-collected if for example, they were only rarely attracted to our particular bait or partial to the forest canopy. In their surveys of moist-semideciduous forests of south-eastern Côte d'Ivoire, Fermon et al. (2003), for example, trapped Charaxinae more frequently in the forest canopy and open canopy forests versus the forest understory. Alternatively, those species putatively "missing" from our trap collections may be largely restricted to the moist evergreen forest subtype, rather than the moist semi-deciduous forests surveyed here, that together comprise the moist forests of Ghana. Our ongoing investigations of these communities will help resolve which species are truly expected in trap surveys that target the moist semi-deciduous forest understory.

The sensitivity we observed of certain Limenitinae to forest modification was also detected by Fermon et al. (2000) in their surveys of managed and regenerating moist semi-deciduous forests in Côte d'Ivoire. Identifying the traits that predispose species to extinction is a primary goal of conservation practitioners. Species that are intrinsically rare are generally presumed to be more vulnerable to habitat modification, and theoretical vulnerability is often used to identify at-risk taxa (Bossart and Carlton, 2002). But the large majority of seemingly vulnerable *Bebearia* and *Euriphene* revealed from our surveys are actually considered generally common members of Ghana's forest butterfly communities. As illustration, *Bebearia tentyris* was by far the most abundant species at Bobiri Reserve, accounting for a quarter of the nearly 1600 individuals trapped, but barely registered at any of the other sites. Only two of the 19 *Bebearia* or *Euriphene* species trapped are considered rare and one of these derived from a single sacred grove. This apparent sensitivity of abundant, widespread species to habitat modification is an important message that can be easy to overlook when devising conservation strategies and setting priorities.

Our rarefied estimates of species richness indicate the small sacred groves apparently have less conservation value than the larger forest reserves. Such estimates may be biased, however, if samples being compared derive from sites with markedly different species abundance distributions (Magurran, 2004), such that species accumulation curves intersect. The rarefaction curve for Bonwire, at least, is sufficiently steep to suggest that species saturation may exceed that observed for Owabi. But the generally lower observed species richness of sacred grove communities matches theoretical expectations of species-area relationships and underscores why biodiversity conservation necessitates the protection of large contiguous blocks of habitat. Such reductions in species richness and differential vulnerabilities could be a function of area per se (Brown, 1995), or a variety of other potential factors whose influence is more pronounced in smaller versus larger habitat islands, e.g., edge effects (Murcia, 1998). Regardless, the conservation significance of these small relict forests will most likely derive from their role in the broader landscape, e.g., by fostering habitat connectivity or spreading extinction risks across populations.

Many of Ghana's sacred groves are imminently threatened and formal legislation may be necessary to reinforce the community-based protective measures that have fostered their long-term persistence (Ntiamoa-Baidu, 2001). These traditional reserves harbor clues regarding changes to forest communities as the surrounding landscape matrix was converted from one of pristine forest to one where forest habitat is absent. Their study can reveal important environmental correlates of species diversity and lead to predictive models of community dynamics in fragmented landscapes and science-based management decisions. Our hope is that by drawing attention to irreplaceable indigenous reserves, others will be similarly inspired to pursue their study and conservation.

ACKNOWLEDGEMENTS

The senior author gratefully acknowledges funding support from The National Geographic Society. Comments of two anonymous reviewers improved the manuscript.

REFERENCES

Bossart, J. L., and Carlton, C. E., 2002, Insect conservation in America: status and perspectives, *Am. Entomol.* **48:**82-92.

Brown, J. H., 1995, *Macroecology*, Univ. of Chicago Press, Chicago.

Fermon, H., Waltert, M., and Mühlenberg, M., 2003, Movement and vertical stratification of fruit-feeding butterflies in a managed West African rainforest, *J. Insect Conserv.* **7:**7-19.

Fermon, H., Waltert, M., Larsen, T. B., Dall'Asta, U., and Mühlenberg, M., 2000, Effects of forest management on diversity and abundance of fruit-feeding numphalid butterflies in south-eastern Côte d'Ivoire, *J. Insect Conserv.* **4:**175-189.

Gotelli, N. J., and Colwell, R. K., 2001, Quantifying biodiversity: procedures and pitfalls in the measurement and comparison of species richness, *Ecol. Let.* **4:**379-391.

Gotelli, N. J., and Entsminger, G. L., 2003, *EcoSim: Null models software for ecology, Version 7*, http://homepages.together.net/~gentsmin/ecosim.htm.

Hall, J. B., and Swain, M. D., 1981, *Distribution and ecology of vascular plants in a tropical rain forest: forest vegetation of Ghana*, W. Junk Publishers, The Hague.

Hawthorne, W. D., and Abu-Juam, M., 1995, *Forest protection in Ghana*, IUCN/ODA, IUCN Publ., Cambridge.

Larsen, T.B., 2001, *The butterflies of Ankasa/Nini-Suhien and Bia protected area systems in western Ghana, Protected areas development programme, Western Region Ghana*, ULG Northumbrian Ltd.

Lebbie, A. R., Freudenberger, M. S., 1996, Sacred groves in Africa: forest patches in transition, in: *Forest patches in tropical landscapes*, J. Schelhas, and R. Greenberg, eds., Island Press, Washington, D.C., pp.300-324.

Magurran, A. E., 2004, *Measuring biological diversity*, Blackwell Publishing, Malden, Massachusetts.

Murcia, C., 1998, Edge effects in fragmented forests: implications for conservation, *Trends Ecol. Evol.* **10:**538-62.

Ntiamoa-Baidu, T., 2001, Indigenous versus introduced biodiversity conservation strategies: the case of protected area systems in Ghana, in: *African rain forest ecology & conservation*, W. Weber, L. J. T. White, A. Vedder, and L. Naughton-Treves, eds., Yale Univ. Press, New Haven, pp. 385-394.

Weber, W., White, L. J. T., Vedder, A., and Naughton-Treves, L., 2001, *African rain forest ecology and conservation*, Yale Univ. Press, New Haven.

BIOGEOGRAPHY, DIVERSITY AND ECOLOGY OF SAWFLIES IN THE AFROMONTANE REGION (INSECTA: SYMPHYTA)

Frank Koch
Institut für Systematische Zoologie, Museum für Naturkunde, Humboldt-Universität zu Berlin, Invalidenstrasse 43, D - 10115 Berlin, Germany, E-mail: frank.koch@museum.hu-berlin.de

Abstract: In the Afrotropical region the sawfly fauna is, in comparison with other biogeographical regions, except Australia, very poor both in number of species and number of individuals. The Afromontane region is composed of a series of archipelago-like disjunctions. This is especially true for eastern and south-eastern Africa, where these consist of regional mountain systems with specific abiotic factors, and it seems that the mountain-specific sawfly diversity, with its many endemic species, is unique and dependent on the mountain systems' genesis. Species of the *Athalia vollenhoveni*-group are known only from the Afrotropical region. The species of this group seem to prefer the montane and sub-montane region and have more narrow ranges than previously thought. The sawfly diversities of different habitats of four mountain systems in the Drakensberg district, South Africa, were investigated over a two-year period. In total 27 species were recognized, including at least 13 species new to science. Furthermore, a minimum of seven species are endemic to vegetational types of the Drakensberg. Only very few species are widely distributed and can be regarded as eurytopic, because these species occur from the lowlands up to the mountains.

Key words: Symphyta; sawflies; *Athalia;* Afrotropical region; Afromontane region; Drakensberg mountain system; East African Coastal District; distribution; ecology

1. INTRODUCTION

The suborder Symphyta is a paraphyletic assemblage comprising the structurally more “primitive” Hymenoptera (Smith, 1995). The adults of Symphyta share a number of plesiomorphic character states. For instance, there is no "wasp-like" constriction at the base of the abdomen as in ants,

B.A. Huber et al. (eds.), African Biodiversity, 159–166.

wasps and bees which are integrated into the other suborder Apocrita. The abdomen of sawflies is broadly attached to the thorax. Females possess a saw-like ovipositor which facilitates the insertion of eggs into plant tissue, hence the English common name "sawflies". The shape of the ovipositor is of taxonomic significance. The larvae of most sawfly species feed externally on plant foliage.

1.1 Afrotropical sawflies

The taxonomy of the Afrotropical sawflies is partly outdated, incomplete and in some cases incorrect. In the middle of the last century, the taxonomic contributions of Pasteels (1949, 1953) were very important for the knowledge of the distribution of the Afrotropical sawflies. He concentrated on the fauna of the Belgian Congo and the countries adjacent to its eastern border (Rwanda, Uganda). Since then, only very few species have been described.

Little is known about the actual distribution of species, because no taxonomic revisions for the whole Afrotropical region have been undertaken. Only the revision of the Athaliini by Benson (1962), containing all known Afrotropical *Athalia* species, seems to be a very valuable contribution for the identification of these species. But in the meantime it is also necessary to revise this genus.

The identification of the species of the allantine genera *Xenapates* Kirby, *Neacidiophora* Enslin, and the blennocampine genus *Trisodontophyes* Enslin are given according to the revisions by Koch (1995, 1998a, 2001).

Little or nothing is known about host plants, larval stages and phenology of the species in this region (Chevin, 1985; Koch, 2003).

Among other factors, this poor knowledge of Afrotropical sawflies can be explained by the fact that no collections specifically targeting sawflies have been undertaken in Africa.

1.2 The Afromontane region

The Afromontane region in eastern and south-eastern Africa consists of regional archipelago-like mountain systems with specific abiotic factors (White, 1978). These unique islands or biocoenosis contain a high degree of endemic species. The Afromontane flora contains at least 4.000 species, of which a remarkably high proportion, about 75 %, are endemic. Moreover, the Afromontane flora probably had a more complex origin and evolutionary history than any other flora in the world (Matthews et al., 1993).

A problem for the protection of the floristic richness is the considerable pressure on the natural ecosystems, particularly from the timber industry. In many areas extensive afforestation has destroyed most of the natural vegetation, and large tracts of land have been severely invaded by alien plants. From a conservation point of view it is necessary to document the sawfly fauna in the remaining original habitats.

Considering the high degree of plant diversity and endemic richness derived from the mountain systems' genesis, it was expected that the investigated sawfly diversities would also be rich in endemic species in the different mountain systems.

These endemic species are either derivatives of widespread Afrotropical species groups or palaeoendemics of the Oriental origin, and they are of vital significance in order to reconstruct historical evolutionary lineages and distribution patterns.

2. MATERIAL AND METHODS

For the presented results it was necessary to examine sawfly material in South African as well as European museums. These data were supplemented by personal collections made between 1992 and 2003.

Most of the material was sampled with a continually increasing number of Malaise-traps. In addition, sweeping with a net was used for selective collecting.

The field work was performed each time over a period of approximately six to eight weeks, mostly during or after the rainy season, from November to March. Several National Parks in Malawi, Zimbabwe and South Africa were visited, each for a period of approximately 5 to 8 days.

3. RESULTS AND DISCUSSION

3.1 Taxonomic and biogeographical investigations

In the examined sawfly material examples for both biogeographical and phylogenetical studies have been found. The genera *Paranetroceros* Koch and *Neoxenapates* Forsius, with one species in each genus, are known as palaeoendemics from Mount Mulanje and the surrounding mountains in Malawi. From *Paranetroceros mirabilis* (Forsius) we currently only know the holotype collected in 1913 (Koch, 1998b). Also from *Neoxenapates affinis*

Table 1. Sawfly diversity in the study areas of the Drakensberg mountain system; W: Wolkberg, S: Soutpansberg, L: Lekgalameetse, B: Blyderevier

	W	S	L	B
Argidae				
Arge laeta (Du Buysson, 1897)		X	X	X
Arge stuhlmanni (Kohl, 1893)		X		
Tenthredinidae				
Athalia concors Konow, 1908	X			
A. gessi Koch, 2003	X		X	X
A. guillarmodi Benson, 1956	X		X	
A. cf. *himantopus* Klug, 1834	X	X	X	
A. incomata Konow,1908	X	X	X	X
A. mashonensis Enslin, 1911	X		X	X
A. ustipennis Mocsáry, 1909	X	X	X	
Athalia spec. nov. 1	X	X	X	X
Athalia spec. nov. 2		X		
Athalia spec. nov. 3	X			
Athalia spec. nov. 4	X			X
Athalia spec. nov. 5	X			
Athalia spec. nov. 6	X			
Athalia spec. nov. 7	X			
Athalia spec. nov. 8	X			
Athalia spec. nov. 9	X			
Athalia spec. nov. 10	X			
Neacidiophora brevicornis Pasteels, 1954	X		X	X
Xenapates similis Benson, 1939			X	X
Distega spec. nov. 1	X		X	
Distega spec. nov. 2	X	X		
Distega spec. nov. 3			X	
Trisodontophyes afra Konow, 1907			X	X
T. triplicata Forsius, 1934			X	
Dulophanes natalensis Forsius, 1931		X	X	
Σ	19	9	16	9

(Forsius) eleven specimens are reported from the beginning of the last century (Koch, 1996). A personal field trip to Mount Mulanje in 2001 with the aim of recording these species, proved unsuccessful. A reason for the absence of these species could be the extensive destruction of the primary forest and the later reforestation with *Pinus* on large tracts of land including the mountain top.

A similar example for species with one or very few records is *Trisodontophyes afra* (Konow). In contrast to the other two species, this one belongs to a widely distributed genus in the Afrotropical region (Koch, 2001), and was collected on Mount Kilimanjaro in 1907. For almost 100 years this species has not been rediscovered since its original description from the place of origin.

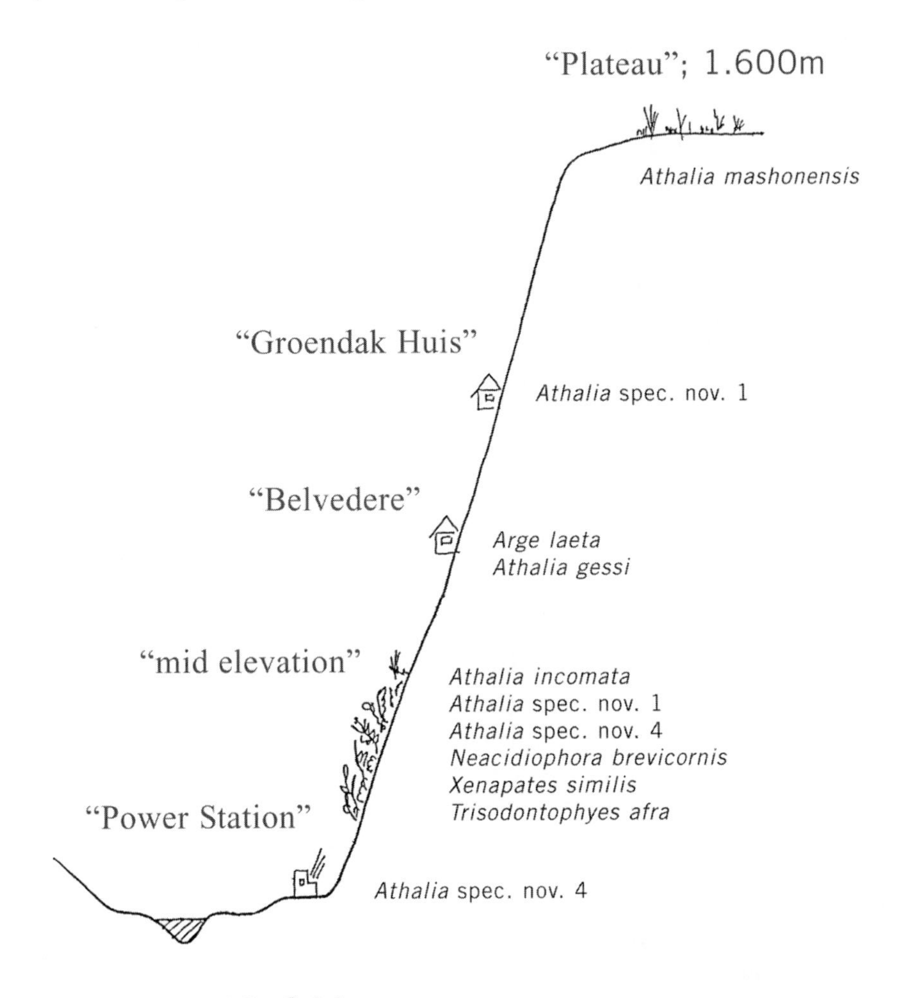

Figure 1. Schematic cross-section through the side canyon at Belvedere in the Blyderevierspoort Canyon Nature Reserve with the species collected in the different investigated habitats.

During field work in 2001 this species was found on Mount Mulanje and in the Lekgalameetse Nature Reserve in the Limpopo Province as well as in the Blyderevierspoort Canyon in the Mpumalanga Province of South Africa. From a biogeographical point of view this species is endemic in the mountains but has a disjunct distribution.

Species of the genus *Athalia* Leach occur in the Palaearctic, Oriental and Ethiopian region, including the Madagascan subregion. According to Benson (1962) for the Ethiopian region, including the Madagascan subregion, six species-groups with 32 species and three subspecies are known. From current data, the number of expected species of this region is significantly higher than before.

One of these Afrotropical species-groups, which is named after *Athalia vollenhoveni* Gribodo has been revised (Koch, in press), and contains ten species. These species are distributed only in the Sahelian, East African, and South African subregions, and seem to prefer the Afrotropical mountain systems. Several species have smaller ranges than previously supposed. For instance, Benson (1962) stated that *A. vollenhoveni* is known from Ethiopia, Kenya, Uganda and Tanzania [Tanganyika]. However, this species is distributed only in Ethiopia and in the Arabian Peninsula (Yemen). The reduction of the ranges is substantiated by misidentifications (Koch, in press). Some species inhabit very small, isolated ranges such as *A. atripennis* Benson from Mount Elgon in Kenya. Specimens of *A. atripennis* were additionally reported by Benson (1962) from the Aberdare Range in Kenya, but the examined specimens were misidentifications too. These examples show the necessity of taxonomic revisions as a basic tool for ecological conclusions as well as biogeographical and phylogenetic hypotheses.

3.2 Investigations in the Drakensberg mountain system

White (1978) has grouped the islands of the Afromontane archipelago into seven regional mountain systems. The southernmost is the Drakensberg system, stretching from the Soutpansberg to Knysna in the Eastern Cape Province. The study areas of the continuous field work belong to the north-eastern Transvaal Escarpment. The investigations were concerned specifically with the following areas: Wolkberg-area, Lekgalameetse Nature Reserve, and Blyderevierspoort Canyon in the Drakensberge and Mount Lajuma in the Soutpansberg.

Up to now, 27 species have been recorded, including at least 13 species, that are new to science (Table 1). All new species are revised and will be described and named in forthcoming revisions. Furthermore, a minimum of seven species of *Athalia* are endemic to vegetational types of the Drakensberg. Two species, *Athalia* spec. nov. 1 and *Athalia* spec. nov. 3, are described from the Drakensberg system, but they show disjunctions to the Chimanimani system in the Eastern Highlands of Zimbabwe. Only *Athalia incomata* Konow and *Arge stuhlmanni* (Kohl) are widely distributed and may be called eurytopic, because these species occur from the lowlands up to the mountains.

Remarkable for its disjunct distribution is *Athalia* spec. nov. 2. This species was discovered in the isolated areas of Mount Mulanje and Soutpansberg and at the south-eastern coast of KwaZula-Natal by the Indian Ocean. This pattern of distribution is likely a result of climatic changes during the Pleistocene. The whole area of its distribution is part of the East African Coastal District.

In the Blyderevierspoort Canyon, sawfly diversity has been investigated in relation to altitude, and in different habitats. The cross-section through the side canyon shows the habitats with the historical power station at the bottom beside the tributary (Fig. 1).

The vegetation behind the power station is relatively poor, and only one *Athalia* species was collected. The habitat "mid elevation" contains many plant species, and correspondingly the sawfly diversity found with six species was distinctly higher than in the other habitats. The so-called "Belvedere"-place resembles a garden with a cultivated lawn and ornamental plants. The "Groendak Huis" is located in a habitat dominated by fern. In both habitats, the sawfly fauna with a total of three species seems to be very poor (Fig. 1). On the "plateau" it is hot and dry, and these microclimatic conditions are suboptimal for the life cycle of sawflies.

Nevertheless, it is worth mentioning that *Lippia javanica* (Burm.f.) Spreng. (Verbenaceae) is distributed on the "plateau", and often it is possible to sample specimens of *Athalia mashonensis* on the flowers of these shrubs while they are eating pollen (Koch, 2003). So far the record of *A. mashonensis* was expected and has now been confirmed.

In summary it can be said for the genus *Athalia* with its 17 reported species, that in contrast to the other sawfly genera, it seems that the southern part of the Afromontane region is a "hot spot" of biodiversity.

ACKNOWLEDGEMENTS

The author thanks C. D. Eardley, Plant Protection Research Institute, Pretoria, I. G. and R. Gaigher, Lajuma Research and Environmental Education Centre, Soutpansberg, for their support in South Africa. Sincere thanks go to D. R. Smith, Washington, and my colleague, J. Dunlop, Berlin for critical review of this manuscript. The German Research Foundation (DFG) is gratefully acknowledged for providing a research grant (445 SUA-113/8/1-1 and 445 SUA-113/8/0-2). Additionally, this study was founded by the National Research Foundation, South Africa.

REFERENCES

Benson, R. B., 1962, A revision of the Athaliini (Hymenoptera: Tenthredinidae), *Bull. Br. Mus. (Nat. Hist.) Ent.* **11**:335-382.

Chevin, H., 1985, Contribution à la faune entomologique du Burundi I. Hymenoptera Symphyta, *Nouv. Revue Ent.* **2**:197-208.

Koch, F., 1995, Die Symphyta der Äthiopischen Region. 1. Gattung: *Xenapates* Kirby, 1882 (Hymenoptera, Tenthredinidae, Allantinae), *Deutsche entomol. Z., N.F.* **42**:369-437.

Koch, F., 1996, Die Symphyta der Äthiopischen Region (Insecta, Hymenoptera): Die Gattung *Neoxenapates* Forsius, 1934 (Tenthredinidae, Allantinae), *Deutsche entomologische Zeitschrift, Neue Folge* **43:**123-128.

Koch, F., 1998 a, Die Symphyta der Äthiopischen Region. Gattung *Neacidiophora* Enslin, 1911 (Insecta: Hymenoptera: Tenthredinidae: Allantinae), *Entomologische Abhandlungen Staatliches Museum für Tierkunde Dresden* **58:**83-118.

Koch, F., 1998 b, Die Symphyta der Äthiopischen Region. Neue Gattungen und Arten aus der Verwandtschaft der *Neacidiophora*- und *Netroceros*-Gruppe (Hymenoptera: Symphyta: Tenthredinidae: Allantinae), *Entomological Problems* **29:**1-17.

Koch, F., 2001, Die Symphyta der Äthiopischen Region. Gattung *Trisodontophyes* Enslin, 1911 (Insecta: Hymenoptera: Tenthredinidae: Blennocampinae), *Entomologische Abhandlungen Staatliches Museum für Tierkunde Dresden* **59:**261-301.

Koch, F., 2003, A new species of *Athalia* Leach (Hymenoptera: Symphyta: Tenthredinidae) from southern Africa, with notes on phenology, *Cimbebasia* **18:**19-30.

Koch, F., in press, The Symphyta of the Ethiopian Region. Genus *Athalia* Leach, 1817 - *Athalia vollenhoveni*-group (Hymenoptera: Symphyta: Tenthredinidae), in: *Sawfly Research: Synthesis and Prospects*, A. Taeger and S. M. Blank, eds., Verlag Goecke and Evers, Keltern.

Matthews, W. S., van Wyk, A. E. and Bredenkamp, G. J., 1993, Endemic flora of the north-eastern Transvaal Escarpment, South Africa, *Biological Conservation* **63:**83-94.

Pasteels, J., 1949, Tenthredinidae (Hymenoptera, Tenthredinoidea), *Exploration du Parc National Albert, Mission G. F. de Witte (1933-1935). Bruxelles* **60:**1-105.

Pasteels, J., 1953, Prodromes d'une fauna des Tenthredinoidea (Hymenoptera) de l'Afrique noir. I. Argidae, *Mém. Soc. Entomologique de Belgique* **26:**1-128.

Smith, D. R., 1995, The sawflies and woodwasps, in: *The Hymenoptera of Costa Rica*, P. E. Hanson and I. D. Gauld, eds., Oxford University Press, Oxford, New York, Tokyo, pp. 157-177.

White, F. (1978): The Afromontane Region, in: *Biogeography and ecology of southern Africa*, M. J. A. Werger, ed., Junk Publishers, The Hague 1, pp. 463-513.

DIVERSITY, DISTRIBUTION PATTERNS AND ENDEMISM OF SOUTHERN AFRICAN BEES (HYMENOPTERA: APOIDEA)

Michael Kuhlmann
Westfälische Wilhelms-University Münster, Institute of Landscape Ecology, Robert-Koch-Str. 26, D-48149 Münster, Germany; E-mail: kuhlmmi@uni-muenster.de

Abstract: An analysis of southern African bee diversity patterns based on 420 species reveals a bipolar pattern with highest species diversity located in the arid west of South Africa and the relatively moist east of the country. A detailed investigation of the distributions of 59 *Colletes* species shows a congruence of distribution areas of most species with seasonality of precipitation, a pattern that can be assumed for the majority of bee species. A total of 32 (sub)genera with about 256 species is strictly endemic to southern Africa with an exceptionally high number of endemic species (95%) strictly confined to the winter rainfall area.

Key words: Southern Africa; bees; Apoidea; *Colletes;* biodiversity; biogeography; endemism

1. INTRODUCTION

With an estimated 30,000 species worldwide, bees are the most important pollinators of angiosperms (Corbet et al., 1991; Proctor et al., 1996). Pollinators are essential for conservation of biodiversity in general and for food security because the majority of crop plants are pollinated by insects, especially by bees (API, 2003). Southern Africa and especially the Cape floristic region are renowned as a centre of phytodiversity of global importance but the bee fauna and pollinator-flower relationships are poorly investigated (Whitehead et al., 1987). Eardley (1989) found that at least 1600 species or one half of all bees described from subsaharan Africa occur in southern Africa, resulting in high species diversity similar to other known

B.A. Huber et al. (eds.), African Biodiversity, 167–172.

centres of bee diversity of the same size. However, detailed biogeographic information is largely lacking. In the present work a preliminary analysis of patterns of bee diversity, distribution and endemism in southern Africa is presented.

2. MATERIALS AND METHODS

For a general analysis of diversity patterns in southern Africa, published and personal distribution records of 420 bee species of 50 genera were mapped on a 2° x 2° grid. For *Colletes*, the largest investigated genus, area maps were created for species with at least three records and species showing similar distributions are joined together in one map. The resulting patterns were compared with climatic factors (Schulze, 1997), phytogeography / -diversity, biomes and vegetation (Cowling et al., 1997).

3. RESULTS AND DISCUSSION

3.1 Diversity patterns

On a global scale phytodiversity as well as bee diversity show an uneven distribution with plants best represented in moist tropical and subtropical regions (Barthlott et al., 1999). The centres of bee diversity are in xeric parts of the world with preference for mediterranean climates (Michener, 1979, 2000). The arid and semiarid west of South Africa is the only place on earth where a centre of bee diversity coincides with a phytodiversity hotspot (Capensis). This is interesting because many bee species are specialised (oligolectic) flower visitors suggesting that coevolutionary processes might be involved in speciation of both bees and flowers. In southern Africa the distribution of bee species diversity (Fig. 1) suggests a bipolar pattern with highest diversity located in the arid west of South Africa and the relatively moist east of the country. This pattern is influenced by the lack of investigations in the central and northern parts of South Africa as well as in large parts of Namibia, Zimbabwe and especially Botswana and Mozambique. Nevertheless, Eardley (1989) pointed out that the observed pattern very probably reflects the real situation.

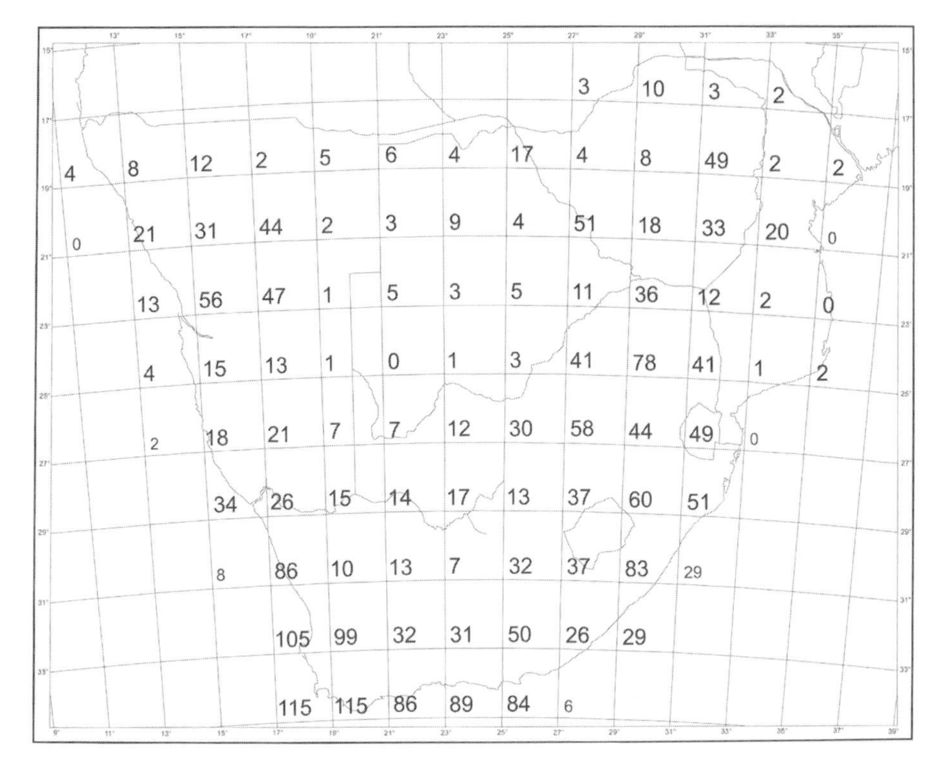

Figure 1. Number of bee species in southern Africa recorded on a 2°x 2° grid based on 420 species of 50 genera.

3.2 Distribution patterns

The species diversity pattern of the genus *Colletes* is largely identical to the one found for all investigated genera (Fig. 1). Thus, *Colletes* is used as a model to analyse distributions in more detail. For 59 of the 101 *Colletes* species known from southern Africa, a sufficient number of records were available to create distribution maps. The distribution of most species show in principle a congruence with seasonality of precipitation. The majority of *Colletes* (34 species, 58%) are limited to the winter rainfall area, some of them partly invading the southern Cape region with rain all year (Fig. 2). Four more widely distributed species of this group have their emphasis in the winter rainfall area or show disjunct distributions. Fourteen species (24%) are basically restricted to the early to mid summer rain area, partly invading adjacent areas and the southern Cape region with rain all year (Fig. 3). For the remaining 11 species a preference is not clearly recognisable. Six of them (10%) seem to have a preference for the late to very late summer rainfall area and the remaining five species (8%) are wide spread on the

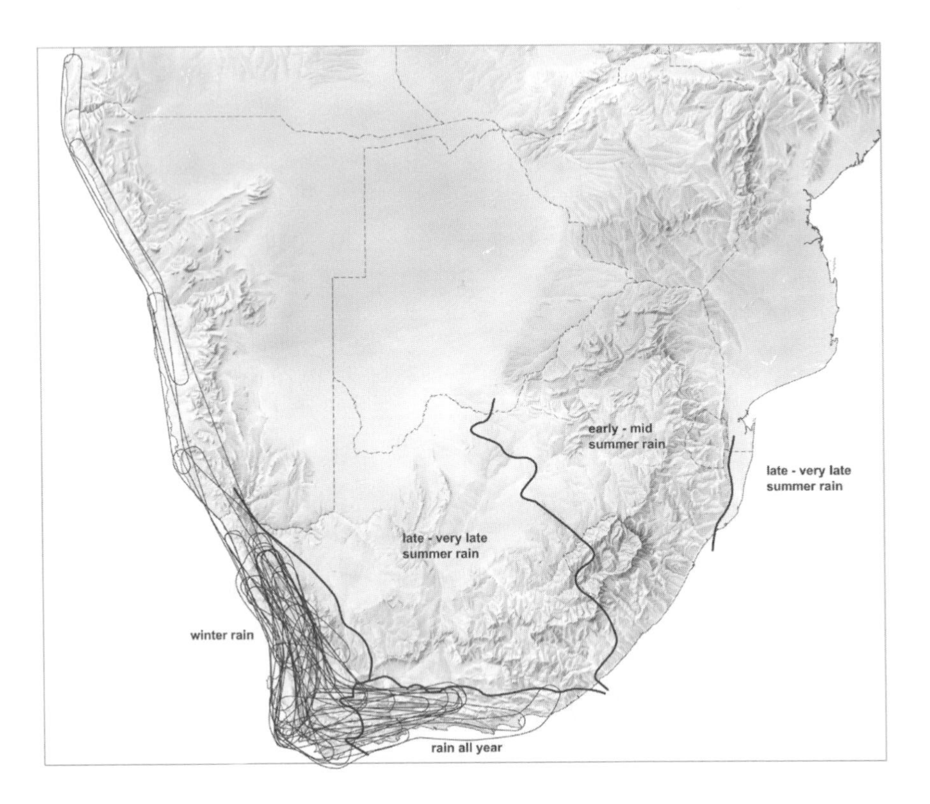

Figure 2. Distribution areas of 30 *Colletes* species that are strictly confined to the winter rainfall area and partly invade the southern Cape region with rain all year (climate data after Schulze, 1997).

subcontinent. Interestingly, the southern Cape region with rain all year does not have a specific fauna. No coincidence of bee distribution patterns is obvious for biome types, vegetation, phytogeographical regions or centres of phytodiversity. Patterns like the ones shown for *Colletes* can be assumed for most bee species in southern Africa.

3.3 Endemism

Of the 1600 bee species that occur in southern Africa about 85% are endemic (Eardley, 1989). Based on Michener (2000) an analysis of distribution patterns at generic level revealed that 32 (sub)genera with a total of 256 described species are strictly confined to this region. Especially the xeric winter rainfall area in western South Africa is a centre of endemism. About 95% of the bees that occur here are endemics (Eardley, 1989). Among them are some basal taxa of different families like the genera *Scrapter*

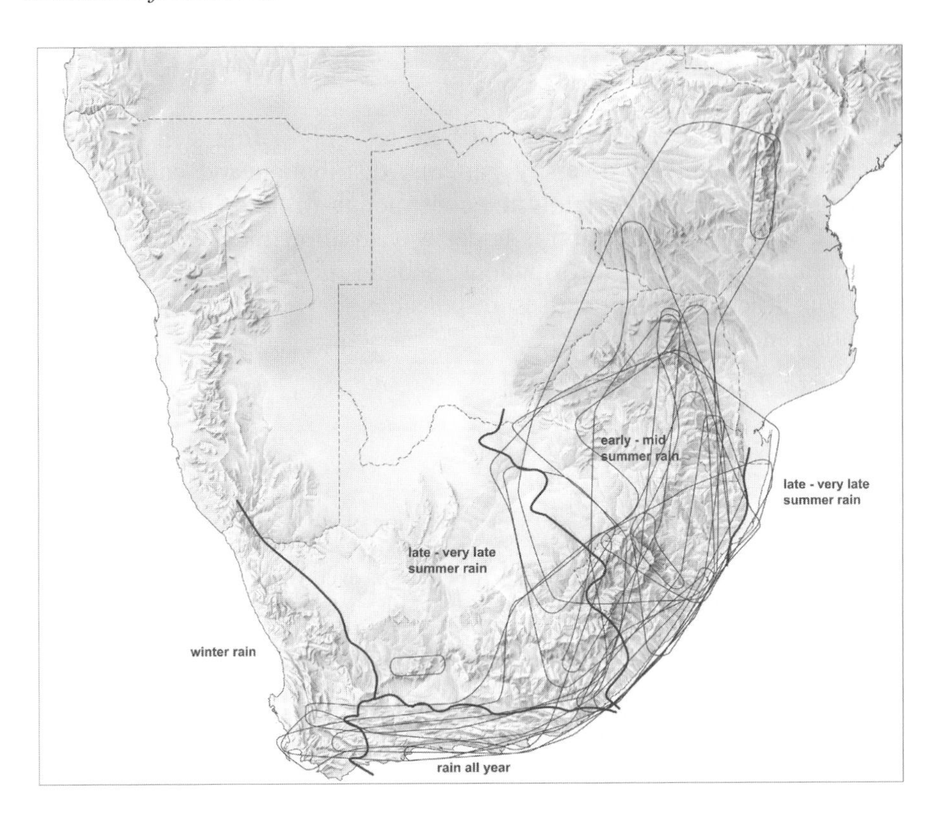

Figure 3. Distribution areas of 14 *Colletes* species that are basically limited to the early to mid summer rain area and partly invade the southern Cape region with rain all year (climate data after Schulze, 1997).

(Colletidae) (Eardley, 1996), *Fidelia* and *Fideliopsis* (Engel, 2002), *Afroheriades* (Peters, 1978) and *Aspidosmia* (Peters, 1972) (Megachilidae) that are of special interest for the understanding of bee phylogeny and evolution. Intensive faunistic investigations in selected parts of this region reveal a high percentage of undescribed species even in recently revised genera that further emphasise its importance. The high degree of endemicity in the winter rainfall area is presumably linked to the fact that it is largely isolated by relatively moist regions to both the north and the east (Eardley, 1989). The moister eastern summer rainfall area of the subcontinent is not distinctly isolated from tropical Africa, thus, only about 75% of the bee species that occur here are endemic to southern Africa. Many bees of this region extend far north (e.g. East Africa) and often reach the southern extreme of their distribution in the Durban area (South Africa) (Eardley, 1989).

3.4 Future perspectives

The preliminary results presented here are only a first step towards a better understanding of diversity patterns, distribution and endemism of southern African bees. Currently the construction of the "Southern African Bee Data Base" (SouthABees) is under way to allow more comprehensive biogeographic study of distributions and to get further evidence for underlying factors and processes responsible for biogeographic patterns, speciation and endemicity.

REFERENCES

API (African Pollinator Initiative), 2003, Plan of action of the African Pollinator Initiative, African Pollinator Initiative Sekretariat, Nairobi.

Barthlott, W., Biedinger, N., Braun, G., Feig, F., Kier, G. and Mutke, J., 1999, Terminological and methodological aspects of the mapping and analysis of global biodiversity, *Acta Bot. Fenn.* **162:**103-110.

Corbet, S. A., Williams, I. H. and Osborne, J. L., 1991, Bees and the pollination of crops and wild flowers in the European Community, *Bee World* **72:**47-59.

Cowling, R. M., Richardson, D. M. and Pierce, S. M., 1997, *Vegetation of southern Africa*, Cambridge University Press, Cambridge.

Eardley, C. D., 1989, Diversity and endemism of southern African bees, *Bull. Pl. Prot. Res. Inst.* **18:**1-2.

Eardley, C. D., 1996, The genus *Scrapter* Lepeletier & Serville (Hymenoptera: Colletidae), *Afr. Ent.* **4:**37-92.

Engel, M. S., 2002, Phylogeny of the bee tribe Fideliini (Hymenoptera: Megachilidae), with the description of a new genus from southern Africa, *Afr. Ent.* **10:**305-313.

Michener, C. D., 1979, Biogeography of the bees, *Annls. Miss. Bot. Gard.* **66:**277-347.

Michener, C. D., 2000, *The bees of the world*, Johns Hopkins University Press, Baltimore.

Peters, D. S., 1972, Über die Stellung von *Aspidosmia* Brauns 1926 nebst allgemeinen Erörterungen der phylogenetischen Systematik der Megachilidae (Insecta, Hymenoptera, Apoidea), *Apidologie* **3:** 167-186.

Peters, D. S., 1978, *Archeriades* gen. n., eine verhältnismäßig ursprüngliche Gattung der Megachilidae (Hymenoptera: Apoidea), *Ent. Germ.* **4:**337-343.

Proctor, M., Yeo, P. and Lack, A., 1996, *The natural history of pollination*, Timber Press, Portland.

Schulze, R. E., 1997, *South African atlas of agrohydrology and –climatology*, Water Research Commission, Pretoria.

Whitehead, V. B., Giliomee, J. H. and Rebelo, A. G., 1987, Insect pollination in the Cape flora, *S. Afr. Scient. Progr. Rep.* **141:**52-82.

DOES GRAZING INFLUENCE BEE DIVERSITY?

Carolin Mayer
Biocentre Klein Flottbek and Botanical Garden, University of Hamburg, Ohnhorststr. 18, 22609 Hamburg, Germany; E-mail: cmayer@botanik.uni-hamburg.de

Abstract: In Namaqualand, the north-western part of the Succulent Karoo of South Africa, a study was conducted to investigate the influence of livestock grazing on the abundance and diversity of bees (superfamily Apoidea). Bees were collected on adjacent rangeland sites which are characterized by a significant fence line contrast, one site showing effects of heavy grazing. Application of different sampling methods (Malaise and colour plate trapping) reveal different results, indicating that methodological influences are significant. Colour traps, in particular, may provide poor estimates of bee abundance due to their apparent sensitivity to competition from surrounding flowers for insect attraction.

Key words: Apoidea; pollinators; biodiversity; grazing; arid rangelands; Succulent Karoo

1. INTRODUCTION

The majority of the rural people living in Namaqualand, the north-western part of the Succulent Karoo of South Africa, subsist on domestic livestock (Hoffman et al., 1999). Particularly in communally managed areas with open-access systems and no formal property rights, grazing by livestock is considered a threat to biodiversity (Davis and Heywood, 1994). Especially affected are the abundance and diversity of the perennial vegetation (Todd and Hoffman, 1999). Besides other factors, the reduction of phytodiversity could lead to a decrease in the associated pollinating insect fauna, which in turn could deteriorate plant recruitment. My own studies revealed lower fruit set for succulent shrubs growing on heavily grazed land compared to plants growing on an adjacent slightly grazed farm (Mayer, 2004). Bees (superfamily Apoidea) represent one of the most important pollinator groups

B.A. Huber et al. (eds.), African Biodiversity, 173–179.

in the Succulent Karoo (Struck, 1994). This article focuses on the hypothesis that bee diversity is limited on heavily grazed farmland.

2. MATERIALS AND METHODS

The present study is part of an ongoing PhD thesis within the southern Africa botanical subproject (S06) of *Biota* (Biodiversity Monitoring Transect Analysis in Africa).

The study sites are situated in Namaqualand, an arid winter rainfall area which belongs to the Succulent Karoo Biome (Rutherford and Westfall, 1986). Close to Leliefontein (Kamiesberg), *Biota* established a pair of standardized 1km^2 Biodiversity Observatories. One Observatory is situated on the commercial farm Remhoogte (No 25, NW-corner of the km^2: S:30,935; E:18,276) while the second one was set up on the communal land of Paulshoek (No 24, NW-corner: S:30,386; E:18,276). Between the adjacent sites a fence-line contrast is apparent. The communal land shows signs of degradation due to heavy grazing by goats, sheep and donkeys. Plant cover is reduced and an unpalatable shrub, *Galenia africana* (Aizoaceae), has become dominant (Todd and Hoffman, 1999).

The presented results originate from field trips from August to October in 2001 to 2003. In all three years of study, two Malaise traps (MT) were set up on either side of the fence-line at 100 and 500m for the flowering season. In 2002 and 2003, every 2nd to 4th day (14 times), five sets of water-filled colour plates (CT) (white and yellow) were placed at different distances from the fence and left for the day. For the results presented here, data from all distances were pooled for each study site. Specimens have been determined to species level except bees collected in 2003 that have been differentiated into morphospecies.

Diversity was calculated with the Brillouin index, HB, which is similar to the Shannon index but should be used when species were recorded selectively with CT (Mühlenberg, 1989).

For further comparison, Evenness (E) was calculated, and Sorensen's quotient (QS) was used to estimate similarity between the sites (Mühlenberg, 1989).

Plant cover as well as average plant height was estimated on 10x10m plots around the CT-sets (i.e. five plots per distance). Within these plots, inflorescence numbers per plant species were counted for every CT-sampling day. Display area (DA) of inflorescences was calculated from $r_{inflorescence}^2 \times \pi = DA$.

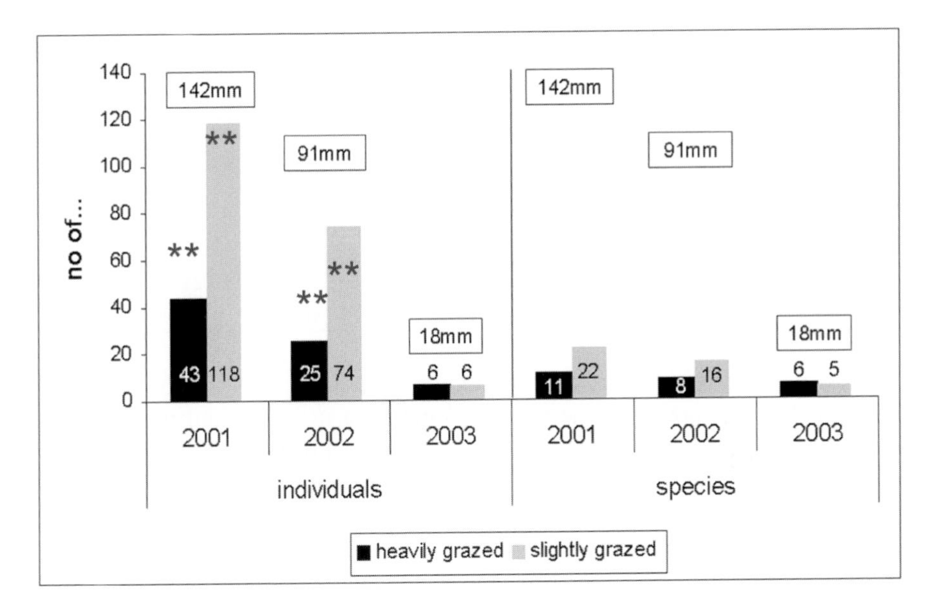

Figure 1. Malaise traps – number of individuals (left) and species (right) of bees caught in subsequent years. Rainfall [mm] from January to July is indicated in shaded fields (not to scale). Significant differences are indicated with **.

3. RESULTS AND DISCUSSION

Bee abundance, species richness (Fig. 1) and diversity (Table 1), as estimated by MTs, were higher on the slightly grazed site in 2001 and 2002, supporting the hypotheses of depauperated bee diversity on overgrazed land. A significant departure of the observed frequencies from those expected from a homogeneous distribution could be demonstrated for individual numbers of bees (χ^2=34,9; df=1 and χ^2=24,8; df=1; p<0,01). The number of bee individuals is significantly influenced by the grazing regime as well as the year in which they were caught (two-way-ANOVA: F=177,62; df=1,1 for year, F=765,57; df=1,1 for grazing regime at p<0,05). However, in 2003, the number of bee individuals and species respectively dropped by about 90% to only 12 specimens in total, with abundance, richness and diversity being virtually equal (Fig. 1 and Table 1). Year 2003 was extremely dry with only 18 mm of precipitation until August (compared to 142 mm and 91 mm in 2001 and 2002, data provided by M.T. Hoffman). The drought postponed the main flowering season for about four weeks to late September and could explain poor insect activity. The observed similarity between the heavily (hgr) and slightly (sgr) grazed site could therefore be neglected as irrelevant due to an exceptional year with insufficient bee abundance.

Table 1. Bee diversity (HB= Brillouin's diversity, E=Evenness) compared on differently grazed sites (hgr=heavily, sgr=slightly grazed) with different trapping methods (MT=Malaise Trap, CT=Colour Trap) in subsequent years. QS=Sorensen's index of community similarity.

	HB		E		QS[%]
	hgr	sgr	hgr	sgr	
MT 2001	0,85	1,72	0,42	0,62	24,24
MT 2002	1,02	1,61	0,60	0,66	25,00
MT 2003	1,10	0,96	1,00	0,94	0,00
CT 2002	2,86	2,50	0,76	0,71	55,77
CT 2003	3,13	3,15	0,74	0,74	73,99

However, looking at results obtained with CTs, the whole picture is inverted. Most amazing is the threefold increase in individual numbers and doubling of species in the dry year 2003 (Fig. 2). Furthermore, in both years more bee specimens (χ^2=15,1; df=1; and χ^2=11,56; df=1; p<0,01) and in 2002 also species were caught on the heavily grazed site, though diversity is only slightly higher or equal (Table 1).

Extremely low similarity between the bee assemblages obtained with different methods may account for these contrasting results (Table 2). Obviously, MTs catch a different set of species to CTs, although one must consider the higher sampling effort conducted with CTs.

Like the decrease of bee individuals and species in MT in 2003, the increased capture of bees in CTs is supposed to be a consequence of low

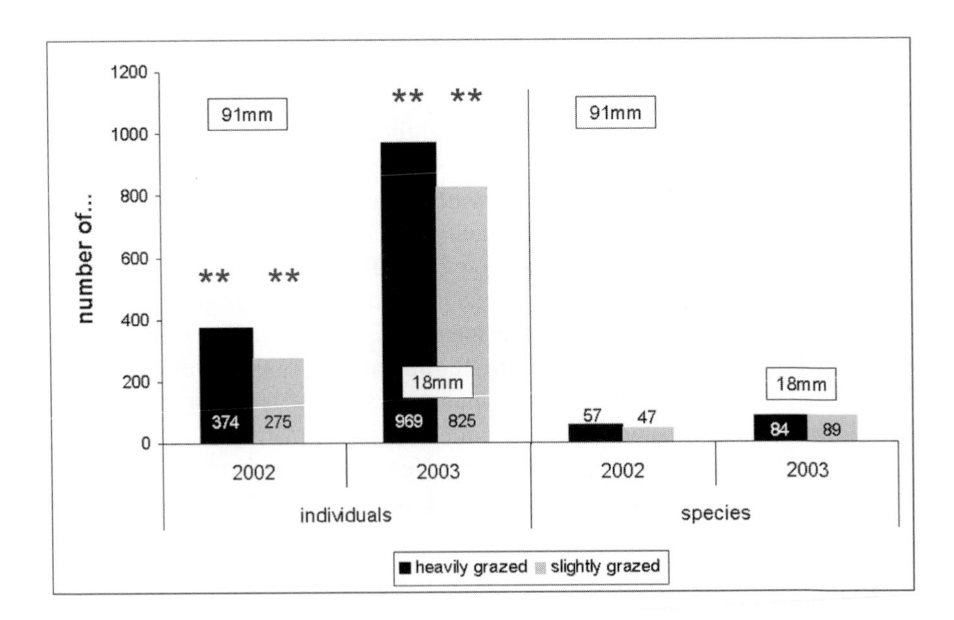

Figure 2. Colour traps – number of individuals (left) and species (right) of bees caught in subsequent years. Rainfall [mm] from January to July is indicated in shaded fields (not to scale). Significant differences are indicated with **.

Table 2. Community similarity indicated with Sorensen's quotient (QS) between bee catches with Malaise traps (MT) and colour traps (CT) for each site (hgr=heavily, sgr=slightly grazed)

	common species	QS[%]
CT & MT 2002 hgr	4	12,31
CT & MT 2002 sgr	6	19,05
CT & MT 2003 hgr	3	5,77
CT & MT 2003 sgr	5	9,71

rainfall and hence late flowering. The attraction of CTs in a flower poor environment is apparently very high. Insect numbers in CTs dropped immediately with the onset of flowering. Number of bee species caught in CTs was negatively correlated with flower display area (Spearman's rank correlation: hgr: R^2=0,48; p<0,001; sgr: R^2=0,29; p<0,001; n=36 each). It seems therefore that CTs compete with flowering plants for the attraction of insects. Additionally conducted insect counts during transect walks on the other hand showed significant increase of insect numbers with increasing flower display.

Insect attraction to CTs by sight also seems to be the decisive factor for contrary results to MTs concerning influence of grazing. On the heavily grazed site, CTs could much easier be detected with plant cover and height being significantly lower (Mann-Whitney-U: Z= -3,01; p<0,003 and Z= -3,26; p<0,002; n=15 each). Flower display area was significantly smaller as well (Mann-Whitney-U: Z= -3,15; p<0,002; n=36). Therefore CTs had a larger catchment area and a greater appeal for flower searching insects explaining higher individual and species numbers of bees caught. But even if this richer bee fauna on the heavily grazed site is an artefact of methodology, it still indicates that there is at least no general lack of potential pollinators. For single plant species though less flower visits as well as visitor species were observed under intensive grazing (Mayer, unpublished data).

Concerning community similarity (Table 1), QS signals that the two sites support different communities according to a threshold lower than 75% (Cowling et al., 1989). On the other hand, both sites share three of the four most common species (adding up >50% of individuals) and differences in species composition are caused by subrecedent species (<1% of individuals). At the moment no good indicator species amongst bees seem to occur, unlike in other pollinator groups, e.g. monkey beetles (Hopliini) (Colville et al., 2002). The higher similarity in 2003 is again assumed to be a consequence of low rainfall, food shortages and therefore greater foraging distances.

4. CONCLUSION

Though MT-results support the hypothesis of depauperated bee assemblages on degraded rangeland, CT-results lead one to assume the opposite. Thus, no firm conclusion can be reached. Recently, it has been argued to carry out studies on pollinator disruption preferably at community level, not on single species (Harris and Johnson, 2004). However, the presented results show that interpretation might be difficult with different effects levelling each other out. Furthermore, it is advisable to consider influences of the methods applied.

ACKNOWLEDGEMENTS

I would like to thank Prof Dr N. Jürgens (Univ Hamburg) and Prof Dr M.T. Hoffman (Univ Cape Town) for supervision. Bee identification was kindly done by Dr M. Kuhlmann, who also added comments on this paper, and K. Timmermann (Univ Münster). Thanks to G. Pufal, C. Rickert, N. Dreber and J. Brüggemann for field assistance and insect mounting. I also thank V. Claasen for accommodation as well as the community of Paulshoek for their hospitality. Northern Cape Nature Conservation Services gratefully provided research permits. This project is imbedded in *Biota* Africa (www.biota-africa.org) which is sponsored by the German Federal Ministry of Education and Research (*Bmbf* 01LC0024).

REFERENCES

Colville, J., Picker, M. D., et al., 2002, Species turnover of monkey beetles (Scarabeidae: Hopliini) along environmental and disturbance gradients in the Namaqualand region of the succulent Karoo, South Africa, *Biodivers. Conserv.* **11**:243-264.

Cowling, R. M., Gibbs Russell G. E., et al., 1989, Patterns of plant species diversity in southern Africa, in: *Biotic diversity in southern Africa: Concepts and conservation*, B. J. Huntley, ed., Cape Town, Oxford Univ.Press, pp.19-50.

Davis, S. D., and Heywood, V. H., 1994, *Centres of plant diversity: a guide and strategy for their conservation*. Oxford, Oxford Univ. Press.

Harris, L. F., and Johnson, S. D., 2004, The consequences of habitat fragmentation for plant-pollinator mutualisms, *Insect Sci. Appl.* **24(1)**:29-43.

Hoffman, M. T., Cousins B., et al., 1999, History and contemporary land use and the desertification of the Karoo, in: *The Karoo. Ecological patterns and processes*. W. R. J. Dean and S. J. Milton, eds., Cambridge, Cambridge University Press, pp. 257-273.

Mayer, C., 2004, Pollination services under different grazing intensities, *Insect Sci. Appl.* **24(1)**:95-103.

Mühlenberg, M., 1989, *Freilandökologie*. Heidelberg, Quelle & Meyer.

Rutherford, M. C., and Westfall, R. H., 1986, Biomes of Southern Africa - An objective categorization, *Mem. Bot. Surv. S. Afr.* **54:**1-98.
Struck, M., 1994, A check-list of flower visiting insects and their host plants of the Goegab Nature Reserve, Northwestern Cape, South Africa, *Bontebok* **9:**11-21.
Todd, S. W., and Hoffman, M. T., 1999, A fence-line contrast reveals effects of heavy grazing on plant diversity and community composition in Namaqualand, South Africa, *Plant Ecology* **142:**169-178.

THE PHOLCID SPIDERS OF AFRICA (ARANEAE: PHOLCIDAE): STATE OF KNOWLEDGE AND DIRECTIONS FOR FUTURE RESEARCH

Bernhard A. Huber
Zoological Research Institute and Museum Alexander Koenig, Adenauerallee 160, 53113 Bonn, Germany, E-mail: b.huber.zfmk@uni-bonn.de

Abstract: Pholcids are among the dominant web-building spiders in many tropical and subtropical areas of the World. They occupy a wide variety of habitats ranging from leaf-litter to tree canopies, and several species occur in caves and in close proximity to humans. This chapter summarizes the present state of knowledge about pholcids in Africa. A total of 17 genera and 162 nominal species are currently described from Africa. High levels of diversity and endemism are recorded for Madagascar, Eastern Africa, and Southern Africa. A brief overview is given on those genera that have not been worked on for several decades and that are most in need of revision.

Key words: Pholcidae; Araneae; Africa; biogeography; diversity; *Smeringopus*; *Pholcus*; *Smeringopina*; *Crossopriza*; *Leptopholcus*

1. INTRODUCTION

1.1 Abundance and diversity of African pholcids

A recent inventory of spiders in the East African Uzungwa Mountain forests (Sørensen, 2003; Sørensen et al., 2002) found pholcids to be the most abundant spider family. In a sample of 14,329 adult specimens, there were 4,319 pholcids, followed by linyphiids (2,025) and theridiids (1,338) (L. Sørensen, Copenhagen, pers. commun.), and pholcids included the first and second most abundant species in the area sampled. At the same time,

B.A. Huber et al. (eds.), African Biodiversity, 181–186.

pholcids are one of the most diverse spider families. The fact that most of this diversity is concentrated in tropical and subtropical regions has historically biased species numbers (Huber, 2003b), but recent revisions have dramatically changed our view. For example, revisions of African taxa previously assigned to *Spermophora* Hentz have increased species numbers more than 5-fold (Huber, 2003a, b, c). At the same time, circumstantial evidence (patchiness of collection sites, low numbers of shared species among sites) suggests that only a small percentage of species actually occurring has been collected. It has been estimated that the present number of about 800 species worldwide may represent no more than maybe 10% of the actual diversity (Huber, 2003b).

1.2 Natural history of pholcid spiders

Several species of pholcids are ubiquitous in human buildings around the world, ranging from regional anthropophilics to cosmopolitans like *Pholcus phalangioides* Fuesslin. Most of these species have characteristically long legs, they spend most of the time in their webs in corners and on the ceiling, and vibrate when disturbed. Several synanthropics are widespread in Africa, namely *Artema atlanta* Walckenaer, *Crossopriza lyoni* (Blackwall), *Smeringopus pallidus* (Blackwall), *Pholcus phalangioides*, *Physocyclus globosus* (Taczanowski), and the smaller *Modisimus culicinus* (Simon) and *Micropholcus fauroti* (Simon). Their impact on disease transmitting insects may be substantial, but this has been studied in only one case (Strickman et al., 1997).

Other species occupy a wide range of habitats in a variety of ecosystems. In arid regions, pholcids occur under rocks, in cavities and crevices, and in caves. In forests, they are found in the leaf litter, in webs between buttresses and twigs of trees reaching up to the canopy, and on the underside of preferably large leaves. Many species are cryptic. For example, those occurring between buttresses are typically dark, those on the underside of leaves tend to be light greenish.

Substantial research has been done on a few species occurring in temperate regions as well as on some Neotropical species, but the biology of African pholcids remains largely unknown.

2. RECENT PROGRESS

Recent revisions of the small, six-eyed pholcids previously assigned to *Spermophora* have resulted in some progress regarding the taxonomy, distribution patterns, and phylogenetic relationships of these taxa in Africa (Huber 2003a, b, c). Most African species are not congeneric with the type species of *Spermophora* (which is Asian). Moreover, there are several regions that combine high species diversity with high levels of endemism, notably Madagascar, Eastern Africa and Southern Africa. In Madagascar, two genera occur (*Zatavua* Huber, *Paramicromerys* Simon), both of them endemic to the island and widely distributed on it, but apparently restricted to largely undisturbed forests (Huber, 2003a). In Eastern Africa, the species-rich genus *Buitinga* Huber is restricted to the humid area delimited by the Somali and Sudanese deserts to the north, the Malagasy rain shadow to the south, and the central African plateau to the west. All known species have limited distributions, most being known from the type locality only (Huber, 2003b). At the same time, some of these species are among the dominant web-building spiders in the Eastern Arc montane rainforests (Sørensen, 2003; Sørensen et al., 2002). In Southern Africa, another species-rich genus (*Quamtana* Huber) is most diverse in eastern South Africa, with a distribution that follows quite exactly the area with more than 600 mm mean annual precipitation. The genus has a wide distribution, with some representatives occurring as far north as Cameroon (Huber, 2003c). However, Western African pholcid diversity remains largely unknown, mainly due to inadequate collecting. The same is true for the Ethiopian mountains and the Congo basin.

3. DIRECTIONS FOR FUTURE RESEARCH

This section will concentrate on the most apparent gaps in our knowledge about the taxonomy, distribution, and relationships of African pholcids. Included are Madagascar and the Comoro Islands, excluded are the Canary and Cabo Verde Islands, Socotra, and the Seychelles. Apart from taxonomic, phylogenetic and biogeographic studies, pholcids offer excellent opportunities for research on adaptation due to their diverse habitats, on their significance for the control of disease transmitting insects, and on a variety of ecological topics. Most pholcids are easily maintained in the laboratory, making them ideal spider objects for in-depth single-species studies.

Figure 1A shows the known African genera with current species numbers. Marked in grey are those genera that have not been revised recently, and these shall be the focus of this section. Figure 1B gives a historical perspective, showing all records of these five genera in Africa. With 95% of all records having been published before 1960, it becomes evident that our knowledge about these genera is quite outdated. The scarcity of records underlines our ignorance about species distributions and levels of

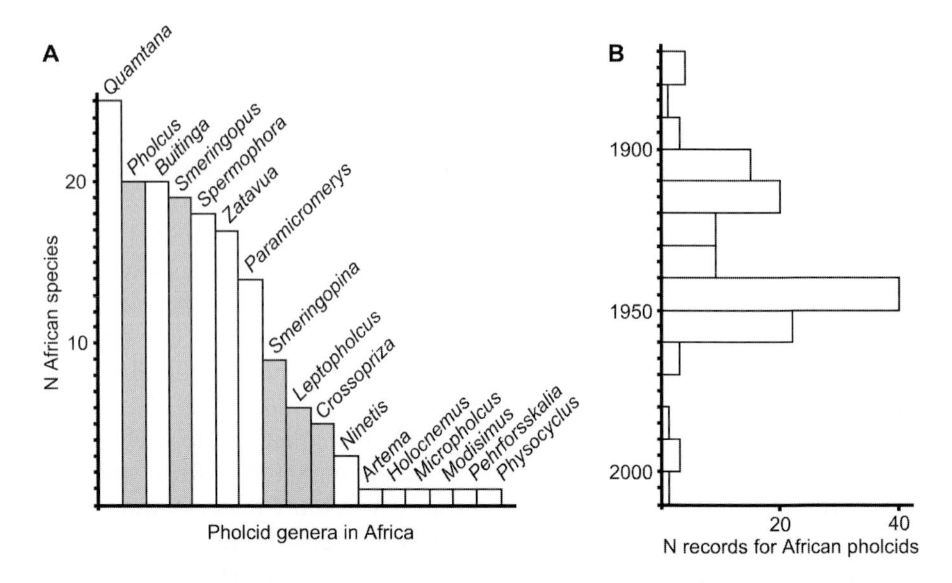

Figure 1. A. Pholcid genera in Africa, with current species numbers. Marked in grey are those genera that are most in need of revision. B. Publication dates of all records published for the five genera marked in grey in Fig. 1A.

endemism: 46% of the species included in these five genera are known from the type locality only. Several species are known from females only, but males are often necessary for species determination in pholcids. With this level of knowledge, interpretation of distribution maps of genera is difficult, but rough estimates can be provided (Fig. 2) and reveal interesting patterns.

At an alpha-taxonomic level, much remains to be discovered. Recent revisions of African *Spermophora* and related taxa has multiplied the number of known species by a factor of 5.2. If a similar increase occurs in the five genera under consideration, this would amount to over 200 new species.

Pholcus Walckenaer is a cosmopolitan (largely Old World) genus, with highest diversity on the Canary Islands and in Asia. In Africa, a single Mediterranean species faces 17 sub-Saharan nominal species. The single record from Madagascar is dubious (one female specimen in a house), and the genus is apparently also absent from southern Africa (Huber, 2003c). The cosmopolitan *P. phalangioides* has probably not originated from Africa but from the Palaearctic. Whether sub-Saharan *Pholcus* represent a monophyletic clade is unknown.

Smeringopus Simon has a quite different distribution. It is largely

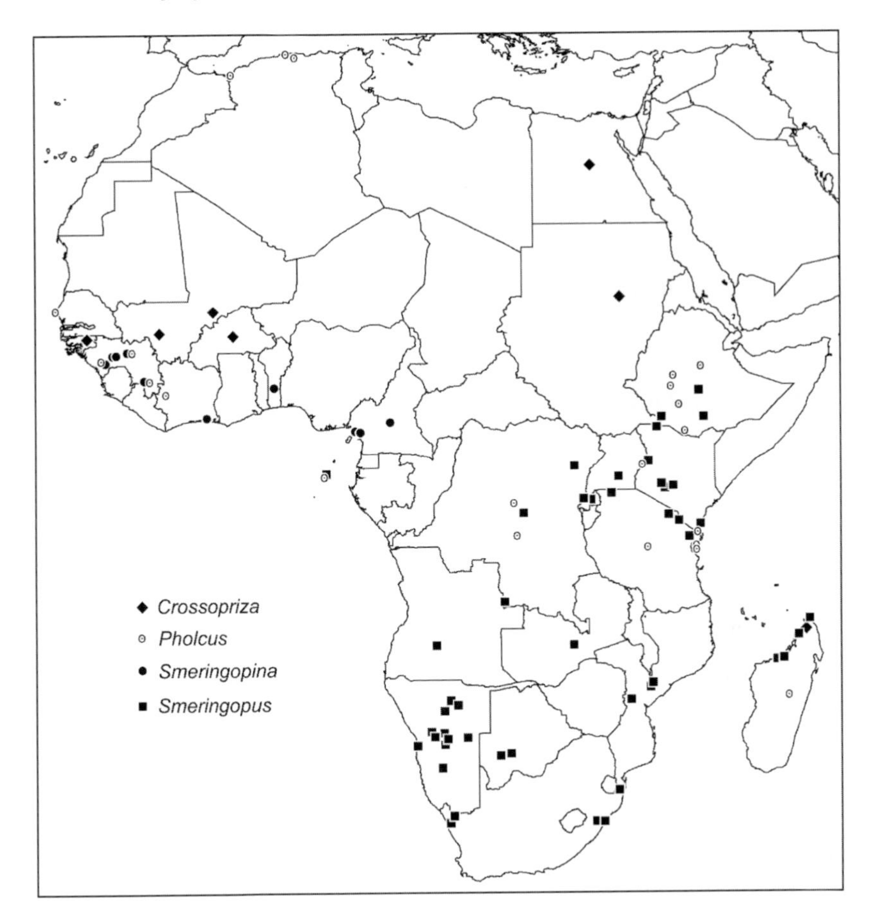

Figure 2. Known distributions of pholcid genera in Africa that are most in need of revision. For *Leptopholcus* see Huber et al. in press.

endemic to Africa, with 18 nominal species restricted to Southern and Eastern Africa (Kraus, 1957), and one species in Yemen (a single East Asian species is either misplaced or a synonym). *Smeringopus natalensis* is anthropophilous (Lawrence, 1947, 1967; Huber, 2001), and *S. pallidus* has followed humans around the world (Kraus, 1957).

Smeringopina Kraus was established for some species originally placed in *Smeringopus* as well as some new species (Kraus, 1957). The nine known species are restricted to Western Africa, and here the genus seems to occur in forests only. Thus, despite the similar names, *Smeringopus* and *Smeringopina* have disjunct distributions, very different ecologies, and are probably not even closely related (Huber, 1995). However, no representative of *Smeringopina* has ever been included in a phylogenetic analysis.

Crossopriza Simon may have similar ecological requirements as *Smeringopus*, but while the latter occurs in Southern and Eastern Africa,

Crossopriza (5 species in Africa) is restricted to the north, with further species on the Arabian Peninsula. The single species from Madagascar is known from one juvenile specimen collected in a house, and is thus probably a synonym of the synanthropic *C. lyoni*.

Leptopholcus Simon has a worldwide distribution, and the six African species occur widely in sub-Saharan Africa (Huber et al. in press). Most or all species appear restricted to relatively humid areas where they live cryptically on the underside of large leaves.

ACKNOWLEDGEMENTS

This study was initially funded by the Austrian Science Foundation (FWF). I thank an anonymous referee for helpful comments.

REFERENCES

Huber, B. A., 2001, The pholcids of Australia (Araneae: Pholcidae): taxonomy, biogeography, and relationships, *Bull. Amer. Mus. Nat. Hist.* **260:**1-144.

Huber, B. A., 2003a, Cladistic analysis of Malagasy pholcid spiders reveals generic level endemism: Revision of *Zatavua* n. gen. and *Paramicromerys* Millot (Pholcidae, Araneae), *Zool. J. Linn. Soc.* **137:**261-318.

Huber, B. A., 2003b, High species diversity in one of the dominant groups of spiders in East African montane forests (Araneae: Pholcidae: *Buitinga* n. gen., *Spermophora* Hentz), *Zool. J. Linn. Soc.* **137:**555-619.

Huber, B. A. 2003c, Southern African pholcid spiders: revision and cladistic analysis of *Quamtana* n. gen. and *Spermophora* Hentz (Araneae: Pholcidae), with notes on male-female covariation, *Zool. J. Linn. Soc.* **139:**477-527.

Huber, B. A., Pérez G., A., and Baptista, R. L. C., in press, First records of the genus *Leptopholcus* (Araneae: Pholcidae) for continental America, *Bonner Zool. Beitr.*

Kraus, O., 1957, Araneenstudien 1. Pholcidae (Smeringopodinae, Ninetinae), *Senck. biol.* **38(3/4):**217-243.

Lawrence, R. F., 1947, A collection of Arachnida made by Dr. I. Trägardh in Natal and Zululand (1904--1905), *K. Vet. Vitterh. Samh. Handl. (ser. B)* **5(9):**3-41.

Lawrence, R. F., 1967, A new cavernicolous Pholcid spider from the Congo, *Rev. Suisse Zool.* **74(4):**295-300.

Sørensen, L. L., 2003, Stratification of the spider fauna in a Tanzanian forest, in: *Arthropods of tropical forests: spatio-temporal dynamics and resource use in the canopy,* Y. Basset, V. Novotny, S. Miller, and R. Kitching, eds., Cambridge University Press, Cambridge, UK, pp. 92-101.

Sørensen, L. L., Coddington, J. A., and Scharff, N., 2002, Inventorying and estimating spider diversity using semi-quantitative sampling methods in an afrotropical montane forest, *Environ. Entomol.* **31:**319-330.

Strickman, D., Sithiprasasna, R., and Southard, D., 1997, Bionomics of the spider, *Crossopriza lyoni* (Araneae, Pholcidae), a predator of Dengue vectors in Thailand, *J. Arachnol.* **25:**194-201.

WEST AFRICAN FISH DIVERSITY – DISTRIBUTION PATTERNS AND POSSIBLE CONCLUSIONS FOR CONSERVATION STRATEGIES

Timo Moritz and K. Eduard Linsenmair
Lehrstuhl für Tierökologie und Tropenbiologie, Theodor-Boveri-Institut, Biozentrum, Am Hubland, Universität Würzburg, D-97074 Würzburg, Germany
E-mail: moritz@biozentrum.uni-wuerzburg.de

Abstract: West African fish diversity was investigated in regard to distribution patterns and species composition of different ichthyofaunal provinces. The coastal rivers region covers only a fraction of West Africa, but harbour 322 of West African's fish species, with 247 restricted to this area and 129 restricted even to small ranges (here: smaller than 40.000 km²). The central rivers fauna, however, comprises 194 fish species, with 119 endemics and only 33 restricted to small areas. While habitat protection is of primary importance in the Guinea regions for preserving its fish diversity there, the actual threats in the Sahelo-Sudan region are different, where additional regulations on fishing techniques are necessary to conserve the diversity of the West African fishes.

Key words: ichthyofaunal provinces; fish diversity; Guinea regions; Sahelo-Sudan region

1. INTRODUCTION

The ichthyofauna of Subsaharan Africa is under heavy threat (Stiassny, 1996) and conservation strategies for fish biodiversity are needed. This article deals with the distribution patterns of West African fish species and suggests regional conservation strategies. West Africa is bordered to the north by the Sahara, to the west and south by the Atlantic Ocean. Only on its

B.A. Huber et al. (eds.), African Biodiversity, 187–195.

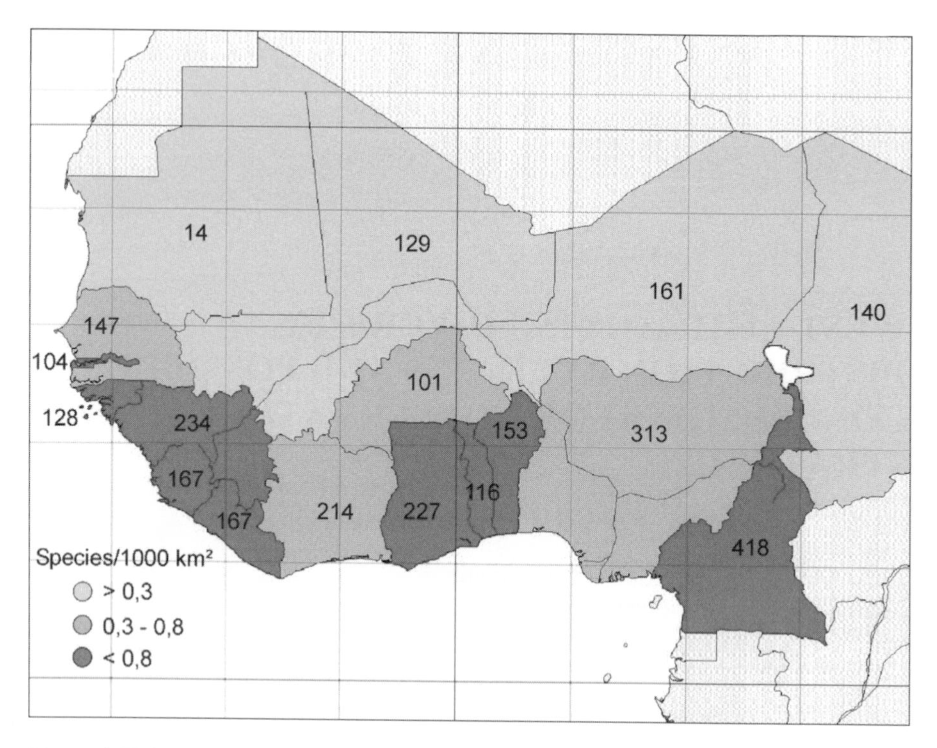

Figure 1. Fish species, freshwater and brackish, in West Africa. Absolute numbers are given for each country. Coloration indicates fish species density per area. Data based on FishBase (Froese and Pauly, 2004).

east side it is flanked by river systems (the Nile and Congo). The main basins in West Africa are the Senegal, Niger, Volta and the Chad basin. Furthermore there are numerous coastal rivers of different dimension.

2. MATERIALS AND METHODS

The actually available data from Fishbase (Froese and Pauly, 2004) on the species richness for the countries of the considered region were plotted on a map of West Africa (Fig. 1). The higher fish diversity in the south and south-east is evident, but for a more detailed picture this illustration is too imprecise. To obtain a finer resolution of the distribution patterns a grid 100 x 100 km was plotted over West Africa. The data on the occurrence of freshwater fish were taken from Daget and Iltis (1965), Roman (1966), Lévêque et al. (1990, 1992), Dankwa et al. (1999), and from personal observations. The received distribution patterns were compared in regard to environment. The so grouped fish species were compared by size and endemism.

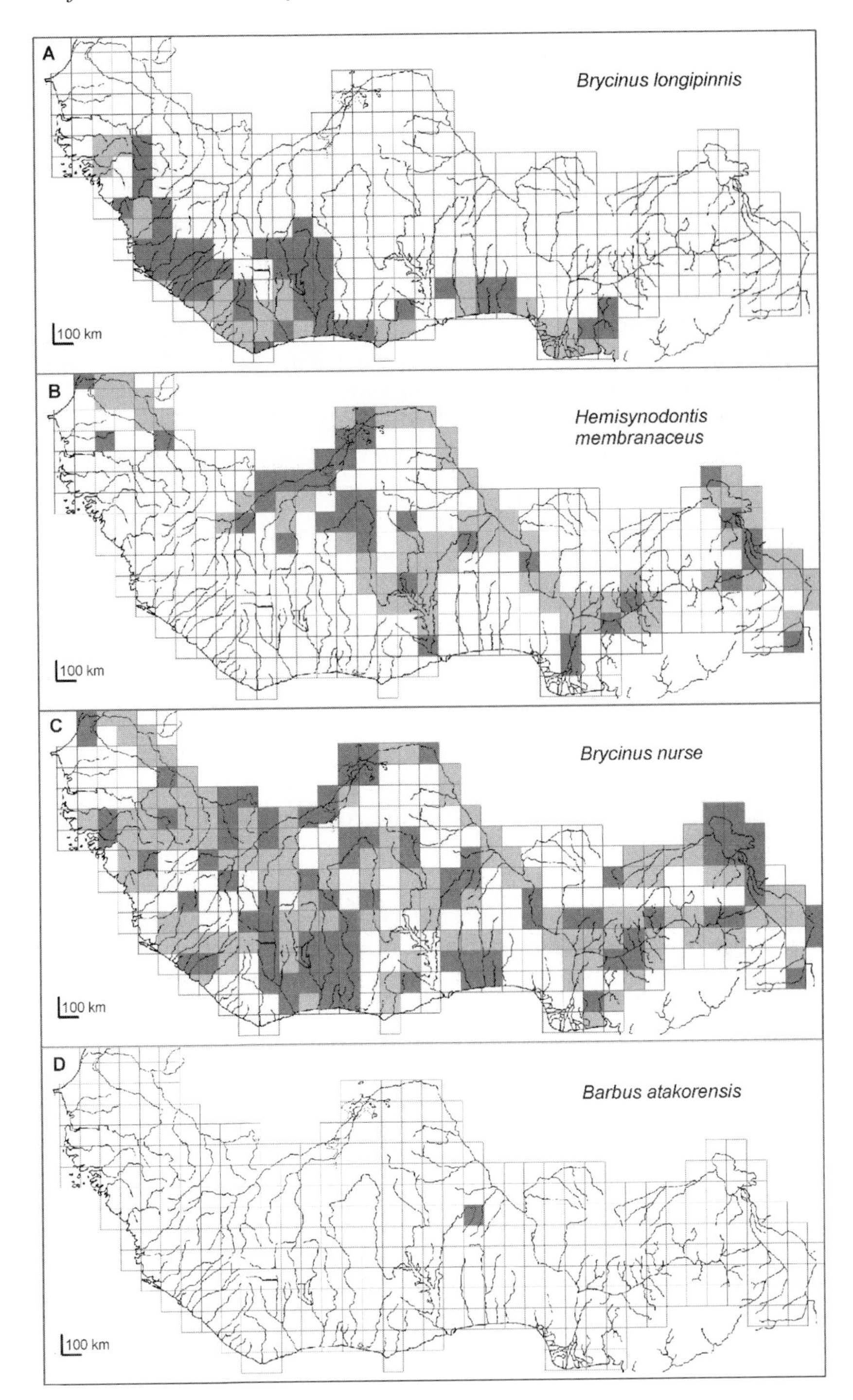

Figure 2. Characteristic distribution patterns of West African fishes. Dark shaded areas indicate records, light shaded areas point to most likely occurrence.

3. RESULTS

The distribution of some species using the 100 x 100 km grid was exemplarily displayed in Figure 2. By the comparison of the distribution maps four major distribution patterns emerged:

(1) Fishes restricted to the southern and/or south-eastern coastal rivers, e.g. *Brycinus longipinnis* (Fig. 2A), *Tilapia guineensis, Polypterus palmas, Ichthyborus besse.*

(2) Fishes restricted to the larger central river systems, above all the Senegal, Niger, Volta and Chad, e.g. *Hemisynodontis membranaceus* (Fig. 2B), *Clarotes laticeps, Gymnarchus niloticus.*

(3) Fishes occurring all over West Africa, e.g. *Brycinus nurse* (Fig. 2C), *Synodontis schall, Clarias gariepinus, Lates niloticus, Hepsetus odoe.*

(4) Fishes restricted to small areas, here to four or less grid cells of 100 x 100 km, i.e. an area of 40,000 km² or smaller, e.g. *Barbus atakorensis* (Fig.2D), *Irvineia voltae, Aphyosemion monroviae.*

The first distribution pattern can be further subdivided into an eastern and a western part with the Volta basin forming the approximate borderline, as some species occur only in one of these two parts.

All together 441 freshwater fish species are known from West Africa (Fig. 3A) (based on Lévêque et al., 1990, 1992): 322 in the coastal rivers region and 194 in the central rivers region (Fig. 3B,C). To point out the differences in species composition of this regions, a further classification was made: 247 species are restricted to the coastal rivers (Fig. 3D) and only half as many, 119, to the much larger region of the central rivers (Fig. 3E). Furthermore there is a third group of 75 fish species that are distributed over both regions (Fig. 3F). Noticeable differences concern the occurrence of Ostariophysi, which include the Characiformes, Cypriniformes and Siluriformes. They contribute a 50% of the coastal region endemics and also 65% of the species restricted to the central rivers region, whereas the cyprinodontiforms constitute 20% of the coastal endemics and just 8% of the central rivers endemics. In absolute numbers this works out to be 50 Ostariophysi to 9 cyprinodontiform species.

Comparing the fishes of the coastal and central rivers in regard to maximum body size (Table 1) shows that the species confined to the coastal

Table 1. Comparison of maximum size (total length) in mm of West African fishes

	coastal rivers	central rivers	both river types
average	95	147	210
minimum	16	23	30
maximum	870	1670	1800
25% quantil	55	56	71
75% quantil	190	383	505
species number	247	119	75

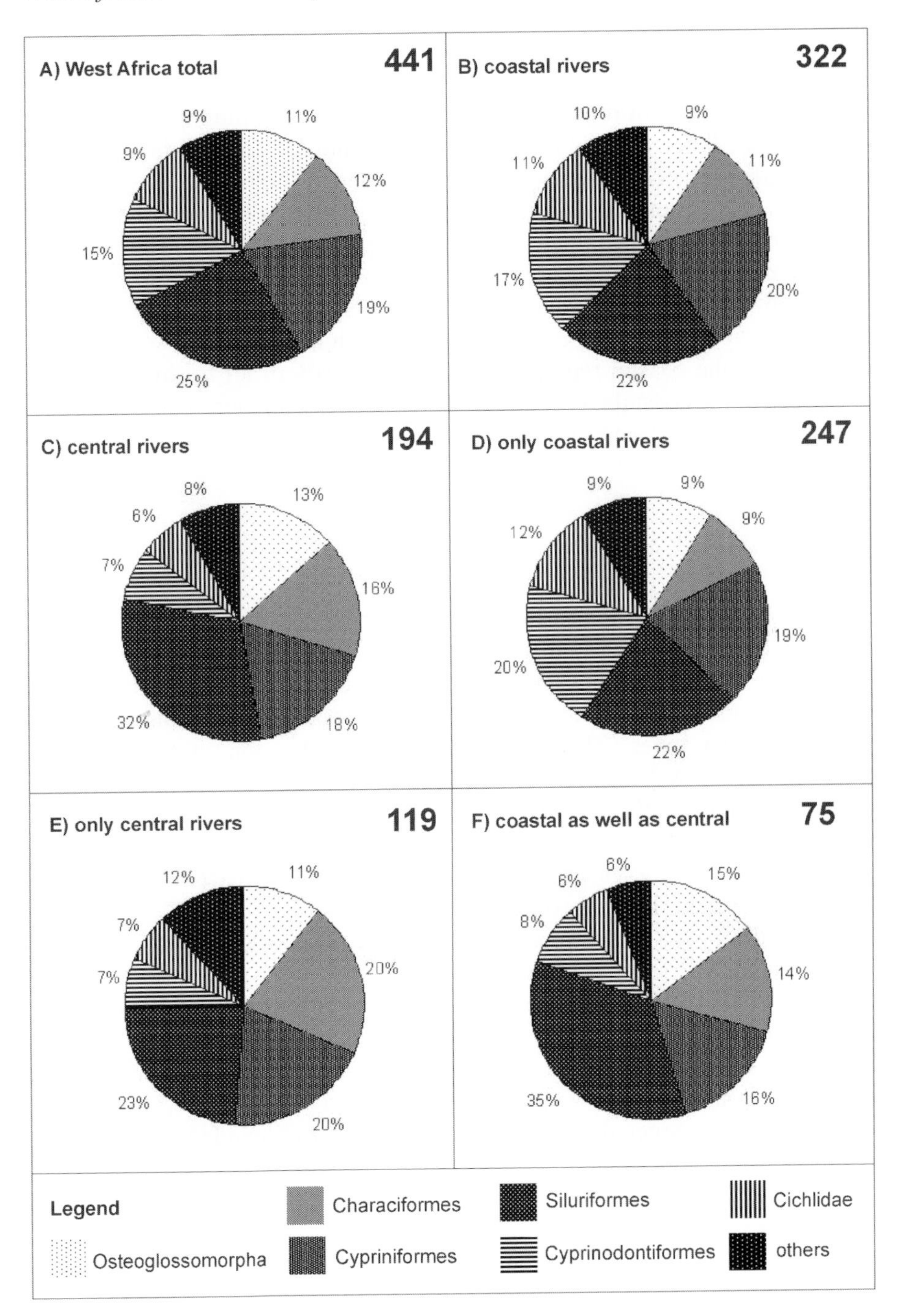

Figure 3. Species composition in West Africa. The bold number indicates the species number for each region.

zones are the smallest, with an average maximum body size of only 9.5 cm. In the group restricted to the central rivers this size is 14.7 cm, and the species living in both regions possess an average maximum body length of 21.0 cm. The latter value is 2.0 cm higher than the 75% quantil in the coastal rivers endemics. Thus species with a wide distribution are bigger than species with restricted distribution. Further differences between the regions become evident if we compare them with regard to endemism. Here species are called endemic, if their area of distribution is smaller than 40,000 km². The central rivers region harbours 33 such species. About half of them belong to two genera, *Barbus* and *Synodontis*. In contrast there are 129 such endemics in the coastal regions. Consequently 40% of all freshwater fishes occurring in the coastal rivers are restricted to areas smaller than 40,000 km² and only 17% in the central rivers region.

4. DISCUSSION

The evaluation of the distribution of West African fishes resulted in four major distribution patterns, of which the first two patterns are related to river length and at the same time roughly to the vegetation types: short rivers with rain forest plus humid savanna, and larger rivers on the other hand with less humid savanna types. This corresponds to the division in Guinean and Sudanian rivers by Daget and Iltis (1965). Taking into account additionally the subdivision of the first pattern into a western and eastern part, the patterns one and two reflect approximately the three ichthyofaunal provinces, which are generally accepted for West Africa (Roberts, 1975; Daget and Durand, 1981; Lévêque, 1997):

(1) The Upper Guinea region, from the Kogon River in Guinea to the Volta in Ghana.

(2) The Lower Guinea region, in the south-eastern part of West Africa continuing to Central Africa. Borders from the Guinea regions are differently drawn by different authors, and so often only parts east from the Cross River are regarded as the Lower Guinea region (Lévêque, 1997), alternatively the Niger delta and parts west of it are included to the Lower Guinea region (Roberts, 1975).

(3) The Sahelo-Sudan region, covering all the rest of West Africa, is continuous with the Nile system and therefore often called the Nilo-Sahelo-Sudan region. The typical vegetation types of this region are different forms of savanna.

Further characteristics of the rivers in Guinea, respectively the Sahelo-Sudan region, have been noted by Welcomme and de Merona (1988) and are listed with additional remarks in Table 2. The third major distribution

Table 2. Comparison of the coastal and central rivers in West Africa

	coastal rivers	**central rivers**
size	Short	long, large basin
vegetation	rain forest	different savanna types
flood regime	relatively stable (reservoir type)	pronounced
floodplains	Restricted	broad
fish species	322	194
endemits	129	33
<40,000 km²	40%	17%
examples	Cavally, Pra, Kogon, Mono	Niger, Volta, Senegal

pattern, the widely distributed species, include fishes, which occur in both regions. The relative species number of catfishes is more pronounced in this group compared to the composition of the species of the entire West African subcontinent, and the cyprinodont species number is smaller, but nevertheless every taxonomic group considered here has members in this group of widespread species. Due to their large distribution, they might have a high adaptability to differing biotic and abiotic conditions. Further evidence for this adaptability comes from relict species in oases. Some 8000 years ago in the Holocene, the Sahara had a much more humid climate than nowadays, but still today one can find 15 fish species there (Lévêque, 1990), and only four of them do not belong to the group of species distributed over all West Africa.

Hugueny (1989) showed, that species richness in West African rivers is positively related to area, i.e. the surface of a river basin. But, as shown here, the many small rivers of the Guinea regions have much more species, especially endemic species, in a much smaller area, than the few large central rivers, which basins cover the greatest part of West Africa. The Guinea regions are heavily endangered, as the coastal regions are the ones with the highest population density, and much of the West African rain forest, representing these rivers' natural environment, is already destroyed. Thus habitat protection is the most important means for conserving fish diversity in the Guinea regions. Protection areas have to be chosen by several criteria, e.g. degree of endemism, size area, or local species richness, which is positively correlated to regional species richness in this region (Hugueny and Paugy, 1995). The fishes of the Sahelo-Sudan region, however, are adapted to severe conditions as seasonal changes are more pronounced in the savanna. Most species in this region have high offspring numbers and these rivers are known to be very productive (Welcomme, 1976, 1979). So these fishes seem to be "pre-adapted" for human impact, e.g. habitat changes or exploitation of fish stocks. But some fishing techniques should be prohibited and the present regulations have to be controlled more extensively, due to their high impact on diversity and productivity: e.g. blocking rivers completely with nets could fish out large species within a few years, since they mature only to a certain size; the unsustainability of poisoning fish is obvious. Further, protection areas within river courses are appropriate, from which fishes can recolonize overexploited areas, e.g. national parks should be such areas. Sadly, in West African national parks there are extensive fisheries controlled mostly only in theory, blocking rivers is a common fishing method, and also poisoning has become more and more frequent. But nevertheless, one should not abandon hope that efforts to protect biodiversity and habitats in West Africa can be successful, e.g. the governmental regulations on mesh size for freshwater fisheries are widely accepted by West African fishermen.

ACKNOWLEDGEMENTS

This project is supported by BMBF, FZ01L0017 BIOTA West. Two anonymous referees provided valuable comments.

REFERENCES

Daget, J., and Durand, J.-R., 1981, Poissons, in: *Flore et faune aquatiques de l'Afrique Sahelo-Soudanienne*, J.-R. Durand, and C. Lévêque, eds., Ed. ORSTOM, *Doc. Tech.* **45:**687-771.

Daget, J., and Iltis, A., 1965, Poissons de Côte d'Ivoire (eaux douces et saumâtres), *Mém. I.F.A.N.*, **74:**1-385.

Dankwa, H. R., Abban, E. K., and Teugels, G. G., 1999, Freshwater fishes of Ghana, *Annls Mus. r. Afr. centr. (Sc. Zool.)*, **283:**1-53.

Froese, R., and Pauly, D., 2004, *FishBase*, World Wide Web electronic publication. www.fishbase.org, version (03/2004).

Hugueny, B., 1989, West African rivers as biogeographic islands: species richness of fish communities, *Oecologia* **79:**236-243.

Hugueny, B., and Paugy, D., 1995, Unsaturated fish communities in African rivers, *Am. Nat.* **146:**162-169.

Lévêque, C., 1990, Relict tropical fish fauna in Central Sahara, *Ichthyol. Explor. Freshwaters* **1:**39-48.

Lévêque, C., 1997, *Biodiversity dynamics and conservation: the freshwater fish of tropical Africa*, Cambridge University Press, Cambridge.

Lévêque, C., Paugy, D., and Teugels, G. G., 1990, Faune des poissons d'eaux douces et saumâtres de l'Afrique de l'Oues, Tome 1, Ed. ORSTOM, *Coll. Faune tropicale* **28:**1-384.

Lévêque, C., Paugy, D., and Teugels, G. G., 1992, Faune des poissons d'eaux douces et saumâtres de l'Afrique de l'Ouest, Tome 2, Ed. ORSTOM, *Coll. Faune tropicale* **28:**387-902.

Roberts, T. R., 1975, Geographical distribution of African freshwater fishes, *Zool. J. Linn.* Soc. **57:**249-319.

Roman, B., 1966, Les poissons des hauts-bassins de la Volta, *Annls Mus. r. Afr. centr.*, in-8°, **150:**1-191.

Stiassny, M. L. J., 1996, An overview of freshwater biodiversity: with some lessons from African fishes, *Fisheries* **21**(9):7-13.
Welcomme, R. L., 1976, Some general and theoretical considerations on the fish yield of African rivers, *J. Fish Biol.* **8:**351-364.
Welcomme, R. L., 1979, *Fisheries ecology of floodplain rivers*, Longman, London.
Welcomme, R. L., and de Merona, B., 1988, Fish communities in rivers, in: *Biology and ecology of African freshwater fishes*, C. Leveque, M. N. Bruton, and G. W. Ssentongo, eds., Ed. ORSTOM, Collection Travaux et Documents **216:**251-276.

DISTRIBUTION AND POPULATION DENSITY OF THE BLACK-EARED MALAGASY POISON FROG, *MANTELLA MILOTYMPANUM* STANISZEWSKI, 1996 (AMPHIBIA: MANTELLIDAE)

David R. Vieites[1]*, Falitiana E. C. Rabemananjara[2], Parfait Bora[2], Bertrand Razafimahatratra[2], Olga Ramilijaona Ravoahangimalala[2], Miguel Vences[1]
[1] *Institute for Biodiversity and Ecosystem Dynamics, Zoological Museum, University of Amsterdam, Mauritskade 61, 1092 AD Amsterdam, The Netherlands*
E-mail: vences@science.uva.nl; vieites@inicia.es
[2] *Université d'Antananarivo, Département de Biologie Animale, Antananarivo, Madagascar*
* *Present address: Museum of Vertebrate Zoology and Department of Integrative Biology, 3101 Valley Life Sciences Building, University of California, Berkeley, CA 94720-3160, USA*

Abstract: We provide first data on the natural history of a poorly known species of frog from Madagascar, the black-eared Malagasy poison frog, *Mantella milotympanum*. Although this species has been intensively collected for the pet trade, not even one precise locality was published until 2003. We here provide further distribution records north and south of the known locality Fierenana, but the encountered populations showed variable colour and patterns intermediate between *M. milotympanum* and *M. crocea*, thus supporting the hypothesis that these are conspecific colour morphs, and that *M. milotympanum* might be a junior synonym of *M. crocea*. Intensive fieldwork at one site next to Fierenana, in February 2003, yielded some data on population structure and density. Snout-vent lengths ranged from 16-24 mm, weights from 0.4-1.4 g in adults, with only few subadults and no juveniles found. The population density, estimated by mark-recapture, was about 470 individuals per ha, which is a quite high density, taking into consideration that this population had probably been under commercial exploitation in the past. We propose that some of the forests in the Fierenana area should be included in the planned extension of Madagascar's network of protected areas, but a controlled and sustainable exploitation should be allowed in these reserves in order to gain the support of local communities.

B.A. Huber et al. (eds.), African Biodiversity, 197–204.

Key words: Amphibia; Anura; *Mantella milotympanum*; Distribution; Conservation; Pet trade; Madagascar

1. INTRODUCTION

The genus *Mantella* Boulenger, 1882 consists of about 15 species endemic to Madagascar (Vences et al., 1999), named Malagasy poison frogs. All of them are small, colourful and usually diurnal frogs, some resembling the neotropical frogs of the family Dendrobatidae, which also exhibit the presence of alkaloid skin toxins (Daly et al., 1996). Together with the tomato frogs (genus *Dyscophus* Grandidier, 1872), Malagasy poison frogs are the amphibian species of Madagascar most requested by the international pet trade, mainly due to their attractive colourations (Behra, 1993; Andreone and Luiselli, 2003). Although all *Mantella* are now included on the Appendix II of the Convention on the International Trade in Endangered Species (CITES), and the numbers exported from Madagascar amount to several thousand individuals per year, almost nothing is known about natural history and distribution and ecology of most species.

The black-eared Malagasy poison frog (*Mantella milotympanum* Staniszewski, 1996) was described by a pet breeder and its status as a separate species is uncertain (see Vences et al., 1999). Although it almost has a uniformly orange-red colour, quite similar to that of the golden *Mantella*, *M. aurantiaca* Mocquard, 1900, it appears to be much more closely related to another species, *M. crocea* Pintak & Böhme, 1990, as evidenced by molecular data (Vences et al., 2004). Because the taxonomy of *Mantella* species was largely unclarified before the revision of Vences et al. (1999), no reliable trade statistics are available for *M. milotympanum* which likely was confounded in the exportation lists with *M. aurantiaca.* From our own observations it is clear that this species is regularly collected by the local pet trade network and exported.

The Global Amphibian Assessment (GAA) workshop in Gland, Switzerland, held in September 2003, classified *M. milotympanum* as Critically Endangered and emphasized the need of further studies on its distribution and natural history (Andreone et al., in press). In this paper we provide new data on the natural distribution of *M. milotympanum*, in addition to the single locality given in Vences et al. (2004), and report some information on population structure and density of one formerly exploited population.

2. METHODS

Field expeditions were directed in 2003 to the Fierenana region to locate *Mantella* populations. In order to find the populations of these patchily distributed frogs, in addition to the one mentioned by Vences et al. (2004), part of the team explored the area south of Fierenana, whereas the area north of Fierenana was separately accessed. The population at a site locally known as Sahamarolambo, close to the village of Fierenana, had been visited in January 2002 by M.Vences; at this site, a campsite was established and more intensive fieldwork carried out from 17-22 February 2003. According to local collectors, this population had been intensively exploited in the past but not any more in the last three years. A parcel of ca. 0.6 ha was then defined in the forest. This parcel bordered on (a) a drift fence line, (b) altered habitat consisting mainly of sun-exposed savanna, (c) a footpath (ca. 2-3 m wide) and our campsite. Although especially the footpath, but also the other limits, certainly are no absolute barriers to *Mantella* dispersal, we believe that their movements across these structures were very limited during the short period in between our sampling and that this population can therefore be considered as a "closed population" for the purpose of mark-recapture analysis. During one week, five to seven people systematically carried out mark-recapture experiments. Frogs were marked by toe-clipping of one finger and immediately released after treating wounds with antiseptic. Recaptured animals did not show any infection caused by the marking procedure. At each sampling date, the number of marked individuals was recorded and all unmarked individuals in the follow-up sample were marked and released. We used these data to calculate a weighted mean population estimate (Seber, 1973), and its standard error (Begon, 1979).

Table 1. Known localities of *Mantella milotympanum* and colour morphs intermediate between *M. milotympanum* and *M. crocea*

Locality	Colour of specimens	GPS coordinates	Altitude (m)
Sahamarolambo next to Fierenana	Typical for *M.milotympanum*; orange-red with black spots on tympanum and nostril	18°32'36"S 48°26'56"E	948
North of Fierenana	Greenish, *M.crocea*-like	18°16'10"S 48°29'03"E	1060
Savakoanina	Pattern like in typical *M.milotympanum*, but greenish-yellow colour	18°36'44"S 48°24'30"E	959
Andriabe	Some specimens like typical *M.milotympanum*; others with *M.crocea*-like pattern but red colour	18°36'46"S 48°19'34"E	1047

From all collected individuals we measured snout-vent length (SVL) to the nearest mm using a dial calliper, and weighted them to the nearest 0.01 g using a digital balance. Allometric growth was tested by a regression analysis of SVL (X) and weight (Y), using the formula $Y = b X^a$, where b is a constant and a is the allometry coefficient (Huxley, 1972). A reliable distinction between sexes was not possible in the field, as in many Malagasy poison frogs, especially outside the breeding season.

3. RESULTS

3.1 Distribution and habitat

The first known populations of *Mantella milotympanum* had been located in the Fierenana region, between Moramanga and Ambatondrazaka, Central-Eastern Madagascar. We discovered three additional *Mantella* populations in the area between 18°16'S and 18°36'S, and 48°19'E and 48°29'E (Table 1). All were in humid tropical areas between 900 and 1100 m a.s.l., in swamp *Pandanus* forests. Whereas the population of Andriabe contained specimens with the typical colour and pattern of *M. milotympanum*, and others of *M. crocea*-like pattern but an orange-red colour, the populations north of Fierenana and at Savakoanina consisted of yellow-green and green coloured specimens.

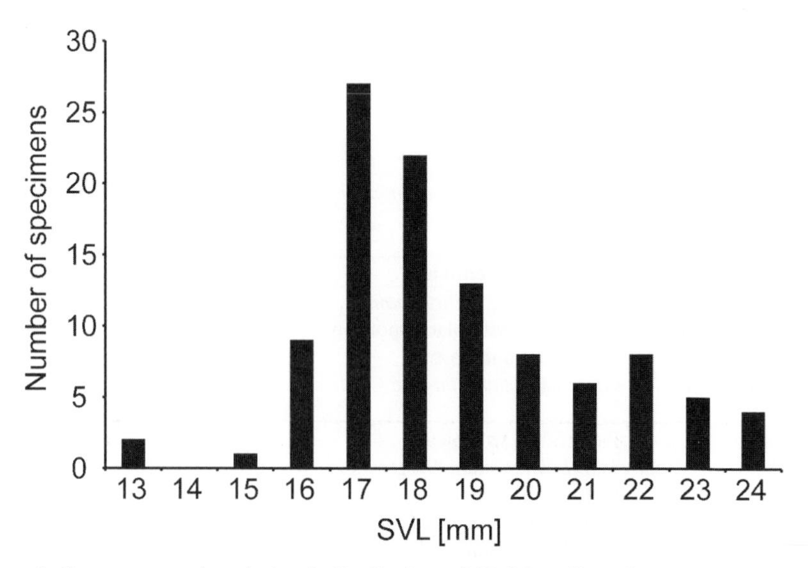

Figure 1. Snout to vent length (mm) distribution of 99 *Mantella milotympanum* specimens captured in February 2003 in Fierenana population. No recaptures were considered.

Specimens were distributed in the forest, being locally more abundant in accumulations of leaf litter under *Pandanus* screw palms and places with dense litter cover close to the swamps. No nocturnal activity was observed.

3.2 Natural history

Calling males were abundant in Sahamarolambo and many other places between Sahamarolambo and Fierenana in January 2002, but not in February 2003, indicating a possible strong seasonality of the reproductive season in this species. Distribution of size classes (Fig. 1) shows that the studied sample consisted mainly on adults. Only three animals of less than 15 mm were found which we here consider as probable subadults. No juveniles were located in the study area despite intensive searches in potential suitable areas like swamps and their immediate surroundings. Adult mean SVL (± standard deviation, minimum and maximum in parentheses) was 19.3 ± 2.2 mm (16.3-24.3mm, n = 99). Adult mean weight was 0.7 ± 0.2g (0.4-1.4g, n = 99). No external character allowed us to distinguish between males and females in the field. Positive allometric growth was observed between SVL and weight ($y=0.0005X^{2.4501}$, $r^2= 0.82$, $P< 0.001$, Fig. 2).

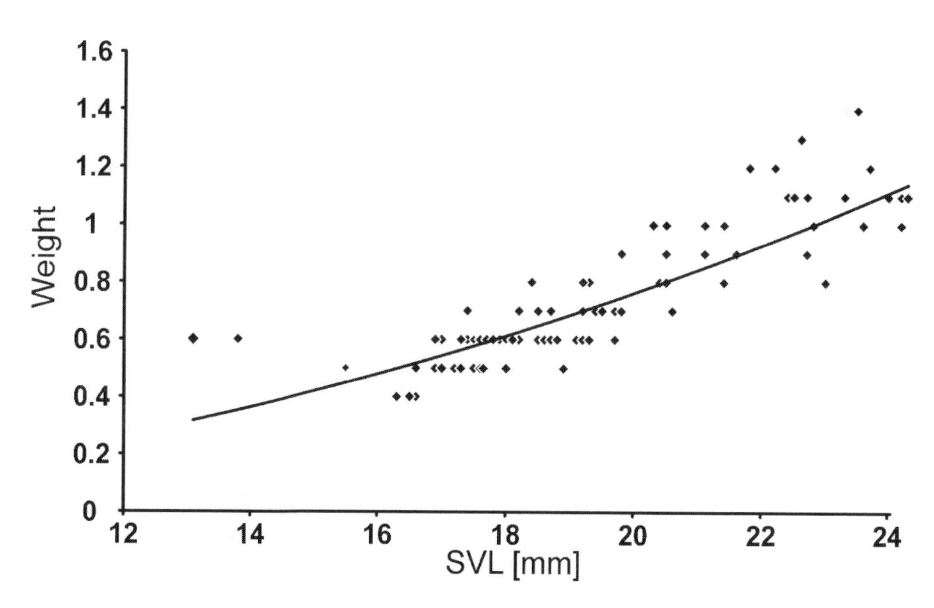

Figure 2. Scatterplot of the relation between body weight (g) and snout-vent length (mm) in the animals captured.

Table 2. Estimate of population size of adult *Mantella milotympanum* in the Sahamarolambo population. (1) Simple Petersen estimate (Petersen, 1896); (2) weighted mean estimate (Seber, 1973), ± standard error (see Begon, 1979) based on the data of the last two sampling days

	18 Feb.	19 Feb.	20 Feb.	21 Feb.	22 Feb.
Marked in population (a)	0	43	91	105	119
Captured (b)	43	48	19	28	30
Marked in captured sample (c)	0	3	5	14	8
(1) Population a (b+1)/(c+1)	--	527	303	203	410
(2) Population Σ a (b+1)/ Σ (c+1)					283±62

3.3 Population estimate

Using the weighted mean estimate (Seber, 1973), the number of frogs in the study parcel was 283±62 (Table 2), with a confidence interval of 222-345 individuals. In relation to the sampled area (0.6ha) this estimate gives a rough density of about 470 specimens per ha.

4. DISCUSSION

Our data support the hypothesis that "typical" *M. milotympanum* occur only next to Fierenana, and the variable and intermediate populations identified corroborate the close relationships, and possible conspecificity, of *M. milotympanum* and *M. crocea*. Populations of the *Mantella milotympanum /M. crocea* complex inhabit a wider area than previously known, including populations north of Fierenana. Surveys in the Zahamena Reserve, further north, revealed that populations of this complex also are present in that area (F. Rabemananjara, per. obs.), although their specific identity still needs to be verified. However, the colour morph typical for *M. milotympanum* appears to be more restricted to the Fierenana area itself. Our preliminary data suggests that the typical habitat of all these populations is constituted by forested edges of large inland swamps at mid-altitudes (900-1100 m a.s.l.), or by swamps inside primary or degraded rainforest areas.

According to our observations, the reproduction of *M. milotympanum* takes place at the beginning of the rainy season (December - January), and after that, the adults spend some weeks active, probably gaining fat reserves for aestivation. After this, their activity is probably less intensive. This coincides with data from exporters who confirmed that it is not possible to obtain this species from local collectors during a long period in the dry season.

The Sahamarolambo population of *M. milotympanum* has been exploited for several years, but not during the three previous years prior to our fieldwork. Our estimate of the adult population during this period gave a relatively high number of specimens per hectare. We consider the information on regular past collecting as relatively reliable since it was confirmed by various professional collectors and locals. Almost certainly these frogs are characterized by a high yearly reproductive output and rate of recruitment, although no precise data are available so far. We therefore speculate that even heavy collecting does not necessarily lead to serious threats for *Mantella milotympanum*, although more data on the population dynamics of this and other species are needed for definitive statements.

The original habitat around Fierenana is being destroyed at very high rates by small-scale slash- and burn-agriculture, and this truly constitutes a high threat to *Mantella* populations. Although the black-eared poison frog may not constitute a species distinct from *M. crocea* (Vences et al., 2004), its distinct colour convergence with *M. aurantiaca* provides a very interesting model for studies on the evolution of aposematism, emphasizing the interest in conserving this locally restricted form. Considering recent efforts of increasing the network of Madagascar's protected areas, the area around Fierenana certainly qualifies for an improved level of protection of natural habitat. However, the fact that populations of *M. milotympanum* apparently are able to withstand strong collection activities and to survive in slightly degraded habitat (as *M. aurantiaca;* see Vences et al., 2004) indicates that a certain degree of controlled exploitation of natural resources in this area could be sustainable and not in contradiction with the proposed conservation efforts.

ACKNOWLEDGEMENTS

We are grateful to Marta Puente, Meike Thomas, and Wouter zum Vörde Sieve Vörding for their help in the field. Euan Edwards provided invaluable logistic assistance and companionship. The work in Madagascar was made possible by a cooperation accord of the Institute for Biodiversity and Ecosystem Dynamics, University of Amsterdam, Netherlands, the Département de Biologie Animale, Université d'Antananarivo, and the Association Nationale pour la Gestion des Aires Protegées, Madagascar. We acknowledge financial support by grants of the University of Vigo to DRV, and of the Volkswagen Foundation. We are especially grateful to the BIOPAT foundation that supported the expeditions to Fierenana.

REFERENCES

Andreone, F., Cadle, J. E., Cox, N., Glaw, F., Nussbaum, R. A., Raxworthy, C. J., Stuart, S. N., Vallan, D., and Vences, M., in press, A complete species review of amphibian extinction risks in Madagascar: results from the Global Amphibian Assessment, *Conserv. Biol.*

Andreone, F., and Luiselli, L. M., 2003, Conservation priorities and potential threats influencing the hyper-diverse amphibians of Madagascar, *Ital. J. Zool.* **70:**53-63.

Begon, M., 1979, *Investigating Animal Abundance. Capture-recapture for Biologists,* Arnold, London.

Behra, O., 1993, The export of reptiles and amphibians from Madagascar, *Traffic Bull.* **13**(3):115-116.

Daly, J. W., Andriamaharavo, N.R., Andriantsiferana, M., and Myers, C.W., 1996, Madagascan Poison Frogs (Mantella) and Their Skin Alkaloids, *Am. Mus. Novit.* **3177:**1-34.

Huxley, J. S., 1972, *Problems of Relative Growth*, Dover, New York.

Petersen, C.G.J., 1896, The yearly immigration of young plaice into Limfjord from the German sea, *Report of the Danish Biological Station* **6**. Copenhagen, Denmark.

Seber, G.A.F., 1973, *The Estimation of Animal Abundance*, Griffin, London.

Vences, M., Glaw, F., and Böhme, W., 1999, A review of the genus *Mantella* (Anura, Ranidae, Mantellinae): taxonomy, distribution and conservation of Malagasy poison frogs, *Alytes* **17**(1-2):3-72.

Vences, M., Chiari, Y., Raharivololoniaina, L., and Meyer, A., 2004. High mitochondrial diversity within and among populations of Malgasy poison frogs, *Mol. Phylogenet. Evol.* **30:**295-307.

AMPHIBIAN AND REPTILE DIVERSITY IN THE SOUTHERN UDZUNGWA SCARP FOREST RESERVE, SOUTH-EASTERN TANZANIA

Michele Menegon[1] and Sebastiano Salvidio[2]
[1] *Museo Tridentino di Scienze Naturali, via Calepina 14,C P.393, I-38100 Trento, Italy*
[2] *DIP.TE.RIS. - Dipartimento per lo Studio del Territorio e delle sue Risorse, Corso Europa, 26, I-16132 Genova, Italy*

Abstract: The Udzungwa Scarp Forest Reserve is one of the largest forest patches within the Udzungwa Mountains of South-central Tanzania. It covers an area of about 200 km^2 on the south eastern slopes of Udzungwa range.
Since 1998 five sites in the southern part of the Udzungwa Scarp Forest Reserve, ranging from 400 to 1900 m a.s.l., have been surveyed and a list of Reptiles (33 species) and Amphibians (36 species) is given. Among them 21.7% and 53.6% of the species are endemic or near endemic to the Udzungwa Mountains and to the Eastern Arc Mountains respectively. Concern is raised for the preservation of this unique highland forest-dependent fauna in the light of continuing habitat alteration.

Key words: Eastern Arc; Udzungwa mountains; Udzungwa Scarp; checklist; Amphibians; Reptiles

1. INTRODUCTION

The Udzungwa Scarp Forest Reserve covers part of the south-eastern slopes of Udzungwa Mountains, which contain large forest reserves within the "Eastern Arc Mountains and Coastal forests of Kenya and Tanzania" biodiversity hotspot. It covers an area of about 200 km^2 and an elevation range of about 1,850 m, between 250 and 2,100 m. The Reserve is part of the Eastern Arc chain of isolated blocks of mountains stretching from Southeast Kenya through south central Tanzania. The abrupt eastern slopes of these mountains are covered by dense rainforest, and are well known for their

B.A. Huber et al. (eds.), African Biodiversity, 205–212.

extremely elevated level of biodiversity (Howell, 1993; Burgess et al., 1998; Myers et al., 2000; Newmark, 2002). In spite of this, few massifs have been sampled adequately for biodiversity, and many others are even lacking checklists (Howell, 2000). This contribution presents a preliminary list of herpetological species based on literature and original records collected by the authors in five forest sites between 1998 and 2004.

2. STUDY SITE

The Udzungwa Scarp Forest Reserve lies in south eastern Tanzania, Kilolo and Kilombero Districts, Iringa and Morogoro Regions. All species herein reported have been recorded both within the Udzungwa Scarp Forest Reserve and in neighbouring areas, these latter are a mosaic of scattered cultivated fields and highly disturbed forest patches, surrounding several villages at altitude of 1,500 - 1,900 m a.s.l.

Rainfall in the study area is unimodal in pattern and ranges from 1,800 and 3,000 mm per year (Shangali et al., 1998). Overall, the study sites are located at different elevations giving a sampled altitudinal range 400 – 1900 m. In all but one case (Kihanga), sampling sites were characterised by the presence of closed canopy rainforest, open areas and ecotone both of anthropogenic and natural origins.

Table 1. Location and main characteristics of the sampling sites within the Udzungwa Scarp Forest reserve. D = day search; N = night search; PFT = pit fall traps with drift fences

Sampling site	UTM coordinates	Elevation range m asl	Main vegetation Type	Main habitats investigated	Collecting methods*
Site 1 W of Chita	36L0818688 9056144	700-1000	Submontane rain forest	closed canopy forest, open wetland, ecotone	D, N, PFT
Site 2 SE of Ihimbo	36L0818777 9060770	1400-1600	Montane rain forest	closed canopy forest, open wetland, ecotone	D, N
Site 3 'Mkalazi'	36L0828439 9070696	1100-1200	Submontane rain forest	closed canopy forest, small wetland	D, N
Site 4 'Kihanga'	36L0828078 9073188	1800-1900	Montane rain forest	closed canopy forest	D-N-PFT
Site 5 'Mkaja'	36L0826838 9076662	1800	Open montane wetland	bamboo forest, open wetland, ecotone	D, N
Village belt		1500-1900	Farmland	Synantropic habitats	Opportunistic

Five forest vegetation types are present in the Udzungwa Scarp (Shangali et al. 1998) but just four were encountered at the sampling sites: lowland rainforest (300-800 m a.s.l.); submontane rainforest (700-1,400 m a.s.l.); montane rainforest (1,400-1,800m a.s.l.) and mountain bamboo forest (Table1).

3. METHODS

Five sites were sampled extensively (see Fig. 1), mainly by SSS (systematic sampling surveys) and by means of pit fall traps with drift fences. Searches were conducted both during the day and by night to sample the highest number of species. Further records were obtained from local people living in the villages at the forest edge, from opportunistic searches and from literature. Only direct records have been considered for running the analysis of similarity (Legendre et al., 1998). All collected specimens were

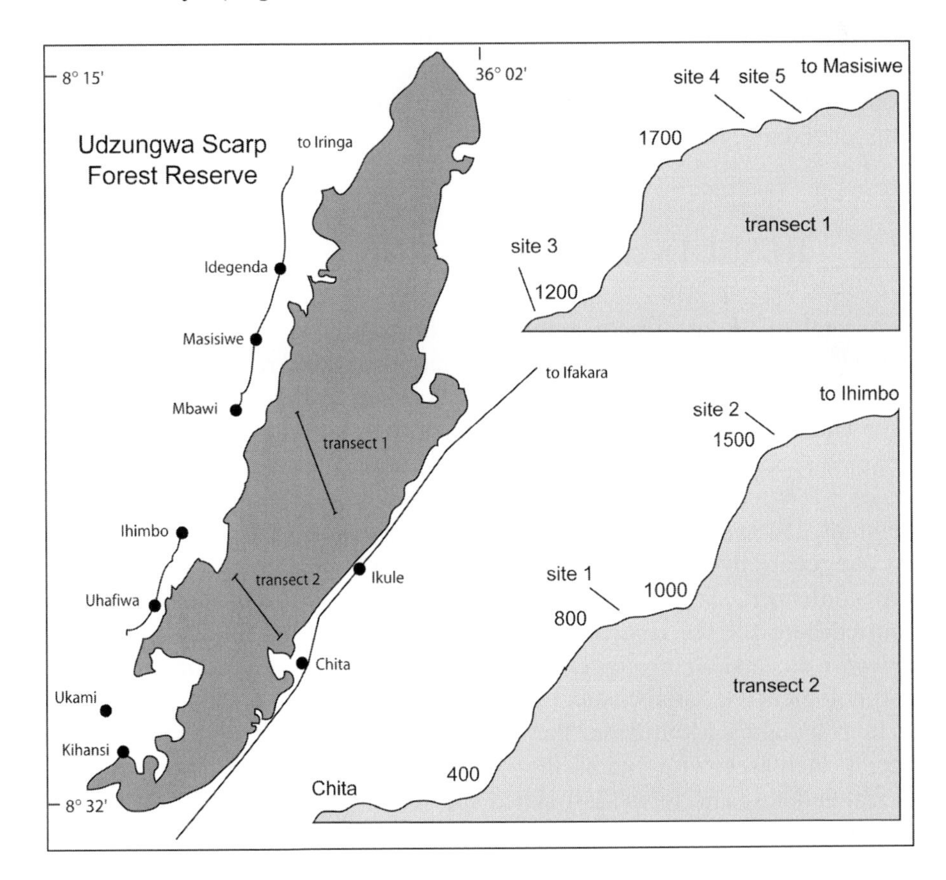

Figure 1. Location of the sampling sites within the Udzungwa Scarp Forest Reserve.

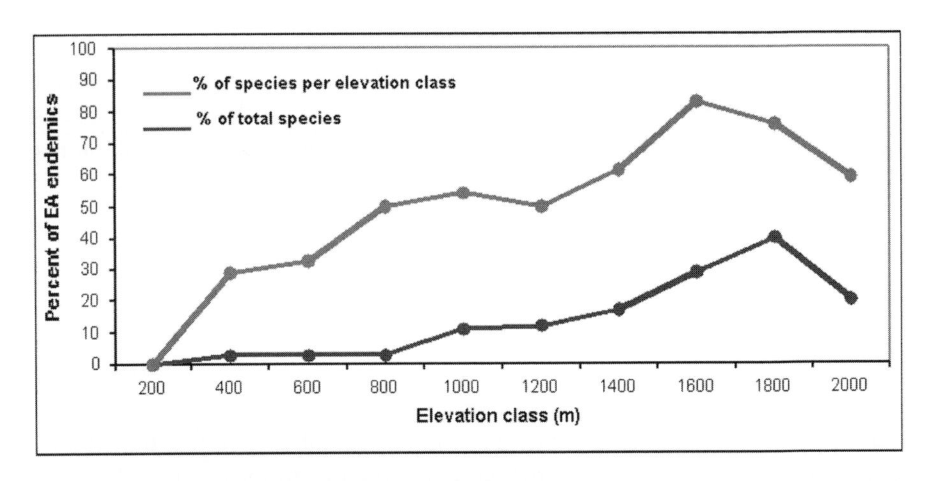

Figure 2. Curves representing the percentage of Eastern Arc endemics and of total species recorded per elevation class in the Udzungwa Forest Reserve (data shown here are the pooled records from the 5 sites reported in table 1 and from opportunistic sampling).

deposited in the collections of the Dept. of Zoology and Marine Biology, University of Dar es Salaam, Dar es Salaam, Tanzania, and specimens, photographs and sound recordings in the Museo Tridentino di Scienze Naturali, Trento, Italy.

4. RESULTS AND DISCUSSION

A total of 36 species of amphibians belonging to six families and 33 species of reptiles belonging to seven families were recorded (Table 2). Among them 21.7% and 53.6% of the species are endemic or near endemic to the Udzungwa Mountains and to the Eastern Arc Mountains respectively.

Some species have never been reported before or were known only from one or a few records from the Udzungwa Mountains. *Cnemaspis dickersoni* (Schmidt, 1919) and *Dipsadoboa werneri* (Boulenger, 1897) were cited from just one record by Perret (1986) and Rasmussen (1997) respectively, and are here confirmed. *Scelotes uluguruensis* (Barbour & Loveridge, 1928) is reported here for the first time from the Udzungwa Mountains. Moreover *Afrixalus* sp. and *Arthroleptis* sp. are believed to be undescribed taxa on morphological and bioaocustical grounds.

In Udzungwa Mountains, the percentage of endemic or near endemic species increases with increasing elevations (Fig. 2). This suggests that several endemic and forest-associated species of the Udzungwa Scarp could be cool-adapted species *sensu* Poynton (2000). According to the Bray Curtis indexes of similarity (Fig. 3) the lowest elevation site (site 1, W of Chita)

Table 2. List of the Udzungwa Scarp Forest Reserve herpetofauna. Species marked with an asterisk are cited in the bibliography or have been collected opportunistically in the village belt surrounding the Forest Reserve or outside the sampling sites, either by local people or by authors. Remaining species have been collected at the sampling sites by the authors and have been considered for running the similarity analysis

	Site 1	Site 2	Site 3	Site 4	Site 5
AMPHIBIA					
ANURA					
ARTHROLEPTIDAE					
Arthroleptis affinis (Ahl, 1939)	+	+	+	+	+
Arthroleptis reichei (Neiden, 1910)		+	+	+	
Arthroleptis stenodactylus Pfeffer, 1893	+		+		
Arthroleptis xenodactyloides (Hewitt, 1933)	+	+	+	+	+
Arthroleptis sp. n.			+		+
Arthroleptis sp.			+		
BUFONIDAE					
Bufo brauni (Neiden, 1910)			+		
Nectophrynoides tornieri (Roux, 1906)	+		+		
Nectophrynoides poyntoni Menegon, Salvidio & Loader, 2004			+		
Nectophrynoides viviparus (Tornier, 1905)				+	+
Nectophrynoides asperginis Poynton, Howell, Clarke & Lovett, 1998*					
Nectophrynoides wendyae Clarke, 1988		+			
HYPEROLIDAE					
Afrixalus morerei Dubois, 1985					+
Afrixalus uluguruensis (Barbour & Loveridge, 1928)		+		+	+
Afrixalus sp. n.		+			
Hyperolius kihangensis Schiøtz & Westergaard, 1999				+	
Hyperolius puncticulatus (Bocage, 1895)	+	+	+	+	+
Hyperolius mitchelli (Loveridge, 1953)			+		
Hyperolius cf. spinigularis (Stevens, 1971)				+	+
Hyperolius sp.				+	
Leptopelis barbouri Ahl, 1929		+	+	+	+
Leptopelis parkeri Barbour & Loveridge, 1928			+	+	
Leptopelis uluguruensis Barbour & Loveridge, 1928	+		+	+	
Leptopelis vermiculatus (Boulenger, 1909)		+			+
Leptopelis flavomaculatus (Günther, 1864)	+				
Phlyctimantis keithae (Schiøtz, 1975)*					
MICROHYLIDAE					
Probreviceps macrodactylus (Nieden, 1926)	+		+	+	
Probreviceps rungwensis (Loveridge, 1932)		+		+	
Spaeleophryne methneri (Ahl, 1924)*					
RANIDAE					
Afrana angolensis (Bocage, 1866)			+	+	+
Arthroleptides yakusini Channing, Moyer & Howell, 2002	+		+	+	
Strongylopus fuelleborni (Nieden, 1910)					+
Phrynobatrachus uzungwensis (Grandison & Howell, 1983)		+	+	+	
Phrynobatrachus parvulus (Boulenger, 1905)				+	
Phrynobatrachus rungwensis Loveridge, 1932				+	

Table 2. (continued)

	Site 1	Site 2	Site 3	Site 4	Site 5
Apoda					
Scolecomorphidae					
Scolecomorphus kirkii Boulenger, 1883		+		+	
REPTILIA					
Squamata					
Gekkonidae					
Cnemaspis uzungwae Perret, 1986	+		+		
Cnemaspis dickersoni (Schmidt, 1919)	+				
Scincidae					
Melanoseps uzungwensis Loveridge, 1942				+	
Melanoseps loveridgei Brygoo & Roux-Estève, 1982		+		+	
Scelotes uluguruensis (Barbour & Loveridge, 1928)	+				
Trachylepis varia (Peters, 1867)		+		+	
Trachylepis maculilabris (Gray, 1845)	+				
Chamaeleonidae					
Bradypodion oxyrhinum (Klaver & Böhme, 1988)				+	
Chamaeleo werneri (Tornier, 1899)				+	
Chamaeleo tempeli (Tornier, 1899)*					
Chamaeleo laterispinis (Loveridge, 1932)*					
Rhampholeon moyeri Menegon, Salvidio & Tilbury, 2002		+	+	+	
Rhampholeon brevicaudatus (Matschie, 1892)	+				
Typhlophidae					
Rhinotyphlops nigrocandidus Broadley & Van Wallach, 2000				+	
Colubridae					
Lycophidion capense jacksoni Boulenger, 1893	+				
Lycophidion uzungwense (Loveridge, 1932)				+	
Duberria lutrix shirana (Boulenger, 1894)		+		+	
Buhoma procterae (Loveridge, 1922)				+	
Crotaphopeltis tornieri (Werner, 1897)			+	+	
Thelotornis kirtlandii (Hallowell, 1844)				+	
Thelotornis mossambicanus (Bocage, 1895)	+		+		
Dipsadoboa werneri (Boulenger, 1897)		+			
Philothamnus macrops (Boulenger, 1895)			+		
Philothamnus hoplogaster Günther, 1863*					
Philothamnus semivariegatus (Smith, 1847)*					
Natriciteres variegata (Peters, 1854)			+		
Dasypeltis medici (Bianconi, 1859)			+		
Atractaspididae					
Atractaspis aterrima Günther, 1863*					
Viperidae					
Causus defilippii (Jan, 1862)*					
Atheris ceratophorus (Werner, 1895)		+		+	
Adenorhinos barbouri (Loveridge, 1930)*					
Bitis arietans (Merrem, 1820)*					
Bitis gabonica (Duméril & Bibron, 1854)*					

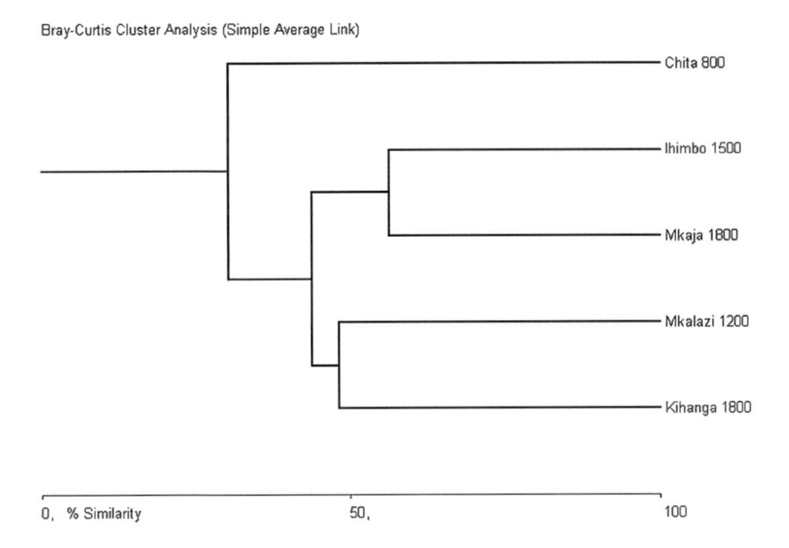

Figure 3. Similarity cluster among the five forest sites within the Udzungwa Scarp Forest Reserve.

seems to be the most discordant from the other sites, where relatively few endemics and more widespread and low altitude species are present. This indicates a marked species turnover between the lower part of the slopes and the submontane and montane layers, which are rich in specialised highland fauna. The firm trend of taxonomic turnover between the lowland and highland fauna of Tanzanian Mountains has been highlighted by Poynton (2003) and Loader et al. (2004). The remarkable similarity between sites 5 'Mkaja' and 2 'Ihimbo' might be related to the presence of extensive open wetlands and ecotone at both sites. In spite of the altitudinal differences (see Fig. 1), this indicates the presence of a specialised amphibian fauna, associated with open and ecotone habitat types. These habitats are completely absent or reduced in sites 4 'Kihanga' and 3 'Mkalazi', where a closed canopy forest is the main vegetation type.

Some of the strictly endemic species recorded for the Forest Reserve show a very restricted distribution area, often limited to a single site or valley (e.g., *Nectophrynoides wendyae, N. poyntoni, N. asperginis, Hyperolius kihangensis, Afrixalus* sp.), the complex distribution pattern and the exiguity of some species population deserve more research and conservation efforts. Large patches of undisturbed montane forests are still present in the Eastern Arc Mountains, while the lowland and sub-montane zones of the forest have suffered from extensive losses and fragmentation due to human disturbance (Newmark, 2002). In this context, the great size and altitudinal range of natural forest cover of this reserve, the extremely high level of strictly endemic and near endemic species, rank this reserve as one of the East-African sites with highest conservation priority.

ACKNOWLEDGEMENTS

Special thanks are due to many people who helped us in many ways. Kim H.Howell, David Moyer, John C.Poynton, Donald Broadley, Simon P.Loader, Charles Msuya, Colin McCarthy, Jens B.Rasmussen and Martin Pickersgill. Moreover Wilirk Ngalason, Francesco Rovero and Liana Trentin shared survey days with us. Finally we would like to thank the Tanzania Commission for Science and Technology (COSTECH) for providing permission to study Reptiles and Amphibians (permit NO 2002-039-ER-98-13); the CITES management authority in Tanzania and the Wildlife Division, who issued the necessary permits for the export of the specimens.

REFERENCES

Burgess, N. D., Fieldsä, J., and Botterweg, R., 1998, Faunal importance of the Eastern Arc Mountains of Kenya and Tanzania, *J. East Afr. Nat. Hist.* **87:**37-58.

Howell, K. M., 1993, Herpetofauna of the Eastern African forests, in: *Biogeography and Ecology of the Rain Forests of Eastern Africa*, J. C. Lovett and S. K. Wasser, eds., Cambridge University Press, New York, pp. 173-202.

Howell, K. M., 2000, An overview of East African Amphibian studies, past, present and future: a view from Tanzania, *Afr. J. Herpetol.* **49**(2):147-164.

Legendre, P., and Legendre, L., 1998, *Numerical Ecology,* 2nd English ed., Elsevier, pp. 853.

Loader, S. P., Poynton, J. C., and Mariaux, J., 2004, Herpetofauna of Mahenge Mountain, Tanzania: a window on African biogeography, *Afr. Zool.* **39**(1):71-76.

Lovett, J. C., 1993, Eastern Arc moist forest flora, in: *Biogeography and Ecology of the Rain Forests of Eastern Africa*, J. C. Lovett and S. K. Wasser eds., Cambridge University Press, New York. pp. 33-55.

Menegon, M., Salvidio, S., and Loader, S., 2004, Five new species of *Nectophrynoides* Noble 1926 (Amphibia Anura Bufonidae) from the Eastern Arc Mountains, Tanzania, *Trop. Zool.* **17**(1):97-121.

Myers, N., Mittermeyer, R. A., Mittermeier, C. G,. da Fonseca, G.A.B., and Kent, J., 2000, Biodiversity hotspots for conservation priorities, *Nature* **403:**853-858.

Newmark, W. D., 2002, Conserving Biodiversity in East African Forests, a Study of the Eastern Arc Mountains. *Ecological Studies 155.* Springer-Verlag, Berlin, pp. 197.

Perret, J. L., 1986, Révision des espèces africaines du genre *Cnemaspis* Strauch, sous-genre *Ancylodactylus* Müller (Lacertilia, Gekkonidae), avec la description de quatre espèces nouvelles, *Rev. Suisse Zool.* **93**(2):457-505.

Poynton, J. C., 2000, Evidence for an Afrotemperate amphibian fauna, *Afr. J. of Ecol.* **49**(1): 33-41.

Poynton, J. C., 2003, Altitudinal species turnover in southern Tanzania shown by anurans: some zoogeographical considerations, *Syst. Biodivers.* **1**(1):117-126.

Rasmussen, J. B., 1997, Afrikanske slanger (12) *Dipsadoboa werneri, Nordisk Herpetologisk Forening* **40**(6):173-177.

Shangali, C. F., Mabula, C. K., and Mmari, C., 1998, Biodiversity and human activities in the Udzungwa Mountain Forests, Tanzania. 1. Ethnobotanical survey in the Uzungwa Scarp Forest Reserve, *J. Est Afr. Nat. Hist.*, **87:**291-318.

INTERPRETING MORPHOLOGICAL AND MOLECULAR DATA ON INDIAN OCEAN GIANT TORTOISES

Justin Gerlach
University Museum of Zoology Cambridge, Downing Street, Cambridge CB2 3EJ, U.K.
E-mail: jstgerlach@aol.com

Abstract: Since 1877 it has been assumed that human exploitation led to the extinction of all Indian Ocean giant tortoises except for the Aldabran *Dipsochelys dussumieri*. A taxonomic review in 1998 proposed that two further species survived in captivity. Some recent molecular studies question the validity of these taxa. All available evidence is re-examined in a discussion of the identity of the Seychelles giant tortoises.
Morphological data identifies three living morphotypes (*D. dussumieri, D. arnoldi* and *D. hololissa*), data on wild and captive growth distinguishes these from the results of abnormal growth. Molecular data is less clear cut with conflicts between different studies. RAPDs identify the morphotypes although microsatellites identify very little population structuring. Different mtDNA genes conflict, with ND4 sequences relating to the different morphotypes but cytochrome *b* failing to do so.

Key words: *Cylindraspi; Dipsochelys;* Mascarenes; molecular taxonomy; morphology; Seychelles

1. INTRODUCTION

Many island systems are well known for the adaptive radiations that have occurred on them. The identification of species within these has been particularly difficult with many morphologically similar but isolated species and disjunct distribution patterns obscuring species relationships. The advent of molecular taxonomy analyses has enabled many cryptic species to be recognised and novel phylogenies constructed, although not without

B.A. Huber et al. (eds.), African Biodiversity, 213–219.

controversy (e.g. Wilson et al., 2000; Omoto et al., 2004). These techniques have recently been applied to western Indian Ocean reptiles (Radtkey, 1996; Austen and Arnold, 2001; Austen et al., 2003), including the tortoise species. These include the giant tortoise genus *Dipsochelys* which was formerly present on Madagascar and many of the Seychelles islands, now restricted to the Seychelles. The taxonomy of Indain Ocean giant tortoises has been highly contentious, with 28 described taxa. The most recent review of *Dipsochelys* (Gerlach and Canning, 1998a) recognised four species (three extant). Although this review has been followed by several authors, some recent molecular based publications have failed to support these conclusions (Austen et al., 2003; Palkovacs et al., 2003). The data concerning the identity of *Dispochelys* tortoises and the conflicts between morphological and molecular identities are reviewed below.

2. EXTERNAL MORPHOLOGY

Morphometric analysis of 354 tortoises (Gerlach, 2004a) supports the accuracy of the descriptions in Gerlach and Canning (1998a). Principal component analysis distinguishes four recent morphotypes: *D. arnoldi, D. daudinii, D. dussumieri* and *D. hololissa*. Males are clearly separated (4.3% overlap between morphotypes) whilst females have 8.8% overlap (Gerlach, 2004a). Analysis of the effects of diet on morphology confirms that this can cause gross distortion or scute pyramiding (Gerlach, 2004b). This impact on the carapace is not reflected in plastron morphology, scute proportions, depressions or notches between scutes. Thus even where captive diet provides a confounding factor, these characters remain taxonomically useful (Gerlach, 2004b).

Dipsochelys onotogeney identifies consistent differences between the morphotypes in coloration, shape and scute proportions that are not due to differences in incubation or parentage (Gerlach and Bour, 2003). Hatchling external morphology is also distinctive, with principal component analysis identifying morphological groupings (Gerlach and Bour, 2003).

3. OSTEOLOGY

The skeletons are highly conservative; variable elements (principally the skulls, femora and dorsal vertebrae) fall into four recent groups corresponding to morphotypes (Gerlach and Canning, 1998). At least some characters defining these groups could result from environmental factors (principally diet) but examination of skeletons of *D. dussumieri* with carapace morphology typical of severe metabolic bone disease shows that these have abnormally porous bones and rough outer surfaces typical of bones undergoing calcium resorption, however, no other abnormal features could be detected (Gerlach, 2004b). This suggests that the external appearance of the morphotypes and their associated osteological characters are of taxonomic significance. *Dipsochelys arnoldi* also has an exceptional arrangement of neural bones (Gerlach, 1999).

4. MOLECULAR DATA

The first molecular study used randomly amplified polymorphic DNA (RAPD) (Gerlach and Canning, 1998b). The limited data available was summarised graphically, indicating the correspondence of three genetic and morphological groups. Cluster analysis supports the initial conclusions with the recognition of three groupings corresponding to the three morphotypes. The only discrepancy is with individuals 4 and 16 which are identified morphometrically as *D. hololissa* but cluster with *D. dussumieri* in the RAPD data. Acceptance of these as *D. hololissa* leads to a minor modification of the interpretation; that the Aldabra population (*D. dussumieri*) represents a subset or descendant of the Seychelles species, *D. hololissa* (Fig. 1).

Microsatellite data should distinguish population structuring even in taxa with low genetic variability (Paetkau et al., 1995), but a study of microsatellites in *Dipsochelys* showed no clear population structuring

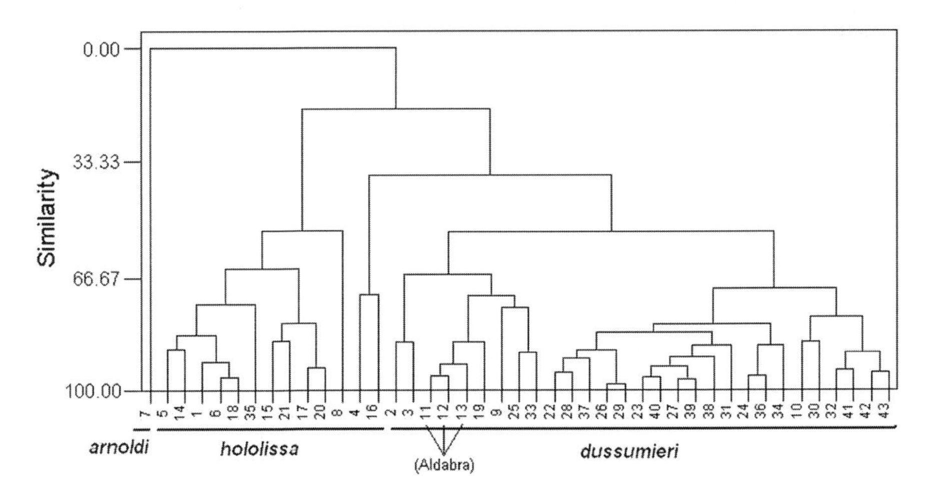

Figure 1. Cluster analysis of RAPD data, individuals are assigned taxon names based on morphometric analysis. The three individuals labelled Aldabra are of known Aldabran origin, all others are of unknown origin.

(Palkovacs et al., 2003) and was unable to resolve the relationships, with only 48% agreement between genotype and morphotype assignments. This assumed the identity of some individuals which had not been directly examined and analysed morphometically, and the results may be distorted by any misidentifications. Restriction of the analysis to directly examined individuals produces slightly different results: the proportion of correctly identified individuals is increased to 56%. Levels of allele differentiation and gene flow indicated by Wright's F_{ST} are not significant overall (F_{ST}=0.0252, P>0.05), although differentiation between *D. arnoldi* and *D. dussumieri* is significant (F_{ST}=0.0467, P<0.05) and the N_m values of gene flow are lowest between this pairing (5.07). The use of the island model in these estimates of N_m and F_{ST} has been criticised due to reservations concerning their biological realism (Whitlock and McCauley, 1999; Neigel, 2002), however, coalescence methods (Beerli, 2002) yield similar results with significant gene flow from *D. hololissa* to *D. dussumieri* (N_m=3.867), but little flow elsewhere other (0.150-1.516).

In the mitochondrial genome the control region is invariant in all living *Dipsochelys* (Palkovacs et al., 2003), indicating a common origin or very recent divergence. Cytochrome *b* sequences form six haplotypes (Austen et al., 2003; unpubl. GenBank sequences). These have been reported to indicate that the greatest diversity is found on Aldabra and that purported Seychelles tortoises are descended from this population (Austen et al., 2003). This interpretation, however, assumes all tortoises of unknown origin to be Aldabran. Classing these as 'unidentified' results in a different interpretation, with the greatest diversity being found in *D. hololissa* and *D. arnoldi*, and Aldabran tortoises being a slightly divergent subset of this grouping (Fig. 2). Living individuals comprise five haplotypes, although only one is retained on Aldabra. The more rapidly evolving ND4 gene (Macey et al., 2004) distinguishes the three living taxa, suggesting a very recent origin of the Aldabra tortoises from *D. hololissa* and a much deeper separation between these species and *D. arnoldi* (Cunningham et al., in prep.).

5. DISCUSSION

Morphological analysis identifies four recent groupings in *Dipsochelys*, of which three are extant (*D. arnoldi, D. dussumieri* and *D. hololissa*). The suggestion that these morphotypes result from dietary effects cannot be supported by the evidence. Growth patterns of captive bred juveniles, reared under identical conditions, indicate that the distinct morphotypes are maintained by underlying genetic factors (Gerlach and Bour, 2003).

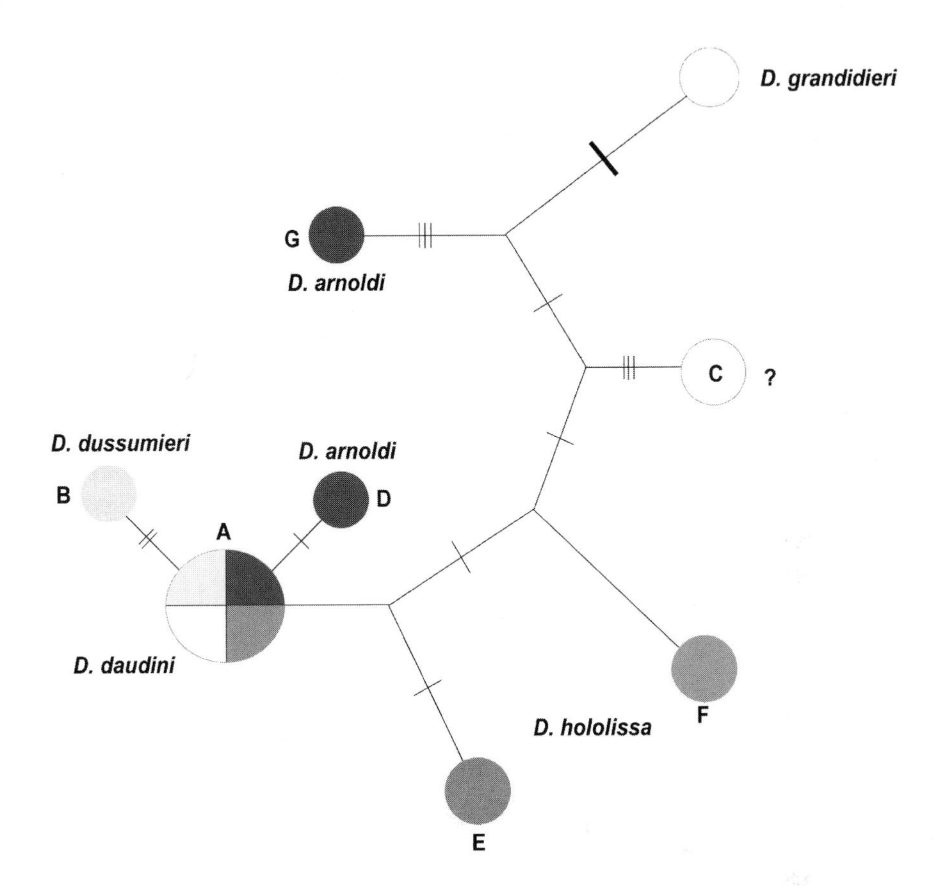

Figure 2. Cytochrome *b* haplotypes in all sampled *Dipsochelys*.

Molecular evidence appears ambiguous with an invariant control region, partial separation of cytochrome *b* haplotypes and three ND4 lineages. This can largely be explained by the different rates of evolution within different parts of the mitochondrial genome; in this case the ND4 sequences being more informative than the cytochrome *b* sequences due to their more rapid evolution. Conflicts between some of the mitochondrial data and the RAPD analysis have been suggested to be the result of limitations of RAPD methods or recent divergence, with the RAPD results reflecting functional genome changes rather than accumulated neutral mutations (Palkovacs et al., 2003). Although there are reservations about the interpretation and reliability of RAPD studies, the close clustering of samples of known Aldabran origin within the *D. dussumieri* grouping (Fig. 1) may indicate that this analysis does provide an accurate representation of whole genome relationships, and may thus be reliable. Microsatellite data suggest a possible synthesis of these data from a Malagasy tortoise colonising the granitic Seychelles islands within the last million years. This may have diverged into an inland, browsing *D. arnoldi* and a coastal, grazing *D. hololissa*. Ecological separation would have provided only a partial genetic barrier, allowing introgression and preventing clear distinction by most genetic methods. Reduced sea-levels 18,000 years ago enabled the coastal morphotype (*D. hololissa*) to colonise Aldabra. The Aldabran population became morphologically distinctive by virtue of the founder effect, its genes remaining a sub-set of those of *D. hololissa*. This pattern agrees with the data indicating strong gene flow from *D. hololissa* to *D. dussumieri.*

Molecular phylogenies of other Testudines are more clearly resolved although conflicts between morphological and molecular data have been found in *Cylidnraspis*, *Testudo, Cuora* and *Graptemys* due to recent divergence under strong selection pressures (Lamb et al., 1994; Kuyl et al., 2002; Stuart and Parham, 2004) which would not be easily detected in neutral genetic markers (Bowie, 2003).

Under the scenarios presented above the living *Dipsochelys* morphotypes are Evolutionarily Significant Units. Distinctions between taxonomic levels for such ESUs are academic as the distinct and diverging populations are the truly important units in evolutionary biology and biodiversity conservation. The three living *Dipsochelys* taxa all merit conservation attention as unique evolutionary groups. Conservation of the low genetic diversity Aldabran *D. dussumieri* population alone is not sufficient to ensure the preservation of the past evolutionary history of these animals or their future evolutionary potential. Whether they are considered full species or just varieties, all three taxa require conservation action.

ACKNOWLEDGEMENTS

I am grateful to Roger Bour, Claudio Ciofi and Eric Palkovacs for helpful discussions and comments, and to an anonymous reviewer for very constructive comments.

REFERENCES

Austen, J. J. and Arnold, E. N., 2001, Ancient mitochondrial DNA and morphology elucidate an extinct island radiation of Indian Ocean giant tortoises (*Cylindraspis*), *Proc. R. Soc. Lond. B* **268:**2515-2523.

Austen, J. J., Arnold, E. N. and Bour, R., 2003, Was there a second adaptive radiation of giant tortoises in the Indian Ocean? *Mol. Evol.* **12:**1396-1402.

Beerli, P., 2002, MIGRATE. Version 1.5. [http://evolution.geneticswashington.edu/lamarc.html]

Bowie, R. C. K., 2003, *Birds, molecules, and evolutionary patterns among Africa's islands in the sky*, Unpublished PhD thesis, University of Cape Town.

Gerlach, J., 1999, Distinctive neural bones in *Dipsochelys* giant tortoises, *J. Morphol.* **240:**33-38.

Gerlach, J., 2004a, *Giant Tortoises of the Indian Ocean*, Edition Chimaira, Frankfurt am Main.

Gerlach, J., 2004b, Effects of diet on the systematic utility of the tortoise carapace, *Afr. J. Herpetol.* **53:**77-85.

Gerlach, J., and Bour, R., 2003, Morphology of hatchling *Dipsochelys* giant tortoises, *Radiata* **12:**11-20.

Gerlach, J., and Canning, L., 1998a, Taxonomy of Indian Ocean giant tortoises (*Dipsochelys*), *Chel. Cons. & Biol.* **3:**3-19.

Gerlach, J., and Canning, L., 1998b, Identification of Seychelles giant tortoises, *Chel. Cons. & Biol.* **3:**133-135.

Honda, M., Yasukawa, Y., Hirayama, R., and Ota, H., 2002, Phylogenetic relationships of the Asian box turtles of the genus *Cuora* sensu lato (Reptilia: Bataguridae) inferred from mitochondrial DNA sequences, *Zool. Sci.* **19:**1305-1312.

Kuly, A. C. van der, Ballasina, D. L. P., Dekker, J. T., Maas, J., Willemsen, R. E. and Goudsmit, J., 2002, Phylogenetic relationships among the species of the genus *Testudo* (Testudinea: Teastudinidae) inferred from mitochondrial 12S rRNA gene sequences, *Mol. Phylogenet. Evol.* **22:**174-183.

Lamb, T., Lydeard, C., Walker, R. B., and Gibbons, J. W. 1994, Molecular systematics of map turtles (*Graptemys*), *Syst. Biol.* **43:** 543-559.

Macey, J. R., Papenfuss, T. J., Kuehl, J. V., Fourcade, H. M., and Boore, J. L., 2004, Phylogenetic relationships among amphisbaenian reptiles based on complete mitochondrial genomic sequences, *Mol. Phyl. Evol.* **33:**22–31.

Neigel, J. E., 2002, Is F_{ST} obsolete? *Conserv. Gene.* **3:**167-173.

Omoto, K., Katoh, T., Chichvarkhin., A., and Yagi, T., 2004, Molecular systematics and evolution of the "Apollo" butterflies of the genus *Parnassius* (Lepidoptera: Papilionidae) based on mitochondrial DNA sequence data, *Gene.* **326:**141-7.

Paetkau, D., Calvert, W., Stirling, I., and Strobeck, C., 1995, Microsatellite analysis of population structure in Canadian polar bears, *Mol Ecol.* **4:**347-354.

Palkovacs, E. P., Maschner, M., Ciofi, C., Gerlach, J., and Caccone, A., 2003, Are the native giant tortoises from the Seychelles really extinct? *Mol. Ecol.* **12:**1403-1414.

Radtkey, R. R., 1996, Adaptive radiation of day-geckoes (*Phelsuma*) in the Seychelles Archipelago: a phylogenetic analysis, *Evolution* **50:**604–623.

Stuart, B. L., and Parnham, J. F., 2004, Molecular phylogeny of the critically endangered Indochinese box turtle (*Cuora galbinifrons*), *Mol. Phylogenet. Evol.* **31:**164-177.

Whitlock, M. C., and McCauley, D. E., 1999, Indirect measures of gene flow and migration: $Fst \neq 1/(4Nm + 1)$, *Heredity* **82:**117-125.

Wilson, P. J., Grewal, S. K., Lawford, I. D., Heal, J.N.M., Granacki, A. D., Pennock, D., Theberge, J. B., Theberge, M. T., Voigt, D. R., Waddell, W., Chambers, R. E., Paquet, P. C., Goulet, G., Cluff, D., and White, B. N., 2000, DNA profiles of the eastern Canadian wolf and the red wolf provide evidence for a common evolutionary history independent of the gray wolf, *Can. J. Zool.* **78:**2156-2166.

MOLECULAR SYSTEMATICS OF AFRICAN COLUBROIDEA (SQUAMATA: SERPENTES)

Zoltán Tamás Nagy[1], Nicolas Vidal[2], Miguel Vences[3], William R. Branch[4], Olivier S.G. Pauwels[5], Michael Wink[1] and Ulrich Joger[6]

[1] *Institute of Pharmacy and Molecular Biotechnology, University of Heidelberg, INF 364, 69120 Heidelberg, Germany, E-mail: lustimaci@yahoo.com*

[2] *Department of Biology, 208 Mueller Laboratory, Pennsylvania State University, University Park, PA 16802, USA*

[3] *Institute for Biodiversity and Ecosystem Dynamics, Zoological Museum, University of Amsterdam, POB 94766, 1090 GT Amsterdam, The Netherlands*

[4] *Bayworld, POB 13147, 6013 Humewood, South Africa*

[5] *Department of Recent Vertebrates, Institut Royal des Sciences Naturelles de Belgique, Rue Vautier 29, 1000 Brussels, Belgium*

[6] *State Natural History Museum Braunschweig, Pockelsstrasse 10, 38106 Braunschweig, Germany*

Abstract: Phylogenetic relationships between African representatives of the Colubroidea have been relatively little investigated. In this paper, DNA sequences of three marker genes were used to identify phylogenetically relevant groups. Viperids represent a basal clade among Colubroidea. The two monophyletic families Elapidae and Atractaspididae are nested within the paraphyletic family Colubridae. African colubroid snakes are found within the cosmopolitan subfamilies Colubrinae and Natricinae, the mainly African subfamilies Lamprophiinae, Psammophiinae and Pseudoxyrhophiinae, and also within the families Elapidae and Atractaspididae. The clade comprising Lamprophiinae, Psammophiinae, Pseudoxyrhophiinae, Elapidae and Atractaspididae is likely to have an African origin, whereas the Colubroidea probably originated in Asia.

Key words: Africa; Serpentes; Colubroidea; phylogeny; mtDNA; *c-mos*

B.A. Huber et al. (eds.), African Biodiversity, 221–228.

1. INTRODUCTION

The African snake fauna has remained one of the biggest challenges for systematic herpetology. High diversity in general and low representation in collections have been the main obstacles for comprehensive analyses. For these reasons, the African taxonomic sampling used in previous molecular systematic investigations (e.g., Heise et al., 1995; Kraus and Braun, 1998; Gravlund, 2001) was insufficient. Moreover, those studies used mitochondrial genes only.

In a collaborative effort, we increased the taxonomic sampling among African colubroids, and used a combination of genes which have already been shown to be suitable at this level of diversification (Nagy et al., 2003).

2. MATERIALS AND METHODS

We tried to include representatives of as many afrotropical genera of colubrid and atractaspidid snakes as possible. A choice of African elapid and viperid taxa were also included, as well as some Eurasian representatives of the genera *Hemorrhois*, *Psammophis*, *Telescopus*, *Platyceps*, *Ptyas* and *Vipera*. Three taxa were used as hierarchical outgroups in the phylogenetic analyses: a scolecophidian snake (*Rhinotyphlops schlegelii*) and two henophidians (*Sanzinia madagascariensis* and *Calabaria reinhardtii*).

Three molecular markers known to have different evolutionary and functional characteristics were amplified in polymerase chain reactions and then sequenced directly (see Nagy et al., 2003 for details): (1) the entire mitochondrial cytochrome *b* gene; (2) part of the mitochondrial 16S rRNA gene; (3) part of the nuclear *c-mos* gene. The partition homogeneity test was performed for the combined set of sequences.

For phylogenetic reconstructions, maximum parsimony (MP) analyses were carried out with PAUP* 4.0b10 (Swofford, 2002), as well as Bayesian inference of phylogeny (BI) using MrBayes 3.0b4 (Huelsenbeck and Ronquist, 2001). For likelihood based methods, an appropriate model of nucleotide substitution was inferred with Modeltest 3.06 (Posada and Crandall, 1998).

All DNA sequences – including collection data of the investigated specimens – were deposited in GenBank and will be published later.

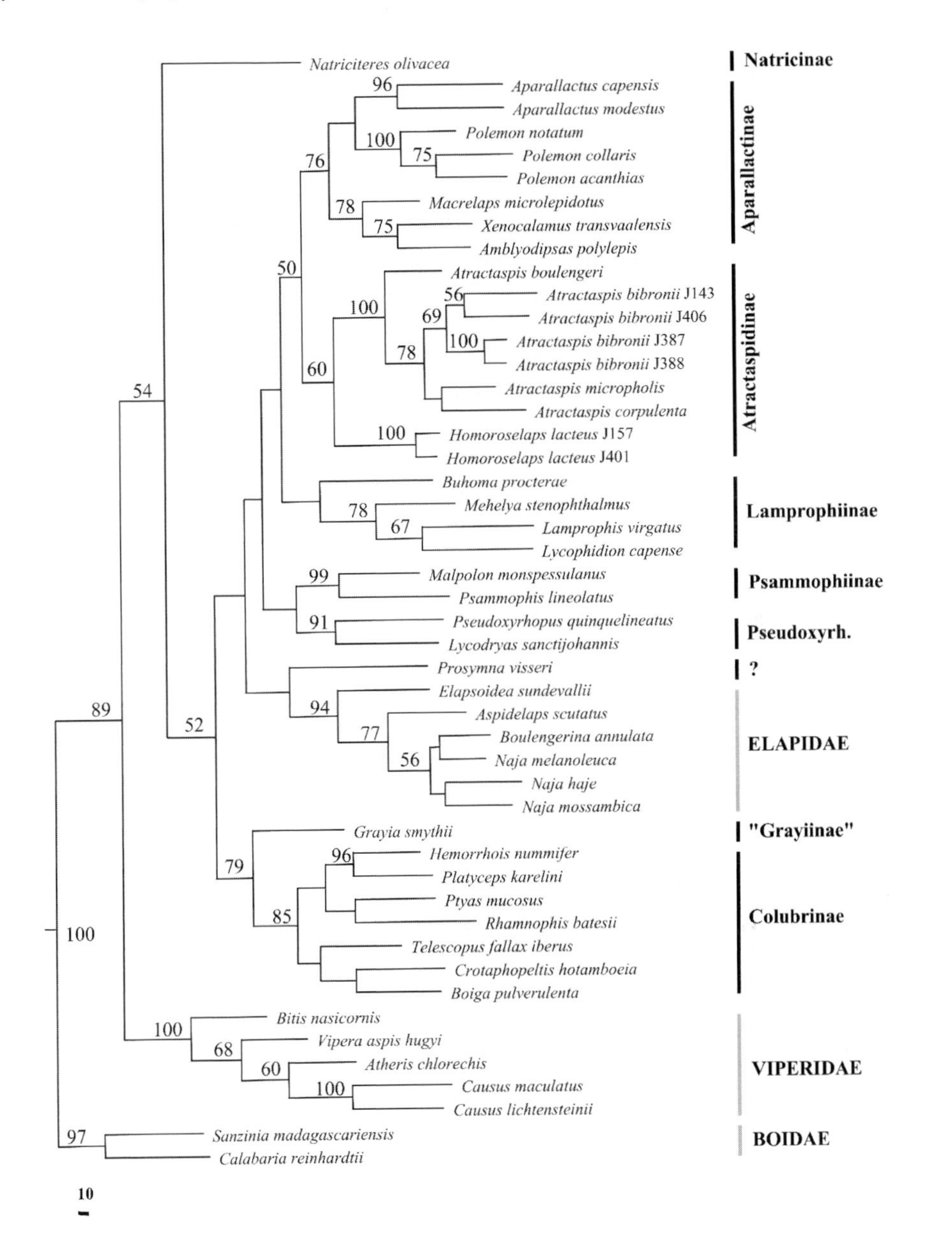

Figure 1. Maximum Parsimony phylogram of African Colubroidea, based on cytochrome *b*, 16S rRNA and *c-mos* sequences (one most parsimonious tree, heuristic search, TBR branch swapping). The most basal outgroup *Rhinotyphlops* is not shown. Numbers are bootstrap values in percentages and given for clades that received support values over 50% (1000 replicates). The actually recognized taxa on family-level are capitalized and indicated with grey bars. Subfamilian taxa are indicated with black bars. *Incertae sedis* are marked with question marks.

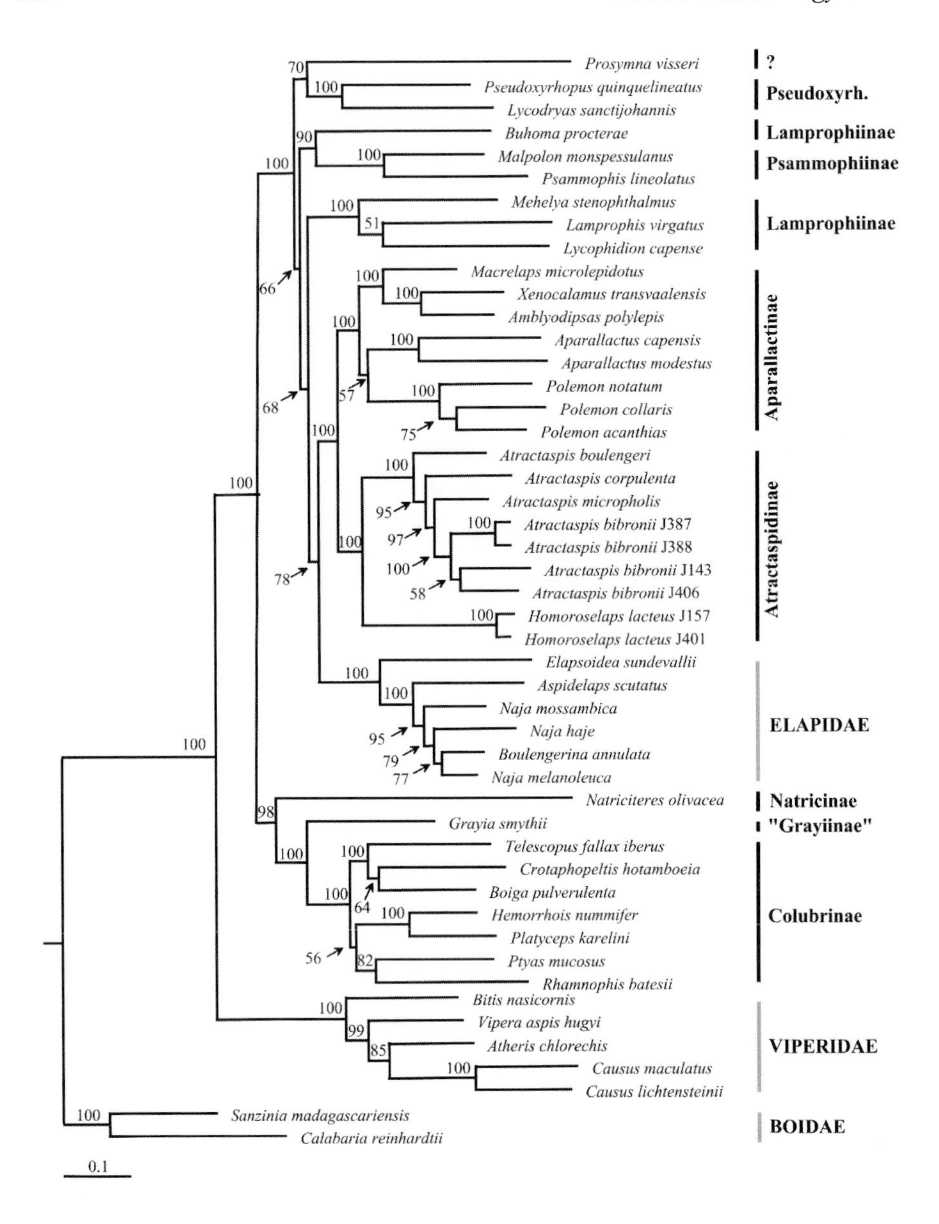

Figure 2. Bayesian Inference phylogram of African Colubroidea, calculated on the nucleotide substitution model GTR+I+G (one million generations, sampling of every 100th generation, the first 500 of 10000 trees were discarded). Numbers are clade credibility values in percentages and given for clades that received support values over 50%. For further details, see figure caption of Fig. 1.

3. RESULTS

The partition homogeneity test did not contradict the congruence of the three investigated genes (p=0.742), thus all sequences were used in the combined analyses. The combined data set comprises 2186 bp for 49 taxa (after the exclusion of the hypervariable regions of 16S rRNA).

The most important result of our phylogenetic analyses is that the colubrid snakes are paraphyletic – even considering African taxa only. Colubrids form a monophyletic unit only with the inclusion of elapid and atractaspidid snakes. The family Viperidae appears to be the sister taxon of this entire group. This observation is in agreement with recent molecular studies (Cadle, 1994; Knight and Mindell, 1994; Dowling et al., 1996; Slowinski and Lawson, 2002; Vidal and Hedges, 2002a, b).

The overall topology of the combined trees (MP: Fig. 1 and BI: Fig. 2) is well supported by high bootstrap and posterior probability values, respectively. Moreover, repetitions of the Bayesian analysis were in excellent agreement with the first one. The main difference between the results of the two reconstruction methods are found (a) in the positions of Elapidae and Atractaspididae among colubrid subfamilies, and (b) in the phylogenetic status of the subfamily Lamprophiinae. A high degree of congruence is found (a) in the monophyly of the Viperidae, Elapidae, Atractaspididae, Psammophiinae, Pseudoxyrhophiinae, and Colubrinae and (b) in the relationships within those clades.

The summarizing tree of all combined analyses is shown in Figure 3.

4. DISCUSSION

4.1 A diversified picture of African venomous snake evolution

The basal position of viperids is congruent with results of other authors (see Results). However, the assumption of earlier studies (e.g. Cadle, 1987; Lenk et al., 2001) that the genus *Causus* represents the most basal lineage within Old World viperids (based on a presumably primitive condition of scalation, unusually round pupils and the morphology of the venom apparatus) is not supported by our data.

The solenoglyphous dentition displayed by the genus *Atractaspis* is not homologous to the solenoglyphous dentition displayed by the viperids. According to our results (Figs. 1 and 2), Atractaspididae are monophyletic and consist of two monophyletic subfamilies, Atractaspidinae and

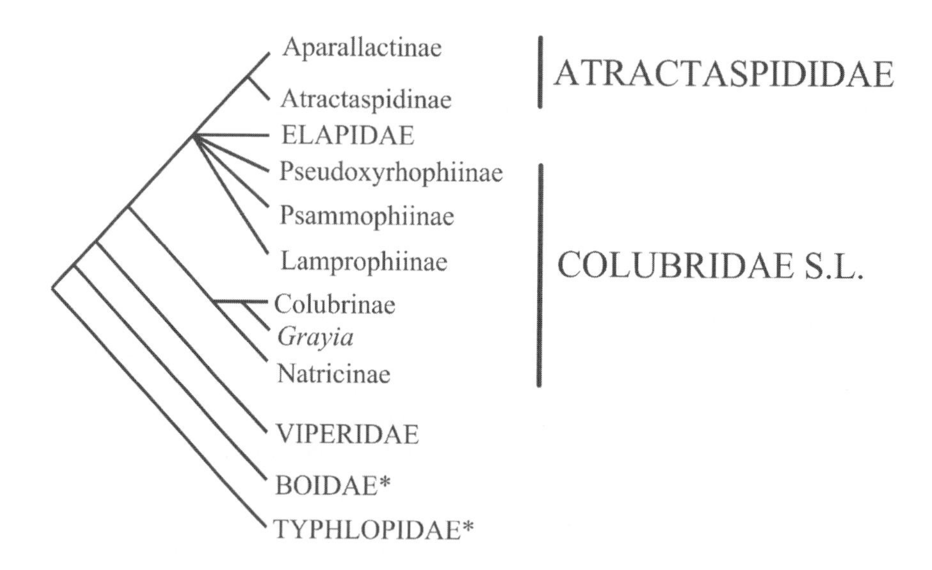

Figure 3. Overall (summarizing) tree showing the general phylogeny of African Colubroidea, based on combined analyses of cytochrome *b*, 16S rRNA and *c-mos* sequences. Asterisks mark outgroups.

Aparallactinae. Within Aparallactinae, the genera *Macrelaps*, *Xenocalamus* and *Amblyodipsas* form a strongly supported monophyletic group. Within Atractaspidinae, the genus *Homoroselaps* is the sister group to the genus *Atractaspis*. Moreover, we found considerable intraspecific variation within *Atractaspis bibronii*. Initially, *Homoroselaps* was considered to be a member of the nominative genus of the family Elapidae, *Elaps*. McDowell (1968) pointed to large morphological differences between elapids and *Homoroselaps* and postulated that the latter was rather a colubrid. Underwood and Kochva (1993) put it back into the family Elapidae. Slowinski and Keogh (2000) found *Homoroselaps* to be basal to the elapids, but unfortunately did not include atractaspidids in their study. Our investigations confirm earlier results (Vidal and Hedges, 2002a) that *Homoroselaps* is part of the Atractaspidinae.

We analysed only African representatives of the family Elapidae. With this limitation, elapids appear monophyletic. However, the relationships of elapids within our African colubroid data set remain unresolved. In the BI phylogram (Fig. 2), Atractaspididae are their sister group, while in the MP tree (Fig. 1), the enigmatic genus *Prosymna* appears closer, but with no significant support. Within African elapids, *Naja* species occupy a derived position, which is concordant with results by Slowinski and Keogh (2000). Moreover, *Naja* appears paraphyletic, as the aquatic cobra *Boulengerina annulata* branches off from within *Naja*. In accordance with results by

Slowinski et al. (1997), the genus name *Boulengerina* should therefore be synonymized with the genus name *Naja*.

4.2 Different origins of the diversity of the African colubrid snake fauna

In our trees, a clear separation of African colubrids into two major groups is evident. One of them contains only Old World subfamilies of African dominance. Species of the subfamily Psammophiinae are found in Africa (including Madagascar) and Asia. The Pseudoxyrhophiinae occupy Madagascar and Socotra (Nagy et al., 2003). The Lamprophiinae include African species only, with the exception of the Asiatic genus *Psammodynastes* (Vidal and Hedges, 2002a). The fact that the afrotropical Atractaspididae and Elapidae branch off from the same origin as this assemblage indicates that it is of African origin.
On the other hand, the two other colubrid subfamilies including African members, Colubrinae and Natricinae, occur both in the Old and New Worlds. An Asian origin of these two lineages is generally assumed (e.g. Rage et al., 1992). According to our study and Vidal and Hedges (2002a), the position of the enigmatic African endemic genus *Grayia* as sister group to the colubrines is strongly supported. Presently, more comprehensive studies are in progress which may lead to the recognition of a new subfamily including the genus *Grayia* only.

ACKNOWLEDGEMENTS

The authors thank Wolfgang Böhme, Frank Glaw and Michele Menegon for providing samples. The manuscript benefited from useful comments provided by Dan Mulcahy and an anonymous reviewer.

REFERENCES

Cadle, J. E., 1987, Geographic distributions: problems in phylogeny and zoogeography, in: *Snakes: ecology and evolutionary biology*, R. A. Siegel, J. T. Collins, and S. S. Novak, eds., MacMillan, New York, pp. 77-105.

Cadle, J. E., 1994, The colubrid radiation in Africa (Serpentes: Colubridae): phylogenetic relationships and evolutionary patterns based on immunological data, *Zool. J. Linn. Soc.* **110:**103-140.

Dowling, H. G., Haas, C. A., Hedges, S. B., and Highton, R., 1996, Snake relationships revealed by slow evolving proteins: a preliminary survey, *J. Zool.* **240:**1-28.

Gravlund, P., 2001, Radiation within the advanced snakes (Caenophidia) with special emphasis on African opistoglyph colubrids, based on mitochondrial sequence data, *Biol. J. Linn. Soc.* **72:**99-114.

Heise, P. J., Maxson, L. R., Dowling, H. G., and Hedges, S. B., 1995, Higher-level snake phylogeny inferred from mitochondrial DNA sequences of 12S rRNA and 16S rRNA genes, *Mol. Biol. Evol.* **12:**259-265.

Huelsenbeck, J. P., and Ronquist, F., 2001, MrBayes: Bayesian inference of phylogenetic trees, *Bioinformatics* **17:**754–755.

Knight, A., and Mindell, D. P., 1994, On the phylogenetic relationship of Colubrinae, Elapidae, and Viperidae and the evolution of front-fanged venom systems in snakes, *Copeia* **1994:**1-9.

Kraus, F., and Brown, W. M., 1998, Phylogenetic relationships of colubroid snakes based on mitchondrial DNA sequences, *Zool. J. Linn. Soc.* **122:**455-487.

Lenk, P., Kalyabina, S., Wink, M., and Joger, U., 2001, Evolutionary relationships among the true vipers (Viperinae) suggested by mitochondrial DNA sequences, *Mol. Phylogenet. Evol.* **19:**94-104.

McDowell, S. B., 1968, Affinities of the snakes usually called *Elaps lacteus* and *E. dorsalis*, *Zool. J. Linn. Soc.* **47:**561-578.

Nagy, Z. T., Joger, U., Wink, M., Glaw, F., and Vences, M., 2003, Multiple colonization of Madagascar and Socotra by colubrid snakes: evidence from nuclear and mitochondrial gene phylogenies, *Proc. R. Soc., Lond.* B **270:**2613-2621.

Posada, D., and Crandall, K. A., 1998, Modeltest: testing the model of DNA substitution, *Bioinfomatics* **14:**817-818.

Rage, J.-C., Buffetaut, E., Buffetaut-Tong, H., Chaimanee, Y., Ducrocq, S., Jaeger, J.-J., and Suteethorn, V., 1992, A colubrid snake in the late Eocene of Thailand: The oldest known Colubridae (Reptilia, Serpentes), *C. R. Acad. Sci. Paris* **314:**1085-1089.

Slowinski, J. B., and Keogh, J. S., 2000, Phylogenetic relationships of elapid snakes based on cytochrome *b* mtDNA sequences, *Mol. Phylogenet. Evol.* **15:**157-164.

Slowinski, J. B., Knight, A., and Rooney, A. P., 1997, Inferring species trees from gene trees: a phylogenetic analysis of the Elapidae (serpents) based on the amino acid sequences of venom proteins, *Mol. Phylogenet. Evol.* **8:**349-362.

Slowinski, J. B., and Lawson, R., 2002, Snake phylogeny: evidence from nuclear and mitochondrial genes, *Mol. Phylogenet. Evol.* **24:**194-202.

Swofford, D. L., 2002, *PAUP*. Phylogenetic Analysis Using Parsimony (* and other methods), Version 4.0b10*, Sinauer, Sunderland, MA.

Underwood, G., and Kochva, E., 1993, On the affinities of the burrowing asps *Atractaspis* (Serpentes: Atractaspididae), *Zool. J. Linn. Soc.* **107:**3-64.

Vidal, N., and Hedges, S. B., 2002a, Higher-level relationships of caenophidian snakes inferred from four nuclear and mitochondrial genes, *C. R. Biologies* **325:**987-995.

Vidal, N., and Hedges, S. B., 2002b, Higher-level relationships of snakes inferred from four nuclear and mitochondrial genes, *C. R. Biologies* **325:**977-985.

PHYLOGENY AND BIOGEOGRAPHY OF MALAGASY DWARF GECKOS, *LYGODACTYLUS* GRAY, 1864: PRELIMINARY DATA FROM MITOCHONDRIAL DNA SEQUENCES (SQUAMATA: GEKKONIDAE)

Marta Puente[1], Meike Thomas[2], and Miguel Vences[3]

[1] *Departamento de Bioloxía Animal, Laboratorio de Anatomía Animal, Universidade de Vigo, Spain; E-mail: martapuente@tiscali.es*

[2] *Zoologisches Forschungsinstitut und Museum A. Koenig, Adenauerallee 160, 53113 Bonn,Germany; E-mail: meike.thomas@uni-koeln.de*

[3] *Institute for Biodiversity and Ecosystem Dynamics, Zoological Museum, University of Amsterdam, Mauritskade 61, 1092 AD Amsterdam, The Netherlands; E-mail: vences@science.uva.nl*

Abstract: Dwarf geckos, *Lygodactylus*, are small, secondarily diurnal lizards that occur in Africa, South America and Madagascar. Identification of many species is based on scale features only, and is therefore extremely difficult in the field due to their small size. Based on DNA sequences of a fragment of the mitochondrial 16S rRNA gene we studied the relationships among the Malagasy taxa and their differentiation relative to some African *Lygodactylus*. The results do not contradict a division of Malagasy *Lygodactylus* into three main lineages, which had been defined as Oriental, Occidental and Meridional clades. However, while the Oriental lineage (subgenus *Domerguella*) was very strongly corroborated by our analysis, no significant support was found for the other groupings. No evidence for a monophyletic group of the Malagasy taxa, or for a monophyletic *Lygodactylus* to the exclusion of *Microscalabotes* was found. The genetic divergences among Malagasy *Lygodactylus* were surprisingly large, indicating a relatively old age of most species and especially of the major lineages.

Key words: Reptilia; Squamata; Gekkonidae; *Lygodactylus*; *Microscalabotes; Millotisaurus*; *Domerguella*; Madagascar

B.A. Huber et al. (eds.), African Biodiversity, 229–235.

1. INTRODUCTION

Lygodactylus are lizards belonging to the family Gekkonidae. The genus is formed by small sized species, with snout-vent lengths usually below 40 mm, although some African forms can reach 42 mm (Pasteur, 1964). In contrast to many other geckos, they are diurnal and have round pupils. Although other diurnal geckos are often bright colored, most species of *Lygodactylus* have a cryptic colour and pattern. The genus currently contains about 59 species, distributed in subsaharan Africa (36 species) and Madagascar (21 species), as well as in South America (two species often classified in their own genus *Vanzoia*).

In the Malagasy region (including the small Mozambique Channel islands Europa and San Juan de Nova), *Lygodactylus* are widely distributed and populate many different environments, from rainforest to semi-desert and also including high mountain habitats up to 2700 m in the Ankaratra Massif (Glaw and Vences, 1994).

The systematics of Malagasy dwarf geckos has been studied by G. Pasteur in a number of papers (e.g. Pasteur, 1962, 1964, 1965, 1995; Pasteur and Blanc, 1967, 1973, 1991). At present they are considered to be divided into three main lineages, supposed to be monophyletic: the Oriental, Occidental and Meridional clades (Pasteur, 1965). The Oriental clade has been formally described as subgenus *Domerguella* Pasteur, 1964, whereas no name is available for the remaining two clades. The genus *Millotisaurus*, described by Pasteur (1962), was later synonimized with *Lygodactylus*, but the single species *L. mirabilis* was not assigned to any of the clades in this genus (Pasteur, 1995). A further, putatively related genus is the monoytypic *Microscalabotes* Boulenger, 1883, which differs from *Lygodactylus* by the structure of its subdigital lamellae.

We here present preliminary results from a molecular survey of 15 species of Malagasy dwarf geckos, as well as of some African species, in order to test previous assumptions of their phylogeny and biogeography.

2. MATERIALS AND METHODS

Tissue samples were taken from specimens collected in the field and stored in 90-96% ethanol. DNA was extracted using standard salt extraction protocols. A part of the mitochondrial 16S rRNA gene (505 bp aligned) was amplified with the primers 16SAL and 16SBH (Palumbi et al., 1991). Cycle sequencing was perfomed with the primer 16SAL, and the product resolved on an automated sequencer (ABI 3100). Sequences were aligned manually in the program Sequence Navigator, and all hypervariable and gapped positions (altogether 93 characters) excluded. Phylogenetic analysis was performed

Table 1. Localities, voucher numbers and Genbank accession numbers of *Lygodactylus* specimens studied. Collection acronyms used: UADBA, Universite d'Antananarivo, Département de Biologie Animale; ZSM, Zoologische Staatssamlung München; ZMA, Zoological Museum Amsterdam. Some vouchers have not yet been catalogued and are given with their field numbers (MV, collection of M.Vences; FGMV, collection of F.Glaw and M.Vences). All specimens except for the first five originated from Madagascar

Species	**Locality**	**Specimen-Voucher**	**Genbank accession**
Lygodactylus cf. *capensis*	Namibia, Popa falls	not collected	AY653248
Lygodactylus sp. 2	Africa (pet trade)	not collected	AY653249
Lygodactylus sp. 1	South Africa	not collected	AY653277
Lygodactylus gutturalis	Guinea Bissau	tissueWME13	AY653251
Lygodactylus gutturalis	Guinea Bissau	tissueWME30	AY653252
Lygodactylus arnoulti	Mount Ibity	tissue 2000B.31	AY653240
Lygodactylus arnoulti	Mount Ibity	UADBA-MV 2001.501	AY653241
Lygodactylus blancae	Lac Itasy	UADBA-MV 2001.260	AY653245
Lygodactylus blancae	Lac Itasy	ZSM 498/2001	AY653246
Lygodactylus guibei	Vohidrazana	ZMA 19631	AY653250
Lygodactylus heterurus	Sambava	ZSM 388/2000	AY653253
Lygodactylus madagascariensis	Tsaratanana	ZSM 781/2001	AY653254
Lygodactylus madagascariensis	Manongarivo	ZSM-FGMV 2002.778	AY653239
Lygodactylus madagascariensis	Manongarivo	FGMV 2002.722	AY653255
Lygodactylus madagascariensis	Manongarivo	UADBA-FGMV 2002.779	AY653256
Lygodactylus madagascariensis	Montagne d'Ambre	ZSM-FGMV 2002.942	AY653257
Lygodactylus madagascariensis	Manongarivo	ZSM-FGMV 2002.721	AY653258
Lygodactylus miops	Ranomafana	ZSM-FGMV 2002.456	AY653260
Lygodactylus miops	Ranomafana	ZSM-FGMV 2002.458	AY653261
Lygodactylus miops	Ranomafana	ZSM-FGMV 2002.459	AY653262
Lygodactylus miops	Ranomafana	UADBA 20735	AY653263
Lygodactylus mirabilis	Ankaratra	ZSM 388/2000	AY653247
Lygodactylus pauliani	Itremo	ZSM 490/2001	AY653264
Lygodactylus aff. *pictus*	Isalo	ZMA 19595	AY653238
Lygodactylus pictus	Ambositra	MV 2002.778	AY653259
Lygodactylus pictus	Ambositra	ZMA 19532	AY653265
Lygodactylus pictus	Ambositra	ZMA 19535	AY653266
Lygodactylus pictus	Ambositra	ZMA 19536	AY653267
Lygodactylus pictus	Ambositra	ZMA 19537	AY653268
Lygodactylus pictus	Ambositra	ZMA 19538	AY653269
Lygodactylus pictus	Sendrisoa	UADBA-FGMV 2001.608	AY653270
Lygodactylus pictus	Ambositra	ZMA 19531	AY653271
Lygodactylus pictus	Ambositra	ZMA 19534	AY653272
Lygodactylus pictus	Antsirabe	MV 2002.2	AY653276

Table 1. (continued).

Species	**Locality**	**Specimen-Voucher**	**Genbank accession**
Lygodactylus rarus	Ankarana	UADBA-FGMV 2002.892	AY653273
Lygodactylus rarus	Ankarana	ZSM-FGMV 2002.941	AY653274
Lygodactylus sp.	Tsaratanana	ZSM 783/2001	AY653275
Lygodactylus tolampyae	Berara	ZSM 419/2000	AY653278
Lygodactylus tolampyae	Ankarafantsika	tissue 2001C.5	AY653279
Lygodactylus tolampyae	Ankarafantsika	ZSM 501/2001	AY653280
Lygodactylus tuberosus	Toliara	UADBA 21069	AY653281
Lygodactylus tuberosus	Toliara	ZMA 19600	AY653282
Lygodactylus tuberosus	Toliara	UADBA 21073	AY653283
Lygodactylus tuberosus	Toliara	ZMA 19601	AY653284
Lygodactylus tuberosus	Toliara	ZMA 19599	AY653285
Lygodactylus tuberosus	Toliara	FGMV 2002.1589	AY653286
Lygodactylus tuberosus	Toliara	FGMV 2002.1591	AY653287
Lygodactylus tuberosus	Toliara	ZMA 19608	AY653288
Lygodactylus verticillatus	Ifaty	ZMA 19596	AY653289
Lygodactylus verticillatus	Ifaty	FGMV 2002.2013	AY653290
Lygodactylus verticillatus	Ifaty	FGMV 2002.2062	AY653291
Microscalabotes bivittis	Andasibe	tissue 2001A.21	AY653242
Microscalabotes bivittis	Andasibe	tissue 2001A.3	AY653243
Microscalabotes bivittis	Andasibe	tissue 2001B.9	AY653244

using PAUP*, version 4b10 (Swofford, 2002). We calculated a maximum likelihood tree after determining the substitution model best fitting the data using Modeltest (Posada and Crandall, 1998). A full heuristic search was performed. Robustness of nodes was tested using maximum likelihood (100 replicates) and maximum parsimony bootstrapping (1000 replicates); for maximum likelihood, several very similar or identical conspecific haplotypes were excluded from the data set due to computational constraints. A species of the diurnal gecko genus *Phelsuma* was used as the outgroup. See Table 1 for voucher specimens and Genbank accession numbers of DNA sequences.

3. RESULTS

The maximum likelihood phylogram (Fig. 1) largely follows the division into Oriental, Meridional and Occidental clades. However, the bootstrap support for these clades is poor except for the oriental clade (subgenus *Domerguella*), which receives a support of 87% in the ML bootstrapping. *Microscalabotes bivittis* groups with species of the Meridional clade, but this placement is not significantly supported. The African species included are

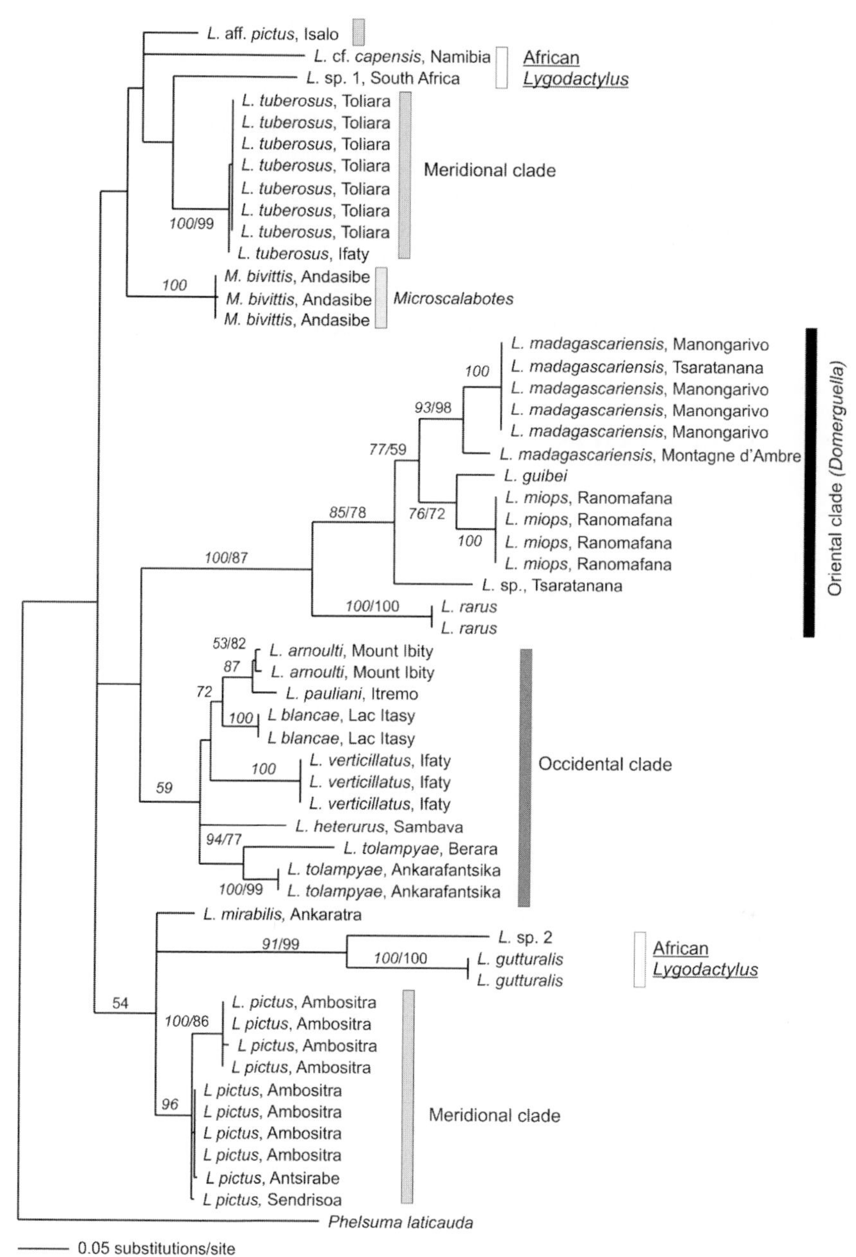

Figure 1. Maximum likelihood phylogram of Malagasy *Lygodactylus*, based on 412 nucleotides of the mitochondrial 16S rRNA gene (hypervariable regions and gapped sites excluded). Numbers at nodes are bootstrap values in percent from a maximum likelihood analysis (100 replicates of a reduced set of taxa, with only single representatives of each species) and maximum parsimony (1000 replicates; in italics).

placed at very different positions of the tree, in both cases nested among the Malagasy taxa. In several species, strong divergences among presumably conspecific individuals were observed. This is true, for instance, among individuals of *Lygodactylus tolampyae* from Berara and Ankarafantsika (12% uncorrected pairwise sequence divergence) and of *L. madagascariensis* from Montagne d'Ambre and Manongarivo/Tsaratanana (10-11%). Within the *L. pictus* population of Ambositra, two divergent haplotypes were found to coexist (divergences of 3%). One individual from Isalo tentatively assigned to *L. pictus* was placed at a large distance from all other included haplotypes of this and other species.

4. DISCUSSION

Our preliminary data indicate that molecular studies bear the potential of clarifying several issues of *Lygodactylus* systematics and biogeography, and to serve as a marker to identify potentially new taxa. However, the gene fragment used here alone is certainly not informative enough to provide a well supported phylogeny for the genus. This poor resolution in our tree might be caused by the high genetic differentiation of these lizards, that possibly is indicative of old ages of most currently recognized species and clades.

Our results did support with sufficiently high bootstrap values the monophyly of the Oriental clade or subgenus *Domerguella*, and interestingly also most internal splits in that clade are reliably resolved. In this group, the two most basal taxa (*L. rarus* and *L.* sp. from Tsaratanana) are distributed in the north and north-west of Madagascar, while the species from the eastern rainforest belt (*L. miops* and *L. guibei*) form a well supported monophyletic group. This indicates that the initial diversification of the Oriental clade may have occurred in the north of Madagascar which also today is the center of diversity of the group (with at least four species: *L. expectatus, L. madagascariensis, L. rarus, L.* sp.)

Several other hypotheses are suggested by our data and need further exploration with more extensive datasets. *Microscalabotes bivittis* was placed with *L. tuberosus* and some African species of *Lygodactylus*, indicating that this monotypic genus might be a synonym of *Lygodactylus.* If the phylogenetic position of the African species within the Malagasy clades was corroborated, it would indicate a complex biogeographic history of the genus, possibly involving multiple and multi-directional dispersal events as postulated for chameleons (Raxworthy et al., 2003).

ACKNOWLEDGEMENTS

We are grateful to Alain Dubois and Annemarie Ohler (MNHN, Paris) who allowed examination of specimens held in their care, to Wolfgang Böhme (ZFMK, Bonn), Franco Andreone (MRSN, Torino), David R. Vieites (Berkeley), and Frank Glaw (ZSM, München) who provided important material. Fieldwork in Madagascar was made possible by a cooperation with the University of Antananarivo (Madagascar) and financially supported by grants of the University of Vigo to M.P., NWO/WOTRO to M. V., and of the Deutscher Akademischer Austauschdienst to M.T. and M.V.

REFERENCES

Glaw, F., and Vences, M., 1994, *A Fieldguide to the Amphibians and Reptiles of Madagascar, 2nd ed.*, Vences and Glaw Verlag, Köln.

Palumbi, S., Martin, A., Romano, S., McMillan, W.-O., Stice, L., and Grabowski, G., 1991, *The Simple Fool's Guide to PCR. Version 2.0,* Priv. published document compiled by S.R. Palumbi, Dept. Zool., Honolulu Univ., Hawaii.

Pasteur, G., 1962, Notes préliminaires sur les lygodactyles (Gekkonidés). III. Diagnose de *Millotisaurus* gen. nov., de Madagascar, *C. R. Soc. Sci. nat. phys. Maroc* **28:**65-66.

Pasteur, G., 1964, Notes préliminares sur les lygodactyles (Gekkonidés). IV Diagnoses de quelques formes africaines et malgaches, *Bull. Mus. nat. Hist. nat.* **36:**311-314.

Pasteur, G., 1965, Recherches sur l'evolution des lygodactyles, lezards afro-malgaches actuels, *Trav. Inst. sci. Chérifien, Ser. Zool., Rabat* **29:**1-132.

Pasteur, G., 1995, Biodiversité et reptiles: diagnoses de sept nouvelles espèces fóssiles et actuelles du genre de lézards *Lygodactylus* (Sauria, Gekkonidae), *Dumerilia* **2:**1-21.

Pasteur, G., and Blanc, C. P., 1967, Les lézards du sous-genre malgache de lygodactyles *Domerguella* (Gekkonidés), *Bull. Soc. Zool. France* **92:**583-597.

Pasteur, G., and Blanc, C. P., 1973, Nouvelles études sur les lygodactyles (Sauriens Gekkonidés). I. Donneés récentes sur *Domerguella* et sur ses rapports avec la phytogéographie malgache, *Bull. Soc. Zool. France* **98:**165-174.

Pasteur, G., and Blanc, P. C., 1991, Un lézard parthénogénétique à Madagascar? Description de *Lygodactylus pauliani* sp. nov. (Reptilia, Gekkonidae), *Bull. Mus. Natl. Hist. nat., Paris, 4° sér.,* **13**:209-215.

Posada, D., and Crandall, K. A., 1998, Modeltest: testing the model of DNA substitution, *Bioinformatics* **14:**817-818.

Raxworthy, C. J., Forstner, M. R. J., and Nussbaum, R. A., 2002, Chameleon radiation by oceanic dispersal, *Nature* **415:**784-786.

Swofford, D.L., 2002, *PAUP*. Phylogenetic Analysis Using Parsimony (* and other methods). Version 4b10*, Sinauer assoc., Sunderland, Massachusetts.

VARIABILITY IN A COMMON SPECIES: THE *LYGODACTYLUS CAPENSIS* COMPLEX FROM SOUTHERN AND EASTERN AFRICA (REPTILIA, GEKKONIDAE)

Beate Röll
Lehrstuhl für Allgemeine Zoologie und Neurobiologie, Ruhr-Universität Bochum, D-44780 Bochum, Germany, Email: roell@neurobiologie.ruhr-uni-bochum.de

Abstract: In live specimens of the dwarf gecko *Lygodactylus capensis* from different localities in Tanzania, Malawi, Zambia, Namibia and South Africa, scalation patterns and colouration of adults and hatchlings were investigated. All specimens examined share the same scalation pattern, but they differ in the number of preanal pores in males. Males from southern Africa generally possess 4 to 5 preanal pores, whereas those from Tanzania possess 7 preanal pores. The general colouration of specimens from southern Africa is brown with a camouflaging pattern of streaks and dots. The extent of the streaks and the number of the spots are quite variable between specimens from different localities and - to a lower extent - between specimens of the same population. Additionally, the colouration strongly depends on internal factors as well as external stimuli, e.g. disturbance. While all specimens from southern Africa share this basic colouration, the Tanzanian specimens are usually coloured grey and possess conspicuous whitish spots with black edges. Their ability to change colour is limited. Hatchlings of all forms except those from Tanzania are again quite similar, with a brownish body and a bright orange-red tail. In contrast, hatchlings of the Tanzanian specimens are dark brown during their first days and become grey subsequently. Additionally, only the undersides of their tails are reddish. The Tanzanian form stands apart from the other populations of *L. capensis* as far as colouration and some morphometric data are concerned. Due to unsatisfactory definitions of East African *Lygodactylus* species, it is not yet clear whether the Tanzanian form is referable to *L.* [*capensis*] *grotei*. Alternatively, it may be a member of the *L. scheffleri* complex.

Key words: southern and eastern Africa; Gekkonidae; *Lygodactylus capensis;* intra-specific variability

B.A. Huber et al. (eds.), African Biodiversity, 237–244.

1. INTRODUCTION

The diurnal dwarf geckos of the genus *Lygodactylus* Gray, 1864 comprise about 60 species. About 22-24 species are natives of Madagascar, of islands close to the Madagascan coast and of two islands in the Strait of Mozambique (Loveridge, 1947; Pasteur, 1965 [1964]; Branch, 1998; Spawls, 2002). Additionally, two species of *Lygodactylus* are found in South America (Brazil and Paraguay) (Vanzolini, 1968; Norman, 1994). However, most of the species - about 33-37 (depending on authors) - occur in sub-Saharan Africa where numerous species live in East Africa and in southern Africa. The geographic ranges of some species are restricted to small areas, e.g. that of the yellow-headed dwarf gecko *L. picturatus* which occurs in coastal areas of Kenya and Tanzania. In contrast, there are some species with a wide area of distribution, e.g. *L. gutturalis*, which occurs from Senegal in western Africa to Sudan and eastern Tanzania in East Africa.

The cape dwarf gecko *Lygodactylus capensis* Gray, 1864 is one of the most common geckos in southern Africa. It occurs from southern Angola to the Cape Province and from the eastern half of southern Africa to Tanzania. Thus, this species is a prime target to study intraspecific variability of specimens of geographically separated populations.

2. MATERIALS AND METHODS

The geckos studied were determined using the morphological characters and descriptions given in the monographs of Loveridge (1947) and Pasteur (1965 [1964]) and in the field guides of Branch (1998) and Spawls et al. (2002). Specimens of *L. capensis* were investigated from different localities in southern and eastern Africa. The localities in southern Africa were: South Africa (Pretoria, Pafuri, Punda Maria, Messina), Namibia (Popa Falls, Palmwag), Zambia (Lusaka) and Malawi (Liwonde). Some specimens were collected and bred in the laboratory.

The specimens of eastern Africa were obtained from the pet trade and had been imported from Tanzania. As Tanzanian specimens of *L. capensis* were usually imported together with *Lygodactylus kimhowelli*, they were probably caught in the geographic range of the latter. *Lygodactylus kimhowelli* is known only from the coastal forest in the vicinity of Amboni Caves, Tanga (Spawls et al., 2002).

In order to facilitate the comparison of the colouration of the animals, both adults and hatchlings were - if possible - photographed using the same grey cardboard as background and the same magnification.

For comparison, standard morphological characters, which are generally used for identification of species within the genus *Lygodactylus*, were examined. These characters were: shape of mental, shape and number of postmentals, number of scansors under the fourth toe, number of scansors under the tail tip, shape and arrangement of subcaudals, number of preanal pores of males. Additionally, the snout-vent lengths (SVLs) were measured.

Hatchlings bred in the laboratory had SVLs of 1.3 to 1.4 cm. Thus, specimens examined in the field were considered as hatchlings if they had a SVL below 1.5 cm.

3. RESULTS

3.1 Morphological characters

Adult specimens of *L. capensis* from almost all localities have SVLs of 3.2 to 3.4 cm (Table 1). Only specimens from Palmwag, Namibia, are smaller with SVLs of 2.8 to 3.0 cm; these specimens are considered as subspecies *L. c. bradfieldi*. The tails of all specimens are generally 1.2 or 1.3 times longer than the cylindrical bodies.

All specimens investigated share the following morphological characters: They all have a mental with distinct lateral clefts, 3 postmentals which have nearly equal sizes, 4 pairs of scansors under the fourth toe, 4-6 pairs of scansors under the tail tip and subequal imbricate subcaudals under original tails.

All males from southern Africa investigated have 4 to 5 preanal pores (Table 1). One male (out of four) from Namibia, Popa Falls, had 6 pores. However, all males from Tanzania have 7 preanal pores. Additionally, males develop a pseudo-escutcheon consisting of shiny, sometimes brownish scales in the preanal area and on the inner sides of the thighs, comparable to the escutcheon of sphaerodactyline geckos. In males of the Tanzanian form, such scales also seem to occur on the underside of the tail base as in males of *Lygodactylus scheffleri*, another Tanzanian species (pers. obs.).

3.2 Colouration

The body of adult specimens is fawn to grey-brown, usually with a dark and a whitish streak on their flanks (Fig. 1). The streaks end before reaching the hindlimbs. Several rows of light spots are arranged longitudinally on the dorsum, on the flanks and on the tail. Belly and throat are whitish to cream,

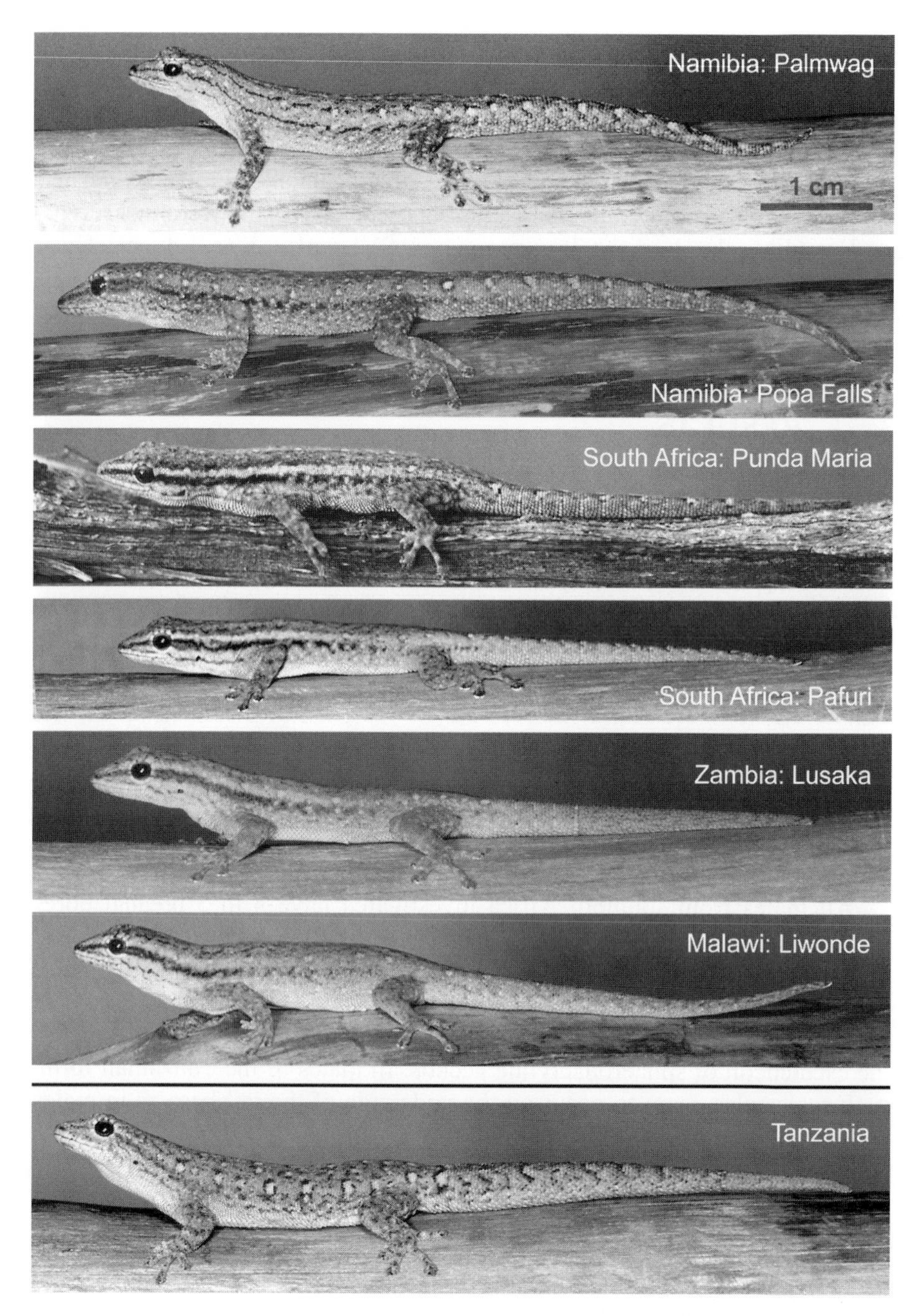

Figure 1. Adult specimens of *Lygodactylus capensis* from different localities in southern and eastern Africa.

Figure 2. Hatchlings of *Lygodactyus capensis* from different localities.

the throat can be stippled with dark dots. The extent of the streaks and the number of the spots are quite variable both between specimens from different localities and also - to a lower extent - between specimens belonging to the same population. For example, specimens from northwestern Namibia - designated as *L. c. bradfieldi* - are characterized by generally only thin dark streaks on the flanks (Fig. 1).

Additionally, the colouration strongly depends on internal factors as well as external stimuli, e.g. disturbance. In the latter case, the animals become darker within a few seconds. In specimens from Pafuri, South Africa, this effect was very pronounced, the whole upper side becoming almost black and the whitish streak on the flanks disappearing completely. Both the sublabials and several scales of the throat became black, too.

All specimens from southern Africa share this basic colouration. The colouration of the Tanzanian specimens differ from that of all other specimens investigated in completely lacking dark and white streaks on the flanks. Additionally, they are usually coloured grey and possess striking whitish spots with black edges on the dorsum and on the flanks. After disturbance they can become brownish. Sometimes - apparently depending on internal factors - Tanzanian animals develop a light brown streak on the flanks extending from the forelimbs to the base of the tail.

Table 1. Morphological characters of *L. capensis* from different localities. SVL: snout-vent length; RSA: Republic of South Africa; NAM: Namibia; MAL: Malawi; TAN: Tanzania; ZAM: Zambia

Localities ***Character***	RSA Pretoria	RSA Pafuri	RSA Punda Maria	RSA Messina	NAM Popa Falls	NAM Palmwag
N adults examined	11	9	3	3	9	13
N hatchlings examined	6	6	-	-	3	5
SVL (cm) of adults	3.2-3.4	3.3-3.4	3.2-3.3	3.2-3.3	3.2-3.5	2.8-3.0
SVL (cm) of hatchlings	1.3-1.4	1.3-1.4	-	-	1.3-1.4	1.2-1.3
N pairs of scansors under tail tip	4-6	5-6	4-5	5	4-5	4-5
N preanal pores in males [N males examined]	4-5 [5]	4-5 [3]	4 [1]	-	4-6 [4]	4-5 [5]

Localities *Character*	ZAM Lusaka	MAL Liwonde	TAN ?
N adults examined	7	6	12
N hatchlings examined	4	4	14
SVL (cm) of adults	3.3-3.4	3.2-3.4	3.2-3.3
SVL (cm) of hatchlings	1.3-1.5	1.3-1.4	1.2-1.4
N pairs of scansors under tail tip	4-5	4-5	4-5
N preanal pores in males [N males examined]	5 [3]	4 [2]	7 [5]

Hatchlings are small and delicate, with SVLs of 1.3 to 1.4 cm (Table 1). Their morphological characters correspond to those of the adult specimens.

Again, the colouration of the hatchlings of all specimens except those from Tanzania is quite similar, with a brownish body and a bright orange-red tail (Fig. 2). In contrast, hatchlings of the Tanzanian specimens are dark-brown during their first days, and only the undersides of the tails are reddish. Thereafter, the hatchlings become as grey as the adults, retaining the ventrally reddish tail for about 3-4 months.

4. DISCUSSION

In this study, 115 live specimens of the dwarf gecko *L. capensis* from different localities in southern and eastern Africa were investigated concerning morphological characters and colouration of both adults and hatchlings. All specimens from southern Africa share the same scalation pattern. The colouration varies between specimens of different localities; however, a basic colour pattern can be recognized for all populations.

The specimens from Palmwag, Namibia, are smaller than those from other populations. Additionally, they differ in the extent of the dark streaks on their flanks. This form is considered a distinct subspecies (*L. c. bradfieldi*) by several authors (e.g. FitzSimons, 1943; Loveridge, 1947; Wermuth, 1965; Werner, 1977; Auerbach, 1987), whereas other authors consider it a full species (e.g. Hewitt, 1932; Pasteur, 1965; Branch, 1998). It seems that *L. c. bradfieldi* - living usually in harsher areas than *L. c. capensis* - is just an ecological dwarf form. Breeding experiments could show whether this reduction in size is genetically fixed.

The Tanzanian form stands apart from all other populations of *L. capensis* as far as colouration and the number of preanal pores are concerned. Due to unsatisfactory definitions of several East African *Lygodactylus* species, it is not yet clear whether the Tanzanian form is referable to *L.* [*capensis*] *grotei* or perhaps to *L. scheffleri*. Further studies, e.g. investigation of mitochondrial DNA of specimens from southern and eastern Africa, will help to clarify the affiliation of the Tanzanian specimens.

ACKNOWLEDGEMENTS

I would like to thank the authorities of South African provinces Mpumalanga and Limpopo for collecting permits. I would also like to thank Wulf Haacke (Transvaal Museum) for extensive information about South African *Lygodactylus*, for access to specimens of some populations and for most valuable advice and help in administrative issues.

REFERENCES

Auerbach, R. D., 1987, *The amphibians and reptiles of Botswana*, Mokwepa Consultants (Pty) Ltd, Gabarone, pp. 295.

Branch, B., 1998, *Field guide to snakes and other reptiles of Southern Africa*, 3rd ed., Cape Town, Struik Publishers (Pty) Ltd, pp. 399.

FitzSimons, V. F., 1943, The lizards of South Africa, *Transvaal Mus. Mem., Pretoria*, **1**:xv-528.

Gray, J. E., 1864, Notes on some new lizards from South-Eastern Africa, with the description of several new species, *Proc. Zool. Soc. Lond.* **34**:58-62.

Hewitt, J., 1932, Some new species and subspecies of South African batrachians and lizards, *Ann. Natal Mus.*, **7**:105-128.

Loveridge, A., 1947, Revision of the African lizards of the family Gekkonidae, *Bull. Mus. Comp. Zool.*, Cambridge (Mass.), **98**(1):1-469.

Norman, D. R., 1994, *Amphibians and reptiles of the Paraguayan Chaco*, Vol. I, San José, C.R.: D. Norman, pp.1-281.

Pasteur, G. 1965 [1964], Recherches sur l'évolution des lygodactyles, lézards afro-malgaches actuels, *Trav. Inst. Scient. Chérif., Sér. Zool.* **29**:1-132.

Spawls, S., Howell, K., Drewes, R. and Ashe, J., 2002, *A field guide to reptiles of East Africa*, Academic Press, San Diego, pp. 543.

Vanzolini, P. E., 1968, Lagartos brasileiros da família Gekkonidae (Sauria), *Arq. Zool. S. Paulo*, **17**:1-84.

Wermuth, H., 1965, Liste der rezenten Amphibien und Reptilien: Gekkonidae, Pygopodidae, Xantusiidae, in *Das Tierreich*, Berlin, pp. 1-246.

Werner, Y. L., 1977, Ecological comments on some gekkonid lizards of the Namib Desert, South West Africa, *Madoqua, Windhoek*, **10**(3):157-169.

MIGRATION WITHIN AND OUT OF AFRICA: IDENTIFYING KNOWLEDGE GAPS BY DATA-MINING THE GLOBAL REGISTER OF MIGRATORY SPECIES

Klaus Riede
Zoologisches Forschungsinstitut und Museum Alexander Koenig, Adenauerallee 160, D-53113 Bonn, Germany, Email: k.riede.zfmk@uni-bonn.de

Abstract: Knowledge gaps for birds migrating within Africa, or visiting the continent either for breeding or wintering, have been identified systematically by data-mining the Global Register of Migratory Species (GROMS). The Register contains a migratory vertebrate species reference list, GIS maps and a literature database with full-text passages on migration. Using the GROMS literature and accessory tables for data mining, knowledge gaps with respect to migration routes and seasonal timing were diagnosed for more than 150 bird species. This analysis was based on a search for text strings such as "migration unknown", "poorly known", "surprisingly little known", etc. The species list of lesser-known migrants generated by data-mining has been published on the GROMS website (www.groms.de), and provides a starting point for future research. The underlying complex query uses Standard Query Language (SQL), requiring a fully documented relational database. Such documentation is often lacking for current biodiversity information systems, which limits their use for complex data mining.

Key words: Migratory birds; intra-African migrants; biodiversity informatics; data-mining

1. INTRODUCTION

Africa provides wintering habitats for billions of birds, mainly from the Palearctic. In addition, its marine shelf areas are important breeding and feeding grounds for marine mammals, seabirds and diadromous fishes. Besides this well-known significance of African ecosystems for inter-

B.A. Huber et al. (eds.), African Biodiversity, 245–252.

continental migrants, there are less well-known intra-African migration phenomena, which seem to depend on rainfall patterns and/or insect outbreaks, e.g. locusts.

The present article exemplifies the potential and problems of data-mining by querying GROMS for African migratory species, particularly searching for knowledge gaps on migration behaviour of non-passerine birds. The GROMS is a cross-sectional database based on a reference list of currently 4,400 vertebrate species and subspecies, combined with 1,100 global distribution maps in GIS format. Besides publication on the World Wide Web (www.groms.de), detailed documentation is available in book format, with a full off-line version of the relational database and the complete GIS map repository on CD-ROM (Riede 2001, 2004a). First results of a global meta-analysis of the entire GIS dataset have recently been published elsewhere (Riede, 2004b). In this paper I will show that such advanced data-mining is only possible by using Standard Query Language (SQL, see Taylor, 2001), together with full documentation of the data model. The result from this text-mining exercise is a document summarising knowledge gaps in African bird migration behaviour, which can be downloaded from the GROMS website (www.groms.de/groms/knowledge_gaps.pdf).

2. METHODS

The definition of migration phenomena, as used in this analysis, is based on cyclical, true migration of more than 100 km (for details, see Riede, 2001, p. 25, and Dingle, 1996).

Full-text citations about migration of all non-passerine birds identified as "migratory" have been entered into the GROMS database, mainly using the "movement" section within species accounts published in del Hoyo et al. (1992-2003). Figure 1 shows the many: many structure of the relational data model, which makes it possible to relate one species with many citations or text passages, or relate one text passage with many species, with several keywords (in this case: migration). Such a data model allows storage of contradictory information from different sources, referring to migration, threat, taxonomic status and other key themes. The data entry module is available on the GROMS CD-ROM (Riede, 2004a).

The recent off-line edition of the GROMS database (Riede, 2004) has been used to identify migratory species occurring in Africa, using the ACCESS (MS-Windows) database query interface to select species according to geographic criteria (GEO-Search: l.c., p. 94).

For text mining, a genuine SQL-statement was generated, using the ACCESS query generator (Fig. 1), searching for text strings such as

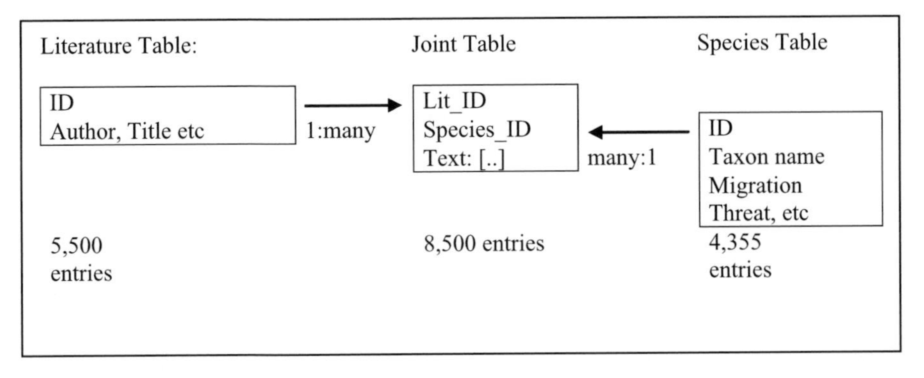


```
SELECT Tab_Arten.Latein, Tab_Arten.Englisch, Tab_Arten.Migration,
Jointab_Art_Lit.Lit_Bezug, Tab_Literatur.Autor_Name, Tab_Literatur.Autor_Vorname,
Tab_Literatur.Coautoren_Namen, Tab_Literatur.Jahr, Tab_Arten.animalgroup,
Tab_Arten.Familie
FROM Tab_Literatur RIGHT JOIN (Tab_Arten INNER JOIN Jointab_Art_Lit ON
Tab_Arten.ID = Jointab_Art_Lit.ID_Art) ON Tab_Literatur.ID = Jointab_Art_Lit.ID_Lit
WHERE (((Jointab_Art_Lit.Theme)=7) AND ((Tab_Arten.Animal_Class)=2) AND
((Jointab_Art_Lit.Lit_Bezug) Like "*  unknown*")) OR (((Jointab_Art_Lit.Theme)=7) AND
((Tab_Arten.Animal_Class)=2) AND ((Jointab_Art_Lit.Lit_Bezug) Like "*    perhaps*")) OR
(((Jointab_Art_Lit.Theme)=7) AND ((Tab_Arten.Animal_Class)=2) AND
((Jointab_Art_Lit.Lit_Bezug) Like "*  little*")) OR (((Jointab_Art_Lit.Theme)=7) AND
((Tab_Arten.Animal_Class)=2) AND ((Jointab_Art_Lit.Lit_Bezug) Like "*     poor*")) OR
(((Jointab_Art_Lit.Theme)=7) AND ((Tab_Arten.Animal_Class)=2) AND
((Jointab_Art_Lit.Lit_Bezug) Like "*  possib*"))
ORDER BY Tab_Arten.animalgroup, Tab_Arten.Familie;
```

Figure 1. Text-mining by using the query generator. Species (Tab_Arten) and literature (Tab_Literatur) table are connected with a many: many relation by a Jointab containing full-text passages. The latter are searched for relevant text strings characteristic for knowledge deficits, which are high-lighted in the statement.

"migration unknown", "poorly known", "surprisingly little known", "thought to be sedentary but...", etc.

3. RESULTS

Searching the GROMS for species migrating within Africa, or migrants visiting the continent either for breeding or wintering, resulted in 1,285 vertebrate species, with a surprisingly high number of migratory fish species from Sub-Saharan Africa (Table 1). The more complex SQL statement described in Figure 1 was necessary to search for birds with insufficiently known migration behaviour. A search for all text entries on bird migration containing text strings such as "unknown", "perhaps", "little", "poor" or "possib" resulted in a list of 349 birds with insufficiently known migration behaviour.

Apus caffer White-rumped swift	Migratory in northernmost and southernmost parts of range. Spanish population present early May to Aug-Oct, some recorded into early Dec, with autumn migration through Straits of Gibraltar mid-Aug to mid-Oct; S African population present Aug-May, mainly absent from S Cape and much reduced farther N within S breeding range Jun-Jul. **Poorly** understood wet-season movements into Sahel may be feature of N sub-Saharan populations. Otherwise resident. Migrates in flocks of up to 100. S African migrants may be transequatorial. Some degree of altitudinal migration in Natal. First record from rabia 1982, and seen at least once subsequently in Tihamah coastal plains, Saudi Arabia, in Mar 1989. Vagrant to Norway (May, Jun) and Finland (Nov).
Caprimulgus climacurus Long-tailed nightjar	**Poorly** known. Nominate race migratory and partially sedentary, some populations moving S after breeding season. Race sclateri **possibly** sedentary and partially migratory. Race nigricans **probably** sedentary. Outside breeding season, range also includes S Ivory Coast, SW Nigeria, S Cameroon, Equatorial Guinea, Gabon, SE Congo (lower Congo river valley), NE Angola (one record Luaco), SE Sudan, SW Ethiopia, W Kenya (sporadic in Turkana and Pokot region) and E Uganda.

Figure 2: Sample output for 2 out of 349 species found searching GROMS with the query listed in Fig. 1. Text particles matching search strings are highlighted.

Examples of two typical text passages are shown in Figure 2; the full list is available as pdf-document at www.groms.de/groms/knowledge_gaps.pdf. Filtering this list for species occurring in Africa still left 107 non-passerine birds from 15 orders (Table 2). Forty-one species are intercontinental migrants, with most of the population visiting the African continent for wintering (Example: Eurasian cuckoo - *Cuculus canorus*). A high number of 42 species shows partial migratory behaviour of a certain subspecies or population (Example: Little green bee-eater - *Merops orientalis*). In these cases, uncertainties with respect to migration mainly refer to difficulties in recognition and tracking of the migratory individuals in the field, where they often mix with the resident part of the population. The fraction of 12 intracontinental migrants contains intra-African migrants with poorly studied, but evidently complex migration patterns (Curry-Lindahl, 1981a, b; example: Pennant-winged nightjar - *Macrodypterix longipennis* [Gould, 1837]). Intracontinental migration often blends with other, less-well defined categories used within the GROMS database, such as "nomadising" (e.g. Tawny eagle - *Aquila rapax* [Temminck, 1828]), "range extension" (e.g. White-backed night heron - *Gorsachius leuconotus* [Wagler, 1827], or

Table 1. Migratory vertebrate species occuring in Africa, according to GROMS.

	Vertebrates	Mammals	Birds	Sea Turtles	Fish
North-	604	58	234	5	307
Sub-Saharan	1.174	79	388	5	702
AFRICA	**1.285**	**93**	**447**	**5**	**740**

Table 2. Number of species of non-passeriform African migrants with insufficiently known migration behaviour within each order (following the systematics used by del Hoyo et al., 1992)

Order	Number of species
Charadriiformes	36
Procellariiformes	13
Gruiformes	12
Falconiformes	10
Caprimulgiformes	9
Cuculiformes	8
Apodiformes	3
Columbiformes	3
Pelecaniformes	3
Anseriformes	2
Ciconiiformes	2
Coraciiformes	2
Strigiformes	2
Podicipediformes	1
Sphenisciformes	1

"possibly migratory" (e.g. Three-banded plover - *Charadrius tricollaris* Vieillot, 1818), altogether comprising an additional 12 insufficiently known migrants.

Table 2 shows that most (36) species with insufficiently known migration behaviour belong to the Charadriiformes, a speciose and highly diverse group comprising 300 migratory species – plovers, lapwings and sandpipers, but also seabirds such as skuas, gulls and auks. A much higher proportion of knowledge deficits is observed within Caprimulgiformes: movements are poorly understood for 9 out of 31 migratory nightjar species. With 8 out of 46, a similarly high proportion is observed in cuckoos (Cuculiformes).

A closer look at some text examples reveals a wide variety of knowledge gaps, related to timing, seasons, food availability and regional distribution. In the following, some typical examples are given. The monotypic white-rumped swift *Apus caffer* (Lichtenstein, 1823) exhibits "...poorly understood wet-season movements into Sahel may be feature of N sub-Saharan populations" (del Hoyo et al., 1999). The Sahel during the wet season is a huge, mostly inaccessible region, and even with better ornithological coverage most specialists will find it difficult to recognise swifts belonging to the northern sub-Saharan populations.

For the little ringed plover *Charadrius dubius* Scopoli, 1786, del Hoyo et al. (1996) noted: "Race *curonicus* migratory, but possibly resident in S breeding areas; W European population migrates across Sahara to tropics;[...]." This is a typical example of unclear delimitation of the frontier between resident and migratory individuals within one population, probably showing clinal variation along a North-South gradient, and possibly variation with annual weather conditions. Weather conditions, and in particular rain, as a trigger for migratory movements is hypothesised for the standard-winged nightjar *Macrodipteryx longipennis* (Shaw, 1796): "An intra-African migrant, movements protracted and possibly influenced by rains, not fully understood. Leaves breeding grounds in southern savannas of W & C Africa from mid-Apr to about Aug and moves N to spend wet season in savannas of Sahel and Sudan." (del Hoyo et al., 1999). In the case of nightjars, it is probably crypsis, together with their nocturnal habits, which hamper research. Even in well-studied regions, such as the Netherlands, recent investigations revealed 50% underestimation of populations by traditional monitoring methods, which could be overcome by application of new techniques, such as song playback (Bult, 2003).

4. DISCUSSION

In summary, most knowledge deficits can be attributed to poorly understood wet season movements. A recent search of the BIOSIS database revealed only a very few papers adding to our understanding of African intra-tropical migration. The extensive study of Anciaux (2000) on migration phenology of a wide range of African non-passerines adds to our knowledge, but is limited to a small study area of inland South Benin. Progress will only be made if the few data from satellite telemetry and ringing are pooled in a more efficient way, using advanced information technologies. Regular monitoring efforts are restricted to a few countries, such as the southern African region (Harrison, 1997), or certain groups, such as waterbirds (Dodman and Diagana, 2003).

Besides the biological results presented here, this small data-mining exercise illustrates that advanced data-mining requires access to entire, fully documented databases, using queries based on Standard Query Language (SQL). This had to be complemented by searching commercial literature databases, such as BIOSIS (www.biosis.org). At present, the Global Biodiversity Information Facility (www.gbif.org) endeavours to harmonise the numerous databases and information systems by harmonising taxonomic standards and designing portals, connecting different databases within "federated systems". However, it will be a challenge to provide user-friendly tools for complex data-mining, which requires deep-drilling into federated systems composed of several biodiversity databases, differing in structure and function.

ACKNOWLEDGEMENTS

Supported by the Secretariat of The Convention on the Conservation of Migratory Species of Wild Animals (UNEP-CMS), Bonn. The development of the GROMS database was supported by the German Federal Agency for Nature Conservation with funds from the German Federal Environment Ministry. Thanks to all members of the GROMS team, and in particular Thomas Lingen and Bedru Sherefa-Muzein, for patiently entering full-text passages on migration from del Hoyo et al. Thanks also to Lynx editions for allowing digital publication of maps and text passages from "Birds of the World".

REFERENCES

Anciaux, M. R., 2000, Approche de la phenologie de la migration des migrateurs intra-africains de l'interieur des terres du sud-benin (plateau d'allada et sud de la depression de la lama). 1. Les non-passeriformes et les non-coraciiformes, *Alauda* **68:**311-320.

Bult, H., 2003, Nachtzwaluwen *Caprimulgus europaeus* onder de rook van Antwerpen, *Limosa* **75:**91-102.

Curry-Lindahl, K., 1981a, *Bird migration in Africa (Movements between six continents). Vol.1*, Academic Press, London.

Curry-Lindahl, K., 1981b, *Bird migration in Africa (Movements between six continents). Vol.2*, Academic Press, London.

del Hoyo, J., Elliott, A., and Sargatal, J. (eds), 1992, *Handbook of the birds of the world. Vol.1: Ostrich to Ducks,* Lynx Edicions, Barcelona.

del Hoyo, J., Elliott, A., and Sargatal, J. (eds), 1994, *Handbook of the birds of the world. Vol.2: New World Vultures to Guineafowl,*- Lynx Edicions, Barcelona.

del Hoyo, J., Elliott, A., and Sargatal, J. (eds), 1996, *Handbook of the birds of the world. Vol.3*: *Hoatzin to Auks*, Lynx Edicions, Barcelona.

del Hoyo, J., Elliott, A., and Sargatal, J. (eds), 1997, *Handbook of the birds of the world. Vol.4: Sandgrouse to Cuckoos,* Lynx Edicions, Barcelona.

del Hoyo, J., Elliott, A., and Sargatal, J. (eds), 1999, *Handbook of the birds of the world. Vol.5: Barn Owls to Hummingbirds*. Lynx Edicions, Barcelona.

del Hoyo, J., Elliott, A., and Sargatal, J. (eds), 2001, *Handbook of the birds of the world. Vol.6: Mousebirds to Hornbills*. Lynx Edicions, Barcelona.

del Hoyo, J., Elliott, A., and Sargatal, J. (eds), 2002, *Handbook of the birds of the world. Vol.7: Jacamars to Woodpeckers*. Lynx Edicions, Barcelona.

Dingle, H., 1996, *Migration - The biology of life on the move*, Oxford University Press, New York.

Dodman, T., and Diagana, C. H., 2003, *African Waterbird Census 1999, 2000 and 2001 - Les Dénombrements d'Oiseaux d'Eau en Afrique 1999, 2000 et 2001*, Wetlands International, Netherlands.

Harrison, J. A., Allan, D. G., Underhill, L. G., Herrmans, M., Tree, A. J., Parker, V., and Brown, C. J., 1997, *The Atlas of Southern African Birds*, Vol.1: Non-Passerines, 2, Johannesburg, South Africa.

Riede, K., 2001, *Global Register of Migratory Species - Database, GIS Maps and Threat Analysis. Weltregister wandernder Tierarten,* Landwirtschaftsverlag, Münster.

Riede, K., 2004a, *Global Register of Migratory Species: From global to regional scales,* Landwirtschaftsverlag, Münster.

Riede, K., 2004b, The "Global Register of Migratory Species" - first results of global GIS Analysis. In: Werner, Dietrich (ed.): *Biological Resources and Migration.* Springer-Verlag, Berlin.

Taylor, A. G., 2001, *SQL for dummies,* Books Worldwide, Indianapolis.

AVIFAUNAL RESPONSES TO LANDSCAPE-SCALE HABITAT FRAGMENTATION IN THE LITTORAL FORESTS OF SOUTH-EASTERN MADAGASCAR

James E. M. Watson
Biodiversity Research Group, School of Geography and the Environment, University of Oxford, United Kingdom, OX1 3BW, Email: james.watson@ouce.ox.ac.uk

Abstract: Madagascar's lowland littoral forests are rich in endemic taxa and considered to be seriously threatened by deforestation and habitat fragmentation. In this study I examined how littoral forest bird communities have been affected by fragmentation at the landscape scale. Bird species composition within 30 littoral forest remnants of differing size and isolation was determined using point counts conducted in October – December in 2001 and 2002. Each remnant was characterised by measures of remnant area, remnant shape, and isolation. Step-wise regression, nestedness analysis, and binomial logistic regression modelling was used to test the relationship between bird species and landscape variables. Bird species richness in remnants was significantly ($p<0.01$) explained by remnant area but not by any measure of isolation or landscape complexity. The bird communities in the littoral forests were significantly ($p<0.01$) nested. The majority of forest-dependent species had significant ($p<0.01$) relationships with remnant area. As deforestation and fragmentation are still significant issues in Madagascar, I recommend that large (>200 ha) blocks of littoral forest are awarded protected status to preserve what remains of their unique bird community.

Key words: Littoral forest; fragmentation; isolation; species-area relationship; birds; island biogeography; Madagascar; nestedness

B.A. Huber et al. (eds.), African Biodiversity, 253–260.

1. INTRODUCTION

The Fort Dauphin (Tolagnaro) region of southeastern Madagascar is arguably one of the most diverse on the island. The region's size (approximately 10000 km^2, or about 1.7% of the total land area of island) is small, but it contains a large variety of habitats, including different forest types, coastal zones, high mountains, and areas of inland freshwater habitat (Goodman et al., 1997). As a result of this habitat variety, the region has one of the highest numbers of bird species of any area on the island; Goodman et al. (1997) found 189 bird species within the area, representing 68% of the birds known in Madagascar. This is impressive when one considers the region's size. However, much of the forested landscape in southeastern Madagascar is fragmented and little is known on how this process affects the region's bird communities.

At the landscape scale, it is believed that dynamics of bird populations are influenced by five interrelated processes associated with habitat loss and subsequent fragmentation: habitat loss, subdivision of habitat, patch isolation, edge effects, and compositional changes to the landscape matrix (Forman, 1995; Lindenmayer and Franklin, 2002). The last four factors are defined as habitat fragmentation – the subdivision of habitat patches – and needs to be distinguished from habitat loss, which is the overall depletion of habitat (Andrén, 1994). In this study I examine the effects of two aspects associated with landscape-scale habitat fragmentation; that is, the effects of sub-division of habitat and patch isolation. The original data for this study can be found in Watson (2004).

2. METHODS

2.1 Study area

The study was conducted in fragmented littoral forest remnants located to the west and north of Fort Dauphin (Fig. 1). The mosaic surrounding these forests include small patches of *Melaleuca* swamp forest and plantations of *Eucalyptus citriodora* Hook and *E. robusta* Blakely, and a dominant heath-type matrix consisting of *Erica* spp. (Ramanamanjato and Ganzhorn, 2001). This region has a subtropical climate with a regional mean annual minimum temperature of 15 °C, mean maximum temperature of 28 °C, and mean annual rainfall ranging from 500 - 3000 mm (Goodman et al., 1997).

Table 1. Definition and summary statistics, with adopted transformations, for landscape structural metrics of 30 littoral forest remnants in southeastern Madagascar. Landscape environmental attributes were determined from a Landsat TM satellite image acquired on 11 November 1999.

Code	Description	Units	Min.	Max.	Mean
Shape Index (SI)	$SI = \frac{A/P}{R/2} x100$ where R is the radius of a circle with area *A* and perimeter *P*	0-1	0.27	0.94	0.707
Edge Density (ED)	Density of littoral forest edge within 500 m of each remnant's edge multiplied by 100	0-1	0.07	0.51	0.28
Patch Density (PD)	Density of littoral forest remnant area within 500 m of each remnant's edge	0-1	0.02	0.38	0.18
Distance to source (DS)	Distance to the nearest block of forest > 1000 ha	M	4813	17502	9965.1
Distance to large remnant (DR)[1]	Distance to the nearest remnant with an area > 100 ha	M	45	3968	903.2
Distance to nearest remnant (DNR)[2]	Distance to the nearest remnant	M	55	1600	132.5
Area of nearest remnant (ANR)[1]	Area of nearest littoral forest remnant	Ha	0.40	464	96.2
Remnant Area (AR)[1]	Total area of the remnant	Ha	0.30	464	57.9

[1]ln transformed; [2]square-root transformed for analyses

2.2 Field methods

Species richness and abundance of birds were determined using the point-count method (Bibby et al., 1992). A total of 90 point count stations were placed in 30 littoral forest remnants. Of the 30 remnants surveyed, 16 remnants were located in the sub-region of Ste. Luce and 14 were located in the sub-region of Mandena. The number of point count stations varied depending on the area of the remnant: one station in remnants < 1 ha; two stations for of 1-10 ha; three for 10-20 ha; four for 20-40 ha; and five for > 40 ha. Each station had a fixed radius of 25 m and stations were located at least 100 m apart to minimize the risk of counting the same individual bird twice. Circles of 25 m radius were selected because this was as far as an observer could reasonably see in littoral forest, and the smallest littoral forest patches in this survey approximated this width. Ten minutes were spent at each station.

Bird surveys were carried out during November and December 2001 and in October and November 2002. Each station was visited twice. Only species sighted within the point count area were recorded as present, calls were used to locate birds and to aid identification. Surveys were confined to the periods 0600 – 1000 hr and 1500 – 1900 hr on days without rain or strong wind.

2.3 Landscape structure data

Eight parameters were used to assess the effects of sub-division of habitat and patch isolation on bird communities in the littoral forest landscape. These were the area of each remnant (AR), distance to nearest large (>1000 ha) block of forest (DS), distance to the nearest remnant > 100 ha (DR), distance to nearest remnant (DRN), area of nearest remnant (ANR), a littoral forest remnant shape index (SI), patch density (PD), and edge density (ED) (Table 1). Measurements from the edge, rather than the centre of the patch, were used in calculating edge and patch density to exclude the area of the patch of interest, because that measurement (remnant area) was included as a landscape metric. All eight variables were determined from a supervised classified Landsat TM satellite image acquired on 11 November 1999 using ArcView GIS software.

2.4 Analysis

A step-wise regression analysis (holding P to enter =0.05 and P to remove =0.1) was used to explore the relationships of species richness (dependent variable) with the landscape mosaic variables recorded for each fragment in order to build the best predictive model for each richness value. The variables ‘AR’ and ‘DR’ were logarithmically transformed while the parameter ‘DRN’ was square-root transformed prior to analyses (Table 1). As multicolinearity was found between the two independent variables ‘SI’ and ‘AR’, I removed the variable ‘SI’ from the regression, as it explained less of the variance in both analyses (Soulé et al., 1988).

Binomial logistic regression was used to identify landscape variables associated with the probability of occurrence of individual species. Logistic regression was used instead of linear regression because of the low number of species found in many patches. Logistic regressions were modelled using a likelihood ratio test based on the presence/absence data for littoral forest bird species in all 30 littoral forest remnants. All regression analyses were calculated using the statistical package SPSS (Kinnear and Gray, 2000). To evaluate the existence of non-random patterns of bird species distribution across fragments, I tested for the existence for nestedness using the Nestedness Temperature Calculator (Atmar and Patterson, 1993).

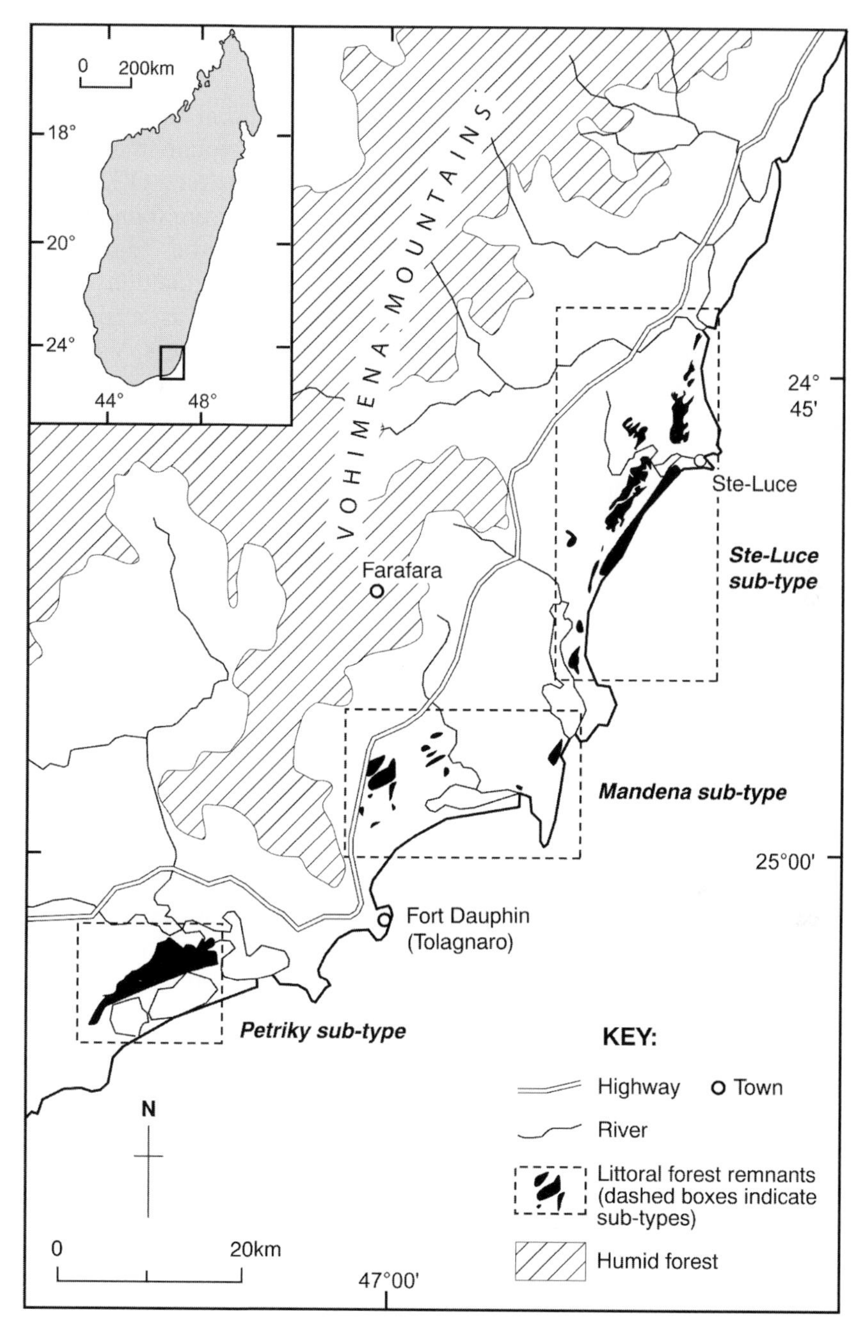

Figure 1. The location of littoral forest remnants surveyed in southeastern Madagascar. The matrix surrounding littoral forest remnants includes swamp forest, *Melaleuca* forest, plantations of *Eucalyptus citriodora* and *E. robusta,* and heath-type vegetation consisting predominately of *Erica* spp. Forest extent was based on the figure presented in Ramanamanjato et al. (2002).

3. RESULTS

In total, 70 bird species were found in the 30 littoral forest remnants. Species-accumulation curves show that the point count methodology adequately captured species richness in all 30 remnants (see O'Dea et al., 2004; Watson, 2004 for further details). Sixteen species were found in fewer than three point count stations and as a result, were treated as 'rare' species and were excluded from the likelihood ratio and incidence function analyses. The step-wise regression model selected remnant 'AR' as a significant ($p<0.01$) factor that explained variation in total species richness. Variance in species richness accounted by the other variables in the step-wise regression was negligible.

The distribution of species among littoral forest remnants was not random as species composition showed a high degree of nestedness ($T^{o}_{observed} = 15.84$, $T^{o}_{random} = 69.31$, $SD = 2.74$, $p<0.001$), highlighting a predictable sequence of species addition. When individual landscape variables were considered in determining the probability of occurrence of individual species, remnant area explained a significant ($p<0.05$) portion of the deviance in the logistic models for 20 species.

4. DISCUSSION

This study showed that the size of littoral forest remnants played a very important role in determining species richness. This marked relationship between total species richness and remnant size was not unexpected (e.g. Watson et al., 2001; Beier et al., 2002). The most likely reason why the littoral forest system has a very strong bird species-area relationship is the lack of forest-dependent species found in small littoral forest remnants, and that the species composition of small fragments was a nested subset of the assemblage found in larger fragments. The high degree of nestedness found in this study not only shows that small remnants are unable to support many species of birds but also suggests that species experience local extinction in a non-random, predictable manner. However, as there are no base-line data on the species richness of littoral forests before fragmentation, it is impossible to determine the extent of species loss in small littoral forest remnants.

No measure of remnant isolation or landscape complexity was related to bird species richness in littoral forest remnants. There are several possible explanations for the relative lack of importance of isolation in this data set. It may be that the bird species found in the study area are so mobile that they are not affected by isolation to the extent that other taxa are, and as such, immigration may not be an issue (Margules et al., 1982). Those species that are not highly mobile may be able to disperse through the landscape matrix, again removing an isolation effect (Andrianarimisa et al., 2000). As the dispersal capabilities of Malagasy birds are unknown, it is impossible to resolve this issue. It is important to recognise that the lack of an isolation effect in this study may not be generalisable to the island's avifauna, as it may reflect the relatively limited range of isolation in the present study (cf. Whittaker, 2000).

5. CONSERVATION IMPLICATIONS

The most conspicuous outcome related to bird conservation is the poor capability of small littoral forest patches to maintain forest-dependent bird species. I found that 72% of the forest-dependent species assessed had a statistically significant relationship with remnant area. A further 12 forest-dependent species, removed from the individual and community analyses due to their rarity, were found solely in the largest littoral forest remnants. If these results turn out to be indicative of how Malagasy bird species are coping with fragmentation in other forested habitats, then a whole suite of species that are considered either 'common' or 'widespread' may become threatened by ongoing processes of forest fragmentation and loss. On a regional scale, these findings suggest that approximately 70% of forest-dependent bird species will be absent from sites composed of small patches of less than 10-20 ha. To preserve the existing forest-dependent bird community, I recommend that as much continuous forest be conserved as soon as possible and, if this not possible, I recommend the protection of a number of large core littoral forest areas in Mandena, Ste. –Luce, and Petriky (> 200 ha).

ACKNOWLEDGEMENTS

This study was funded by the Rhodes Trust and Hertford College, University of Oxford. I thank QMM (QIT Madagascar Minerals) for generously providing logistic support for this research. I thank Steve Goodman and an anonymous referee for providing useful criticisms of drafts of the manuscript and Ruth Ripley and Robert Whittaker for statistical advice.

REFERENCES

Andrén, H., 1994, Effects of habitat fragmentation on birds and mammals in landscapes with different proportions of suitable habitat: a review, *Oikos* **71**:355-366.

Andrianarimisa, A., Bachmann, L., Ganzhorn, J. U., Goodman, S. M., and Tomiuk, J., 2000, Effects of forest fragmentation on genetic variation in endemic understory forest birds in Central Madagascar, *Journal für Ornithologie* **141**:152-59.

Atmar, W. and Patterson, B.D., 1993, The measure of order and disorder in the distribution of species in fragmented habitat, *Oecologia* **96**:375-392.

Beier, P., Drielen, M. V., and Kankam, B. O., 2002, Avifaunal collapse in West African forest fragments, *Conserv. Biol.* **16**:1097-1111.

Bibby, C. J., Burgess, N. D., Hill, D. A., and Mustoe, S. H., 1992, *Bird Census Techniques* Academic Press, London.

Forman, R. T., 1995, *Landscape mosaics*. Cambridge University Press, Cambridge.

Goodman, S. M., Pidgeon, M., Hawkins, A. F. A. and Schulenberg, T. S., 1997, The birds of southeastern Madagascar, *Fieldiana: Zoology*, new series **87**:1-132.

Kinnear, P. R., and Gray, C. D., 2000, *SPSS for windows made simple,* Psychology Press Ltd, Hove.

Langrand, O., 1990, *Guide to the Birds of Madagascar,* Yale University Press, London.

Lindenmayer, D.B., and Franklin, J.F., 2002, *Conserving forest biodiversity*, Island Press, London.

Margules, C. R., Higgs, A. J., and Rafe, R. W., 1982, Modern biogeographic theory: are there any lessons for nature reserve design? *Biol. Conserv.* **24**:115-128.

O'Dea, N., Watson, J. E. M. & Whittaker, R. J., 2004, Rapid assessment in conservation research: a critique of avifaunal assessment techniques illustrated by Ecuadorian and Madagascan case study data, *Divers. Distrib.* **10**:55-63.

Ramanamanjato, J. B., and Ganzhorn, J. U., 2001, Effects of forest fragmentation, introduced *Rattus rattus* and the role of exotic tree plantations and secondary vegetation for the conservation of an endemic rodent and a small lemur in littoral forests of southeastern Madagascar, *An. Cons.* **4**:175-183.

Ramanamanjato, J.B., McIntyre, P.B. & Nussbeum, R.A., 2002, Reptile, amphibian, and lemur diversity of the Malahelo Forest, a biogeographical transition zone in southeastern Madagascar, *Biodivers. Conserv.*, **11**:1791-1807.

Soulé, M. E., Bolger, D. T., Alberts, A. C., Wright, J., Sorice, M., and Hill, S., 1988, Reconstructed dynamics of rapid extinctions of Charparral- requiring birds in urban habitat islands, *Conserv. Biol* **2**:75-92.

Watson, J. E. M., Freudenberger, D., and Paull, D., 2001, An assessment of the Focal-species Approach for conserving birds in variegated landscapes in southeastern Australia, *Cons. Biol.* **5**:1364-1373.

Watson, J. E. M., 2004, *The effects of habitat fragmentation on birds: illustrations from Madagascan and Australian case studies.* Unpublished D.Phil Thesis, School of Geography and the Environment, University of Oxford.

Whittaker, R. J., 2000, Scale, succession and complexity in island biogeography: are we asking the right questions? *Glob. Ecol. and Biog.* **9**:75-87.

USE OF SATELLITE TELEMETRY AND REMOTE SENSING DATA TO IDENTIFY IMPORTANT HABITATS OF MIGRATORY BIRDS (*CICONIA CICONIA* (LINNAEUS 1758))

Birgit Gerkmann and Klaus Riede
Alexander Koenig Research Institute and Museum of Zoology, Adenauerallee 160, 53113 Bonn, Germany, E-mail: birgit.gerkmann@gmx.de, k.riede.zfmk@uni-bonn.de

Abstract: This study is using new technologies such as satellite telemetry and remote sensing data to identify important habitats of migratory birds (*Ciconia ciconia*). Previously collected telemetry data are being used for further analysis to localise regions of frequent stopovers during migration and wintering. An index has been developed to calculate migration and staging periods semi-automatically from the satellite dataset. African land cover maps were used to characterise these staging sites. First results show the importance of wintering sites in Chad and Sudan, where the preferred habitats consist of savanna, grassland and cropland.

Key words: migratory birds; satellite telemetry; remote sensing; GIS; conservation

1. INTRODUCTION

Sufficient food resources and adequate habitats in breeding and wintering areas of migratory birds are of major importance for successful migration and reproduction.

Loss of staging and feeding sites, together with direct anthropogenic impacts such as pesticide use and hunting, are considered principal negative factors affecting migratory bird populations (Berthold, 2000). For the White Stork, population declines are due to deteriorating habitat conditions in breeding areas (Dallinga and Schoenmakers, 1989), while pesticide use and

B.A. Huber et al. (eds.), African Biodiversity, 261–269.

hunting have been identified as the main negative impacts during migration and in wintering areas (Schulz, 1995).

Previous studies on three crane species (*Grus leucogeranus, G. vipio, G. monacha*) already demonstrated that the use of remote sensing data can be combined with migration data, thereby helping to identify important staging areas and habitats (e.g. Minton, 2003; Fujita, 2004). Especially the use of satellite telemetry has improved our understanding of migration considerably, due to higher accuracy compared to ringing data, together with continuous temporal and spatial data availability throughout the migration route. This digitised data stream can be transferred directly into a Geographical Information System (GIS). A project carried out by the Max Planck Institute for Ornithology (Vogelwarte Radolfzell) moved investigations using satellite telemetry ahead by tracking more than 100 White Storks (*Ciconia ciconia*) for more than a decade, starting in 1991 (Berthold et al., 2002; Van den Bossche et al., 2002). Although the White Stork is already a well-known species, the analysis of these satellite data revealed new insights, which improved our understanding considerably (e.g. Berthold et al., 1997; Berthold et al., 2001).

The main topic of this paper is the development of semi-automatic procedures facilitating data-cleaning and analysis of these telemetry data. The improved data repository can then be used for intersection with other datasets (e.g. land cover, precipitation regime), thereby helping to understand the factors affecting migration routes.

2. METHODS

2.1 Datasets and data cleaning

This study uses satellite telemetry data from 66 White Storks, all equipped with transmitters at their breeding sites in northeast Germany, in northeast and northwest Poland and in Israel between 1991 and 2003 (Van den Bossche et al., 2002). Satellite data were verified by Kaatz (2004) localising White Storks on the ground with a Global Positioning System (GPS) from 1991 to 2003 in Loburg, Saxony-Anhalt, Germany. He observed considerable differences between satellite and GPS data, and developed criteria to select high-accuracy data (Kaatz, pers. comm. 1999). In addition we assumed data from storks resting on the ground were more reliable. Because most storks rest after sunset and before sunrise, it was possible to extract reliable data by setting a temporal threshold (Kaatz, 2004). By application of these criteria as a filter on the entire dataset of 68764 locations, unreliable data were eliminated. Examples of migration routes have been published as animated satellite tracks on the internet - http://www.groms.de/groms/Ciconia_Info.html.

2.2 Indices for classification of migration behaviour

Identification of staging areas requires differentiation between periods of migration and staging. To facilitate this classification, a directional index I_{dir} based on significant variables such as the distance flown and change of direction between consecutive datapoints was calculated:

$$I_{dir} = (r \times t)/s$$

with r = change of direction [degree]; t =days after change of direction and s = distance flown after change of direction [km]. Major changes of flight direction (bearing), together with a short daily distance flown, indicate a "nomadising" or foraging behaviour of the bird during staging periods, and resulted in Indices >1. Indices during obvious migration periods were below 1, so that a threshold of 1 was set for semi-automatic classification of staging (>1) and migration (<1) periods.

However, some ambiguities were observed, for example during short foraging flights at constant bearing. Therefore an improved index based entirely on distances was applied. This improved "distance index" I_{dist} is used in "random walk" models. It has already been applied to analysis of *Grus vipio* movements by Fujita (2004), who placed the distance actually flown between breeding and wintering sites in relation to the geographical distance (i.e. the shortest distance on the Earth´s surface) between the start and end locations. This ratio varies with directionality: during migration with constant bearing, the flown distance nearly equals the geographical distance between every 3rd location, resulting in an I_{dist} close to 1. Taking into account small deviations, the threshold was set to 1.1. An important additional criterion for a reliable classification was the average daily travel distance, which was set to less than 50 km for nondirectional movement (between point 1 and 3), or a minimal distance of 50 km for directional movement (migration). All criteria together are summarised in Table 1.

Applying all criteria to the entire dataset allows semi-automatic classification of staging areas – most points were classifiable (Fig. 1 and Table 2). Once identified, staging areas can be superimposed with remote sensing data using a Geographical Information System (GIS).

2.3 Remote sensing data

Already classified data of the "Global Land Cover Characteristics

Table 1. Criteria for the classification of (non-)directional movement

Mean distance flown[km/day]	Geographic distance between 1st and 3rd location resp.	I_{dist}*	Classification
<50	<50 or >50	>1.1	Nondirectional movement - foraging or staging
<50	<50 or >50	<1.1	unclassifiable
>50	>50	<1.1	Directional movement - migration
>50	>50	>1.1	unclassifiable

* I_{dist} = (distance flown)/ (geographic distance)

Database" (GLCC) (Loveland et al., 2000) were used for a preliminary analysis of land cover classes at the staging areas identified. The dataset is based on NOAA-AVHRR data from April 1992 to May 1993 and has a resolution of 1 km. Several maps with already classified data are available, showing different classification systems (Loveland et al., 2000; Brown, Loveland et al. 1999). In addition, the most commonly used dataset "USGS land use" based on the U.S. Geological Survey´s classification (Brown et.

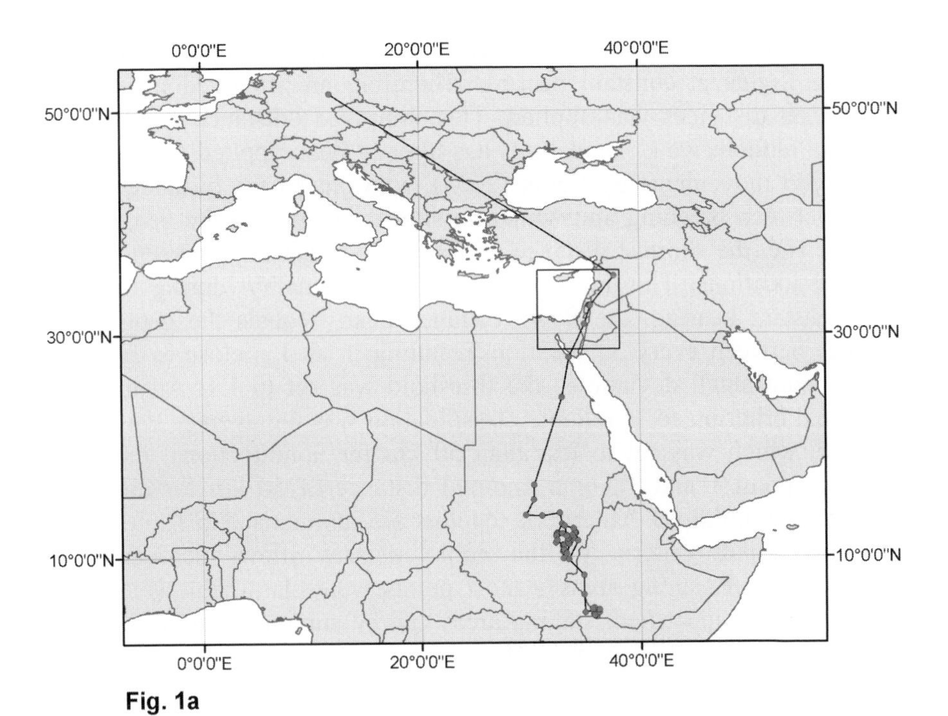

Figure 1a. Migration route of White Stork tag number 14549 at different scales. Entire migration route from 01/09/94 (breeding) to 14/01/95 (wintering).

al., 1999) was used. However, these data exhibit accuracy problems, especially at regional scale.

3. RESULTS AND DISCUSSION

3.1 Satellite tracks of White Stork migration - data cleaning and classification

Data cleaning, as described in the methods section, was applied to the huge input dataset of 68,764 locations. The result was a more accurate dataset for 57 birds, consisting of 4990 locations. Improved accuracy was a prerequisite for the following detailed analysis. As a first step, the directional index (I_{dir}) was calculated and compared with the distance index (I_{dist}). Figures 1b and 1c show the results of classifying sections of the migration route for White Stork tag number 14549 (1994/1995; Fig. 1a).

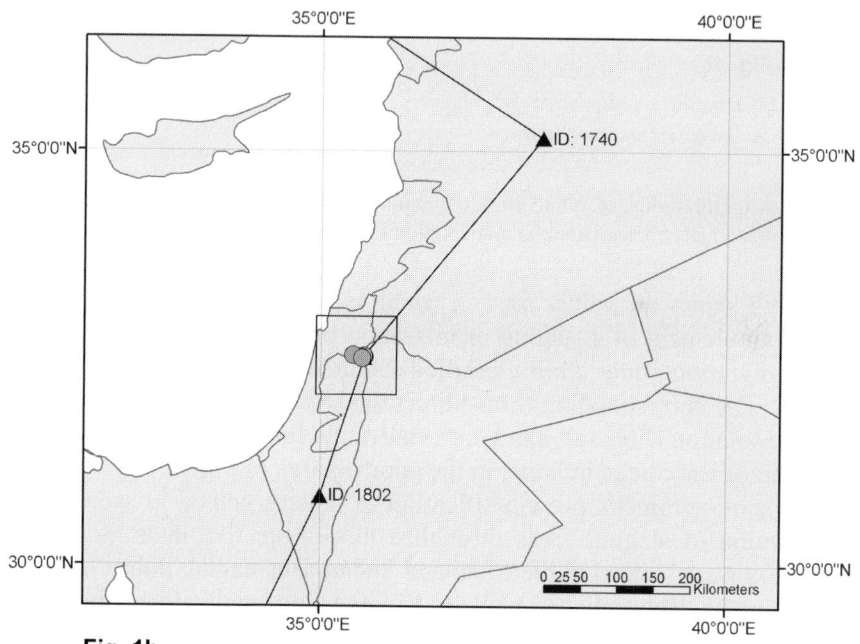

Figure 1b. Migration route of White Stork tag number 14549 at different scales. Migration data from the Middle East at a higher resolution showing locations of nondirectional movement (circles) and directional movement (triangles), the area outlined by a rectangle has been further enlarged in Fig. 1c. ID numbers indicate the points listed in Table 2, listing values for I_{dist}.

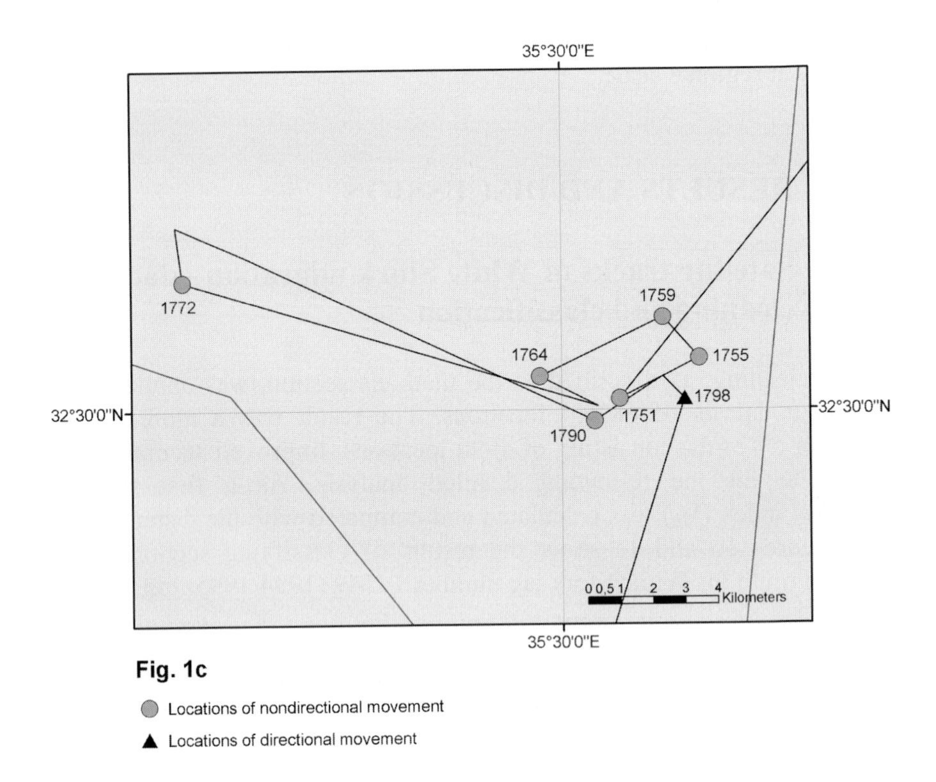

Figure 1c. Migration route of White Stork tag number 14549 at different scales. Detailed plot of staging area in Bet Shean, Israel (area marked in Fig. 1b).

Table 2 shows the values for I_{dist}, which made it possible to classify each point as an element of a staging or migration period; 53 out of 83 locations of this migration route were classified using the criteria presented in the methods. The corresponding (non-)directional points become clearly visible at high resolution (Fig. 1c), showing small-scale foraging movements within the region of Bet Shean in Israel in the zoomed area of Figure 1b.

This semi-automatic pre-classification of indices helped to speed up the identification of staging areas through a more objective method. Most of these areas were located in the Chad and Sudan. But staging points have also been found in Botswana as well as isolated points all along the eastern migration route. These results are coincident with important wintering areas of the White Stork already identified by (Van den Bossche et al., 2002).

Classification and indexing of staging sites allowed automatic extraction of all staging sites. This reduced dataset was used to analyse the distribution of staging patterns, together with further GIS analysis by overlaying remote sensing data.

Table 2. Distances, indices and classification as (non-)directional movement for locations shown in Fig. 1b and 1c

ID	Date (1994)	Lat	Long	Distance of successive locations (km)	Mean daily distance (km)	Distance between 1st and 3rd location = shortest path (km)	I_{dist}	Classification N= non-directional/ staging D= directional/ migration
1740	12.09.	35.16	37.718	359.05	179.53	356.89	1.01	D
1751	14.09.	32.503	35.516	2.40	2.40	2.70	1.46	N
1755	15.09.	32.514	35.538	1.54	1.54	4.17	1.25	N
1759	16.09.	32.525	35.528	3.65	3.65	3.16	1.71	N
1764	17.09.	32.509	35.494	1.75	1.75	9.73	1.35	N
1772	19.09.	32.535	35.395	1.68	1.68	1.90	2.04	N
1790	23.09.	32.497	35.509	2.23	2.23	2.44	1.27	N
1795	24.09.	32.509	35.528	0.87	0.87	193.58	1.00	D
1798	25.09.	32.503	35.534	193.08	193.08	554.51	1.00	D
1802	26.09.	30.830	34.996	362.73	362.73	760.32	1.00	D

3.2 Identification of habitats - overlay with remote sensing data

The overlay of staging data based on I_{dir} with the USGS land cover map gave the pixel value (= land cover class out of 24 classes) for every location. Most of the points were found in regions classified as Savanna (49%). Further land cover classes with many locations were Grassland (14%) and Dryland Cropland and Pasture (13%). The habitats identified coincide with the preferences of the White Stork in its breeding areas – grassland and green crops (Nowakowski, 2003), grassland (Böhning-Gaese, 1992) or arable fields (Ozgo and Bogucki, 1999). However, contrary to earlier observations of storks in Africa (Mullié et al., 1995), no preference for wetlands could be found. A possible explanation could be an under-representation of wetlands in the land cover map (Loveland et al., 2000).

Although the maps comprise only the time span from April 1992 to May 1993, while the locations of the White Stork range from 1991 to 2003, it was assumed that main patterns of land cover are still valid.

4. OUTLOOK

In future work, staging areas will be classified more accurately by taking into account the duration of staging and data from already identified Important Bird Areas (Fishpool and Evans, 2001).

For a more detailed analysis of habitats used, more accurate land cover maps (Global Land Cover – GLC 2000: Mayaux, 2003) and MODIS NDVI (data with a higher spatial and temporal resolution, 250 m, nearly every day, http://modis.gsfc.nasa.gov/about/index.html) will be used. For example, the MODIS vegetation index (NDVI) can be used to derive information about vegetation growth and precipitation regime. This will facilitate detection of possible correlations between migration pathways and external factors, such as precipitation or food abundance. Furthermore, accuracy and reliability of the remote sensing data will be examined on the ground during field studies at selected sample sites.

ACKNOWLEDGEMENTS

We would like to thank the "Deutsche Bundesstiftung Umwelt" (DBU) foundation for funding the PhD thesis of B. Gerkmann (AZ 20003/465), and Prof. Berthold of the Max Planck Institute for Ornithology (Vogelwarte Radolfzell) for generously providing the complete dataset of White Storks, as well as Prof. G. Menz, Dr. Michael Kaatz, Prof. B.-U. Meyburg and many others for supporting this PhD thesis. We thank an anonymous referee for valuable criticism of a first draft of this manuscript.

REFERENCES

Berthold, P., Van den Bossche, W., et al., 1997, Satelliten-Telemetrie der Jahreswanderung eines Weißstorchs *Ciconia ciconia* und Diskussion der Orientierungsmechanismen des Heimzugs, *J. Ornithol.* **138:**229-234.

Berthold, P., 2000, Vogelzug - Eine aktuelle Gesamtübersicht, Wissenschaftliche Buchgesellschaft, Darmstadt.

Berthold, P., Van den Bossche, W. et al., 2001, Der Zug des Weißstorchs (*Ciconia ciconia*): eine besondere Zugform auf Grund neuer Ergebnisse [German*], J. Ornithol.* **142:**73-92.

Berthold, P., Van den Bossche, W., et al., 2002, Long-term satellite tracking sheds light upon variable migration strategies of White Storks (*Ciconia ciconia*), *J. Ornithol.* **143:**489-495.

Böhning-Gaese, K., 1992, Zur Nahrungsökologie des Weißstorchs (*Ciconia ciconia*) in Oberschwaben: Beobachtungen an zwei Paaren, *J. Ornithol.* **133**(1):61-71.

Brown, J., Loveland, T. et al., 1999, The Global Land-Cover Characteristics Database: The Users´ Perspective, *Photogramm. Eng. Remote Sens.* **65**(9):1069-1074.

Dallinga, J. and Schoenmakers, S., 1989, Population changes of the White Stork *Ciconia ciconia* since the 1850s in relation to food resources, in: Weißstorch - White Stork Proc. I Int. Stork Conserv. Symp., G. Rheinwald, Ogden and H. Schulz (eds), *Schriftenreihe des DDA.* **10:**231-262.

Fishpool, L. D. and Evans, M, 2001, *Important Bird Areas in Africa and associated islands - Priority sites for conservation*, Pisces Publications and BirdLife International, Newbury and Cambridge, UK.

Fujita, G., Hong-Liang G, et al., 2004, Comparing areas of suitable habitats along travelled and possible shortest routes in migration of White-naped Cranes *Grus vipio* in East Asia, *Ibis* **146:**461-474.

Kaatz, M., pers. comm. (1999) Studie zur Neubewertung von ARGOS-Satellitenkoordinaten, unveröffentlicht, pp. 26.

Kaatz, M., 2004, Der Zug des Weißstorchs *Ciconia ciconia* auf der europäischen Ostroute über den Nahen Osten nach Afrika, Dissertation, Martin-Luther-Universität, Halle - Wittenberg, pp. 163.

Loveland, T., Reed, et al., 2000, Development of a global land cover characteristics database and IGBP DISCover from 1 km AVHRR data, *Int. J. Remote Sensing* **21**(6 & 7):1303-1330.

Mayaux, P., Bartholomé, W., et al., 2003, The Land Cover Map for Africa in the Year 2000, European Commission Joint Research Centre.

Minton, J, Halls, J, et al., 2003, Integration of satellite telemetry data and land cover imagery: a study of migratory cranes in northeast Asia, *Trans. GIS* **7**(4)**:**505-528.

Mullié, W., Brouwer, J. et al., 1995, Numbers, distribution and habitat of wintering White Storks in the east-central Sahel in relation to rainfall, food and anthropogenic influences, Internat. Symp. on the White Stork (Western Population), Basel.

Nowakowski, J., 2003, Habitat structure and breeding parameters of the White Stork *Ciconia ciconia* in the Kolno Upland (NE Poland), *Acta Ornithol.* **38**(1)**:**39-46.

Ozgo, M. and Bogucki, Z., 1999, Home range and intersexual differences in the foraging habitat use of a White Stork (*Ciconia ciconia*) breeding pair, in: *Weißstorch im Aufwind? - White Storks on the up?* Proceedings, Internat. Symp. on the White Stork, H. Schulz (eds), Hamburg, - NABU (Naturschutzbund Deutschland e.V.), Bonn, pp. 481-492.

Schulz, H. 1995, Zur Situation des Weißstorchs auf den Zugrouten und in den Überwinterungsgebieten, in: *Proceedings of the International Symposium on the White Stork (Western Population),* Basel (Switzerland), Schweizerische Vogelwarte Sempach (Switzerland).

Van den Bossche, W., Berthold, P., et al., 2002, Eastern European White Stork Populations: Migration Studies and Elaboration of Conservation Measures, Bundesamt für Naturschutz, Bonn, pp. 197.

CONSERVATION PRIORITIES AND GEOGRAPHICAL VARIATION IN FLYCATCHERS (AVES: PLATYSTEIRIDAE) IN THE DEMOCRATIC REPUBLIC OF CONGO

Michel Louette
Royal Museum for Central Africa, B-3080 Tervuren, Belgium
E-mail: michel.louette@africamuseum.be

Abstract: This paper highlights representatives of the bird family Platysteiridae in the DR Congo which merit special conservation attention because of the following distinctive characteristics: (1) morphology (plumage coloration) that is unusual within family: sexual monomorphism or polymorphism in the male; (2) ecological specialisations: restricted to the forest canopy; occurring near the forest floor; restricted to the submontane zone; or restricted to a single vegetation type.
These species occur mainly outside nature reserves, and thus urgently require additional protection. The morphological diversity of *Batis erlangeri* in DR Congo is also discussed and a colonisation history of its southern population is suggested.

Key words: Platysteiridae; Africa; DR Congo; conservation; geographical variation

1. INTRODUCTION

I will develop here two points from my preparation of the chapter on the Platysteiridae for 'Handbook of birds of the World' (Louette, in press 2006), a bird family with peculiar morphology, occurring throughout sub-Saharan continental Africa.

Fishpool and Evans (2001) identified 'Important Bird Areas' (IBAs) in all African countries. Species protection appears to be best ensured in legally protected areas (World Heritage Sites, National Parks, Hunting Reserves,

B.A. Huber et al. (eds.), African Biodiversity, 271–277.

etc.). However, not all IBAs (nor zones of greatest biodiversity) are confined to protected areas. BirdLife International (2000) also categorised threat status for individual bird species according to rarity (small range, low population or both) and decreasing population size. Here, I advocate the introduction of the criterion "unusual characteristics", to be taken into account for conservation prioritization. This criterion may include morphological, ethological or other ecological characteristics. The Platysteiridae exhibits its greatest diversity in the equatorial region. Fourteen out of the 29 species constituting the family occur in the DR Congo (see Appendix); none of these is listed as of conservation concern by BirdLife International (2000). Seven species have restricted ranges (with a global breeding range of less than 50,000 km^2) or can be categorized as biome-restricted, and thus qualify for special attention (Demey and Louette, 2001).

2. MATERIAL AND METHODS

I studied more than 2,000 specimens held at the Royal Museum for Central Africa, Tervuren, Belgium.

As a general rule in birds, colourful (typically male) plumage is connected to territorial activity and cryptic (typically female) plumage to predator avoidance. Trade-offs occur: not all related bird taxa are sexually dimorphic to the same extent and in some cases, females are also colourful (as in all Platysteiridae). Even in those cases, however, plumages can be classified as 'masculine' or 'feminine'. Particularities in social organisation (such as competition for partner) are a plausible explanation. The majority of species in the family exhibit a 'typical' sexual dimorphism consisting of a black breast band in males and a brown breast band in females. Some members, however, have "unusual characteristics" in plumage colouration, some with atypical, masculine, females. In other regions of Africa another taxon has an atypical (feminine) male, but its female is 'hyperfeminine', and still another taxon has a typical female and a 'semi-feminine' male. The role of these plumages in social behaviour is still unknown and may yield important ethological information.

Whereas most Platysteiridae are rather eurytopic forest birds, some have a more specialist ecology: several occur in the forest canopy or just above the forest floor; others are restricted to the submontane zone or to a single vegetation type.

3. CONSERVATION

Four of the species occurring in DR Congo, *Megabyas flammulatus*, *Dyaphorophyia blissetti*, *D. castanea* and *D. tonsa*, are restricted to the Guinea-Congo Forests biome (Demey and Louette, 2001) but are widespread (Urban et al., 1997; Louette, in press 2006) and will not be discussed further here.

Dyaphorophyia concreta is not listed in any conservation category, but has peculiar morphological and ecological characteristics. In marked contrast to other Platysteiridae it has mainly metallic bottle green upperparts and bright yellow underparts (the greenish and yellow colours are shared only with the related species *D. blissetti*). Ventrally some specimens are washed with chestnut of variable intensity, and there is an as yet unexplained plumage colour polymorphism in males in DR Congo (and Uganda): a yellow-bellied form and a chestnut-bellied form (see Prigogine, 1971). This is a bird living just above the forest floor, but this ecology does not explain its peculiar morphology. It probably occurs in several protected areas in the east of DR Congo.

The DR Congo (collecting) localities of the remaining three biome-restricted species, *Batis diops, B. ituriensis* and *B. margaritae,* are shown in Figure 1, together with the most important protected areas.

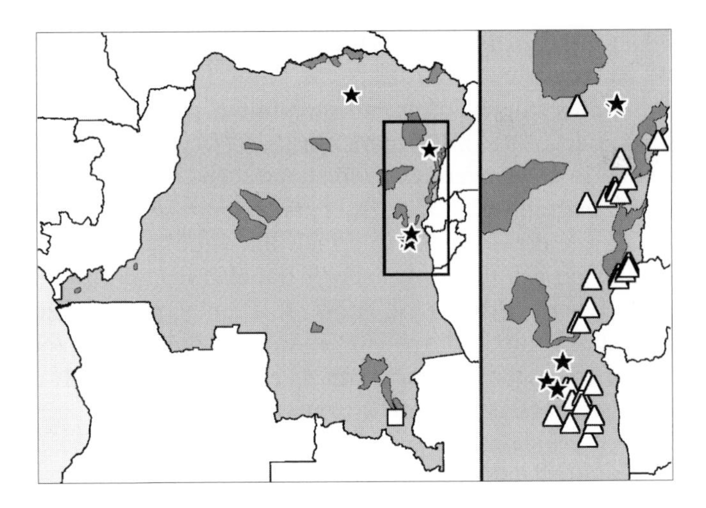

Figure 1. The most important protected areas in the DR Congo and localities within the country for three rare Platysteiridae species: *Batis diops* (triangles), *B. ituriensis* (stars) and *B. margaritae* (square).

The restricted-range species *Batis diops* is monomorphic masculine, except for its iris colour, which is yellow in the male and reddish in the female. The social structure of this species, which is no doubt connected to this plumage, is as yet unknown. It occurs throughout the montane evergreen forests of different types in the Albertine Rift in the altitudinal range of 1340–3300 m. It is thus localised, and fortunately locally still relatively common, mainly because these higher altitudes are relatively inaccessible to people for the time being. It occurs at the higher elevations in several protected areas.

Batis ituriensis is apparently a monomorphic masculine species, but its possible sexual dimorphism is still under discussion. Its white loral spot is indeed variable in size and this may be related to sex, the birds with very small spots being females. But contrary to *B. diops*, the iris is yellow in both sexes. It is known from a few localities only: in the northern part of its range it occurs in the mid and upper strata of lowland forest, whereas in the southern part it is found in submontane forests between 950–1300 m. Altitudinal factors seem to be paramount in its distribution, but also vegetation type and it is not often recorded in primary forest (see e.g. Rossouw, 2001). It belongs to a group of submontane specialists occurring at the lower levels of the Albertine Rift for which Bober et al. (2000) showed that they are of greatest conservation concern, because these submontane habitats are only partly included in protected areas. It has been observed, but not collected in Kahuzi-Biega National Park (Demey and Louette, 2001).

Batis margaritae exhibits a typical sexual dimorphism, with a black breast band in the male and a rufous one in the female. *Batis m. kathleenae* is the taxon restricted to northernmost Zambia and extreme southern DR Congo (Schouteden, 1971). It occurs mostly in *Cryptosepalum* forest and to a certain extent in other evergreen forest and man-induced thickets, possibly as a seasonal migrant (R. Dowsett, pers. comm. 2004). It uses the mid stratum, i.e. lower down in the tree than the widespread and eurytopic *B. molitor*, which can use the open canopy but also occurs lower down in the vegetation elsewhere (Urban et al., 1997; R. Demey, pers. comm. 2004), and is thus a potential competitor (as is *Platysteira peltata)*, a factor to be taken into consideration in combination with its restricted habitat. It is not known to occur in any protected area in DR Congo.

4. GEOGRAPHICAL VARIATION

In most recent works (e.g. Urban et al., 1997), *Batis erlangeri* is treated as a subspecies of *B. minor* but elsewhere (Louette, in press 2005), I give reasons for considering it a separate species, because it is significantly larger

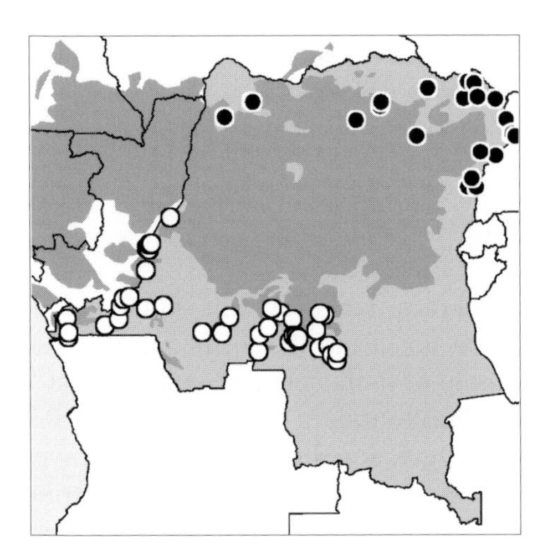

Figure 2. Equatorial forest range in the DR Congo (simplified from the WCMC global forest cover map in www.wcmc.org.uk/protected_areas/pavl/parc.htm) and localities within the country for *Batis erlangeri*: nominate race (full circles), race *congoensis* (open circles).

in size and darker in the mantle, but less black on top of the head. The females have decidedly more olive wash than *B. minor.* Figure 2 shows that *B. erlangeri*, a woodland bird, occurs north of the equatorial forest and reaches the northernmost areas of DR Congo; there is also an isolated population in southern DR Congo (and adjoining areas). My investigation of the standard bird measurements, however, reveals geographic variation: small but statistically significant differences in male tail length and bill length in both sexes, indicating incipient speciation (Louette, in press 2005). I therefore recognize the subspecies *congoensis* for the southern birds. It is thought that this *congoensis* population arrived from northern woodlands during a period with a geographical gap in the rainforest along the Ubangi river corridor in northwestern DR Congo, a scenario accepted for some other woodland birds (see Prigogine and Louette, 1983, for this hypothesis in relation to *Dendropicos goertae*). Whereas there is no clue as to the biological meaning of these differences, the low tail/wing ratio of *B. e. congoensis* is quite remarkable. This ecomorphological trait also appears in the look-alike species *B. molitor* (an ecological vicariant), whose population in south-eastern DR Congo also has a much lower tail/wing ratio than its populations elsewhere in southern and eastern Africa (Louette, 1987). Despite being very similar in plumage pattern and colouration, there appears to be no close phylogenetic relationship between the two species *B. molitor* and *B. erlangeri*, as they are vocally quite different. Both have the 'standard' sexually dimorphic plumage, with the main morphological difference a brown throat spot in the female *B. molitor* (absent in the female *B. erlangeri*) but they differ in the number of postjuvenile moult episodes: two moults transform the juvenile rusty wing stripe into a white (adult) one in *B. erlangeri*, while in *B. molitor* three moults are necessary to reach that stage (L. Verstraelen, pers. comm. 2003).

ACKNOWLEDGMENTS

I thank Hans Beeckman for providing botanical information, Robert Dowsett for the relevant text from the forthcoming 'The Birds of Zambia', Benoît Hardy for the map of protected areas and Liesbet Verstraelen for data on the moult in some Platysteiridae. Assistance at the RMCA was given by Danny Meirte, Alain Reygel, Jos Snoeks and Wim Tavernier. Adrian Craig and Ron Demey, who refereed the paper, contributed much to its improvement.

REFERENCES

BirdLife International, 2000, *Threatened birds of the World*, Lynx Edicions and BirdLife International, Barcelona and Cambridge (UK).

Bober, S. O., Herremans, M., Louette, M., Kerbis Peterhans, J. C., and Bates, J. M., 2001, Geographical and altitudinal distribution of birds endemic to the Albertine Rift, Proceedings 10th Pan African Ornithological Congress, Kampala, Uganda. *Ostrich Supplement* **15:**189-196

Demey, R., and Louette, M., 2001, Democratic Republic Congo, in L.D.C. Fishpool and M.I. Evans, eds., Important Bird Areas in Africa and associated islands: Priority sites for Conservation. Newbury and Cambridge, U.K.: *Pisces Publications and BirdLife International (BirdLife Conservation Series N° 11)* pp. 199-218.

Fishpool, L. D. C., and Evans, M. I., 2001, eds. Important Bird Areas in Africa and associated islands: Priority sites for Conservation, Newbury and Cambridge, U.K.: *Pisces Publications and BirdLife International (BirdLife Conservation Series N° 11).*

Louette, M., 1987, Additions and corrections to the avifauna of Zaïre (1), *Bull. Brit. Orn. Cl.* **107:**137-143.

Louette, M., (in press, 2005), The Western Black-headed Batis *Batis erlangeri*, a separate species consisting of two subspecies, *Bull. Afr. Bird Cl.*

Louette, M., (in press, 2006), Platysteiridae, in J. del Hoyo, A. Elliot and J. Sargatal, eds., *Handbook of Birds of the World.* Lynx, Barcelona.

Prigogine, A., 1971, Les Oiseaux de l'Itombwe et son Hinterland, Volume I. *Ann. Sc. Zool.* **185**. Musée Royal de l'Afrique Centrale. Tervuren.

Prigogine, A., and Louette, M., 1983, Contacts secondaires entre les taxons appartenant à la super-espèce *Dendropicos goertae*, *Gerfaut* **73:**9-83.

Rossouw, J. D., 2001, New records of uncommon and poorly known species for Ugandan National Parks and Forest Reserves. *Scopus* **21:**23-34.

Schouteden, H., 1971, La Faune Ornithologique de la province du Katanga, *Doc. Zool.* **17**. Musée Royal de l'Afrique Centrale, Tervuren.

Urban, E. K., Fry, C. H., and Keith, S. eds., 1997, *The Birds of Africa*, Vol. 5, Academic Press, London.

APPENDIX

The 14 Platysteiridae species occurring in DR Congo. Those with * are discussed in the Conservation section of this paper; those with § in the Geographical variation section.

Megabyas flammulatus Verreaux, 1855
Bias musicus (Viellot, 1818)
Batis diops Jackson, 1905 *
B. ituriensis Chapin, 1921 *
B. margaritae Boulton, 1934 *
B. erlangeri Neumann, 1907 §
B. minulla (Barboza du Bocage, 1874)
B. molitor (Küster, 1850) §
Dyaphorophyia concreta (Hartlaub, 1855) *
D. blissetti Sharpe, 1872
D. castanea (Fraser, 1842)
D. tonsa Bates, 1911
Platysteira cyanea (Müller, 1776)
P. peltata Sundevall, 1850

RARE WEAVERS (AVES: PLOCEIDAE): ARE SOME *PLOCEUS* SPECIES HYBRIDS?

Adrian J.F.K. Craig
Department of Zoology & Entomology, Rhodes University, Grahamstown, 6140, South Africa

Abstract: Two central African forest weavers, *Ploceus aureonucha* and *P. flavipes*, are known only from < 10 specimens each since 1920, and < 3 sightings each in the past 40 years. A re-examination of the available data raised the possibility that these taxa might be hybrids. For *P. flavipes*, the postulated parental species are *P. nigerrimus* X *P. albinucha*. For *P. aureonucha*, the parental species could be *P. nigerrimus* X *P. tricolor*, but it seems more likely that *P. aureonucha* represents an unrecognised plumage stage of *P. tricolor*. Resolution of the status of these two species is relevant to assessments of biodiversity, and conservation priorities, in the Ituri region.

Key words: Ploceidae; Africa; taxonomy; hybrids

1. INTRODUCTION

The final volume of the landmark handbook, "Birds of Africa" has just been published, and this will for many years be the standard source for information on African birds. Thus it will be used in calculating regional biodiversity, and in assessing the conservation status of species and their habitats. However, these data are always subject to revision. After completing the species accounts for many of the Ploceidae, I have re-examined the data for some of the lesser-known species, and now question my earlier acceptance of the status of two rare taxa: *Ploceus aureonucha* and *Ploceus flavipes* (Craig, 2004). Why are there so few records of these species, and little consistency in the appearance of the available specimens? Clearly there are three possible conclusions: (1) that these are valid species, and that their variation is age and/or sex-related; (2) that they are hybrids

B.A. Huber et al. (eds.), African Biodiversity, 279–286.

between other known species; (3) that they represent unrecorded or aberrant plumage stages of previously described species.

Hybridization traditionally represented a 'problem' for the biological species concept, but is accepted more easily under other views such as the phylogenetic species concept. Recently Grant and Grant (1992, 1996, 1997) have re-examined the frequency and possible evolutionary role of hybridization between populations of wild birds. The African Ploceidae appear to meet some of the suggested criteria for conditions where hybrids might be expected to occur most often, such as areas of sympatry where one species is rare (Randler, 2002).

2. MATERIALS AND METHODS

In preparing text for "Birds of Africa" I personally examined museum specimens of all species, wrote plumage descriptions, and took standard measurements: wing length and tail length to the nearest mm using a stopped rule as issued to bird ringers; tarsus length, bill length from the base of the bill against the skull, bill depth and width at the anterior border of the nostril – these to 0.1 mm using vernier callipers. While my measurements undoubtedly differ from those taken by other authors, they do provide a basis for my own comparisons between species. Following discussions after I had presented my paper in Bonn, I visited the Africa Museum in Tervuren again, and re-examined *Ploceus flavipes*, *P. albinucha*, *P. nigerrimus* and *P. tricolor* in the company of Dr Michel Louette.

3. RESULTS AND DISCUSSION

3.1 *Ploceus aureonucha*

This taxon was described by Sassi (1920) from specimens collected in the Ituri region of the Democratic Republic of Congo by Grauer. The type and two other specimens are in the Natural History Museum in Vienna; a fourth specimen initially identified as belonging to this species was later assigned to *Ploceus flavipes* by Stresemann (1925). I have not seen the material in Vienna, but Dr Walter Sontag sent me photographs of these specimens, and Dr Anita Gamauf provided measurements; I took my own measurements of three specimens in the American Museum of Natural History in New York. Three specimens were collected in 1910, two in 1921, and one in 1926. In 1986 sight records were reported, including "a flock of up to 60 birds" (Collar et al., 1994), although perhaps for this canopy species the phrase should be reworded as "a flock, some of which were tentatively identified as *P. aureonucha*". There was a subsequent report of a pair feeding two young in the same region of the Ituri forest (Collar et al., 1994).

The plates in Fry et al. (2003) and in Sinclair and Ryan (2003) show an interesting contrast, with the former showing a much lighter-coloured bird, and the latter a dark basic plumage. The illustration in Fry and Keith (2004) is not a good match for the description in the text, and since there are no specimens of this species in the British Museum at Tring, I presume that the artist relied on secondary sources for his representation.

Clearly, in plumage characters *P. aureonucha* is most like *P. tricolor* but darker. Thus if one were to cross *P. tricolor* with another species, losing some of the coloured areas and producing a darker overall appearance, which weavers come into question? It should also be a forest species, which is known to occur sympatrically with *P. tricolor*. An all-black bird seems the best choice. Black forest weavers include females of the two *Malimbus* species *cassini* and *rubricollis*, but these are forest canopy birds, breeding solitarily after constructing a complex nest with the involvement of other group members (Fry and Keith, 2004), and not likely to associate with a colonial ploceid such as *P. tricolor* when nesting. This would leave two black *Ploceus* species, *albinucha* and *nigerrimus*. In the region from which specimens of *P. aureonucha* have been collected, *P. nigerrimus* is common, whereas *P. albinucha* and *P. tricolor* are both at the fringe of their ranges. Mixed colonies of *P. nigerrimus* and *P. tricolor* have been reported from both West and Central Africa (Bannerman, 1949; Chapin, 1954), and a comparison of plumage characters and measurements (Table 1) suggests that this is a plausible combination.

However, there is the third possibility, namely that *P. aureonucha* is in fact merely a plumage stage of another species. Immature specimens of *P. tricolor* in the Africa Museum at Tervuren immediately recalled the specimens of *P. aureonucha* that I had seen in New York, and the photographs from Vienna. I have always respected Chapin's judgement, and re-reading his account of these two species, I found that he had seen no immature specimens of *P. tricolor* and that two specimens of *P. aureonucha* which he collected were from a flock, apparently all subadult birds (Chapin, 1954). Thus his acceptance of *P. aureonucha* as a valid taxon was based on one specimen, assessed as an adult male. The clustering of specimens, and the sight records, would certainly support the idea of misidentification rather than rare hybrids, for which one might expect only one specimen to be collected at a time.

Table 1. Comparative coloration (of breeding male birds) and measurements of *P. aureonucha* and possible 'parent' species. Measurements are given as ranges of male and female specimens combined (since the sexing is uncertain), with mean values in brackets.

Character	*P. aureonucha* n = 6 (3 ♂, 3 ♀)	*P. tricolor* n = 24 (12 ♂, 12 ♀)	*P. nigerrimus* n = 24 (12 ♂,12 ♀)
Bill colour	Black	Black	Black
Leg colour	Brown	Brown	Brown
Iris colour	Brown	Red	Golden yellow
Head colour	Black	Black	Black
Nape colour	Yellow collar, orange-brown edge	Yellow collar	Black
Mantle colour	Black with some yellow streaks	Black	Black
Rump colour	Black	Black	Black
Wing colour	Black	Black	Black
Tail colour	Black	Black	Black
Chest colour	Maroon	Chestnut brown	Black
Belly colour	Charcoal grey	Chestnut brown	Black
Wing length	79-82 (80.5)	78-91 (84.6)	75-92 (80.4)
Tail length	42-44 (43.0)	46-55 (50.5)	47-58 (52.2)
Tarsus length	17.7-19.1 (18.4)	17.3-20.8 (18.9)	19.0-22.0 (20.8)
Bill length	20.0-20.5 (20.2)	18.8-21.5 (20.2)	19.1-23.0 (20.8)

3.2 *Ploceus flavipes*

The type specimen was brought to Chapin by one of his hunters. It is a carefully sexed female bird in all black plumage, shot in the forest canopy, and now in the American Museum of Natural History. Chapin (1916) noted the "dull yellow legs" and "pinkish nasal tubercle". There are nine specimens in all (Louette, 1988); I have not examined those in Brussels (Prigogine, 1960) and in Stockholm. I have seen the type specimen, the Vienna specimen, and three specimens in Tervuren (male, female and unsexed). Because of its all black plumage, this species has at times been placed in the genus *Malimbus* (Sibley and Monroe, 1990), and also in the monotypic genus *Rhinoploceus* on the strength of the unusual nasal structure (Gyldenstolpe, 1924; Wolters, 1975-1984). The distinctiveness of the nasal tubercle has probably been over-emphasized, when compared with a range of other ploceids, and this may not be an important synapomorphy. However, yellow legs appear to be a unique character within the Ploceinae. In the area from which *P. flavipes* has been recorded, *P. nigerrimus* is common and widespread, while *P. albinucha* is uncommon and at the eastern limits of its distribution. Adults of both sexes of *P. flavipes* are uniform black, while apparent immatures are unmarked greenish, which is quite dissimilar to the pattern of adult female and immature *P. nigerrimus*,

Table 2. Comparative coloration (of breeding male birds) and measurements of *P. flavipes* and possible 'parent' species. Measurements are given as ranges of male and female specimens combined (since the sexing is uncertain), with mean values in brackets.

Character	*P. flavipes* n = 5 (2 ♂, 3 ♀)	*P. albinucha* n = 24 (12 ♂, 12 ♀)	*P. nigerrimus* n = 24 (12 ♂,12 ♀)
Bill colour	Black	Black	Black
Leg colour	Yellow	Brown	Brown
Iris colour	Pale yellow	White	Golden yellow
Head colour	Black	Black	Black
Nape colour	Black	Black, white bases	Black
Mantle colour	Black	Black	Black
Rump colour	Black	Black	Black
Wing colour	Black	Black	Black
Tail colour	Black	Black	Black
Chest colour	Black	Black	Black
Belly colour	Matt black	Matt black	Black
Wing length	75-83 (78.6)	74-85 (78.9)	75-92 (80.4)
Tail length	40-52 (46.6)	42-51 (46.5)	47-58 (52.2)
Tarsus length	16.5-19.0 (17.9)	16.7-19.0 (17.7)	19.0-22.0 (20.8)
Bill length	16.5-20.0 (18.5)	18.5-20.6 (19.4)	19.1-23.0 (20.8)

but more like the uniform dull brownish or greyish of immature *P albinucha*. The specimens were collected in 1910, 1913, 1921, 1950, 1952, 1953, 1959; this spread over many years could fit the pattern expected from occasional hybridization. Since then there have been two sight records, in 1990 and 1994 (Collar et al., 1994).

3.3 Characterizing hybrids

From studies of hybrid hummingbirds (Graves, 1990), antbirds (Graves, 1992), manakins (Graves, 1993) and wood warblers (Graves, 1996), there do seem to be predictable patterns in the phenotypes of bird hybrids. Graves (1996) concluded that the measurements of hybrids should fall within the cumulative range of measurements of a large sample of the parental species, and that plumage characters often resemble those of one or the other parental species rather than a blending or intermediate condition. However, he notes that novel plumage characters are known from some hybrids.

3.4 Hybrid Ploceidae

Uncritical compilations of bird hybrids in the literature have drawn most of their records from avicultural sources, where identification of the species involved, especially the female partner, is often uncertain. There are still relatively few documented field records. Chapin (1954) collected a male bird which he described as a hybrid *P. taeniopterus* X *P. melanocephalus* and also several hybrid *P. nigerrimus* X *P. cucullatus*. I have seen the specimens (AMNH, New York), and support his assessment; there are further hybrid specimens of *P. nigerrimus* X *P. cucullatus* at Tring and Tervuren. The taxon "*Ploceus victoriae*" described from Kampala, Uganda (Ash, 1986) is clearly a hybrid as suggested by Louette (1987), and I consider it most likely to be a hybrid *P. castanops* X *P. melanocephalus*. Other apparent hybrids between these two species have been reported from mixed colonies in Uganda (Fry and Keith, 2004).

Why should we expect hybridization amongst the Ploceidae? Randler (2002) concluded that key factors in hybridization are that one of the species is rare, and that there are errors in mate recognition based on female choice. Two bird families in which many interspecific hybrids have been described from natural populations are the Paradisaeidae, with > 20 confirmed combinations (Frith and Beehler, 1998), and the Trochilidae, with c. 50 different combinations reported (G. Graves, pers. comm.). Both these groups have polygynous mating systems, where females select a mating partner. Thus in weavers we might expect hybrids to occur in species which overlap near the range limits of one partner, which often occur in mixed colonies, and where males display within their territories to any non-breeding plumaged bird that enters the area. These criteria are met at many breeding sites of polygynous *Ploceus* species, and field workers in Africa should be alert to the possible occurrence of hybrids, although these will only be recognisable with any confidence if they are male birds.

4. CONCLUSIONS

We need a critical assessment of the status of taxa which are poorly represented in museum collections, and may consequently be interpreted as rare species of special conservation concern. The two weaver species discussed in this paper may serve as case studies. No final conclusions are possible at this stage. I consider that *Ploceus flavipes* could be a hybrid, if the novel character of yellow legs could arise during a cross between the only plausible parental species, *P. albinucha* and *P. nigerrimus*. However, we may yet rediscover yellow-footed weavers when research can resume in the Ituri region. For *Ploceus aureonucha* a hybrid origin from *P. tricolor* and *P. nigerrimus* is plausible, but I now consider it more likely that this taxon in fact comprises subadult plumage stages of *P. tricolor*, which could be elucidated by careful study of all the available material on this species.

ACKNOWLEDGEMENTS

This study began in museum collections, and may ultimately be resolved there. I am indebted to all those who have assisted me during my visits: Michel Louette (Africa Museum, Tervuren, Belgium), Phil Canning, Joel Cracraft, Bob Dickerman, Mary Lecroy, Lester Short (American Museum of Natural History, New York, USA), Robert Prŷs-Jones (British Museum of Natural History at Tring, UK), Gary Graves (National Museum of Natural History, Washington DC, USA). Anita Gamauf (Vienna), Gary Graves (Smithsonian Institution), Michel Louette (Tervuren) and Walter Sontag (Vienna) generously provided additional information and useful discussion. I also thank the organisers of the Symposium on African Biodiversity at the Museum Alexander Koenig in Bonn for the opportunity to present these speculative ideas at the conference. These studies have been supported by grants from the National Research Foundation of South Africa, Rhodes University and the Smithsonian Institution. Collaborative research in France, funded by the CNRS (France) and the NRF (South Africa) and supported financially by the Université de Rennes 1, gave me the opportunity to attend the conference in Bonn.

REFERENCES

Ash, J. S., 1986, A *Ploceus* sp. nov. from Uganda, *Ibis* **128:**330-336.

Bannerman, D. A., 1949, *The birds of tropical West Africa.* Vol. VII, Oliver & Boyd, Edinburgh & London.

Chapin, J. P., 1916, Four new birds from the Belgian Congo, *Bull. Amer. Mus. Nat. Hist.* **35:**3-29.

Chapin, J. P., 1954, The birds of the Belgian Congo. Part IV, *Bull. Amer. Mus. Nat. Hist.* **75B:**1-846.

Collar, N. J., Crosby, M. J. and Stattersfield, A. J., 1994, *Birds to watch 2. The world list of threatened birds,* BirdLife International, Cambridge.

Craig, A.J.F.K., 2004, Family Ploceidae: buffalo-weavers, sparrow-weavers, weavers, bishops and widowbirds, in: *The birds of Africa.* Vol. VII, C. H. Fry and S. Keith, eds, Christopher Helm, London, pp. 104, 180-181.

Frith, C. B. and Beehler, B. M., 1998, *The birds of paradise*, Oxford University Press, Oxford.

Fry, C. H. and Keith, S., eds, 2004, *The birds of Africa.* Vol. VII, Christopher Helm, London.

Grant, P. R. and Grant, B. R., 1992, Hybridization of bird species, *Science* **256:**193-197.

Grant, P. R. and Grant, B. R., 1996, Speciation and hybridization in island birds, *Phil. Trans. R. Soc. Lond. B* **351:**765-772.

Grant, P. R. and Grant, B. R., 1997, Hybridization, sexual imprinting and mate choice, *Am. Nat.* **149:**1-28.

Graves, G. R., 1990, Systematics of the 'green-throated sunagels' (Aves: Trochilidae): valid taxa or hybrids? *Proc. Biol. Soc. Wash.* **103:**6-25.

Graves, G. R., 1992, Diagnoses of a hybrid antbird (*Phlegopsis nigromaculata* X *Phlegopsis erythroptera*) and the rarity of hybridization among suboscines, *Proc. Biol. Soc. Wash.* **105:**834-840.

Graves, G. R., 1993, A new hybrid manakin (*Dixiphia pipra* X *Pipra filicauda*) (Aves: Pipridae) from the Andean foothills of eastern Ecuador, *Proc. Biol. Soc. Wash.* **106:**436-441.

Graves, G. R., 1996, Hybrid wood warblers, *Dendroica striata* X *Dendroica castanea* (Aves: Fringillidae: Tribe Parulini) and the diagnostic predictability of avian hybrid phenotypes, *Proc. Biol. Soc. Wash.* **109:**373-390.

Gyldenstolpe, N., 1924, Zoological results of the Swedish expedition to central Africa 1921. Vertebrata. I. Birds, *K. Svenska Vetensk. Akad. Handl.*, ser. 3, vol. 1, no. 3: 1-326.

Louette, M., 1987, A new weaver from Uganda? *Ibis* **129:**405-407.

Louette, M., 1988, Additions and corrections to the avifauna of Zaïre (2), *Bull. Brit. Orn. Cl.* **108:**43-50.

Prigogine, A. 1960. Le male de *Ploceus flavipes* (Chapin), *Rev. Zool. Bot. Afr.* **61:**364-365.

Randler, C., 2002, Avian hybridization, mixed pairing and female choice, *Anim. Behav.* **63:**103-119.

Sassi, M., 1920, Zwei neue Weber aus Mittelafrika, *Ornith. Monatsber.* **28:**81.

Sibley, C. G. and Monroe, B. L., 1990, *Distribution and taxonomy of birds of the world,* Yale University Press, New Haven.

Wolters, H. E., 1975-1984, *Die Vogelarten der Erde,* Paul Parey, Hamburg.

HISTORICAL DETERMINANTS OF MAMMAL DIVERSITY IN AFRICA: EVOLUTION OF MAMMALIAN BODY MASS DISTRIBUTION IN AFRICA AND SOUTH AMERICA DURING NEOGENE AND QUARTERNARY TIMES

Manuel Nieto[1], Joaquín Hortal[1], Cayetana Martínez-Maza[1], Jorge Morales[1], Edgardo Ortiz-Jaureguizar[2], Pablo Pelaez-Campomanes[1], Martin Pickford[3], José Luis Prado[4], Jesús Rodríguez[1], Briggite Senut[2], Dolores Soria[1], and Sara Varela[1]

[1]*Museo Nacional de Ciencias Naturales (CSIC); José Gutiérrez Abascal, 2. Madrid 28006, Spain*

[2]*Museo Paleontológico Egidio Feruglio; Trelew, Argentina*

[3]*Collège de France & Muséum National d'Histoire Naturelle; 8 Rue Buffon, 75005 Paris, France*

[4]*Facultad de Ciencias Sociales; Universidad Nacional del Centro; Olavarría, Argentina*

Abstract: Local mammalian communities in Africa present the highest species richness in the world, only paralleled by some communities in the Oriental biogeographic region. Differences in mammalian species richness are especially outstanding when compared with South American communities, despite their similar latitudinal position and regional species richness. Recent study has shown that these differences are not only related to contemporary determinants but also to biogeographic-historic factors, which acted on the composition of the regional pool of species. One of the main differences in composition between the two regions relates to the high diversification of large mammals in Africa, which greatly contributes to the high values of local community richness in this region. The absence of extant large mammals in the South American region has been proposed to result from Pleistocene-Holocene extinctions, which affected large mammals all over the world. However, a gradual pattern of decrease in the abundance of large mammal species can be appreciated in almost all regions except Africa since the late Miocene and through the Pliocene. To test these hypotheses we compare the patterns of macromammal body mass distribution - at regional and local scales - in the two regions over the past 20 million years and relate the observed changes to major geological events.

B.A. Huber et al. (eds.), African Biodiversity, 287–295.

Key words: Mammals; communities; species richness; body mass; Neogene; Quaternary; Africa; South America

1. INTRODUCTION

Most of the richest mammal communities in the world are found in Africa south of the Sahara desert. Moreover, on average African mammal communities are richer than those of any other biogeographic region (*sensu* Cox, 2001) except the Oriental one (see Fig. 1). Such differences in species richness are especially striking when African communities are compared to South American ones, despite their similar area, latitudinal position, landscapes (see Vivo and Carmignotto, 2004 and references therein) or richness of their regional pool (Africa: 861 species; South America: 777; data from Nowak, 1999). These differences have been related to differences in the abundance of medium and large mammal species (see Cristoffer and Peres, 2003; Vivo and Carmignotto, 2004), that is, to differences in the composition of their regional pools. While Africa has an abundant fauna of large mammals, such species are almost absent from South America. Several hypotheses based on historical processes have been proposed to explain these differences. Most of them establish that the differences are due to the disappearance of the large mammals in South America either during the Great American Biotic Interchange, the megafaunal extinctions of the Pleistocene, or even as late as the middle Holocene. Other authors have reported that a pattern of decrease in the abundance of large mammal species since the late Miocene can be observed throughout all continents except Africa, making this continent a refuge of high large mammal diversity and posing the question of what makes Africa so special.

In the present contribution we will compare the evolution of mammal sizes in Africa and South America during the last 20 million years. Our aim is to describe the pattern of change in body size in the mammal faunas in the two continents and to find out if the available data support any of the previous hypotheses. Before analyzing the fossil data, we test the premises about the recent faunas on which they are based on. That is, we test whether differences in the abundance of large mammals are responsible for the differences in local richness between continents, and particularly between Africa and South America. We also test whether these differences are related to historical causes or if they can be explained by ecological, contemporary causes such as climate, productivity or others. Both analyses of recent fauna are preliminary since our intention is simply to confirm that these conditions are met and that the fossil record can help to understand the recent patterns of species richness distribution. Concerning the fossil record, we first compare the size distribution of the regional pools of South America and

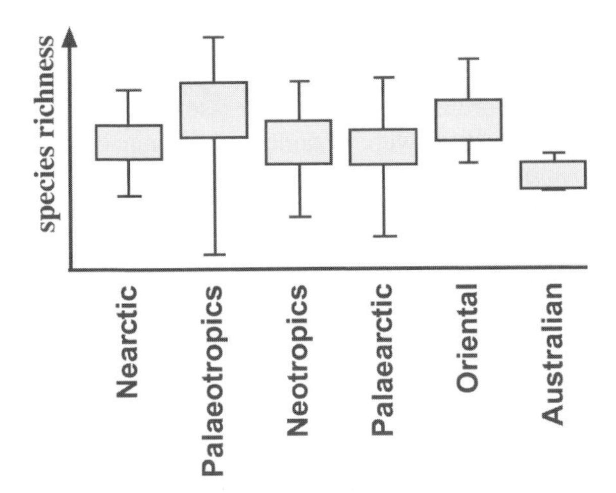

Figure 1. Box & Whisker plot of the local mammal richness (number of species) in the different biogeographic regions. Total number of localities=236; Palaeotropics=26; Neotropics=27; Oriental=10; Australian=8; Nearctic=92; Palaearctic=73. Locality data obtained from the bibliography.

Africa to determine their similarities and differences and the period when today's differences were established. We finally explore how variations in the regional pool of the different body sizes affect the species richness of local faunas in Africa.

2. MATERIALS AND METHODS

Recent data were compiled from the bibliography and existing databases. Lists of species for the different localities were obtained from the bibliography following criteria specified in Rodríguez (2001). Body mass data for recent species were obtained mainly from Nowak (1999) and to a lesser degree from other bibliographic sources. Fossil data was reduced to two subregions, East Africa in Africa and South America south of the Equator, where enough information and detailed reviews were available. Data on the generic diversity in East Africa is taken from Pickford and Morales (1994) and in Southern South America from Pascual et al. (1996). There are marked differences in the time intervals as well as in the taxonomy and sampling units used in both continents, however we feel that they will not substantially affect the obtained patterns of change though it would clearly affect absolute numbers. Body mass data comes from the bibliography and anatomical comparisons. Only macromammal species, i.e. over 1 kg, were considered. Body masses were assigned to the following categories modified from Andrews et al. (1979): C=1-10 kg; D=11-45 kg; E=46-90 kg; F=91-180 kg; G=181-360 kg; H=361-1000 kg; and I>1000 kg (A and B correspond to masses <1 kg). Each genus was assigned to one or several categories according to the body masses of its species. Data from individual African localities were obtained from faunal lists of localities in Pickford (1986) and other bibliographic sources.

Today's differences in local species richness of large and small mammals between continents were assessed by means of ANOVA tests. Analyses were replicated for species ≥45 kg and <45 kg using as the dependent variable $\log_{10}$ of local richness of species below and over 45 kg; as factor, the biogeographic region; and locality area as a covariate. The Tuckey honest test was used as a post-hoc test to look for significant differences between continents.

The analyses of the effects of contemporary-ecologic factors and biogeographic-historic factors on local species richness of large and small mammals were performed by means of partial regressions using GLM (see Legendre and Legendre, 1998). Mammal data comes from checklists of 86 localities (36 from Africa and 50 from South America; see Rodríguez, 2001). Environment was measured as the 6 axis extracted from monthly values of Cloudiness, Precipitation and Temperature (12x3=36 variables; Leemans and Cramer, 1991), and annual ETP (Deichman and Eklundh, 1991; GIS database from http://www.grid.unep.ch/data/grid/). Analyses were performed for local species richness of mammals over and under 45 kg and for all mammal species in the communities.

Patterns of body mass distribution during the last 20 Ma were estimated from the body mass of all genera present in each of the considered times periods in South America and Africa. For genera with body masses in more than one category, the value for each category was calculated as the correspondent fraction (i.e. if three categories are present in a genus, each category is assigned a value of 0.33 for that genus). Genus values for each category in a time period are added and the obtained value is used as a proxy of the richness of that category during that time lapse. Body mass values for the different categories are plotted against time. Millions of years in Africa and local biochronologic units (SALMAs) in South America (according to Pascual et al., 1996) were used as time units.

Kendall's Tau was used to analyze the correlation between regional pool and local richness of the different body mass categories in Africa. Regional richness was estimated at the generic level, while local richness was measured at the species level to maximize the information of each locality.

Table 1. Tuckey HSD test (post-hoc test) for species both over (≥) and under (<) 45 kg. PAR: Palaearctic; OR: Oriental; NTR: Neotropics; NAR: Nearctic; AUS: Australian; PTR: Palaeotropics. Significant differences between regions in bold type (p<0.05). Dependent variable: Log_{10} of local richness of species below and over 45 kg. Factor: biogeographic region. Locality area included as a covariate

Species ≥ 45 kg					
	PAR	OR	NTR	NAR	AUS
PAR					
OR	.231				
NTR	.108	**.001**			
NAR	.688	.671	**.004**		
AUS	.088	**.004**	.415	.037	
PTR	**.000**	.187	**.000**	**.000**	**.000**

Species < 45 kg					
	PAR	OR	NTR	NAR	AUS
PAR					
OR	.655				
NTR	.997	.842			
NAR	.999	.606	.991		
AUS	.686	.113	.535	.719	
PTR	**.003**	.862	.015	**.002**	.011

3. RESULTS

ANOVA analyses verify that large mammal richness plays a relevant role in the differences in local richness between continents (F=20.54; p<0.001). ANOVA analysis also detects significant differences for the species under 45 kg (F=6,542; p<0.001). However, differences between paleotropical and neotropical regions are highly significant for richness of species over 45 kg (Tuckey test, p<0.001) while only marginally significant for sizes under this body mass (p=0.015; see Table 1).

Concerning the factors affecting species richness, variation partitioning analysis shows the important role of biogeographic factors related to the history and topography of each region in determining local richness of large mammals (Fig. 2). While the effect of Environment was higher when considering all mammal species or those under 45 kg, the effect of Region (Paleotropical vs Neotropical), as well as the region-mediated environmental influence, were prominent for species over 45 kg. In fact, the results confirm that biogeographic factors including history can be considered more relevant than environment in determining the differences in local richness of large mammal species in South America and Africa.

The evolution of the body mass composition in the regional pool in South America and Africa are presented in Figure 3. Both continents reveal an

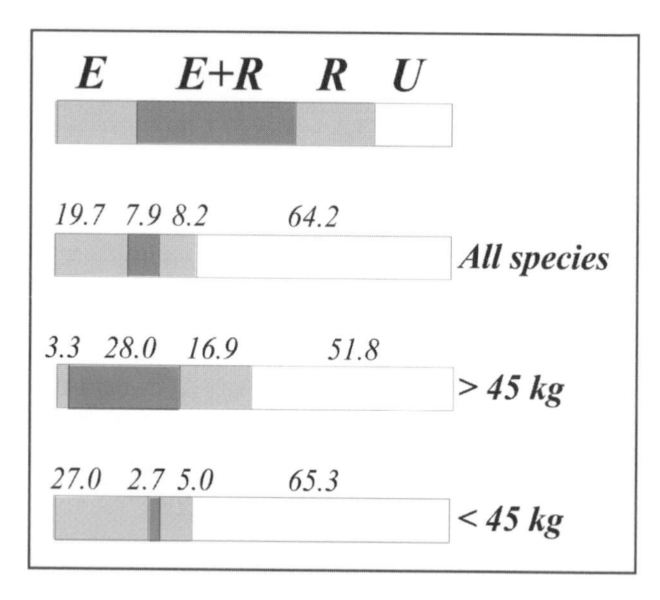

Figure 2. Bars represent the separate influence over local species richness of Environment (E), Region (R), their joint effects (E+R), and the unexplained variation (U). While the effect of Environment was higher when considering either all mammal species, or only those below 45 kg, the effect of Region (Paleotropical vs Neotropical), as well as the region-mediated environmental influence, were much higher when only species of equal or more than 45 kg were considered.

increase of large mammal species richness until ca 7 Ma. Later, South America suffered a dramatic reduction of richness during the late Miocene and Pliocene and, after a partial recovery, almost all genera over 200 kg became extinct during the Pleistocene extinctions. Meanwhile, Africa suffered a decrease in proboscidean abundance, compensated by an increase in genera between 90 and 1000 kg. Thus, present regional differences in body mass composition may be linked to the Pleistocene megafaunal extinction suffered by South America, but also to a long-term increase of large mammals in Africa (between 90 to 1000 kg).

Lastly, Kendall's Tau analysis of the relationship between regional pool and local richness in the different body mass categories in Africa shows that local richness of mammals under 100 kg (categories C, D and E; see Table 2) is not correlated with their regional pool, while mammals over 100 kg (categories F, G, and I) do present a significant correlation between richness at both scales.

Category H does not show correlation, probably due to scarcity of taxa in the African record. Similar results have been obtained analyzing recent faunas (data not shown) pointing out that changes in the regional pool of

Table 2. Kendall's Tau analyses of the correlation between regional pool and local richness in Africa. Data on regional richness of genera from Pickford & Morales (1994); local richness of species taken from faunal lists of localities in Pickford (1986) and other bibliographic sources. Cat.=body mass category; N=number of comparisons. Significant correlations marked in bold type (p<0,01)

Cat.	N	Kendall's Tau	Z	p-level
C	20	0,00559	0,0345	0,973
D	20	-0,24809	-1,5294	0,126
E	20	0,34464	2,1245	0,034
F	**20**	**0,47363**	**2,9196**	**0,004**
G	**20**	**0,49373**	**3,0436**	**0,002**
H	20	-0,22636	-1,3954	0,163
I	**20**	**0,49303**	**3,0392**	**0,002**

large mammals have major effects on local richness while these effects are more restricted when they affect medium and small sized taxa.

4. DISCUSSION AND CONCLUSIONS

In agreement with previous authors (see Vivo and Carmignotto, 2004 and references therein), we have shown that biogeographic differences in the regional pool of large species of Africa and South America may underlie the higher richness of African mammal communities. Our results also agree with previous authors (see op. cit.) in the leading role of historical factors in determining the differences in large species abundance. In this sense, we have shown that while the abundance of "small" mammals (below 45 kg) in a community depend on environmental factors, the richness of "large" species mainly depends on biogeographical factors.

Concerning the fossil record of the body mass distribution from Africa and South America, our results do not allow us to fully accept or deny any of the hypotheses proposed. Trends of diversification and extinction are apparent since the late Miocene in both continents, although the Pliocene climatic-environmental changes due to the Andean Diaguita diastrophic phase and Pleistocene extinctions meant major events in South America. However, the main differences can be observed in the generic richness of the body mass categories. That is, while Pliocene and mainly Pleistocene extinctions caused the disappearance of all mammals over 200 kg in South America, it was restricted to mammals over 1000 kg in Africa and was coupled with an increase in the richness of the rest of the large mammals. The effects of these regional differences on local richness depend on the body mass category affected (at least in Africa; see Fig. 3). Local richness of large species (species over 100 kg) is correlated to regional richness,

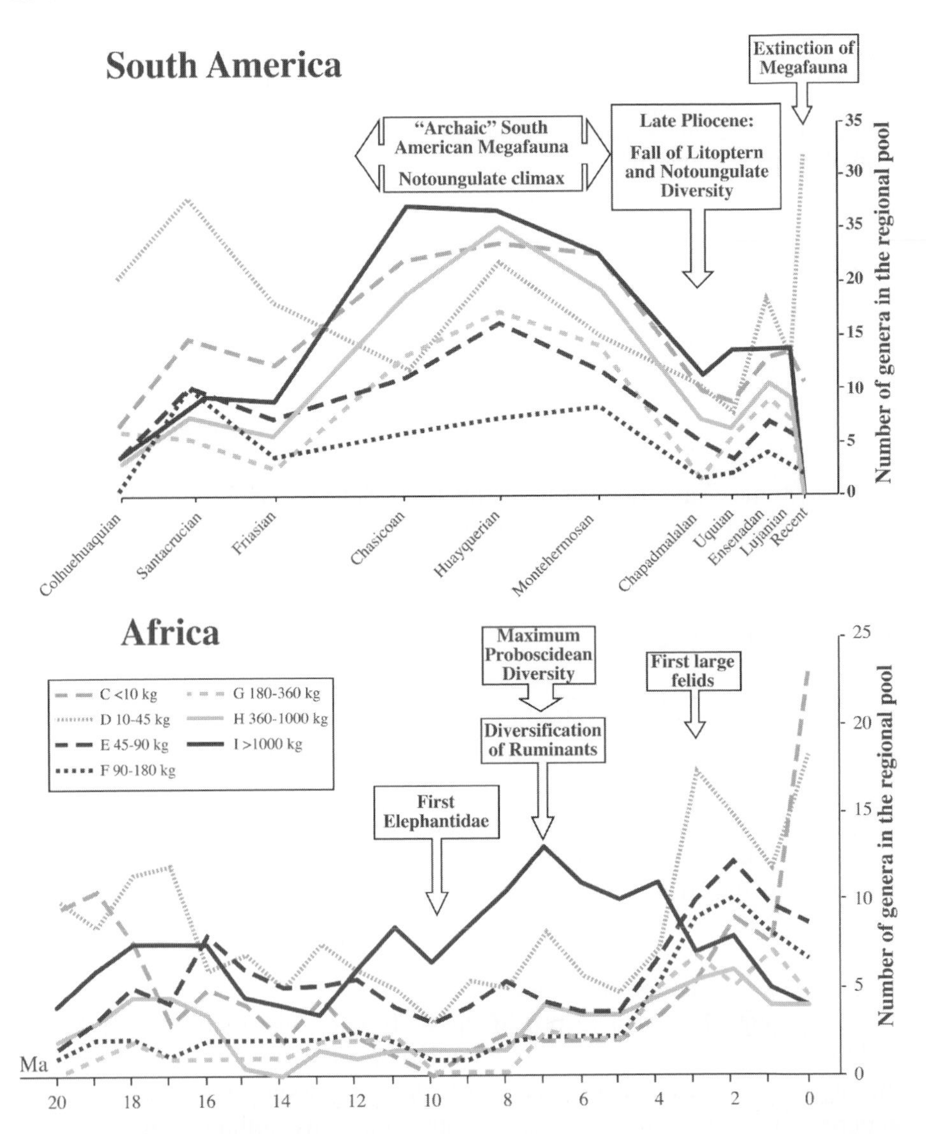

Figure 3. Numbers of genera in the considered body mass categories in East Africa and Southern South America during the last 20 million years. Time units: millions of years in Africa and local biostratigraphic units (SALMAs) in South America (dating of the different SALMAs according to Pascual et al., 1996). Codes for body mass categories detailed in the graph.

while the local richness of small mammals seems to be independent of the regional pool. Thus, processes affecting the regional pool of large species have a deep effect on local communities through time.

In summary, our results show that African high local richness is related to a restricted megafaunal extinction (only species over 1000 kg) and to a unique history of diversification of large mammals (between 200-1000 kg).

The effects of these processes are specially relevant due to strong correlation between Local and Regional Richness in species over 100 kg.

ACKNOWLEDGEMENTS

During the preparation of this manuscript, one of its authors, Dr. Dolores Soria, died in a car accident. We dedicate this contribution to her memory; we shall never forget her. This work was supported by the Spanish D.G.I. projects REN2001-1136/GLO, PB98-0513, BX X 2000-1258-CO3-01 and BTE 2002-00410. Thanks to Maite Alberdi and an anonymous referee for their helpful comments.

REFERENCES

Andrews, P., Lord, J. M., and Evans, E. M. N., 1979, Patterns of ecological diversity in fossil and modern mammalian faunas, *Biol. J. Linn. Soc.* **11:**177-205.

Cox, C. B., 2001, The biogeographic regions reconsidered, *J. Biogeogr.* **26:**511-523.

Cristoffer, C., and C. A. Peres., 2003, Elephants vs. Butterflies: the ecological role of large herbivores in the evolutionary history of two tropical worlds, *J. Biogeog.* **30:**1357–1380.

Deichmann, U., and Eklundh, L., 1991, *Global Digital Data Sets for Land Degradation Studies: a GIS Approach*, UNEP/GEMS and GRID, Nairobi, Kenya.

Leemans, R., and Cramer, W. P., 1991, *The IIASA Database for Mean Monthly Values of Temperature, Precipitation and Cloudiness of a Global Terrestrial Grid*, IIASA, Laxenburg, Austria.

Legendre, P., and Legendre, L., 1998, *Numerical Ecology*, 2nd English ed., Elsevier, Amsterdam.

Nowak, R. M., 1999, *Walker's Mammals of the World*, Johns Hopkins Univ. Princeton, USA.

Pascual, R.; Ortiz-Jaureguizar, E., and Prado, J. L., 1996, Land Mammals: Paradigm for Cenozoic South American Geobiotic Evolution, *Münch. Geowiss Abh.* **30:**265-319.

Pickford, M., 1986, Cainozoic palaeontological sites of western Kenya, *Münch. Geowiss Abh.* **8:**1-151.

Pickford, M., and Morales, J., 1994, Biostratigraphy and palaeobiology of East Africa and the Iberian Peninsula, *Palaeogeogr. Palaeoclimat. Palaeoecol.* **112:**297-322.

Rodríguez, J., 2001, Structure de la communauté de mammifères pléistocenes de Gran Dolina (Sierra de Atapuerca, Burgos, Espagne), *L'Anthropologie* **105:**131-157.

Vivo, M., and Carmignotto, A. P., 2004, Holocene vegetation change and the mammal faunas of South America and Africa, *J. Biogeogr.* **31:**943-957.

DIVERSITY AND ABUNDANCE OF DIURNAL PRIMATES AND FOREST ANTELOPES IN RELATION TO HABITAT QUALITY: A CASE STUDY FROM THE UDZUNGWA MOUNTAINS OF TANZANIA

Francesco Rovero[1] and Andrew R. Marshall[2]

[1] *Vertebrate Zoology Section, Museo Tridentino di Scienze Naturali, Via Calepina 14, I-38100, Trento, Italy, E-mail: francesco_rovero@yahoo.it*

[2] *Centre for Ecology Law and Policy, Environment Department, University of York, Heslington, York YO10 5DD, United Kingdom, E-mail: andrewrmarshall@hotmail.com*

Abstract: We used line transect counts to collect data on population abundance of primates and small antelopes from three different moist forest blocks within the Udzungwa Mountains of south-central Tanzania. Results show that both diversity and relative abundance (number of primate groups/individual antelopes seen per km walked) are generally higher in the lower part (300-800 m a.s.l.) of Mwanihana Forest (Udzungwa Mountains National Park), than in medium to high altitude and less protected forests (Ndundulu Forest, 1400-2100 m a.s.l., and New Dabaga/Ulangambi Forest Reserve, 1800-2100 m a.s.l.). The latter, in particular, is the most degraded and encroached forest patch, and presents the most impoverished community. The combined effects of altitude, hunting and human-induced alteration of forest cover might account for the differences observed.

Key words: Udzungwa mountains; primates; duikers; census; logging; habitat degradation

1. INTRODUCTION

Primates and forest antelopes are key rainforest animals and documenting the effects of habitat quality on their abundance is crucial for developing conservation strategies. Habitat degradation and hunting by humans are two major threats to primates and forest antelopes. Generally, populations

B.A. Huber et al. (eds.), African Biodiversity, 297–304.

respond to these threats with reduced abundance (reviews in Mittermeier, 1987; Plumptre, 1994; Struhsaker, 1997; Wilkie and Finn, 1990; Cowlishaw and Dunbar, 2000; Hart, 2000). Species diversity might also decrease if threats cause local extinctions (e.g. Chapman and Lambert, 2000). In mountaineous areas, moreover, elevation-related habitat features might account for differences in population abundance (e.g. Marshall et al., in press).

The Udzungwa Mountains of south-central Tanzania (10,000 km^2) are ideal for investigating patterns of abundance across forest blocks, because exceptional diversity is found in a wide variety of forest habitats with different levels of disturbance and broad altitudinal ranges (Dinesen et al., 2001). The Udzungwa Mountains are one of the major forest blocks of the Eastern Arc Mountains, an area of global importance for biodiversity conservation (Myers et al., 2000). Of special importance is the presence of two endemic and threatened monkeys (Udzungwa red colobus, *Procolobus gordonorum* Matschie, 1900, and *Sanje mangabey*, *Cercocebus galeritus sanjei* not yet formally described, and an antelope with a very restricted-range: Abbott's duiker (*Cephalophus spadix* True, 1890). We compare the results on population abundance of primates and forest antelopes from contrasting forest sites to evaluate the effects of habitat degradation, hunting and altitude on community composition and relative density.

2. METHODS

Primate groups and forest antelopes were counted along seven paths located in three forest sites (Fig. 1):

Mwanihana Forest (centred on 7°46'S, 36°43'E; 177 km^2), located in the Udzungwa Mountains National Park (1990 km^2). The forest cover is continuous from 300 to 2100 m a.s.l. Three, 4 km-long line transects were positioned about 6 km apart, to sample a mosaic of habitat types from 300 to 800 m a.s.l., from lowland, deciduous woodland to sub-montane, evergreen moist forest.

Ndundulu Forest (centred on 36°35' E, 07°45' S; 250 km^2) is located in the West Kilombero Scarp Forest Reserve, about 20 km west of Mwanihana Forest. This is a submontane and montane evergreen forest (1400-2100 m a.s.l.), with herbaceous clearings and patches of bamboo and high altitude woodland. Part of the forest has been affected by commercial logging in the past. Two transects (about 3100 m long) were positioned on cartographic lines of latitude 6 km apart through evergreen forest.

New Dabaga/ Ulangambi Forest Reserve (here referred to as NDUFR; centred on 35°55' E, 08°05' S; 37 km^2) is a high altitude (1800-2100 m

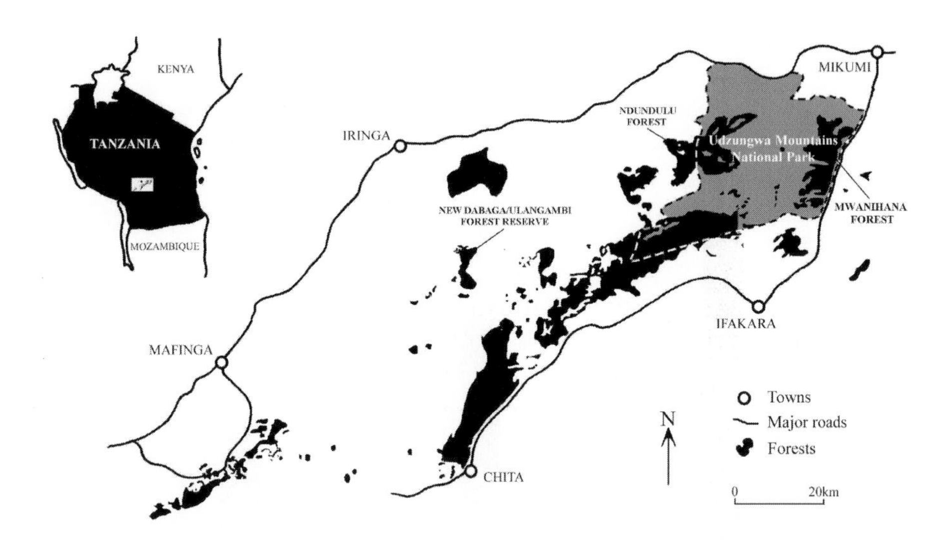

Figure 1. Map of the Udzungwa Mountains, Tanzania, highlighting the major forest blocks and location of the three study sites. Modified from Marshall et al. (in press).

a.s.l.) Forest Reserve consisting of secondary montane evergreen forest. Two transects (about 3100 m long) were placed in the reserve.

These areas are affected by human encroachment to various degrees. Poaching of duikers using snares has been reported in all of these areas. People from adjacent villages are allowed to enter Mwanihana forest for firewood collection: this may be having negative impact on ground foraging species, such as the Sanje mangabey. The highest level of encroachment has occurred in NDUFR, which is surrounded by villages on all sides, and includes animal trapping, hunting, pole-cutting and pit-sawing. Hunting of colobus monkeys has recently been reported (Nielsen, 2002). An index of human forest use is given here as the number of animal traps, pit-sowing sites, timber piles and cut tree stumps counted along transect lines: these were 47.2 per ha in NDUFR and 4 in Ndundulu Forest, while only in one case a snared duiker was found along Mwanihana transects (details in Marshall et al., in press; F. Rovero, unpubl. data).

Mwanihana Forest transects were walked twice per month from July 2002 until January 2003 by FR (14 repetitions per transect). Ndundulu Forest and NDUFR transects were walked approximately once every month during the period January-December 2000 by ARM (10 repetitions per transect; 6 walks conducted early in the survey period were approximately 500 m shorter than the lengths given above). Census lines were walked beginning at 7.00-7.30 h at an average speed of about 1 km h^{-1}, recording all sightings of primates and antelopes. Groups that were heard but not seen were not scored. For each observation, time, species, number of individuals

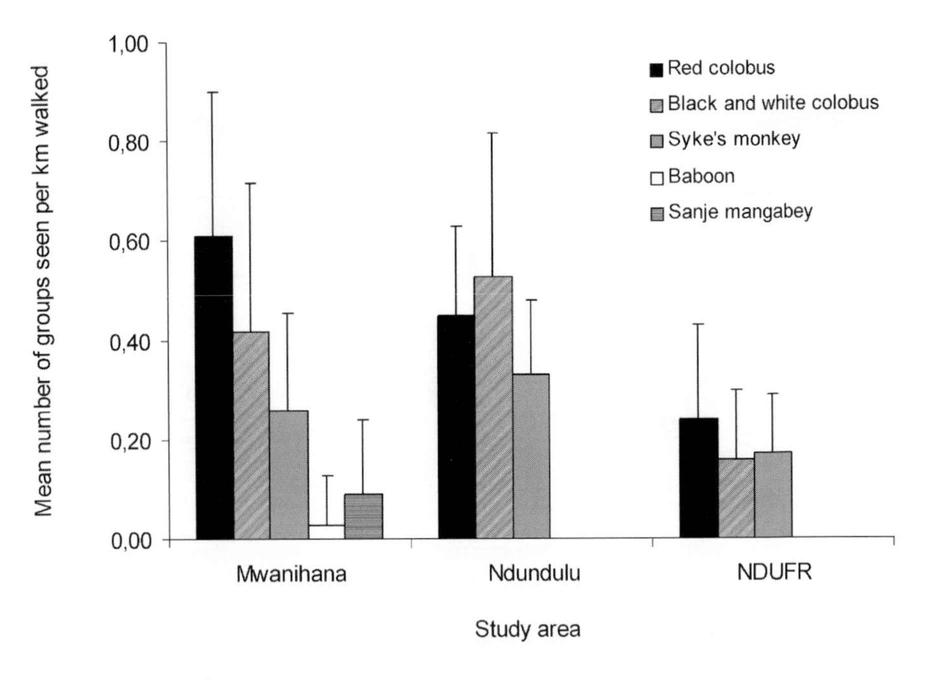

Figure 2. Census results of primate species from line transect walks in three forests of the Udzungwa Mountains, Tanzania. Relative abundance is shown as encounter rate (mean number, and standard deviation, of primate groups seen per km walked).

and position of the observer along the transect were recorded. Because of low antelope sighting rates in Ndundulu Forest and NDUFR, counts of dung piles and forest antelope paths (i.e. sets of footprints) were also made in these two areas. Four 1 km transects were placed through evergreen forest in each of NDUFR and Ndundulu Forest near the transects used for censuses. The number of dung piles within 5 m of each transect and the number of sets of forest antelope footprints crossing each transect were recorded.

The number of primate groups/individual antelope sightings per km walked, defined as "encounter rate" (ER), was computed as an index of relative abundance (e.g. Seber, 1982). We did not compute absolute density because of potential inter-observer differences in distance estimation (e.g. Mitani et al., 2000) and because primate census data often violates many of the assumptions of density estimation (Struhsaker, 1997; Buckland et al, 2001). We compared ER values between the three areas by using Kruskal-Wallis tests followed by post-hoc Mann-Whitney U-tests (Siegel and Castellan, 1988). We applied the same analysis to compare primate group count data, that were collected for the two colobus only (Marshall et al., in press; F. Rovero, unpubl. data). Group counts of red colobus in Mwanihana are reported by Struhsaker et al. (2004). Differences in the number of dung piles and tracks between Ndundulu Forest and NDUFR were tested using Mann-Whitney U-tests.

3. RESULTS

Differences in sighting frequency of primate groups between study areas (Fig. 2) are highly significant for red colobus and black and white colobus, *Colobus angolensis* (Sclater, 1860) (Kruskal-Wallis tests: χ^2=14.23 $p<0.01$, χ^2=10.74 $p<0.01$, respectively), and weakly significant for Syke's monkeys, *Cercopithecus mitis moloney* (Wolf, 1822) (χ^2=5.35, p=0.068). Post-hoc testing reveals that greatest differences are in the reduced abundance of colobines in NDUFR than Mwanihana (U=171, $p<0.001$ and U=235, $p<0.005$ for red colobus and black and white colobus, respectively). Abundance of Syke's monkey is significantly higher in Ndundulu than Mwanihana (U=272, $p<0.05$), and not significantly different between Mwanihana and NDUFR (U=405.5, p=0.7). Group count comparisons show that group size of red colobus decreases from Mwanihana (mean 40.8 individuals, N=31) to Ndundulu Forest (22.5, N=6) to NDUFR (11.8, N=8). Differences were significant (χ^2=19.52, $p<0.001$). Group size of black and white colobus does not vary significantly between Mwanihana (9.4, N=16) and Ndundulu (10.4, N=4; U=23, p=0.44) while it is smaller in NDUFR (6.3, N=6). Overall differences are significant (χ^2=9.77, $p<0.01$). The primate species composition also differs across study sites as Sanje mangabey and yellow baboon, *Papio cynochephalus* (Linnaeus, 1766) were only sighted in Mwanihana.

The only antelope seen regularly during transect walks was Harvey's duiker (*Cephalophus natalensis harveyi* Thomas 1893) in Mwanihana (mean of 0,16 ± 0.17 sightings/km walked, N=25). Other small antelopes known to occur in these forests (blue duiker, *Cephalophus monticola* Thunberg, 1789, suni, *Neotragus moschatus* von Dueben, 1846 and bush duiker *Sylvicapra grimmia* Linnaeus, 1758) were sighted only four times in Mwanihana but the species could not be distinguished during the censuses. The Abbott's duiker was never seen. No antelope sightings were scored during census walks in Ndundulu and NDUFR, although data on tracks and dung piles indicate a greater number of antelope signs in Ndundulu than NDUFR (mean of 60 versus 7 sets of footprints per km of transect and 6 versus 0 dung piles per ha, for Ndundulu and NDUFR, respectively; U=16 $p<0.02$).

4. DISCUSSION

The results show differences across study areas in the diversity and abundance of primates and antelopes, decreasing abundance and diversity being generally coincident with increasing habitat degradation, levels of hunting and altitude. An exception to this trend is in the higher abundance of black and white colobus and Syke's monkeys at Ndundulu than Mwanihana, which is significant for the latter species. This might reflect preference of these monkeys for areas with regenerating vegetation and clearings. The lower abundance of colobines in NDUFR is even more remarkable when considering individual abundance instead of group abundance, confirming earlier suggestions (Struhsaker et al., 2004; Marshall et al., in press). Mangabeys in Ndundulu might be at very low density (Dinesen et al., 2001) and they have never been recorded from NDUFR. Baboons are found almost exclusively in the lower, deciduous part of Mwanihana and their absence from montane forest habitats is not surprising.

The low sighting rates of forest antelopes indicate that transect walks might not be appropriate for censusing elusive or restricted range species (Rovero and Marshall, in press). Moreover, camera-trapping data from Mwanihana indicate that Abbott's duiker, suni and bush duiker are mainly crepuscular and nocturnal, while blue duiker has not been photographed (Rovero et al., in press). The relatively high number of antelope signs for Ndundulu Forest, however, suggests that poor visibility due to dense vegetation might have lowered detection of forest antelopes. Nevertheless, the higher abundance of those antelopes that were seen during censuses in Mwanihana Forest versus NDUFR and Ndundulu is likely to reflect actual differences in density. A lower abundance of antelopes in NDUFR than Ndundulu is also indicated by antelope signs. This coincides with correspondingly high levels of human forest use in NDUFR than in Ndundulu, and probably with less hunting in Mwanihana.

Decreased abundance in relation to habitat degradation and hunting is consistent with other studies on primates and antelopes (e.g. Skorupa, 1986; Dubost, 1979; Wilkie and Finn, 1990; Struhsaker, 1997; Hart, 2000; Cowlishaw and Dunbar, 2000). Our data, however, do not allow determination of the relative effects of such factors on abundance. Moreover, the study compares sites of different altitudes. Elevation-related vegetation features might account for differences such as the relatively higher density of red colobus and antelopes in Mwanihana than Ndundulu. This might be related to energy limitations and/or reduced food density/quality at high elevations (e.g. Caldecott, 1980; see also Marshall et al., in press). In conclusion, our study provides evidence of primate and antelope variation in population abundance and species composition across contrasting forest sites. More research involving ad-hoc selection of study sites will be necessary to independently evaluate the effects that might account for the differences observed.

ACKNOWLEDGEMENTS

We thank two anonymous referees for their precious comments on the manuscript. Permission to conduct the study was granted by the Tanzanian Commission for Science and Technology and by Tanzanian National Parks. FR was supported through a grant from the Margot Marsh Biodiversity Foundation to Drs. Ehardt, Struhsaker and Butynski and through a Rufford Small Grant. ARM collected data for Frontier Tanzania during the Udzungwa Mountains Biodiversity Surveys, involving the Society for Environmental Exploration, the University of Dar es Salaam, and the Udzungwa Mountains Forest Management and Biodiversity Conservation project, supported by DANIDA.

REFERENCES

Buckland, S. T., Anderson, D. R., Burnham, K. P., Laake, J. L., Borchers, D. L., and Thomas, L., 2001, *Introduction to distance Sampling: Estimating Abundance of Biological Populations*, Oxford Univ. Press, New York.

Caldecott, J. O., 1980, Habitat quality and populations of two sympatric gibbons (Hylobatidae) on a mountain in Malaya, *Folia Primatol.* **33:**291-309.

Chapman, C. A., and Lambert, J. E., 2000, Habitat alteration and the conservation of African primates: case study of Kibale National Park, Uganda, *Am. J. Primatol.* **50:**169-185.

Dinesen, L., Lehmberg, T., Rahner, M.C., and Fjeldså, J., 2001, Conservation priorities for the forests of the Udzungwa Mountains, Tanzania, based on primates, duikers, and birds, *Biol. Cons.* **99:**223-236.

Dubost, G., 1979, The size of African forest artiodactyls as determined by the vegetation structure, *Afr. J. Ecol.* **17:**1-17.

Frontier Tanzania, 2001, *New Dabaga/Ulangambi Forest Reserve – Zoological Report,* Report for the Udzungwa Mountains Forest Management and Biodiversity Conservation Project, MEMA, Iringa, Tanzania.

Hart, J. A., 2000, Impact and sustainability of indigenous hunting in the Ituri Forest, Congo-Zaire: a comparison of unhanted and hunted duiker populations, in: *Hunting for sustainability in tropical forests*, Robinson, J.G., and Bennett, E.L., eds., Columbia Univ. Press, New York, pp.106-153.

Marshall, A. R., Topp-Jørgensen, J. E., Brink, H., and Fanning, E., in press, Monkey abundance and social structure in two high elevation forest reserves in the Udzungwa Mountains of Tanzania, *Int. J. Primatol.*

Mitani, J. C., Struhsaker, T. T., and Lwanga, J. S., 2000, Primate Community Dynamics in Old Growth Forest over 23.5 Years at Ngogo, Kibale National Park, Uganda: Implications for Conservation and Census Methods, *Int. J. Primatol.* **21:**269-286.

Mittermeier, R. A., 1987, Effects of Hunting on Rain Forest Primates, in: *Primate Conservation in the Tropical Rain Forest*, C. W. Marsh, and R. A. Mittermeier, eds., A. R. Liss, Inc., New York, pp.109-146.

Myers, N., Mittermeier, R. A., Mittermeier, C. G., de Fonseca, G. A. B., and Kent, J., 2000,. Biodiversity hotspots for conservation priorities, *Nature* **403:**853-858.

Plumptre, A., 1994, The effects of long-term selective logging on blue duikers in the Budongo Forest Reserv,. *Gnusletter* **13:**15-16.

Rovero, F., and Marshall, A.R., in press, Estimating the abundance of forest antelopes by using line transect techniques: a case from the Udzungwa Mountains of Tanzania. *Tropical Zoology*.

Rovero, F., Jones, T., and Sanderson, J., in press, Notes on Abbott's duiker (*Cephalophus spadix* True 1890) and other forest antelopes of Mwanihana Forest, Udzungwa Mountains, Tanzania, as revealed from camera-trapping and direct observations. *Tropical Zoology*.

Seber, G., 1982, *The estimation of animal abundance and related parameters*, Macmillan, New York.

Siegel, S., and Castellan, N .J., 1989, *Nonparametric statistics*, McGraw-Hill International Editions.

Skorupa, J. P., 1986, Responses of rainforest primates to selective logging in Kibale Forest, Uganda, in: *Primates: The road to self-sustaining populations,* K. Benirschke, eds., Springer-Verlag, New York, pp.57-70.

Struhsaker, T. T., 1997, *Ecology of an African Rainforest: Logging in Kibale and the Conflict between Conservation and Exploitation*, Univ. Press of Florida, Gainesville.

Struhsaker, T T., Marshall, A. R., Detwiler, K. M., Siex, K., Ehardt, C. L., Lisbjerg, D. D., and Butynski, T. M., 2004, Demographic Variation Among the Udzungwa Red Colobus (*Procolobus gordonorum*) in Relation to Gross Ecological and Sociological Parameters, *Int. J. Primatol.* **25:**615-658.

Wilkie, D. S., and Finn, J. T., 1990, Slash-burn cultivation and mammal abundance in the Ituri forest, Zaïre, *Biotropica* **22:**90-99.

SMALL MAMMAL DIVERSITY AND REPRODUCTION ALONG A TRANSECT IN NAMIBIA (BIOTA S 07)

Peter Giere and Ulrich Zeller*
Institut für Systematische Zoologie, Museum für Naturkunde, Invalidenstraße 43, 10099 Berlin, Germany, E-mail: Peter.Giere@MUSEUM.HU-Berlin.de
**authors in alphabetical order*

Abstract: As part of an interdisciplinary project on biodiversity (BIOTA South), basic data on small mammal diversity and reproduction from three sampling periods along a transect covering a rainfall gradient in Namibia are presented. A total of 16 species (350 specimens) of Soricidae, Rodentia, and Macroscelidea were captured. Small mammal diversity was highest in the farming area north of Windhoek, whereas it was low in an overgrazed site in the drier south. Reproductive data supports seasonality of reproduction after the rainfall in the north of Namibia but not in the southern sites.

Key words: Small mammal diversity; reproduction; Soricidae; Rodentia; Macroscelidea; semi-arid environments; Namibia

1. INTRODUCTION

Despite extensive research on small mammal in southern Africa (e.g. Eccard et al., 2000), basic data on the distribution and biology of several small mammal species of the area are still in short supply (cf. Matson and Blood, 1994). Yet, besides providing basic data on each of the species concerned, these data are required for understanding interactions between the organism and its environment and to assess the effects of man-made changes to the environment (cf. Hoffmann and Zeller, in press).

Within a larger interdisciplinary project (BIOTA South), this project attempted to assess the diversity of small mammals in the semi-arid ecosystems of Namibia and to study aspects of biotic interrelationship with

B.A. Huber et al. (eds.), African Biodiversity, 305–313.

the fauna and flora. Here, only data concerning small mammal diversity and reproduction with regard to the climatic gradient on the transect in Namibia is used in an initial attempt to fill some of these gaps.

2. MATERIALS AND METHODS

2.1 Study areas

All study areas are located on biodiversity observatories of 1 km^2 in extent. Each observatory is subdivided into 100 one-hectare plots, which ideally are typical and uniform representatives of the respective biome. The observatories form a transect that follows a precipitation gradient from northern Namibia to western South Africa. Here, only the Namibian part of the transect is considered (Fig. 1). Seven observatories were visited in Namibia one to three times between 2001 – 2003. Sampling took place at the end of the Namibian rainy season (March or April). Table 1 provides biotic and abiotic information on the observatories visited.

2.2 Trapping design

Small mammal trapping followed a standard design that was modified from Boye and Meinig (1996). A trapline of 30 Sherman traps (5 m intervals) and three pit fall traps was placed on each 1 ha sampling site. Traps were baited (peanut butter, rolled oats and sardines or banana) an hour before dusk.

In 2002 and 2003, the traps were left open during the night and checked after sunrise. As a standard, two 1 ha plots were sampled for four nights per trapping session.

In 2001, however, the number of nights for trapping varied between the observatories due to logistic restraints. Thus, these data are excluded from statistical analysis. The same applies to additional plots sampled or additional trapping sites used to sample special habitat features. Additional traps to the standard design were placed at bushes, fallen trees, termitaria, etc. These sometimes produced species that had not been trapped previously. However, there was no standard used and thus these data are excluded from statistical analysis but are incorporated as additional information in this report.

Captured small mammals were euthanized using chloroform according to the permit issued by the Ministry of Environment and Tourism. Standard measurements (weight, length of: head-body, tail, ear, and hind-foot) were

Table 1. Characteristics and location of the biodiversity observatories sampled; observatories are arranged from north to south, with colloquial names in quotation marks; for biomes and average annual precipitation, see Mendelsohn et al. (2002)

Observatories	Biome	Average annual precipitation (mm)	Common plant species (BIOTA data)	Land use	Coordinates (NW-corner)	Remarks
Mutompo	north eastern Kalahari woodland	550	*Combretum* spec., *Terminalia sericea, Burkea africana, Aristida stipitata, Guibortia colophosperma*	communal land, low grazing pressure	18°18' 06.4"S 19°15' 30.0"E	Neighbouring Mile 46
Mile 46 Research Station	north eastern Kalahari woodland	550	see Mutompo	research farm, rotational grazing, 25 ha/cattle	18°18' 06.1"S 19°15' 24.4"E	Neighbouring Mutompo
Toggekry 250 "Omatako Ranch"	thornbush shrubland	≤ 400	*Stipagrostis uniplumis, Eragrostis* spec., *Acacia mellifera, Acacia tortilis, Boscia albitrunca, Monechma genistifolia*	commercial farm, cattle, game	21°30' 37.7"S 16°43' 45.2"E	in the vicinity of Otjiamongombe
Otjiamongombe West 44 "Erichsfelde"	thornbush shrubland	≤ 400	see Toggekry	commercial farm ca. 17.5 ha/cattle or game	21°35' 48.7"S 16°56' 41.2"E	in the vicinity of Toggekry
Gellap-Ost 3, Research Station	Karas dwarf shrubland	≤ 150	*Tetragonia schenkii, Rhigozum trichotomum, Stipagrostis* spec.	research farm, rotational grazing, Karakul or Gellaper sheep, low grazing pressure	26°24' 04.2"S 18°00' 17.4"E	fence effect to neighbouring Nabaos
Nabaos 7	Karas dwarf shrubland	≤150	see Gellap-Ost	communal land, goats, donkeys, highly overgrazed	26°23' 58.8"S 17°59' 44.0"E	fence effect to neighbouring Gellap-Ost; bush encroachment (e.g. *Rhigozum trichotomum*) due to overgrazing (Hoffmann and Zeller, in press)
Karios 8	dwarf shrub savanna (Nama Karoo)	< 100	*Euphorbia gregaria, Rhigozum trichotomum, Stipagristis ciliata, Stipagrostis obtusa*	private "Gondwana Cañon Park", game, cattle, low grazing pressure	27°41' 07.9"S 17°48' 05.3"E	

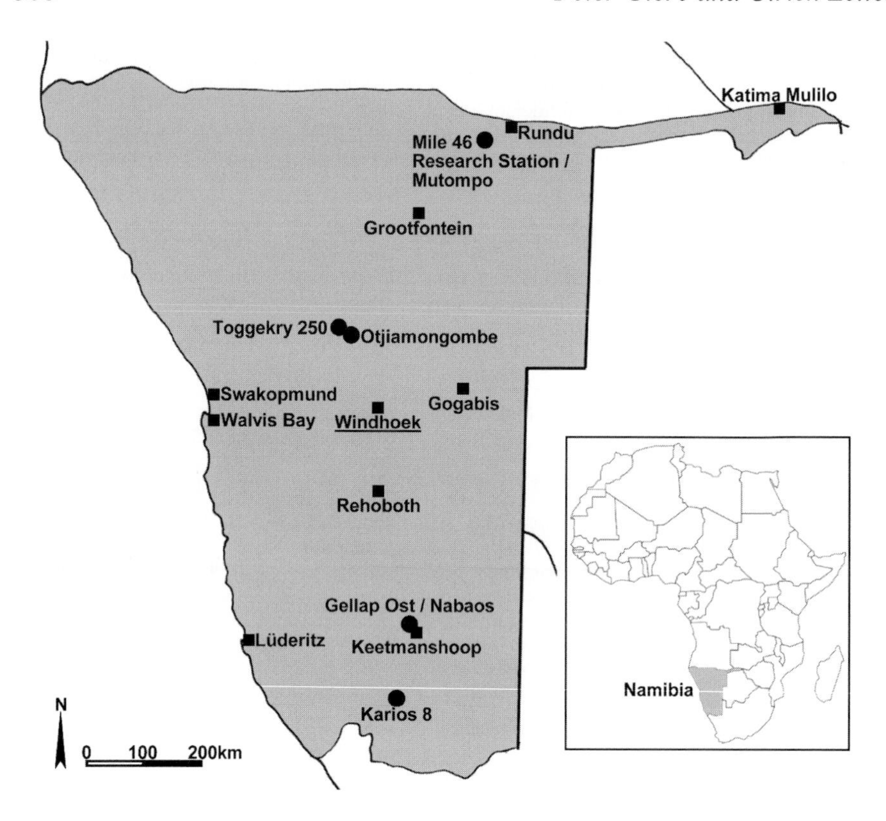

Figure 1. Location of the study sites (solid circles) and major cities (solid squares) in Namibia.

taken before the animal was treated according to one of these protocols: 1. fixation in 6% formalin, storage in 70% ethanol, 2. perfusion with ringer solution and 6% formalin, fixation in 6% formalin and storage in 70% ethanol, 3. skinning, fixation of body in 6% formalin and storage in 70% ethanol.

All sampling was conducted according to the guidelines for the capture, handling and care of mammals by the Animal Care and Use Committee (1998). Specimens are the property of the National Museum of Namibia, Windhoek, but most currently are stored at the Museum für Naturkunde, Berlin.

2.3 Reproductive status

For obtaining accurate information on the reproductive traits of the mammals captured, the female specimens were perfused (see above) whenever possible in the field and measurements of the uteri were taken to classify pregnancies into stages. As early pregnancies cannot be detected macroscopically, female genital systems were embedded in paraffin, sectioned at 10 μm, stained with hematoxyline-eosine and studied microscopically. Later pregnancies were treated histologically for the documentation of fine scale measurements and details of reproductive parameters.

2.4 Species identification

Field identification was re-examined using cranial and body characters and by comparison with specimens from the collection of the Museum für Naturkunde, Berlin. R. Hutterer (Museum Alexander Koenig, Bonn) determined some of the specimens from 2001, which were subsequently used as reference. Since sympatric *Mastomys coucha* and *M. natalensis* can only be distinguished chromosomally and only few specimens were analysed, all were recorded as *Mastomys* sp. The taxonomy follows that of Musser and Carleton (1993).

3. RESULTS

3.1 Small mammal diversity

During the three-year study, 16 species and a total of 350 individuals (rodents, shrews, and elephant shrews) were captured (Table 2). The highest individual numbers were obtained for *Mastomys* (n=119 for 2001-2003), some of which were released after identification according to the collection permit.

Diversity indices were only calculated for the specimens caught with the standard trapping design. Table 3 provides an overview of the Shannon-Wiener indices of diversity (H_s) and the evenness (E_s) calculated for 2002, 2003 and the pooled data for these years. Fewer species (S), with lower abundance, were trapped in the dryer southern regions (Karios: S=3, n=7 and Nabaos: S=2, n=3, respectively). The pooled data for Otjiamongombe (S=7, H_s=1.664, E_s=0.885, n=38) revealed a lower diversity index than in Gellap-Ost (pooled data: S=7; H_s=1.808, E_s=0.929, n=20).

A striking difference is found in the data for neighbouring Gellap-Ost and Nabaos: whereas Gellap-Ost ranked among the most diverse small mammal faunas sampled (see above), Nabaos only had a total of three specimens representing two species for 2002 and 2003 (H_s=0.637, E_s=0.918). There, not a single specimen was caught during 264 trap nights of the standard design in 2003.

Table 2. Species captured on the observatories in 2001 (x), 2002 (●) and 2003 (■). Open symbols refer to species trapped by standard design and thus included in the calculation of the diversity index (Table 3). *gravid females; + early pregnancies

Species	Observatory						
	Mu-tompo	Mile 46	Otjiamon-gombe	Togge-kry	Gellap-Ost	Nabaos	Karios
*Crocidura fuscomurina**		●	x				
Crocidura hirta				x			
Saccostomus campestris+*	□	●□	x○□	x□	○□		
Steatomys pratensis+*			x●	x			
Desmodillus auricularis+*			●		○	○	●
Gerbillurus paeba+*							x○□
Gerbillurus vallinus+*					x□	x○■	x
*Tatera brantsii**		○		x□			□
Tatera leucogaster+*	■	○■	x○□	x□	x□		x□
Aethomys chrysophilus+*			○	x□			
*Aethomys namaquensis**					○□		
*Mastomys sp. *+*			x○□	x□			
Mus indutus+*		●	x○	x	○□		
Rhabdomys pumilio+*			x○	x	□		●
*Thallomys nigricauda**				x			
Elephantulus intufi+*			○□	x□	x		

If all data is taken into account, the highest overall number of species was found in the farming area north of Windhoek (Toggekry: S=11, n=132, Otjiamongombe: S=10, n=132; cf. Table 2), followed by Gellap-Ost (S=8, n=39) in the vicinity of Keetmanshoop. The lowest overall numbers were encountered in Nabaos (S=2, n=7) and Mutompo (S=2, n=4).

3.2 Reproductive status

Overall, 53.3% of all female specimens (n=137) were gravid at capture. These represented all species listed in Table 2 with the exception of *Crocidura hirta*. In the northern observatories, the overall pregnancy rate was higher (Mile 46: 70.0%, n=10; Toggekry: 62.2%, n=45; Otjiamongombe: 53.7%, n=54), whereas it was much lower in the drier south of Namibia (Gellap-Ost: 17.6%, n=17). Too few specimens were captured for analysis of the remaining observatories (Karios: 0.0%, n=5; Nabaos 100.0%, n=3; Mutompo 100.0%, n=3).

Due to the histological examination of uteri, early pregnancies – which may be overlooked by macroscopic analysis alone – were detected in this study (11 species, n=24, Table 2). This, however, contributed to the

Table 3. Shannon-Wiener index of diversity (Hs), evenness (Es), number of species (S) and specimens (n) for small mammals captured on the observatories using the standard trapping design in 2002, 2003 and the pooled data for 2002 and 2003. n.a.: not applicable; - : observatory not sampled

Observatory / S	Year		
	2002: H_s / E_s / n	2003: H_s / E_s / n	2002+2003: H_s / E_s / n
Mutompo / 1	- / - / -	0.000 / n.a. / 4	(0.000 / n.a. / 4)
Mile 46 / 3	1.055 / 0.960 / 5	0.562 / 0.811 / 4	0.937 / 0.853 / 9
Otjiamongombe / 7	1.720 / 0.884 / 24	1.240 / 0.894 / 14	1.664 / 0.855 / 38
Toggekry / 6	- / - / -	1.494 / 0.834 / 33	(1.494 / 0.834 / 33)
Gellap-Ost / 7	1.242 / 0.896 / 6	1.475 / 0.823 / 14	1.808 / 0.929 / 20
Nabaos / 2	0.637 / 0.918 / 3	n.a. / n.a. / 0	0.637 / 0.918 / 3
Karios / 3	0.000 / n.a. / 1	0.868 / 0.790 / 6	0.796 / 0.725 / 7

detection of an overall high variability in the developmental stages of the embryos encountered. For instance, *Aethomys chrysophilus* was found to occur in early and late stages of gravidity in Otjiamongombe in March 2003.

4. DISCUSSION

Unpredictable environments with patchy rainfall and irregular supply of food resources can lead to high variations in small mammal populations (e.g. Christian, 1980). This is reflected in this study as the species and numbers captured in the semi-arid environment of Namibia varied from year to year using a standardized trapping design. This is also expressed in the varying diversity indices, which nevertheless compare to known data (H_s=1.47 for Gellap-Ost in Hoffmann and Zeller, in press). Unpredictable environments, however, can only explain variation in the data of one sampling point between two years, whereas other factors are likely to explain more of the variation seen between the observatories. One of these factors, difference in land use practises, was obvious in neighbouring Gellap-Ost and Nabaos. In the latter, a change in land use has resulted in land degradation within the past decades (Kuiper and Meadows, 2002). Overgrazing and resulting loss of habitat heterogeneity as found in Nabaos resulted in a comparatively low diversity (this study, Hoffmann and Zeller, in press; cf. Joubert and Ryan, 1999) since the availability of shelter (e.g. for breeding and protection) and food items was reduced. Small mammal species richness in southern Africa appears to be largely influenced by plant richness (Andrews and O'Brien, 2000) and grazing has a negative impact on the small mammal fauna (Eccard et al., 2000). Faunal response to degraded land as in Nabaos was found not only in small mammals (this study, Hoffmann and Zeller, in press) but also in arthropods, where species diversity and abundance was lower than in neighbouring, non-degraded Gellap-Ost (K. Vohland, M. Uhlig, A. Hoffmann, Berlin, and E. Marais, Windhoek, pers. comm.).

Interestingly, the diversity index calculated for Gellap-Ost was higher than that for Otjiamongombe and Toggekry, the areas that had the highest overall number of species and the highest abundance. This, however, can be attributed to the fact that in Gellap-Ost only 3 out of 7 species overlapped between 2002 and 2003. The species numbers for individual years and the higher abundance of small mammals in the observatories of the farming area north of Windhoek correspond to the higher grass and herb cover encountered there.

Overall, the species compositions found on the observatories reflect the distribution of small mammals as is compiled in the taxonomic literature (e.g. de Graaff, 1981; Skinner and Smithersm, 1990) or in the few studies available on small mammal diversity of the areas sampled (Matson and Blood, 1994; Hoffmann and Zeller, in press). However, *Gerbillurus paeba*, which according to other authors (Griffin, 1990; Matson and Blood, 1994) is widely distributed in Namibia, was found only in Karios. Another difference to published data is the occurrence of *Mus indutus* in Gellap-Ost, an area that lies outside the distribution map for this species (Skinner and Smithers, 1990; but compare locality 65 in Matson and Blood, 1994).

The high pregnancy rate of small mammals in the sites north of Windhoek emphasised the existing notion that small mammal reproduction is often linked to rainfall – and thus to the availability of sufficient resources for rearing young, especially for altricial species with prolonged lactation. Various triggers for reproductive activities currently are discussed. One of these, the perception of secondary plant compounds as found in fresh greenery, is linked to rainfall (e.g. White and Bernard, 1996). In line with this, a correlation between monthly rainfall and reproduction was stated for *Gerbillurus paeba* in the Karoo (White et al., 1997). The low pregnancy rate in the much drier observatory of Gellap-Ost, however, does not fit into this pattern. There, the rainy season appears to have little influence on the reproductive activities of the mammals sampled and opportunistic behaviour may be more common. In such unpredictable environments, species with brief lactation periods, especially precocial species, may be favoured since the duration of increased energy demand is shorter (cf. Künkele and Trillmich, 1997; Zeller, 1999).

ACKNOWLEDGEMENTS

We are grateful to R. Hutterer, Museum Alexander Koenig, Bonn, for the identification of specimens, to A. Hoffmann (Museum für Naturkunde, Berlin) for many helpful hints and comments on the manuscript, and Seth

Eiseb (National Museum of Namibia, Windhoek) for his help in the field and the loan of specimens. Two anonymous reviewers are acknowledged for improvements to the manuscript. This project was funded by the German Bundesministerium für Bildung und Forschung (BMBF, commission number 01 LC 0024).

REFERENCES

Andrews, P., and O'Brian, E. M. O., 2000, Climate, vegetation, and predictable gradients in mammal species richness in southern Africa, *J. Zool., Lond.* **251:**205-231.

Animal Care and Use Committee, 1998, Guidelines for the capture, handling, and care of mammals as approved by the American Society of Mammalogists, *J. Mammal.* **79:**1416-1431.

Boye, P., and Meinig, H., 1996, Flächenbezogene Erfassung von Spitzmäusen und Mäusen, *Schriftenreihe für Landschaftspflege und Naturschutz / Bundesamt für Naturschutz* **46:**45-54.

Christian, D. P., 1980, Patterns of recovery from low numbers in Namib Desert rodents, *Acta Theriol.* **25:**431-450.

de Graaff, G., 1981, *The rodents of Southern Africa*, Butterworths, Durban.

Eccard, J. A., Walther, R. B., and Milton, S. J., 2000, How livestock grazing affects vegetation structure and small mammal distribution in the semi-arid Karoo, *J. Arid Environ.* **46:**103-106.

Griffin, M., 1990, A review of taxonomy and ecology of gerbilline rodents of the central Namib Desert, with keys to the species (Rodentia: Muridae), in: *Namib ecology: 25 years of Namib research*, M. K. Seely, ed., Transvaal Museum Monograph No. 7, Transvaal Museum, Pretoria, pp. 83-98.

Hoffmann, A., and Zeller, U., in press, Influence of variations in land use intensity on species diversity and abundance of small mammals in the Nama Karoo, Namibia, *Belgian J. Zool.*

Künkele, J., and Trillmich, F., 1997, Are precocial young cheaper? Lactation energetics in the guinea pig, *Physiol. Zool.* **70:**589-596.

Kuiper, S. M., and Meadows, M. E., 2002, Sustainability of Livestock farming in the communal lands of southern Namibia, *Land Degrad. Develop.* **13:**1-15.

Matson, J. O., and Blood, B. R., 1994, A report on the distribution of small mammals in Namibia, *Z. Säugetierkd.* **59:**289-298.

Mendelsohn, J., Jarvis, A., Roberts, C., and Robertson, T., 2002, *Atlas of Namibia. A Portrait of the Land and its People*, Published for the Ministry of Environment and Tourism by David Philip.

Musser, G. G., and Carleton, M. D., 1993, Rodentia: Sciurognathi: Muridae, in: *Mammal Species of the world*, D. E. Wilson, and D. M. Reeder, eds., Smithsonian Institution Press, Washington, D.C., pp. 501-755.

Skinner, J. D., and Smithers, R. H. N., 1990, *The mammals of the southern African subregion,* University of Pretoria.

White, R. M., and Bernard, R. T. F., 1996, Secondary plant compound and photoperiod influences on the reproduction of two southern African rodent species *Gerbillurus paeba* and *Saccostomus campestris*, *Mammalia* **60:**639-649.

White, R. M., Kerley, G. I. H., and Bernard, R. T. F., 1997, Pattern and controls of reproduction of the southern African rodent *Gerbillurus paeba* in the semi-arid Karoo, South Africa, *J. Arid Environ.* **37:**529-549.

Zeller, U., 1999, Mammalian reproduction: origin and evolutionary transformations, *Zool. Anz., Jena* **238:**117-130.

POSSIBLE KARYOLOGICAL AFFINITIES OF SMALL MAMMALS FROM NORTH OF THE ETHIOPIAN PLATEAU

Nina Sh. Bulatova and Leonid A. Lavrenchenko
Severtsov Institute of Ecology and Evolution, Russian Academy of Sciences, Leninsky prospect 33, Moscow, 119071 Russia, E-mail: sevin@orc.ru

Abstract: Representatives of eight genera of Ethiopian rodents were collected and their chromosome characteristics were analyzed for the first time in the northern montane part of the country (between Tana Lake and Mt Guna, altitudes from 1800 to 3800 m a.s.l. at latitudes between 11-12° N). Even preliminary chromosome analysis gives rich empirical materials for the correction of current systematic definitions, interpretation of endemic taxa and evaluation of possible relationships with the already known karyotypic forms in Ethiopia and abroad. As a result of this study, four new karyotypes are manifested and five karyotypic descriptions already known for Ethiopian species are confirmed.

Key words: karyotypes; muroid rodents; endemics; Ethiopia; East Africa

1. INTRODUCTION

Many former studies of rodents taken for karyological identification in Ethiopia have focused on highland areas as the source of biodiversity and endemism in this African country. In this respect, however, a segment of the Ethiopian Plateau advanced far northwards to the marine (Red Sea) African/Asian boundary remained the true "white spot". Indeed, there were no karyological data from the highland areas at latitudes above 10° N. The aim of the present communication is to fill in this gap. The chromosome information for eight murid genera was for the first time obtained from the north-western montane district of the country between Mt Guna and Lake Tana.

B.A. Huber et al. (eds.), African Biodiversity, 315–319.

Animals for chromosome analyses were collected from three sites around Lake Tana - Dega Istefanos island (11°53'N 37°19'E), Bahir Dar (11°36'N 37°23'E), Vanzaye (11°47'N 37°40'E) meant below Tana1, Tana2, Tana3 - and Mt Guna, meant Guna (11°43'N 38°15'E) with an intermediate site, Debre-Tabor (11°55'N 37°57'E) all belonging to the highland area (altitudes from 1800 m ASL for the lake to 3800 m for the mount). A standard colchicine method adapted to field conditions was used. Diploid numbers (2n) and autosomal fundamental numbers (NFa) were counted from quality microphotographs. In sum, samples of minimum 1-2 to maximum 11 specimens were investigated for each species.

In view of a number of karyotypic data reported from Ethiopia in the last twenty years, there were certain expectations regarding the chromosome features of populations of the most abundant rodents distributed up north. It is true for the members of genera *Arvicanthis*, *Mastomys*, *Myomys* where quite expectable species were determined *(A. abyssinicus, A. dembeensis, M. natalensis, M. albipes)*. Nevertheless, the results were not simply predictable in genera *Stenocephalemys, Mus (Nannomys), Lophuromys, Otomys, Tachyoryctes.* The karyotypes revealed in these five genera cannot be attributed to a certain species name and will need further taxonomic satisfaction (Table 1).

2. KARYOLOGICAL RESULTS

Arvicanthis abyssinicus Ruppell, 1842 – *A. dembeensis* Ruppell, 1842. Two different karyotypes are detected among the *Arvicanthis* samples studied in the north. One of them matches well with the formerly known karyotype of the endemic *A. abyssinicus* (1 specimen from Mt Guna), while the other possesses the karyotype of the widely distributed *A. dembeensis* (vicinities of Lake Tana, 8 specimens). Despite the same geographic area, two species are obviously separated in altitudes of their ranges in this northern district as generally everywhere.

Mastomys natalensis Smith, 1834. It is noticed that the 32-chromosome specific karyotype found in three lower sites around Tana Lake seems to be in details similar with one from Gambela, a lowland region in the west of country (Bulatova et al., 2002).

Myomys albipes Ruppell, 1842. A single specimen examined at the higher altitude (Debre-Tabor) above Lake Tana level (two specimens from Dega-Istefanos island) revealed a rare autosomal heteromorphism. The largest pair that looks usually acrocentric in this endemic of Ethiopian

Table 1. Karyotypic characteristics of rodent species studied in the Ethiopian north

Taxon	2n	NFa	Geographic origin and specimens examined
Arvicanthis abyssinicus	62	64	Guna (1♂)
Arvicanthis dembeensis	62	62	Tana 2 (3♂)
			Tana 3 (4♂, 1♀)
Mastomys natalensis	32	54	Tana 1 (1♂, 1♀)
			Tana 2 (2♂)
			Tana 3 (4♂, 3♀)
Myomys albipes	46	50	Tana 1 (2♀)
		51(A/ST) *	Debre-Tabor (1♀)
Stenocephalemys sp.	50	56	Guna (1♂, 2♀)
Mus (Nannomys) setulosus	36	34	Tana 3 (2♀)
Mus (Nannomys) sp.?	36	34	Debre-Tabor (1♂, 1♀)
Lophuromys sp.	70	82	Guna (3♂, 1♀)
			Debre-Tabor (7♂, 4♀)
			Tana 3 (4♂)
Otomys sp.	58	56? 58?	Guna (1♀)
Tachyoryctes cf. *splendens*	48	58	Tana 3 (1♀)

Highlands is heteromorphic (acrocentric/subtelocentric) due to centromeric shift or pericentric inversion (Fig. 1A). Formerly, the rearrangement of a similar type was reported in lesser pair 4 from Bale Mountains National Park (Lavrenchenko et al., 1999).

Stenocephalemys sp. A quite new karyotype is determined in a northern montane sample of *Stenocephalemys* Frick, 1914 (Fig. 1B). Three specimens examined revealed new value 2n=50 versus 2n=54 described in *S. albocaudata* Frick, 1914 and *S. griseicauda* Petter, 1972 from Bale (Lavrenchenko et al., 1999). Very likely the difference between the new and the known karyotypes is based on simple chromosome rearrangements of a fusion type because of the presence of two more bi-armed pairs (large submetacentric and medium metacentric) quite corresponding to a decrease in 2n for two pairs.

Mus (Nannomys) setulosus Peters, 1876. To distinguish between the uniform 36-chromosome karyotypes which form the common evolutionary stock in African (and Ethiopian, respectively) representatives of the genus *Mus*, the procedure of *C*-banding was applied to obtain extra characteristics of non-identified on a routine level acrocentrics comprising the karyotypes of four mice from a lower (Tana 3) and an elevated (Debre-Tabor) sites. Only in two females from Vanzaye (i.e. Tana 3), a standard procedure with Ba hydroxide brought success. Here, the characteristic pattern of C-bands proved that specimens of this latter site might belong to *N. setulosus* according to initial karyotype descriptions from West Africa (Jotterand-Bellomo, 1981). No information on extended heterochromatic segments in several pairs of acrocentrics, diagnostic for this species, was so far reported for Ethiopian *Mus*.

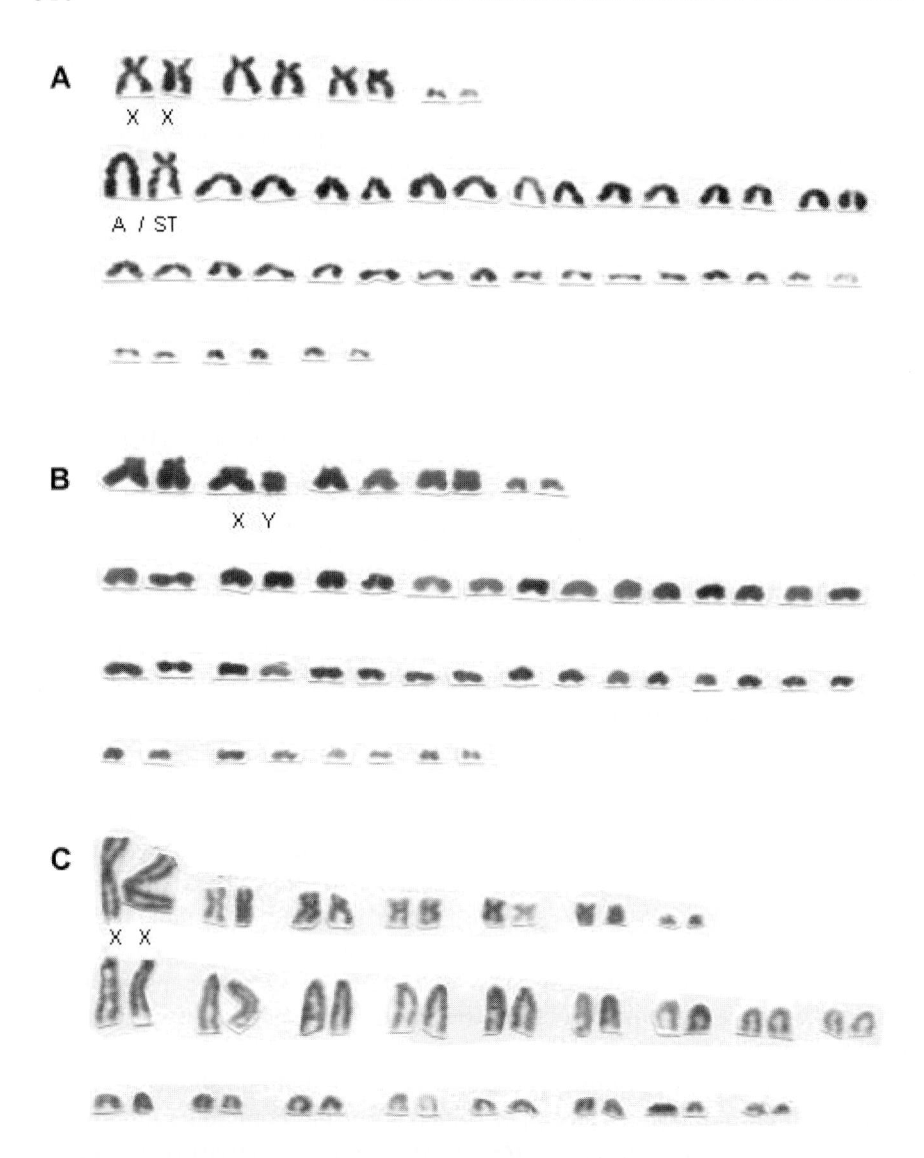

Figure 1. Rodent karyotypes from northern Ethiopia; A. *Myomys albipes*, 2n=46 (Debre-Tabor, female), A/ST – heteromorphic pair; B. *Stenocephalemys* sp., 2n=50 (Mt Guna, male); C. *Tachyoryctes cf. splendens*, 2n=48 (Vanzaye, female); XX, XY – sex chromosomes.

Lophuromys flavopunctatus Thomas, 1888 species-complex. For all the *Lophuromys* specimens studied in the north, the value 2n=70 is only found, contrary to a lesser value 2n=68 typical to all other populations examined along the plateau (Lavrenchenko et al., 1998). Remarkably, no variation was noticed in chromosomes of representatives of all three-altitude levels concerned in this study.

Otomys sp. Uncertain morphology of lesser elements in poor preparations of a single specimen studied from Mt. Guna prevents a true count of Nfa, which may be 56 or 58. In the latter case meaning the bi-armed condition of the smallest pair of the chromosome complement, the karyotype found in the northern part of the plateau is similar to that reported from Bale as *Otomys* sp. B (Lavrenchenko et al., 1997).

Tachyoryctes cf. *splendens* (Ruppell, 1835). The karyotype of a single specimen from the north is characterized by the same 2n=48, but reveals some steady differences from the variable karyotype of *T. splendens*. Diagnostic features of the new karyotype concern not only the increased number of metacentrics – five instead of the usual three, but also the presence of a minute bi-armed pair never previously reported for *T. splendens*. Endemic *T. macrocephalus* as shown from the Bale Massif, differs both in 2n (50) and the absence of a metacentric group, except in the two smallest pairs. As in *T. splendens*, the extra large submetacentric pair of a female karyotype from around Lake Tana shown in Fig. 1C corresponds to XX chromosomes (Aniskin et al., 1997).

ACKNOWLEDGEMENTS

Karyological part of the work was in portions supported by the INTAS grant 01-2163 and the RFBR project 03-04-48924.

REFERENCES

Aniskin, V. M., Lavrenchenko, L. A., Varshavskii, A. A., and Milishnikov, A. N., 1997, Karyotypes and Chromosomal Differentiation of Two *Tachyotyctes* Species (Rodentia, Tachyoryctinae) from Ethiopia, *Rus. J. Genet.* **33:**1079-1084.

Bulatova, N., Lavrenchenko, L., Orlov, V., and Milishnikov, A., 2002, Notes on chromosomal identification of rodent species in western Ethiopia, *Mammalia* **66:**128-132.

Jotterand-Bellomo, M., 1981, Le caryotype et la spermatogenese de Mus setulosus (bandes Q, C, G et coloration argentique), *Genetica* (Neth.) **56:**217-227.

Lavrenchenko, L. A., Milishnikov, A. N., Aniskin, V. M., and Warshavsky, A. A., 1999, Systematics and phylogeny of the genus *Stenocephalemys* Frick, 1914 (Rodentia, Muridae): a multidisciplinary approach, *Mammalia* **63:**475-494.

Lavrenchenko, L. A., Milishnikov, A. N., Aniskin, V. M., Warshavsky, A. A., and Woldegabriel Gebrekidan, 1997, The Genetic Diversity of Small Mammals of the Bale Mountains, Ethiopia, *SINET: Ethiop. J. Sci.* **20:**213-233.

Lavrenchenko, L. A., Verheyen, W. N., and Hulselmans, J., 1998, Systematic and distributional notes on the *Lophuromys flavopunctatus* Thomas, 1888 species-complex in Ethiopia (Muridae – Rodentia), *Bull. Inst. R. Sci. Nat. Belg. Biol.* **68:**199-214.

GEOGRAPHIC VARIATION IN THE WEST AFRICAN SCALY-TAILED SQUIRREL *ANOMALURUS PELII* (SCHLEGEL AND MÜLLER, 1845) AND DESCRIPTION OF A NEW SUBSPECIES (RODENTIA: ANOMALURIDAE)

Anja C. Schunke and Rainer Hutterer
Zoologisches Forschungsinstitut und Museum Alexander Koenig, Adenauerallee 160, 53113 Bonn, Germany, E-mail: a.schunke.zfmk@uni-bonn.de, r.hutterer.zfmk@uni-bonn.de

Abstract: Coat colour variation of *Anomalurus pelii*, the largest and most striking species of scaly-tailed squirrels in Africa, was studied based on the examination of 142 skins. Within a relatively small distribution area from central Liberia to Ghana, three distinct colour morphs (dorsal surface and tail black; dorsal surface black with narrow white margins and white tail; dorsal surface partly black with broad white margins and white tail) were found. The observed variation is not clinal but changes stepwise across the Sassandra and Bandama Rivers which divide the distribution clearly into a western, central, and eastern range. This pattern is weakly paralleled by variation of skull size. Between the three sub-populations there is probably reduced gene flow and they represent relicts of a formerly wider distribution. We assign two available names to the western and eastern and suggest a new name for the central population.

Key words: Anomaluridae; West Africa; colouration; geographic variation; taxonomy

1. INTRODUCTION

Rodents of the family Anomaluridae occur throughout the African rain forest belt. All but one species are able to perform a gliding flight. The extent of distribution differs strongly between species. *Anomalurus derbianus* occurs from Liberia to Tanzania, while *Anomalurus pelii* is

B.A. Huber et al. (eds.), African Biodiversity, 321–328.

restricted to a relatively small area in West Africa (Liberia to Ghana). The dorsal fur of *A. pelii* is generally dark brown to black, with white margins along the edges of the patagia and a white tail (Rosevear, 1969, plate 6). Occasionally specimens without such markings and with a completely black dorsal surface are found. Such a form was described by Matschie (1914) as a separate species, *A. auzembergeri*, which Kuhn (1966) later recognized as a subspecies, *A. pelii auzembergeri*. In the course of current analyses of the geographic distribution of different colour morphs of Anomaluridae (Schunke and Hutterer, in press), we found in *A. pelii* a strict correlation between locality and colouration within a relatively small area. Here we present a detailed analysis of the geographic variation of this species.

2. MATERIALS AND METHODS

2.1 Data basis

142 specimens of *A. pelii* were studied from 11 scientific collections: American Museum of Natural History (AMNH, New York), Natural History Museum (BMNH, London), Field Museum of Natural History (FMNH, Chicago), Liverpool Museum (LIVCM), Museum National d'Histoire Naturelle (MNHN, Paris), Naturhistorisches Museum Basel (NHMB), National Museum of Natural History (NMNH, Washington), Naturhistorisches Museum Wien (NMW), Naturalis/Nationaal Natuurhistorisch Museum (RMNH, Leiden), Zoölogisch Museum Amsterdam (ZMA), and Museum für Naturkunde (ZMB, Berlin). The black and white patterns were analysed from colour slides of skins (Schunke and Hutterer, in press). Information on the sex was either obtained from the skins, or copied from labels, along with body measurements (only field measurements considered). When skulls were available relative age was estimated from tooth wear. Measurements of skulls were taken with a digital caliper to the nearest 0.01 mm. The data basis for the analyses is given in Table 1.

2.2 Statistics

All statistical analyses were performed with SPSS 10.0. Data were tested for normal distribution with the Kolmogorov-Smirnov test. Relationships between skull and body measurements and sex, age, and dorsal colouration were checked with a one-way ANOVA and/or with a Kruskal-Wallis test if variances of the respective characters were not homogeneous. These tests were restricted to specimens of DC2 in order to avoid biases caused by geographic variation.

Table 1. Number of specimens (N) of *A. pelii* used for different character sets

Character	N
Dorsal colouration	142
Ventral colouration	58
Skull	108
Body measurements	36 (+1 from literature)
Sex	64
Age class	117
Locality coordinates	118

3. RESULTS

In an earlier analysis (Schunke and Hutterer, in press) of the variation of coat colouration in *Anomalurus pelii* all specimens were assigned to four groups, based on the amount of white on the dorsal surface (the ventral colouration principally follows the dorsal colouration in the degree of darkness, although less regularly, and is not further considered here). Dorsal colouration (DC) was coded as follows: DC1, completely black; DC2, white margins (maximum width measured) less than one third of pleuropatagia and separated from uropatagia margins; DC3, white margins less than one half of pleuropatagia, connected with uropatagia margins; and DC4, white margins more than one half of pleuropatagia, shoulder frequently also white. The latter two groups (DC3 and 4) are combined as DC3 in the following analyses because they occur sympatrically and because the amount of white varies gradually in the two groups (Schunke and Hutterer, in press). The three groups are restricted to certain areas, with completely black specimens occurring only west of the Sassandra River, individuals with a little white only east of the Bandama River, and specimens with large white markings occurring in between the two rivers. The observed variation of the colour patterns is not clinal but changes stepwise, resulting in three sharply delimitated sub-ranges (Fig. 1), with the lightest specimens between the darker sub-populations. Unfortunately no body or skull measurements were available from specimens collected in the triangle between the Bandama and Nzi Rivers, where specimens of DC2 (RMNH 17828, ZMA 21.329) and DC3 (MNHN 1983-578) apparently co-occur.

3.1 Body measurements

The data set on body measurements is biased towards specimens from the eastern region of the distribution area, with 81% (n=30) showing colouration class DC2 (white markings small). Large white markings (colouration class DC3) are represented by six specimens (16%), and the completely black

Table 2. Arithmetic means of body measurements of *A. pelii* specimens assigned to different dorsal colourations (DC1-3); * data from Kuhn (1966)

Character	DC1*	DC2		DC3	
	♂ (N=1)	♀♀ (N=12)	♂♂ (N=17)	♀♀ (N=4)	♂♂ (N=2)
Total length [mm]	850	888	837	896	895
Tail length [mm]	445	450	423	452	469
Hindfoot length [mm]	75	83	80	88	86
Ear length [mm]	43	46	45	44	48
Weight [g]	1770	1767	1477	1833	1700

morph by 1 specimen (3%) only (data from Kuhn, 1966). The data set used for the analysis consisted of the following standard characters: total length (n=35, partially calculated from head and body plus tail length), tail length (n=35), length of hindfoot (n=35), ear length (n=35) and weight (n=24).

The Kolmogorov-Smirnov test showed that all body measurements were normally distributed. None of the characters listed above showed a significant correlation with age class in the one-way ANOVA and/or the Kruskal-Wallis test. Different results were obtained for the relationship between body measurements and sex. The one-way ANOVA showed a significant correlation between sex and total length ($P = 0.005$), tail length ($P < 0.05$), and weight ($P < 0.01$), while hindfoot and ear length were not significantly correlated with sex. Because of these results the one-way ANOVA for the different colourations were calculated for males and females separately as well as in a combined analysis. However, no statistically significant differences were found, with a single exception (tail length in males, $P < 0.05$). This difference may have been caused by the preparation technique, because the two male specimens with large white markings were both from the AMNH collections. Mean values for females and males are given in Table 2.

3.2 Craniometric data

As in body measurements, data for skulls differ considerably in sample size, with only 3 unsexed skulls available from completely black specimens (3 %), 36 from specimens with a little white (33 %), and 69 from individuals with large white markings (64 %). Skull measurements were more difficult to analyse. From an original data set of 64 measured characters (details to be published elsewhere), most were excluded because they showed either sexual dimorphism or age dependence within the tested sub-sample (DC2), or were not available for all three skulls of the DC1 (completely black) sample. Tooth characters were completely omitted because most showed age dependence. This reduced the data set to 19 characters of which 8 showed

Table 3. Craniometric characters which differ significantly between specimens of *A. pelii* assigned to different dorsal colourations (n.s. = not significant)

Character	Arithmetic mean			*P*-values		
	DC1 (N=3)	DC2 (N=36)	DC3 (N=69)	DC1 vs. DC2	DC1 vs. DC3	DC2 vs. DC3
Sagittal cristae largest width	18.52	18.12	17.08	n. s.	n. s.	$P < 0.005$
Sagittal cristae smallest width	6.73	7.77	6.49	n. s.	n. s.	$P < 0.001$
Incisive foramen length	8.13	7.38	8.03	$P < 0.05$	n. s.	$P < 0.001$
Gum width at first molar	2.90	2.36	2.49	$P < 0.05$	n. s.	n. s.
Choana width	7.54	6.79	7.15	$P < 0.005$	n. s.	$P < 0.001$
Distance between posterior end of zygomatic arcs	35.23	33.42	33.48	$P < 0.005$	$P < 0.001$	n. s.
Infraorbital foramen height	9.56	8.71	8.91	$P < 0.05$	$P < 0.05$	$P < 0.005$
Mandible height at diastema	9.03	8.07	8.19	$P < 0.01$	$P < 0.001$	n. s.

statistically significant differences between the groups in a one-way ANOVA (Tab. 3).

4. DISCUSSION AND CONCLUSION

The morphometric data support the traditionally accepted hypothesis of *Anomalurus pelii* as a single species, although the three sub-populations separated by the Sassandra and Bandama Rivers clearly differ. Most obvious are the entirely black individuals, which also tend to have larger skulls than specimens from east of the Sassandra River. Significant differences in skull characters were also found between the other two populations (Table 3). The most striking difference separating the populations is the proportion of black and white markings on the dorsal surface (Figs. 1, 2). We assume that the current sub-populations represent relicts of a formerly wider distribution and are now stranded in their respective ranges between major river systems. Reduced gene flow and cladogenesis are probable but should be tested by genetic data. The apparent sympatry of DC2 and DC3 in the confluence of the Bandama and Nzi Rivers (Fig. 1) is not well documented; the few specimens might also have been collected on the other side of the respective river.

Two of the three discrete sub-populations of *A. pelii* defined in this paper received formal recognition in the past. The third and previously unrecognized population from central Ivory Coast is named below. In addition, a lectotype is designated for *A. pelii*. The ranges of the three subspecies are shown in Figure 1.

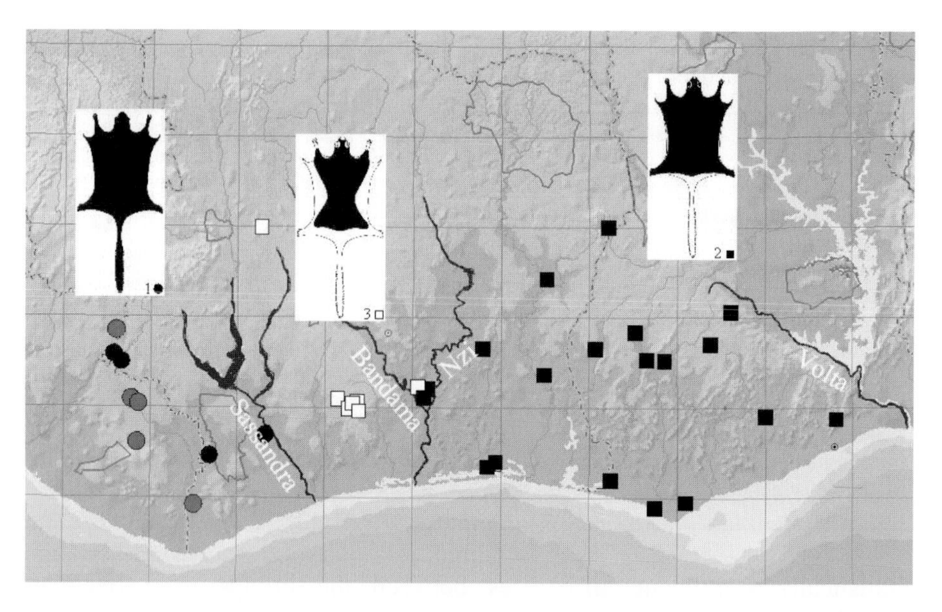

Figure 1. The distribution of *Anomalurus pelii* based on voucher specimens. The three populations defined in this paper are marked by different symbols according to the dorsal colouration. For the western population (*A. p. auzembergeri*) some records from the literature have been added (grey circles).

Anomalurus pelii auzembergeri Matschie, 1914
Holotype: ZMB 18271, male, skin (Fig. 2a).
Type locality: "Bei Patokla am mittleren Cavally, Elfenbeinküste, 150 km vom Meere" [Patokla, Ivory Coast, 05°28' N, 07°19' W].
Diagnosis: Dorsal surface including tail and limbs uniformly black.
Distribution: E Liberia and extreme W Ivory Coast west of Sassandra River. No evidence for its occurrence in Sierra Leone, as previously assumed by Honacki et al. (1982), Corbet and Hill (1991), and Dieterlen (1993).
Specimens examined (9): BMNH 1, MNHN 1, RMNH 1, ZMA 5, ZMB 1.

Anomalurus pelii pelii (Schlegel and Müller, 1845)
Lectotype (here designated): RMNH 26761, male, skin and skull (Fig. 2c).
Paralectotypes: RMNH 26762, presumably also BMNH 50.5.7.1 and LIVCM 1981-270 (see Largen 1985).
Type locality: "bij Daboeram, aan de Goudkust" [Dabocrom, Ghana, 04°57'N, 01°53'W (Grubb et al. 1998)].
Diagnosis: Dorsal surface black, tail and margins of uropatagia white, fore feet completely black or with very little white, white margins less than one third of pleuropatagia, hindfeet at least partly black.
Distribution: Ghana and E Ivory Coast east of Bandama River.
Specimens examined (58): BMNH 14, FMNH 4, LIVCM 1, MNHN 7, NHMB 1, NMNH 16, NMW 1, RMNH 3, ZMA 10, ZMB 1.

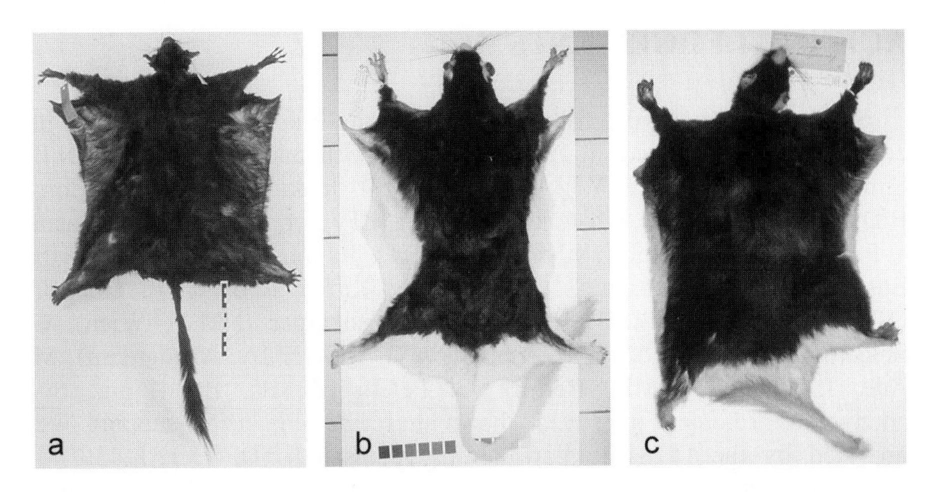

Figure 2. Skins of type specimens in dorsal view. a, *A. p. auzembergeri*, holotype ZMB 18271, photo courtesy V. Heinrich, ZMB; b, *A. p. peralbus* new subspecies, holotype ZMA 21.275; c, *A. p. pelii*, lectotype RMNH 26761.

Anomalurus pelii ***peralbus*** new subspecies

Holotype: ZMA 21.275, female, skin and skull, collected on 20 October 1970 by L.J.R. Bellier (Fig. 2b).

Paratypes: 47 further specimens collected in 1970 by L.J.R. Bellier and in 1973 by J. Vissault at the type locality (ZMA 21.264-270, 21.272-274, 21.276-289, 21.293-315).

Type locality: "Gueboua, Côte d'Ivoire" [Ivory Coast, 05°59' N, 05°41' W].

Diagnosis: Dorsal surface black with broad lateral and caudal margins of white, more than one third of pleuropatagia, tail white, fore feet with little white to broad white markings from feet to shoulders, hind feet completely or almost white. Variation in the amount of white higher than in *A. p. pelii*, but generally all white markings, including margins of patagia, shoulders, nose, and ear basis tend to be larger.

Distribution: Ivory Coast, between rivers Bandama and Sassandra.

Etymology: The subspecies epithet is a combination of the Latin *per* (very) and *albus* (white).

Specimens examined (75): AMNH 4, MNHN 2, ZMA 69.

Comment: The conservation status of Pel's scaly-tailed squirrel has been assessed by the IUCN as Lower Risk/ near threatened (Baillie, 1996). We are not able to add substantial new data, as most of the specimens examined by us were collected decades ago. In a forest in eastern Ivory Coast, *Anomalurus pelii* constituted 1.5% of the bush meat yield (Caspary, 2000). Given the rapid rate of forest degradation in West Africa, most populations are likely to be threatened. This may be particularly true for the new subspecies. Surveys to assess the current status of all populations are urgently needed.

ACKNOWLEDGEMENTS

Staff members of the following institutions kindly allowed access to collections and assisted in many ways: P. Brunauer, B. Randall, N.B. Simmons (AMNH, New York), D. Hills, P.D. Jenkins (BMNH, London), B.D. Patterson (FMNH, Chicago), C. Fisher (LIVCM, Liverpool), J. Cuisin, J.-F. Ducroz, M. Tranier (MNHN, Paris), U. Rahm, R. Winkler (NHMB, Basel), M.D. Carleton, L.K. Gordon, H.L. Kafka, R.W. Thorington, Jr. (NMNH, Washington), B. Herzig, F. Spitzenberger (NMW, Wien), C. Smeenk (RMNH, Leiden), M. Vences, A.G. Rol (ZMA, Amsterdam), M. Ade, I. Thomas (ZMB, Berlin). G. Peters (ZFMK, Bonn) and two anonymous referees improved the manuscript. Financial support was provided by the Deutsche Forschungsgemeinschaft (DFG, HU 430/1-1, 430/1-2). Visits to BMNH and MNHN were funded by the TMR Programme of the European Commission (Bioresource London and Parsyst Paris).

REFERENCES

Baillie, J., 1996, *Anomalurus pelii*, in: 2003 IUCN Red List of Threatened Species, IUCN 2003. <www.redlist.org>

Caspary, H.-U., 2000, *Faune sauvage et la filière viande de brousse au Sud-est de la Côte d'Ivoire*, Weissensee Verlag, Berlin.

Corbet, G. B. and Hill, J. E., 1991, *A world list of mammalian species,* 3rd ed., Natural History Museum Publications and Oxford Univ. Press, London and Oxford.

Dieterlen, F., 1993, Family Anomaluridae, in: *Mammal species of the world: A taxonomic and geographic reference,* D. E. Wilson and D. A. Reeder, eds., Smithsonian Institution Press, Washington, D.C., pp. 757-758.

Grubb, P., Jones, T.S., Davies, A.G., Edberg, E., Starin, E.D. and Hill, J.E., 1998, *Mammals of Ghana, Sierra Leone and The Gambia,* The Trendrine Press, Zennor, St. Ives, UK.

Honacki, J. H., Kinman, K. E. and Koeppl, J. W., 1982, *Mammal species of the world,* Allen Press, Lawrence, Kansas.

Kuhn, H.-J., 1966, *Anomalurus pelii auzembergeri* in Liberia, *J. Mamm.* **47:**334-338.

Largen, M. J., 1985, Taxonomically and historically significant specimens of mammals in the Merseyside County Museums, Liverpool, *J. Mamm.* **66:**412-418.

Matschie, P., 1914, Ein neuer *Anomalurus* von der Elfenbeinküste. *Sitzb. Ges. Naturf. Freunde, Berlin*: 349-351.

Rosevear, D. R., 1969, *The Rodents of West Africa,* Trustees of the British Museum (Natural History), London.

Schlegel, H. and Müller, S., 1843-1845, Bijdragen tot de natuurlijke geschiedenis der Vliegende eekhorens (*Pteromys*) door Herm. Schlegel en Sal. Müller, in: *Verhandeel. over Natuurl. Geschiedn. Nederland. Bezittingen, Zool.*, C. J. Temminck, ed., Natuurkundige Commissie in Ostindie, pp. 103-114.

Schunke, A. C. and Hutterer, R., in press, The variance of variation: Geographic patterns of coat colouration in *Anomalurops* and *Anomalurus* (Mammalia, Rodentia, Anomaluridae). *Bonn. zool. Beitr.*

MIDDLE EAR OSSICLES AS A DIAGNOSTIC TRAIT IN AFRICAN MOLE-RATS (RODENTIA: BATHYERGIDAE)

Simone Lange[1]*, Hynek Burda[1], Nigel C. Bennett[2], Pavel Němec[3]
[1] *Dept. General Zoology, FB 9, University of Duisburg-Essen, D-45117 Essen, Germany*
[2] *Dept. Zoology and Entomology, University of Pretoria, Pretoria 0002, Rep. South Africa*
[3] *Dept. Zoology, Charles University, Viničná 7, CZ-128 44 Praha 2, Czech Republic*
* *E-mail: simone.lange@uni-essen.de*

Abstract: Within the family Bathyergidae (subterranean African mole-rats), *Cryptomys* and *Coetomys* are the most speciose genera. Whereas each species can be characterized karyologically and from mitochondrial gene sequence data genetically, most of them exhibit marked intraspecific polymorphism and an overlap in morphological and biometric traits making them very hard to discern for taxonomic studies. As a consequence, pelage colour, body size and craniometric characters are inappropriate for species diagnosis.
Here we report on a comparative morphological and biometric analysis of the auditory ossicles in seven species of *Coetomys* and one species of *Cryptomys*. While species can be identified by a unique combination of oto-morphological features, there is also a strong correlation between biometry of functionally relevant traits and the habitat. Middle ear biomechanical efficiency and hence hearing sensitivity appears to improve with increasing humidity of the habitat. This trend is evident even when comparing different populations at an intraspecific level.

Key words: functional morphology; middle ear; species diagnosis; *Coetomys*; *Cryptomys*; Bathyergidae

1. INTRODUCTION

Within the family Bathyergidae (subterranean African mole-rats), *Cryptomys* and *Coetomys* are the most speciose genera. Whereas the species can be characterized karyologically and genetically, most of them exhibit

B.A. Huber et al. (eds.), African Biodiversity, 329–337.

marked intraspecific polymorphism including overlap in morphological and biometric traits (for most recent reviews see Bennett and Faulkes, 2000; Faulkes et al., 2004; Ingram et al., 2004; VanDaele et al., 2004). Consequently, pelage colour, body size or craniometric characters are inappropriate for species diagnosis. Following the findings in spalacid mole-rats (Burda et al., 1990), we studied the use of the oto-morphological characters of the middle ear (that can be examined in museum specimens after opening one bulla tympanica) for species diagnosis as well as the effect of habitat humidity upon ear sensitivity in African mole-rats.

2. MATERIALS AND METHODS

Middle ears were examined in eight species of two genera of African mole-rats (Bathyergidae, Rodentia) of different body sizes (skull lengths) occurring in different precipitation regimes: *Coetomys amatus* (CAM), *Coetomys anselli* (CA), *Coetomys damarensis* (CDAM), *Coetomys darlingi* (CDAR), *Coetomys kafuensis* (CK), *Coetomys mechowi* (CM), *Coetomys whytei* (CW) and *Cryptomys pretoriae* (CPRE) (*Coetomys* represents a resurrected genus, the species of which were previously referred to as *Cryptomys*; Ingram et al., 2004). Ears of each adult individual were examined bilaterally. Origin and numbers of the adult specimens involved in our study are given in Table 1. The animals were sacrificed and perfused by 4% paraformaldehyde as part of a larger study to investigate the neuroanatomy of the eight representatives. The heads were kept immersed in 70% ethanol for at least four weeks prior to preparation. The numbers of analyzed ossicles are given in the text.

2.1 Comparative biometry and species diagnosis

Condylo-basal length of the skull and maximum length, width and height of the tympanic bulla were measured using digital callipers. The bulla was opened, middle ear structures were prepared and examined under a stereoscopic binocular (Olympus, Type SZH 10 Research Stereo) and traced using a drawing tubus at different magnifications (15x – 40x), both *in situ* and after removal and isolated *in toto*. Measurements were taken from drawings taking into account the respective magnifications. The following variables were measured: longer and shorter radius of the eardrum, longer and shorter radius of the cross-section of the bony meatus, length of the mallear lever, length of the incudal lever, longer and shorter radius of the stapedial footplate. Mallear and incudal levers were measured as the

Table 1. Ear dimensions, morpho-functional ear variables and precipitation sorted by the mean annual precipitation at the given locality. The table shows means and standard deviations where possible. R – mean annual precipitation (mm/year) (see Material and Methods), CB – condylobasal length (mm), B –volume of bulla tympanica (mm^3), PT – area of pars tensa (mm^2), BS - area of basis stapedis (mm^2), AR - area ratio (MT/BS), ML - mallear lever (mm), IL - incudal lever (mm), LR - lever ratio (ML/IL), TR - transformation ratio (AR x LR)

	R	CB	B	PT	BS	AR	ML	IL	LR	TR
Coetomys whytei (Karonga, Malawi) N=5	1603	36.9 ±1.9	469 ±109	11.6 ±1.4	0.50 ±0.04	23.2 ±1.9	2.4 ±0.1	1.0 ±0.1	2.4 ±0.3	55.7 ±7.6
Coetomys mechowi (Ndola, Zambia) N=9	1185	45.9 ±6.7	942 ±305	14.5 ±2.2	0.67 ±0.07	21.6 ±4.0	2.5 ±0.2	1.2 ±0.1	2.1 ±0.3	45.4 ±13.8
Coetomys darlingi (Chimanimani, Zimbabwe) N=2	1175	36.2 ±3.9	331 ±11	9.7 ±2.3	0.48 ±0.05	20.2 ±2.6	2.0 ±0.1	1.0 ±0.1	2.0 ±0.1	40.4 ±4.3
Coetomys amatus (Serenje District, Zambia) N=1	1132	32.5	208	8.9	0.46	19.3	1.9	1.0	1.9	36.7
Coetomys damarensis (West Zambezi, Zambia) N=4	900	32.2 ±1.7	384 ±49	11.4 ±1.0	0.53 ±0.11	21.5 ±3.4	2.1 ±0.1	1.1 ±0.1	1.9 ±0.1	40.9 ±7.5
Coetomys darlingi (Goromonzi, Zimbabwe) N=8	843	36.7 ±3.4	346 ±63	8.5 ±1.4	0.54 ±0.07	15.7 ±1.9	2.1 ±0.1	1.2 ±0.1	1.8 ±0.1	28.3 ±2.6
Coetomys anselli (Lusaka, Zambia) N=9	822	32.8 ±1.9	380 ±68	8.4 ±1.4	0.52 ±0.10	15.6 ±4.2	2.2 ±0.2	1.0 ±0.3	2.2 ±0.3	32.8 ±10.1
Coetomys kafuensis (Itezhi-Tezhi, Zambia) N=11	787	31.7 ±1.8	363 ±44	8.3 ±1.3	0.53 ±0.07	15.7 ±2.5	2.2 ±0.1	1.1 ±0.2	2.0 ±0.3	31.4 ±8.5
Coetomys pretoriae (Pretoria, South Africa) N=12	777	38.2 ±2.3	319 ±42	9.9 ±1.3	0.58 ±0.04	17.1 ±1.5	2.2 ±0.1	1.1 ±0.1	2.0 ±0.2	34.2 ±3.7
Coetomys damarensis (Dordabis, Namibia) N=10	400	39.2 ±2.8	471 ±96	11.7 ±2.0	0.73 ±0.09	16.0 ±2.6	2.5 ±0.1	1.4 ±0.2	1.8 ±0.2	28.8 ±5.4

perpendicular distances from the rotatory axis to the tip of the manubrium or lenticular apophysis (Fig. 1). The eardrum, meatus and stapedial footplate areas were calculated as ovals, outer bulla volume was calculated as the product of length, height and width. Bulla volume, eardrum area, mallear lever, incudal lever and stapedial footplate area were related to the condylobasal length (Spearman rank correlation, two-tailed; SPSS 11.0).

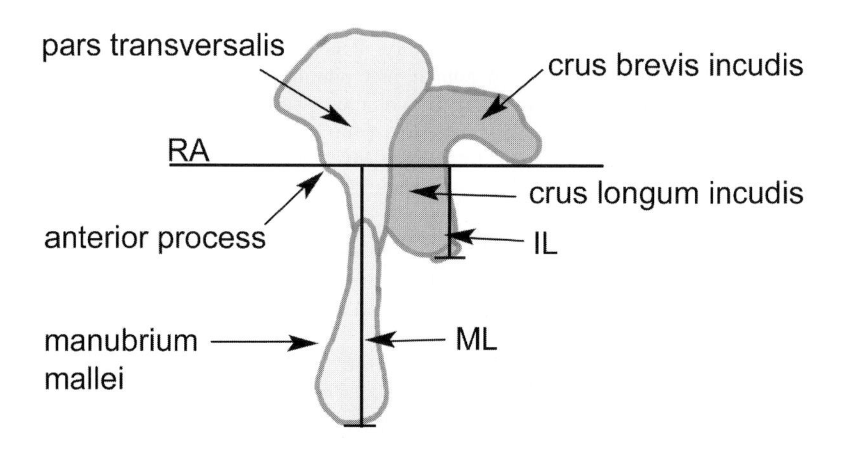

Figure 1. Morphological terminology and measurements of the mallear-incudal-complex, the malleus is light shaded, the incus is dark shaded. (RA – rotatory axis).

2.2 Functional morphology

Movements of the eardrum are transmitted to the mallear-incudal complex and then to the stapes that is inserted into the vestibular window of the cochlea. Through the arrangement of this ossicular chain, pressure and force are amplified. This is accomplished by two mechanisms: the difference between the area of the eardrum and the area of the stapedial footplate leading to the area ratio (AR) and the difference in the length of the mallear and incudal levers leading to the lever ratio (LR; cf. Relkin, 1988). The product of both ratios (transformation ratio, TR) expresses the middle ear efficiency in sound transmission and thus middle ear sensitivity (Wever and Lawrence, 1954; Webster and Webster, 1975). TR was correlated to the mean annual precipitation of the locality (Spearman rank correlation, two-tailed; SPSS 11.0).

2.3 Climatic data

The mean annual precipitation was calculated from monthly values from an average of about 55 years recorded by the nearest climatological station according to the Global Historical Climatological Network database: <www.ncdc.noaa.gov/ol/climate/research/ghcn/ghcn.html>.

3. RESULTS

3.1 Comparative morphology and species diagnosis

Cerumen was found in the bony meatus of the following species: CDAR, CHOT, CW, CM, CA, CK. The tympanic bulla had a volume between 208 mm^3 in CAM (n = 1) and 942 mm^3 in CM (n = 9). The eardrum was nearly circular in all the studied species, with the umbo at the center, no pars flaccida was apparent. The mean area of the pars tensa ranged between 8.4 mm^2 in CA (n = 9) and 14.5 mm^2 in CM (n = 9) (cf. Table 1). The middle ear ossicular chain had a quite massive malleo-incudal-complex, e.g. malleus and incus were solidly fused. The malleus exhibited a reduced pars transversalis, the gonial was missing, the incus was relatively large and massive, and the manubrium mallei and the crus longum incudis were parallel to each other (Fig. 2). The malleus was firmly attached to the bulla wall by a ligament at the back, other connections were loose – auditory ossicles were not fused with the tympanic bone, neither at the mallear head nor at the crus breve incudis, so the middle ears were of the "freely mobile" type. The incudo-stapedial joint was rather loose and the malleo-incudal complex separated easily from the stapes; the annular ligament was generally very weak whereas in CA it was stronger and in some specimens of CK the stapes was strongly connected with the incus.

While bulla volume, eardrum and meatus area were correlated with condylo-basal length, mallear and incudal lever seemed to be independent of body size. A statistical analysis of CW, CM, CDAR - Chimanimani, CAM, CDAM - West-Zambezi, CA and CK revealed that the volume of the tympanic cavity was strongly positively correlated with the condylo-basal length ($r = 0.941$, $p < 0.01$). Eardrum area ($r = 0.600$, $p < 0.01$), the area of the stapedial footplate ($r = 0.528$, $p < 0.01$) and the length of the mallear levers ($r = 0.424$, $p < 0.01$) were less dependent upon the condylo-basal lenght. The length of the incudal levers ($r = 0.144$, $p = 0.38$) was independent of it. The stapedial footplate was absolutely and relatively noticeably small in CAM, CDAR and CW.

The shape of the middle ear ossicles (Fig. 2) was similar, though some species-specific differences were identified. Thus, the head of the malleus was particularly massive in *C. whytei*, whereas in *C. pretoriae* it was conspicuously round. The transitional part between the manubrium and the head of the malleus, where the tensor tympani muscle attaches, was rather straight in *C. whytei* and *C. amatus* whereas in the other species, the anterior process was more (*C. anselli, C. pretoriae*) or less (*C. kafuensis, C. darlingi, C. damarensis* and *C. mechowi*) conspicuous. The shape of the incus was

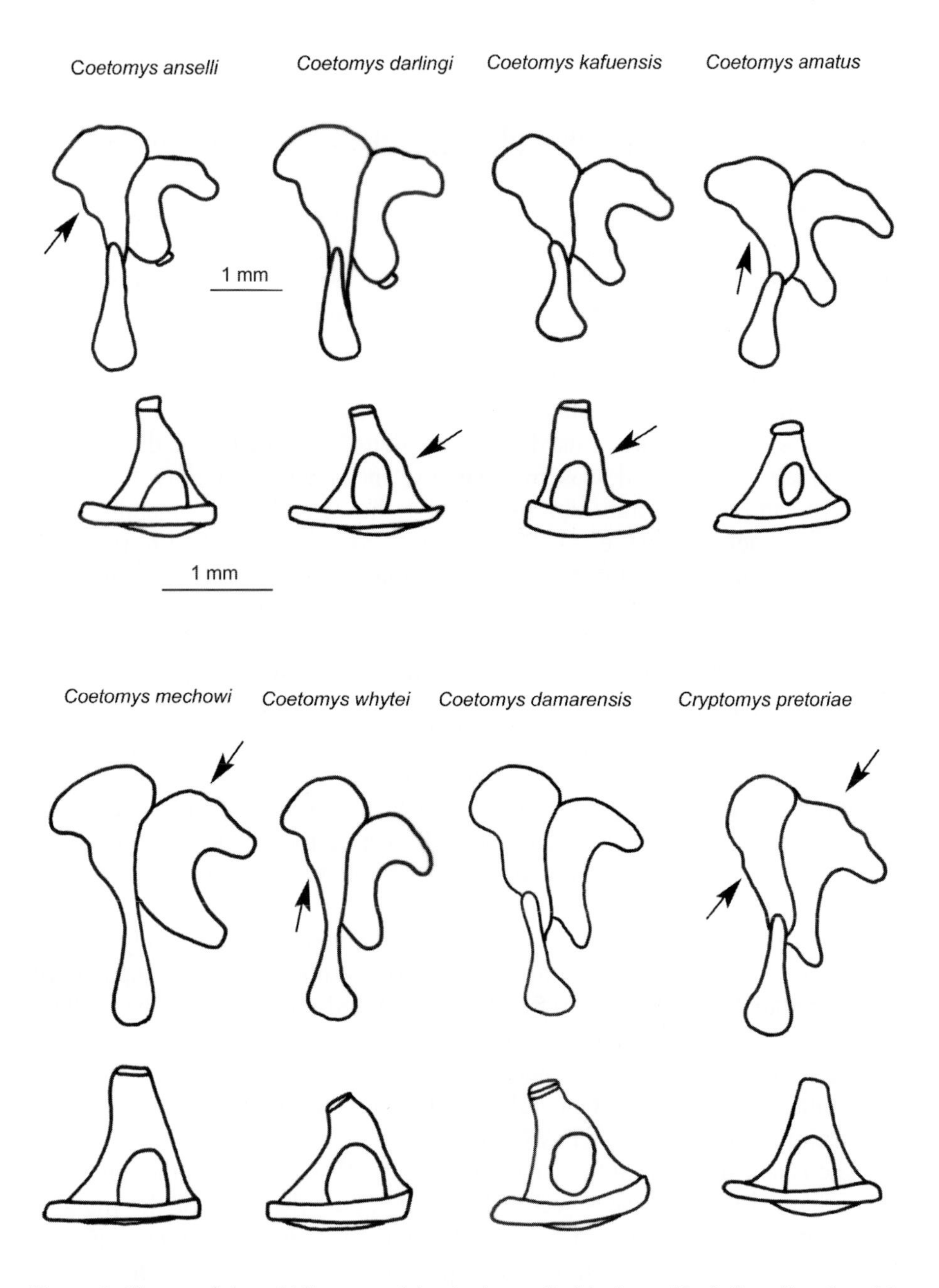

Figure 2. Shapes of the middle ear ossicles in the studied bathyergids (left mallear-incudal complexes from lateral view and left stapes from ventral view).

similar in all the species, in *C. mechowi* and *C. pretoriae* it was more pronounced. The stapes had a normal stirrup-like appearance. In *C. kafuensis* and *C. darlingi*, crus stapedis posterior was broader than crus stapedis anterior, whereas the opposite was true in the other studied species. The attachment of the stapedial muscle was clearly differently located and thus stapedes were shaped very differently. No intraspecific variability was found in the shape of the malleo-incudal-complex whereas the shape of the stapes was more varying. No stapedial artery was found in the studied species. The area of the stapedial footplate was relatively large ranging from 0.46 mm² in CAM (n = 1) to 0.73 mm² in CDAM - Dordabis (n = 10). Middle ear muscles (m. tensor tympani and m. stapedius) were missing or reduced in all the studied species.

3.2 Functional morphology

The lever arm of the malleus amounted to values between 1.9 mm (n = 1) in CAM and 2.5 mm in CM (n = 9) and CDAM - Dordabis (n = 10), that of the incus to values between 1.0 mm in CW (n = 5), CDAR - Chimanimani (n = 2), CAM (n = 1) and CA (n = 9) and 1.4 mm in CDAM - Dordabis (n = 5).

The calculated functional parameters (Table 1) could be correlated to the mean annual precipitation in the area of occurrence of the particular species: With increasing rainfall the transformation ratio (expressing mechanical middle ear efficiency and thus middle ear sensitivity) increased from the driest habitat (CDAM – Dordabis, Namibia) to the most mesic habitat (CW – Karonga, Malawi; $r = 0.480$, $p < 0.01$, $df = 68$, Fig. 3).

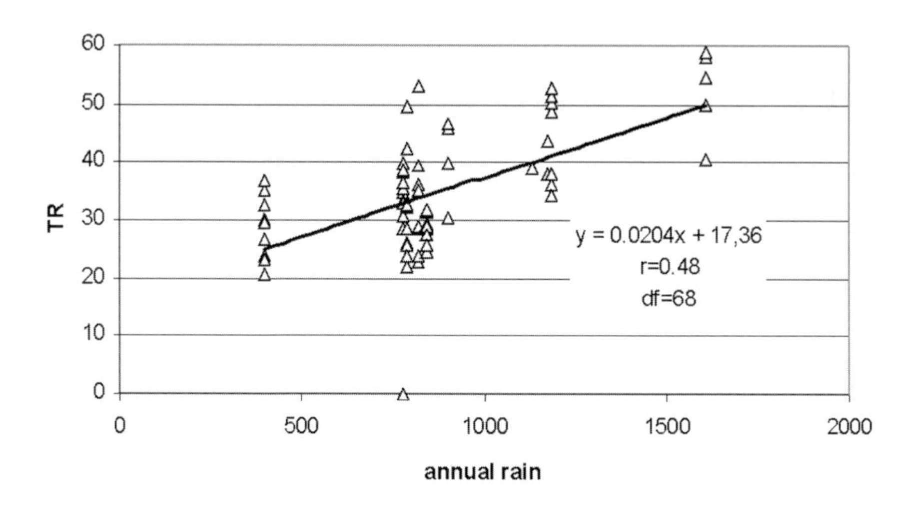

Figure 3. Correlation between annual precipitation (mm/year) and middle ear sensitivity (expressed as transformation ratio, TR).

4. DISCUSSION AND CONCLUSIONS

Both intra- and interspecific comparison of the shape of auditory ossicles revealed that they can provide a valuable tool for species diagnosis (cf. Fig. 2). Each species can be identified by a unique combination of qualitative oto-morphological features. Shape of the malleo-incudal complex is particularly useful since it exhibits no obvious inter-individual variability but clearly differs between species. Species diagnosis based upon the morphology of the middle ear is a simple method that does not require sophisticated and expensive laboratory equipment and can be applied on skulls, requiring only local unilateral and rather negligible damage of the material. Correlation analysis between condylo-basal length on the one hand and the measured (mallear and incudal lever) or calculated ear variables (volume of tympanic bulla, area of the eardum, area of the stapedial footplate) on the other hand showed that most of them are related to body size. However, the incus size (reflected in incudal lever) emancipated from dependency upon the body size. This could indicate the adaptive functional importance of the size (and thus also mass) of the incus.

Additionally, there is a strong correlation between the biometrics of the middle ear and some characteristics of the habitat. Middle ear biomechanical efficiency (determined by the transformation ratio) and hence hearing sensitivity increases with increasing humidity (expressed by mean annual precipitation) of the habitat (even at an intraspecific level). This result contrasts the findings in heteromyids (Webster and Webster, 1975) and in the *Spalax ehrenbergi* species complex (Burda et al., 1990). The biological meaning of this correlation remains obscure. Probably, precipitation has no immediate influence upon acoustics and hearing ablities. Unfortunately, to date the influence of soil type, density and moisture upon the acoustics in burrows is unknown and hearing and its role in sensory ecology across diverse *Coetomys* and *Cryptomys* species was not studied at all. Therefore it would be highly speculative to discuss causal relations.

ACKNOWLEDGEMENTS

This study was supported by a scholarship of the University of Duisburg-Essen (GRAFÖG) to the first author. N.C.B. acknowledges the National Research Foundation for funding and University of Pretoria for animal collection. P.N. was supported by the Grant Agency of the Czech Republic (project no. 206/03/0638) and by the Ministry of Education, Youth and Sport of the Czech Republic (project no. 1131-00004). We wish to thank two anonymous reviewers for helpful comments on the manuscript.

REFERENCES

Bennett, N. C., and Faulkes, C. G., 2000, *African Mole Rats: Ecology and Eusociality,* Cambridge University Press, Cambridge.

Burda, H., Nevo, E., and Bruns, V., 1990, Adaptive radiation of ear structures in subterranean mole-rats of the *Spalax ehrenbergi* superspecies in Israel, *Zool. Jb. Syst.* **117:**369-382.

Faulkes, C. G., Verheyen, E., Verheyen, W., Jarvis, J.U.M., and Bennett, N. C., 2004, Phylogeographic patterns of speciation and genetic divergence in African mole-rats (Family Bathyergidae*), Mol. Ecol.* **13:**613-629.

Ingram, C. M., Burda, H., and Honeycutt, R. L., 2004, Molecular phylogenetics and taxonomy of the African mole-rats, genus *Cryptomys* and the new genus *Coetomys* Gray, 1864, *Mol. Phylogenet. Evol.* **31:**997-1014.

Relkin, E. M., 1988, Introduction to the analysis of middle-ear function, in: *Physiology of the ear*, Jahn & Santos-Sacchi, eds., New York.

Van Daele, P. A. A. G., Dammann, P., Kawalika, M., Meier, J.-L., Van De Woestijne, C., and Burda, H., 2004, Chromosomal diversity in *Cryptomys* mole-rats (Rodentia: Bathyergidae) in Zambia; with the description of new karyotypes, *J. Zool. Lond.*, in press

Webster, D. B., and Webster, M., 1975, Auditory systems of heteromyidae: functional morphology and evolution of the middle ear*, J. Morph.* **146:**343-376.

Wever, E. G., and Lawrence, M., 1954, *Physiological acoustics*, Princeton University Press, Princeton, pp. 454.

COMMUNITY ANALYSIS OF MURIDAE (MAMMALIA, RODENTIA) DIVERSITY IN GUINEA: A SPECIAL EMPHASIS ON *MASTOMYS* SPECIES AND LASSA FEVER DISTRIBUTIONS

Denys, C.[1]*, Lecompte, E.[1,3], Calvet, E.[1], Camara, M.D.[2], Doré, A.[2], Koulémou, K.[2], Kourouma, F.[2], Soropogui, B.[2], Sylla, O.[2], Allali-Kouadio B.[4], Kouassi-Kan, S.[4], Akoua-Koffi, C.[4], ter Meulen, J.[3] and Koivogui, L.[2]

[1] *FRE2695: OSEB – Département Systématique et Evolution – MNHN - Mammifères & Oiseaux, Case courrier 051, 55 rue Buffon, 75 231 Paris cedex 05, France*

[2] *Projet de Recherches sur les Fièvres Hémorragiques en Guinée, Centre Hospitalier Donka, CHU Donka, BP 5680, Conakry, Guinée*

[3] *Institute of Virology, Robert-Koch-Str. 17, 35037 Marburg, Germany*

[4] *Institut Pasteur, BP490, 01 Abidjan, Côte d'Ivoire*

* *E-mail: denys@mnhn.fr*

Abstract: The Murid rodent diversity has been sampled following the 9th Meridian in the east of Guinea from the forest region to Sudanian savannas of southern Mali. This represents the first small mammals survey in North Guinea. Murid diversity patterns have been researched using correspondence analysis and faunal comparisons. A difference between southern forest communities and northern ones is observed. *Mastomys natalensis* is found only in houses in southern Guinea while in the north it is found in all sampled habitats. *M. erythroleucus* is absent from the forest zone, but found from the ecotone forest-savanna to the north, and it seems that this species never enter houses. Implications for Lassa fever transmission are discussed.

Key words: Rodents; *Mastomys*; Lassa; Forest Savanna; Guinea

B.A. Huber et al. (eds.), African Biodiversity, 339–350.

1. INTRODUCTION

Forests of the Guinean block represent one hotspot of biodiversity (Mayr and O'Hara, 1986), but like elsewhere deforestation and anthropogenic pressure may have important consequences both on diversity and disease emergences. Several Lassa fever outbreaks have been recently detected in Guinea (Birmingham and Kenyon, 2001; Richmond and Baglole, 2003) and the multimammate rat, *Mastomys*, has been identified as a natural host of Lassa virus. In Guinea, only forest zones have been explored to date for rodent systematics (Roche, 1971; Heim de Balsac and Lamotte, 1958; Ziegler et al., 2002). None of those studies used recent tools of systematics or surveyed areas affected by anthropogenic factors, whereas in Guinea, the commensal genus *Mastomys*, is potentially represented by three, probably sympatric, cryptic species (see Granjon et al., 1997), whose exact distribution and state of commensalism are still unknown. The exact Lassa virus reservoir species in Guinea is not yet clearly identified (Monath et al., 1974; Demby et al., 2001). Moreover, Lassa fever seems to occur all over the territory but human prevalence varies geographically. In southern Guinea, human seroprevalence may reach 40%, while in the North and coastal region prevalence is lower (from 3 to 5%) (Lukashevich et al., 1993; ter Meulen et al., 1996; Baush et al., 2001). According to White (1986) three different vegetation zones cross Guinea from South to North along a decreasing rainfall gradient. In order to see if *Mastomys* species and murid communities differ from South to North and try to explain the differences in Lassa prevalence, we have undertaken a trapping survey during the dry

Table 1. Geographic coordinates and main vegetation types (after White et al., 1986) of the Guinean localities sampled

Localities	Coordinates	Main vegetation	White zone
Bhoita	N08°05'13 - W08°54'5	dense forest with plantations of utile trees (Coffee-Cacao-Cola-Palm trees)	Ombrophyle forest
Sangassou	N08°36'49 - W10°58'22	dense forest with plantations of utile trees (Coffee-Cacao-Cola-Palm trees)	Ombrophyle forest
Macenta	N08°33'47 - W09°29'20	gallery forest close to plantations	Ombrophyle forest
Maikou	N09°02'04 - W09°02'44	woodland savanna	Transition zone
Franfina	N09°38'04 - W08°56'37	Combretum-Terminalia Woodland	Transition zone
Kodoko	N10°31'14 - W08°58'58	Combretum-Terminalia Woodland	Sudanian
Saourou	N11°30'41 - W09°00'48	Combretum-Terminalia Woodland	Sudanian
Massakoroma	N12°16'15 - W08°45'35	woodland savanna	Sudanian

Table 2. Details of trapping efforts and all taxa success along each village's habitats. TOT = Total number of night traps in each locality

Village	Habitat	Night traps	Trap success
Bhoita	*Houses*	150	16,00%
	Cultures	375	2,70%
TOT: 660	*Forest*	135	16,00%
Macenta	*Cultures*	25	12,0%
TOT: 45	*Forest*	20	55,0%
Maikou	*Houses*	180	0,00%
	Cultures	200	1,00%
	Gallery forest	320	9,38%
TOT : 738	*Woodland*	38	0 %
Franfina	*Houses*	240	29,60%
	Cultures	560	0,71%
	Gallery forest	300	8,00%
TOT: 1330	*Woodland*	230	5,22%
Kodoko	*Houses*	70	14,30%
	Cultures	140	5,00%
	Gallery forest	100	18,10%
TOT: 410	*Woodland*	100	9,00%
Saourou	*Houses*	200	14,50%
	Cultures	200	4,00%
	Gallery forest	200	8,50%
TOT: 800	*Woodland*	200	2,00%
Massakoroma	*Houses*	160	18,75%
	Cultures	320	6,25%
	Gallery forest	100	26,00%
TOT: 640	*Woodland*	160	12,50%

season following a transect along the 9th meridian from the forest block in the south of Guinea, through the ecotone forest - savannas to the south Mali in the Sudanian environment. This aims to better understand the distribution of murid communities along a rainfall and vegetation gradient and by focusing on *Mastomys* species distribution, bring new hypotheses to aid in the understanding of Lassa fever geographical prevalence differences.

2. MATERIALS AND METHODS

Following the 9th meridian, we performed during one month (February 2004) trapping sessions in eight localities separated by about 100 kms from each others (Table 1). Due to logistic problems relative to such a type of survey and in order to catch as many rodents as possible, the number of night traps is variable and the field procedure is not standardized. Trapping success for rodent captures only and trapping details are provided in Table 2 and 3. The only constraint was to sample the same types of habitats in each

Table 3. Rodent faunal list along the transect (Sa = Sangassou, Bh = Bhoita, Mac = Macenta, Mk = Maikou, Fra = Franfina, Kdk = Kodoko, Sao = Saourou, Msk = Massakoroma). Mean trapping success (including other taxa than rodents): 9,91%. Murid Trapping success is the number of individuals of all murid species trapped in each locality divided by the trap-nights. S-W = Shannon-Wiener index calculated here from murid species proportions (P) using the classical formula: H = -SUM(Pi(ln(Pi)), P(i) = number of individuals for each species / total number of individuals per locality

Murid Species	**Sa**	**Bh**	**Mac**	**Mk**	**Fra**	**Kdk**	**Sao**	**Msk**	**P**
Tatera guineae					2		2		4
Thamnomys/Grammomys butingi					1				1
Dasymys rufulus						1			1
Lemniscomys zebra					1	1		4	6
Lemniscomys linulus						2		4	6
Hybomys trivirgatus	1								1
Praomys tullbergi/rostratus	11	7	2	27	19	16	1	17	100
Hylomyscus simus	5								5
Mastomys erythroleucus					1	2	6	23	32
Mastomys natalensis	14	24			74	12	34	38	196
Nannomys minutoides/musculoides		5	1		2	9			17
Nannomys setulosus	6	7	4	2	4				23
Lophuromys sikapusi	3	9	7	5	8				31
Arvicanthis ansorgei								1	1
Taterillus gracilis								5	5
Myomys daltoni				1		3	17	10	31
Other small mammals	8	6	1	6	8	7	2	2	40
Total no. of murid individuals	40	52	14	35	112	46	60	102	460
Murid Diversity (no. of species)	6	5	4	4	9	8	5	8	16
Murid Success %	6,4	7,9	31,1	4,7	8,4	11,2	7,5	15,9	11,7
Shannon-Weaver index	1,55	1,43	1,17	0,74	1,15	1,65	1,09	1,75	1,82

locality like: 1- the village (i.e. houses and reserves), 2- the surrounding cultivated areas, 3- the surrounding woodland savanna, 4- the close gallery-forest or rainforest. Results of rodent trapping are presented here by localities pooled for a broad biogeographic description of murid community patterns and by similar habitats in order to detect compositional changes in murid communities along the transect. The number of traps put in houses depends on the size of the village and of the first night trapping success.The 9th meridian has been chosen because it covers most of the vegetation zones of Guinea and starts in forest, a high Lassa prevalence zone. Species diversity was calculated using the Shannon-Weaver index. Trapped rodents were handled according to published safety and ethic standard rules (Mills et al., 1995), using P3-security masks and gloves. We used a molecular test based on polymorphism in the Cytochrome b gene developed by Lecompte et al. (subm) to assign individuals to the *Mastomys* species. This molecular test as well as sequencing were used to genotype all the *Mastomys* specimens to provide reliable determinations. Other species were identified by classical morphometrics on external measurements and skulls, and the keys of Rosevear (1969) were used. The nomenclature follows Wilson and Reeder (1993). Correspondence analysis (AFC) (Benzecri and Benzecri, 1984) was performed on the contingency tables based on species frequencies per sampled localities using XLSTAT (version 5.1). Using the chi-square distance, this method allows the illustration of the relationship between variables and individuals on the same graph.

3. RESULTS

Mastomys, with 214 individuals, represented 45,78% of the small mammals trapped. *Praomys* (20,08%) was the second most numerous, followed by *Nannomys* (8,02%) and *Lophuromys* (7,02%). The absence of cytogenetical analysis prevented determinations such as sibling species taxonomy, but based on some molecular taxonomy data we can produce a provisional faunal list of 13 rodent genera and 16 species (Table 3). The Shannon-Weaver index is lowest in Maikou and highest in Massakoroma, with no real trend from south to north (Table 1).

Mastomys natalensis is the dominant species in all localities with 10 to 55% contribution except Maikou and Macenta (Table 3). This was the only species trapped inside houses. In the forest region *M. natalensis* was found only inside houses, while in savannas it occurs both in houses and cultivated zones, and was also regularly trapped in all the other sampled habitats. On the contrary, *M. erythroleucus* was never trapped inside houses or in forest

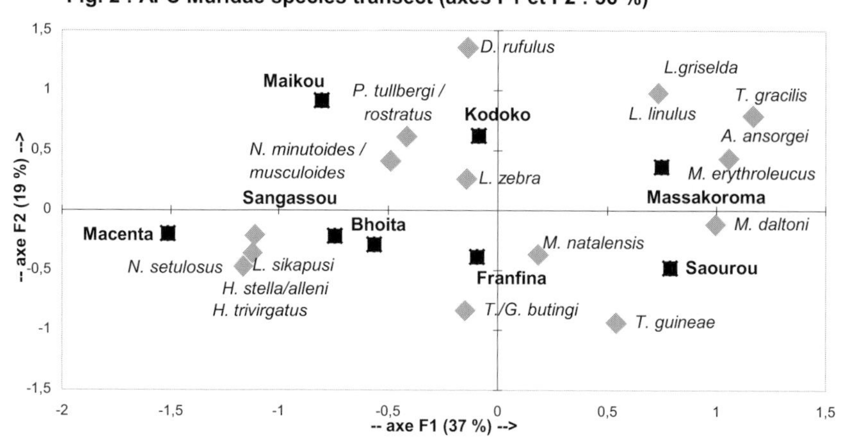

Figure 1. AFC graph of axis 1 and 2 (56%) of the transect Guinea Murid species (8 variables= transect localities, 16 individuals= species), based on Table 2 data.

villages (Table 3) but did occur in cultivated zones or woodlands. While two *Nannomys* species were trapped in Guinea forests, only one was found in Kodoko and none in the northern localities (Saourou and Massakoroma). On the contrary, the southern localities did not provide any *Lemniscomys*, whereas two species were trapped in the northern ones. The northern Guinean locality of Saourou lacks *Arvicanthis* and *Taterillus*.

The correspondence analysis graph (Fig. 1) displays 56% of total variability and a clear differentiation of three faunal groups along axis 1. On the left of the graph, the forest faunas are characterized by *Nannomys, Hybomys, Hylomyscus, Lophuromys* and *Praomys* (Macenta, Bhoita, Sangassou, Maikou). The savanna faunas in the right of the graph, comprise *Myomys, Arvicanthis, Taterillus, Lemniscomys* and *Tatera* (Saourou and Massakoroma). The intermediate localities of Franfina and Kodoko are placed in an intermediate position. Maikou has a strong contribution to axes 1 and 2 due to the absence of *Mastomys*. Figure 1 shows that the presence of *M.erythroleucus* characterizes northern localities while *M. natalensis* is close to the centre of gravity of the graph due to its presence in nearly all localities.

The woodland habitat displays an interesting contrast between southern and northern regions (Fig. 2). In the south, *Lophuromys* and *Nannomys* dominate, in the North *Lemniscomys* and Gerbillidae (mainly *Taterillus)* are more abundant. This habitat has the greatest diversity compared to others but it is interesting to notice that each site has its "dominant" genus: *Lophuromys* in Franfina, *Nannomys* in Kodoko, *Tatera* in Saourou and *Lemniscomys* in Massakoroma. In the same way, *Nannomys, Praomys* and *Lophuromys* dominate in the cultivated zones in the South, whereas *Myomys* and

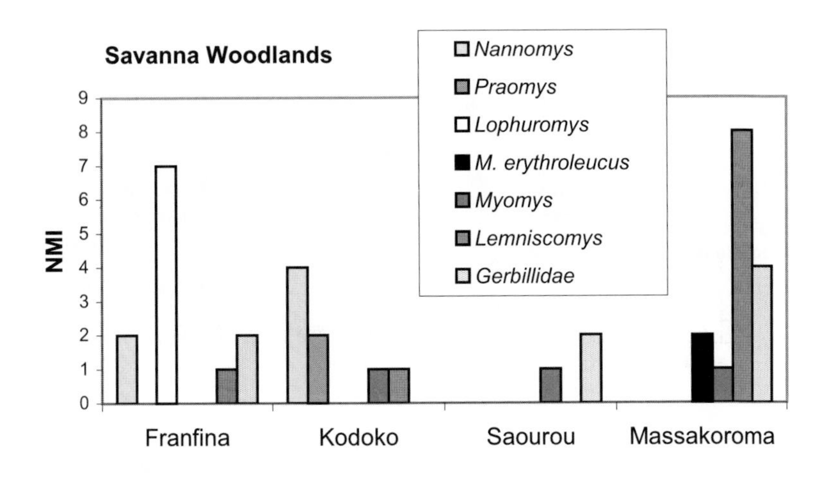

Figure 2. Frequency distribution of the most abundant murid genera in woodland savanna. *Tatera* and *Taterillus* have been grouped together as representatives of Gerbillinae.

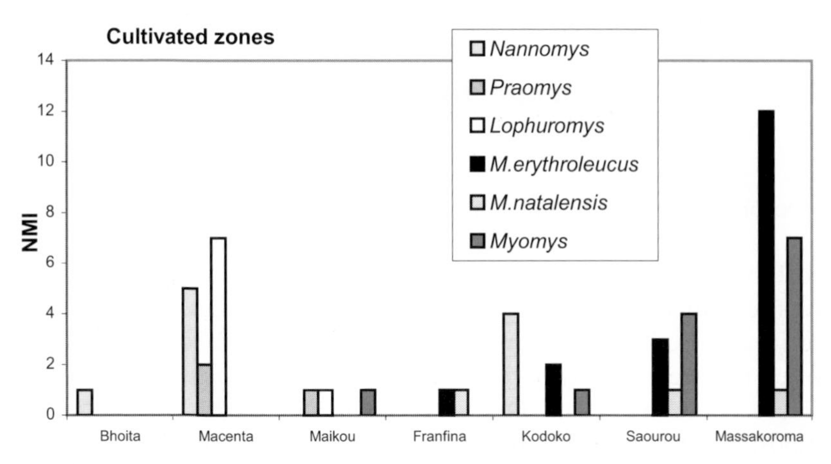

Figure 3. Frequency distribution of *Mastomys* spp. and other murid species in cultivated areas.

Mastomys are predominant in the North (Fig. 3). In this habitat, neither *Tatera, Taterillus* nor *Lemniscomys* were trapped. At all latitudes, *Praomys* is dominant in gallery forests, except in Bhoita where it is replaced by *Nannomys* and *Lophuromys* and in Saourou where *Myomys* is the most abundant (Fig. 4). Gallery forests also yielded *Graphiurus* in Maikou, *Grammomys* in Franfina and *Lemniscomys* in Kodoko.

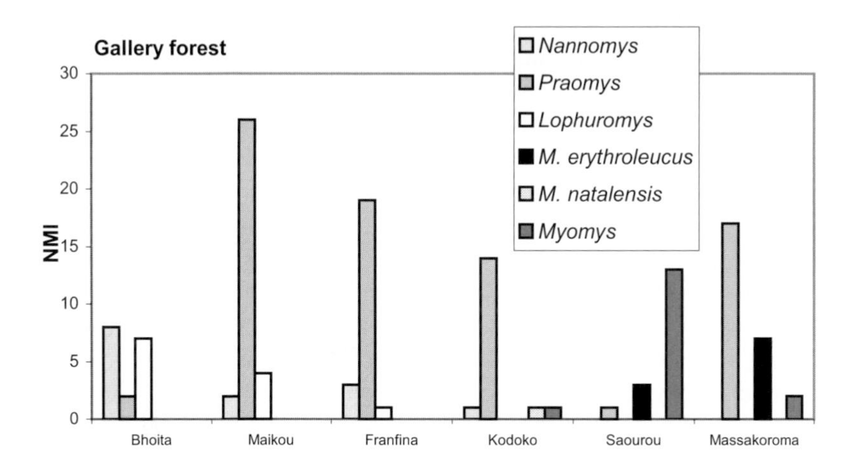

Figure 4. Frequency distribution of *Mastomys* spp. and other main murid species in gallery forests or forests close to villages sampled.

4. DISCUSSION

4.1 Distribution and diversity of rodents along the gradient

Despite rather comparable habitats, a strong contrast is shown between forest and savanna murid communities along the transect. In the geographically intermediate localities of the transition vegetation zone (Table 3), diversity composition but not richness differ from that of the ombrophile forest and of Sudanian savannas, but this faunal change is due to progressive species replacement. Also, *Nannomys* and *Lophuromys* of the South are replaced by *Myomys* and *Lemniscomys* in the North. Such a contrast in composition was already observed in Nigeria (Happold, 1987) and Benin (De Visser et al., 2001). From south to north, the forest-galleries communities change but, in contrast to *Lophuromys*, *Praomys* is found everywhere. These gallery forests have a very low extension (less than 100 m wide on each side of the rivers) and are not connected together. They therefore constitute refuge islands of strong conservation importance (Happold, 1986).

4.2 Distribution of the two *Mastomys* species in Guinea and Lassa fever implications

This study also provides more precision on *M. natalensis* and *M. erythroleucus* distribution and habitat preferences during the dry season in Guinea. According to the literature, *Mastomys* is not a typical inhabitant of forest zones (Dieterlen, 1989) and is catalogued as a savanna species which occurs in savanna-like habitats within rainforest (Happold, 1987). In the southern localities, it was trapped only in villages, confirming its close relationships with man. As in Senegal (Duplantier and Granjon, 1988) *M. natalensis* was trapped inside houses in most villages and *M. erythroleucus* (present only in northern Guinea) preferred cultivated areas. However, in northern Guinea, *M. natalensis* was not strictly commensal, similar to what was found in Tanzania (Leirs, 1995). Moreover, two forms corresponding probably to *M. erythroleucus* and *M. natalensis* were identified in houses in Sierra Leone (McCormick et al., 1987). The literature indicates the presence of *M. erythroleucus*, rather than *M. natalensis*, in the forest zone (Roche, 1971; Adam, 1977), but none of these specimens have been karyotyped or genotyped and misidentification is probable pending an attentive reexamination of these specimens. Moreover, while the presence of *M. erythroleucus* in houses is documented in Senegal and Chad (Duplantier and Granjon, 1988; Granjon et al., 2004) and in Guinea (Kindia prefecture in savanna zone: Calvet, pers. comm.), its absence from southern Guinea during the transect may result from a low sampling or a seasonality effect. Nevertheless, it is probable that *M. natalensis*, the most anthropophilic species, is able to follow humans into houses but not to colonise the surrounding true forest biotopes. The absence of *Mastomys* in Macenta is due to a trapping bias with only two habitats sampled for a night using a low number of traps. In a previous study in the Seredou and Macenta region, Roche (1971) found *Mastomys* in abundance. The absence of *Mastomys* in the Maikou village is quite surprising. Maikou was the smallest village sampled, situated in the Milo river floodplain. An explanation of the *Mastomys* absence could be that, after the flood season, *Mastomys* may not have had time to recolonize the village, or the predation pressure by domestic cats, which were very abundant, has removed rodents.

Because *M. natalensis*, the supposed Lassa virus host, is present in nearly all transect localities, its distribution cannot explain why southern Guinea (Forest region) has a higher human Lassa Fever prevalence than northern Guinea. Our data suggest that the presence of both *M. natalensis* and *M. erythroleucus* in the North may play a role in maintaining lower Lassa prevalence. In forests, *M. natalensis* seems unable (at least in the dry season) to occur outside the village, while in savanna it is found in fields, and even in gallery forests. In south Guinea, the competition with forest species may restrict *M. natalensis* to houses and makes it more sensitive to the virus. In village houses surrounded by forest, *M. natalensis* populations may be concentrated and fragmented, with low inter-population gene exchange reducing resistance, which may favour the virus transmission within rodent families (vertical transmission) or between non related specimens (horizontal transmission) and increase the chances of human contact with urine of rodents and aerosols. This may be at the origin of the higher prevalence observed in forest Guinea. The phylogeography of *M. natalensis* (E.L., in progress) and population genetics may provide a better understanding of this higher prevalence in Guinean forests. Another explanation could be cultural differences: in the southern forest Guinea bush-meat is highly consumed and may be a supplementary factor of prevalence (ter Meulen et al., 1996). However, such rodent consumption is not only a forest behaviour and has been found in savannas zones of Benin (de Visser et al., 2001), Chad and Malawi (pers. obs.). In the North, the more open environments and larger size of cultivated zones provide more feeding possibilities and then allows *M. natalensis* and *M. erythroleucus* to enter in contact outside houses. This may provide the opportunity for the virus to switch from *M. natalensis* to *M. erythroleucus*. This possible dispersion of the virus in a wide population, with loss of a part of the vertical transmission and the higher genetic diversity of populations may decrease the prevalence.

5. CONCLUSION

The results of this first transect along a south-north gradient in Guinea highlights an important difference between forested and savanna regions. Moreover, it appears that, at least in the dry season, *M. erythroleucus* is absent from the forest zone where *M. natalensis* is restricted to houses. In the North, *M. erythroleucus* does not enter houses while *M. natalensis* is also found in all surrounding environments. The absence of *M. erythroleucus* and the confinement of *M. natalensis* to homes, as well as the fragmented nature of its populations in forests may explain the higher Lassa prevalence in southern Guinea. This hypothesis should be tested by further trapping, in different seasons, both in degraded and undisturbed environments all over Guinea. Further investigations in South Guinea are necessary to improve our knowledge of the distribution of the two *Mastomys* species and their relationships with humans. Moreover, investigations on the vector capacity of these two species, i.e. the capacity to act as a vector for Lassa virus, exploring the presence and quantity of Lassa virus in the different excretions (as urine and faeces) and on the transmission modes are also needed. Also in Guinea, similar situations can be present between coastal region and the Fouta Djalon Plateau that will be investigated soon. Lastly, this study stresses the importance of a correct taxonomic analysis, using cytogenetic and molecular techniques, for ecological and virological studies.

ACKNOWLEDGMENTS

This work has been funded by CEE project INCO-DEV ICA4-CT2002-10050 and by a Marie Curie Intra-European Fellowships within the 6th European Community Framework Programe - Guinean Ministry of Health: CHU Donka (Conakry) and by the French CNRS (FRE2695). Many thanks to R. Cornette, C. Houssin, A. Delaprée, E. Gros (S.P.O.T.) for taxidermy preparation, thanks to J. Cuisin and A. Bens for collections management. Thanks to L. Granjon and Karima Mouline (IRD, Mali). We are grateful to two anonymous referees whose comments were very helpful and to B. Huber for editorial assistance.

REFERENCES

Adam, F., 1977, Données préliminaires sur l'habitat et la stratification des rongeurs en forêt de Basse Côte d'Ivoire., *Mammalia*, **41**:283-290.

Bausch, D. G., Demby, A. H., Coulibaly, M., Kanu, J., Goba, A., Bah, A., Condé, N., Wurtzel, H. L., Cavallaro, K. F., Lloyd, E., Binta Baldet, F., Cissé, S. D., Fofona, D.,

Savané, I. K., Tamba Tolno, R., Mahy, B., Wagoner, K. D., Ksiazek, T. G., Peters, C. J., and Rollin, P. E., 2001, Lassa fever in Guinea: I. Epidemiology of human disease and clinical observations, *Vector-Borne Zoonot.* **1**:269-281.

Benzecri, J. P, and Benzecri, F., 1984, *Pratique de l'analyse des données (tome 1),* Bordas Ed., Paris, pp. 456.

Birmingham, K., and Kenyon, G., 2001, Lassa fever is unheralded problem in West Africa, *Nat. Med.* **7**:878.

Demby, A. H., Inapogui, A., Kargbo, K., Koninga, J., Kourouma, K., Kanu, J., Coulibaly, M., Wagoner, K. D., Ksiazek, T. G., Peters, C. J., Rollin, P. E., and Bausch, D. G., 2001, Lassa fever in Guinea: II. Distribution and prevalence of Lassa virus infection in small mammals, *Vector-Borne Zoonot*, **1**:283-296.

De Visser, J., Mensah, G. A., Codjia, J.T.C., and Bokonon-Ganta, A. H., 2001, *Guide préliminaire de reconnaissance des rongeurs du Benin*, Coco Multimedia, Cotonou, Benin, pp. 252.

Dieterlen, F., 1989, Chap.21 – Rodents, in: *Tropical rainforest ecosystems, biogeographical and ecological studies*, M. Leith, and M. J. A. Werger, eds., Elsevier, Amsterdam, Oxford, New York, Tokyo, pp. 383-400.

Duplantier, J.. M., and Granjon, L., 1988, Occupation et utilisation de l'espace par des populations du genre *Mastomys* au Sénégal: étude à trois niveaux de perception, *Sci. Techn. Anim. Lab.*, **13**:129-133.

Granjon, L., Duplantier, J.. M., Catalan, J., and Britton-Davidian, J., 1997, Systematics of the genus *Mastomys* (Thomas, 1915) (Rodentia: Muridae), A review, *Belg. J. Zool.*, **127**(1):7-18.

Granjon, L., Houssin, C., Lecompte, E., Angaya, M., Cesar, J., Cornette, R., Dobigny, G., and Denys, C., 2004, Community ecology of the terrestrial small mammals of Zakouma National Park, Chad, *Acta Theriol.*, **49**(2):215-235.

Happold, D.C.D., 1987, *The mammals of Nigeria*. Clarendon Press, Oxford, pp. 550.

Happold, D.C.D., 1996, Mammals of the Guinea-Congo rainforest, *Proc. R. Soc. Edinburgh*, **104B**:243-284.

Heim de Balsac, H., and Lamotte, M., 1958, La réserve intégrale du Mont Nimba.XV: mammifères rongeurs (Muscardinides et Murides), *Mém. I.F.A.N.* **53**:339-57.

Lecompte, E., Granjon, L., Kerbis-Peterhans, J., and Denys, C., 2002, Cytochrome b-based phylogeny of the *Praomys* group (Rodentia, Murinae): a new African radiation? *C. R. Biologies*, **325**:827-840.

Leirs, H., 1995, Population ecology of *Mastomys natalensis*: implications for rodent control in Africa, in: *Report from the Tanzania-belgium Joint Research Project, BADC, Agricultural editions*, **35**.

Lukashevich, L. S., Clegg, J. S., Sidibe, K., 1993, Lassa virus activity in Guinea: distribution of human antiviral antibody defined using enzyme-linked immunosorbent assay with recombinant antigen, *J. Med. Virol.*, **40**:210-217.

Mayr, E., O'Hara, R., 1986, The biogeographic evidence supporting the pleistocene forest refuge hypothesis, *Evolution*, **40**:55-67.

McCormick, J. B., Webb, P. A., Krebs, J. W., Johnson, K. M., and Smith, E. S., 1987, A prospective Study of the Epidemiology and Ecology of Lassa Fever, *J. Infect. Dis.*, **155**(3):437-444.

Meinig, H., 2000, Notes on the mammal fauna of southern part of the Republic of Mali, west Africa, *Bonn zool. Beitr.*, **49**:101-114.

Mills, J. N., Childs, J., Ksiazek, T. G., Peters, C. J., and Velleca, W. N., 1995, *Methods for trapping and sampling small mammals for virologic testing,* Centers for Disease Control and Prevention. Atlanta, Georgia, pp. 61.

Monath, T. P., Newhouse, V. F., Kemp, G. E., Stzer, H., and Cacciapuoti, A., 1974, Lassa virus isolation from *Mastomys natalensis* rodents during an epidemic in Sierra Leone, *Science*, **185**:263-265.

Richmond, J. K., and Baglole, D. J., 2003, Lassa fever:epidemiology,clinical features,and social consequences, *BMJ,* **327**(7426):1271-1275.

Roche, J., 1971, Recherches mammalogiques en Guinée forestière, *Bull. Mus.Nat. Hist. Nat., Zoologie,* **16**(3):737-782.

Rosevear, D. R., 1969, *The rodents of West Africa.* British Museum of Natural History, London, pp. 604.

ter Meulen, J., Lukashevich, I., Sidibe, K., Inapogui, A., Marx, M., Dorlemann, A., Yansane, M. L., Koulemou, K., Chang-Claude, J., and Schmitz, H., 1996, Hunting of peridomestic rodents and consumption of their meat as possible risk factors for rodent-to-human transmission of Lassa virus in the Republic of Guinea, *American J. Trop. Med. Hyg.*, **55:**661-666.

White, F., 1986, *La végétation de l'Afrique. Mémoire accompagnant la carte de végétation de l'Afrique* Unesco/AETFAT/UNSO. Orstom – Unesco eds., pp. 1- 384.

Wilson, D. E., and Reeder, D. M., 1993, *Mammals species of the world: a taxonomic and geographic reference*, Smithsonian Inst. Press, Washington, pp. 1207.

Ziegler, S., Nikolaus, G., and Hutterer, R., 2002, High mammalian diversity in the newly established National Park of Upper Niger, Republic of Guinea, *Oryx*, **36:**73-80.

EFFECTS OF DIFFERENT LAND USE ON THE PARASITE BURDEN AND MHC CONSTITUTION OF TWO RODENT SPECIES (*GERBILLURUS PAEBA*, *RHABDOMYS PUMILIO*) IN THE SOUTHERN KALAHARI DESERT

Rainer Harf, Götz Froeschke, and Simone Sommer*
Dept. Animal Ecology and Conservation, University of Hamburg, Martin-Luther-King-Platz 3, 20146 Hamburg, Germany
* *corresponding author; E-mail: Simone.Sommer@zoologie.uni-hamburg.de*

Abstract: Human impact causes land degradation with negative effects on both density and genetic variability of animal populations. Recent studies indicate that host genetic diversity plays an important role in buffering populations against widespread epidemics. Within the mammalian immune system the highly variable major histocompatibility complex (MHC) plays a central role in pathogen defence. In this study we used two common African rodent species, the hairy-footed gerbil (*Gerbillurus paeba*) and the striped mouse (*Rhabdomys pumilio*), as model organisms to investigate effects of different land use on (1) vegetation structure and population density, (2) parasite burden and (3) the importance of the MHC-constitution for resistance to parasites by comparing two sites on a private nature reserve and one site on a commercially used farm in the Southern Kalahari/South Africa. Both rodents were significantly more abundant and showed more helminths infections in the two less degraded nature reserve sites, although the MHC diversity was not significantly different between sites. Either higher parasite species richness or the increased host population density in the more diverse nature reserve sites might be responsible for the higher parasite prevalence. In both species significant associations were found between specific alleles and infection status, giving evidence for the significance of the MHC in the defence of pathogens.

Key words: Habitat degradation; immune gene variability (MHC); parasite load; *Gerbillurus paeba; Rhabdomys pumilio*; South Africa

B.A. Huber et al. (eds.), African Biodiversity, 351–361.

1. INTRODUCTION

In Southern Africa anthropogenic impacts have lead to severe environmental changes due to increasing human population and extensive land use (Jeltsch et al., 1997; Seeley, 1998). Selective grazing and high stocking rates of domestic animals have marked effects on the soil and vegetation structures with shrub encroachment as a typical sign of degradation in arid savannahs (Barnard, 1998; Jeltsch et al., 2000). Due to habitat destruction, degradation and anthropogenically induced vegetation changes a lot of animal populations are likely to decrease in size and density. Small populations often suffer from reduction of genetic diversity due to genetic drift and inbreeding effects (Lande, 1994; Dudash and Fenster, 2000; Keller and Waller, 2002). An increasing number of studies indicate that host genetic diversity plays an important role in buffering populations against widespread epidemics (Altizer et al., 2003). The loss of genetic variation can lead to short-term reduction of the immunocompetence and therefore the ability to face pathogens (Bernatchez and Landry, 2003).

Within the mammalian immune system the most important genetic component is a group of closely linked genes known as the Major Histocompatibility Complex (MHC) (Klein, 1986). MHC-encoded glycoproteins bind short peptides from pathogens and present them to T-lymphocytes thereby triggering the appropriate immune responses (Janeway and Travers, 2002). Certain genes within the MHC constitute the most polymorphic loci known in vertebrates. Pathogen-driven selection processes are thought to be involved in the maintenance of diversity at MHC loci (Jeffery and Bangham, 2000; Penn, 2002; Bernatchez and Landry, 2003). Up to now, however, most of the studies investigating the selection processes and role of MHC variability and parasite resistance have been conducted in humans or under laboratory conditions. Studies in free-ranging, wild animal populations are still very limited (Bernatchez and Landry, 2003). Furthermore, ecological parameters have rarely been included into studies on parasites and MHC.

In this study, we used two widely distributed and abundant African rodent species, the hairy-footed gerbil (*Gerbillurus paeba*) and the striped mouse (*Rhabdomys pumilio*), as model organisms to test the hypothesis that populations would suffer from habitat degradation leading to lower immune gene diversity and thus higher parasite prevalence. We investigated effects of different land use on (1) vegetation structure and population density, (2) parasite burden and (3) the importance of MHC-constitution for resistance to parasites.

2. MATERIALS AND METHODS

2.1 Study areas and sample collection

The study took place in June 2002 in the "Southern Dunefield" in South Africa (26°45'39.3"S, 20°36'50.5"E), the driest southwestern part (mean annual precipitation of 200-300 mm) of the Kalahari Desert, approximately 30 km south of the Kgalagadi Transfrontier Park. Animals were live-trapped with Sherman-traps® at three study sites using permanent grids with a 20 m-interval of 140 traps. Two study sites (A and B) were located on a private nature reserve, whereas site C was placed on a commercially used farm. Trapped animals were individually marked, the usual body measurements, stool and tissue samples were taken. The animals were released at their respective trapping sites. In order to get information about habitat degradation, vegetation surveys were conducted around every trap within a radius of 3 m.

2.2 Parasitic screening

For the identification of parasite species a modification of the McMaster method was used (Sloss et al., 1994). This technique is noninvasive and considered as an appropriate method to quantify nematode eggs and thus to evaluate worm burdens in mammals (Paterson et al., 1998; Cassinello et al., 2001). Each faeces sample was homogenized in a flotation-dilution of potassium iodide with a specific density of 1.5 g/ml (Meyer-Lucht, unpublished data). 97 stool samples of *G. paeba* and 76 stool samples of *R. pumilio* were screened for nematode eggs by counting two chambers of McMaster.

2.3 Molecular techniques

Forty individuals of *G. paeba* and 58 individuals of *R. pumilio* were MHC-genotyped. We examined variations of the highly polymorphic 171 bp fragment of the functional important exon 2 of the MHC class II gene DRB coding for the antigen-binding sites. PCR amplification of this region was accomplished using the primer pair JS1 and JS2 (Schad et al., 2004; in press). All identified alleles were sequenced bi-directionally. Details on the molecular techniques are outlined in Sommer et al. (2002) and Sommer (2003).

2.4 Statistical treatment

Expected heterozygosity (gene diversity, Nei, 1987) was used as a measure of genetic variation within each population. For parasite analyses, odds ratio tests were conducted to assess the relative risk of being infected. It is a common test in epidemiological studies to evaluate the exposition of individuals carrying a risk factor. The ratio of the odds of an event occurring in one group is compared to the odds of it occurring in another group by using a 2×2 crossclassification table (Sachs, 1992). All additional statistical tests were performed by using SPSS version 9.0. Calculations are two-tailed and based on a significance level of $\alpha = 0.05$. Bonferroni corrected significance levels were used for multiple comparisons and calculations of pairwise differences (Rice, 1989; Sachs, 1992).

3. RESULTS

3.1 Vegetation and trapping success

The vegetation surveys revealed 39 different plant species on the three study sites (Appendix 1). Sites A and B showed a higher number of plant species than site C (Fig. 1a). Whereas in the private nature reserve 27 (site A) and 31 (site B) species were found, only 12 different plant species were present on the commercially used farm (site C). On sites A and B eight grass

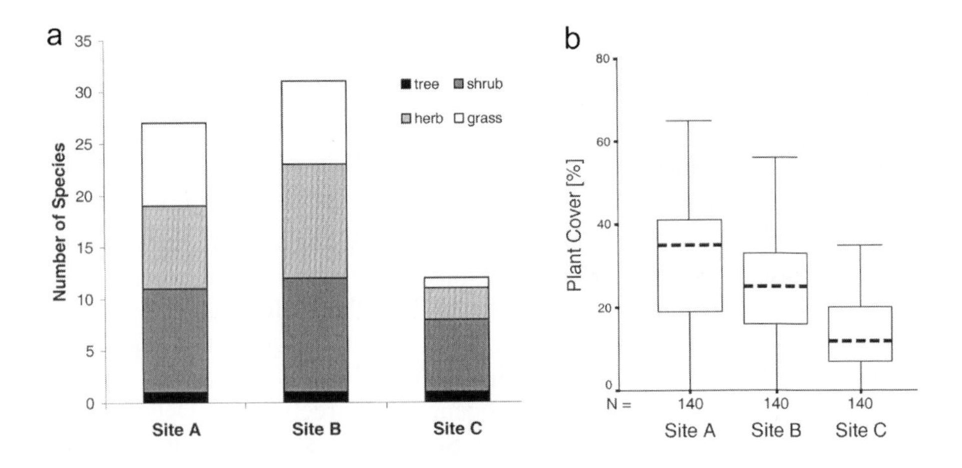

Figure 1. (a) Number of plant species in different growth form-categories and (b) mean plant cover (%) on the three study sites A and B (private nature reserve) and C (commercially used farm). The boxplots indicate medians, quartiles, minima and maxima. N = number of vegetation surveys.

species could be identified, whereas on site C only one grass species (*Stipagrostis amabilis*) was present. Shrubs were the most abundant growth form on site C. The three study sites also differed significantly in their plant cover, which had the lowest value on site C (Kruskal Wallis-test: $\chi^2 = 76.1$; df= 2, p< 0.0001; pairwise MWU-tests: sites A-B: Z= -4.11, p< 0.0001; sites B-C: Z= -6.26, p< 0.0001; sites A-C: z= -7.76, p< 0.0001, all Bonferroni significant) (Fig. 1b).

Further, the three sites revealed significant differences in the number of trapped individuals (*G. paeba*: χ^2= 13.21, df= 2, p= 0.001; *R. pumilio*: χ^2= 9.08, df= 2, p= 0.01). In both species no significant difference in the trapping success was found between the sites A and B, whereas the trapping success was in both species significantly lower on site C (*G. paeba*: site A-C: χ^2= 12.85, p= 0.0003; site B-C: χ^2= 5.99, p= 0.013; *R. pumilio*: site A-C: χ^2= 4.92, p= 0.024; site B-C: χ^2= 8.45, p= 0.0035; all Bonferroni significant) (Fig. 2a).

Differences in the trapping rate can be explained by the plant cover. In both species, the mean plant cover in a 3 m-radius around traps where individuals were caught was significantly higher compared to traps where no individual was captured (MWU-test: *G. paeba*: Z= -5,45; p< 0,0001; *R. pumilio*: Z= -4.47; p< 0,0001) (Fig. 2b).

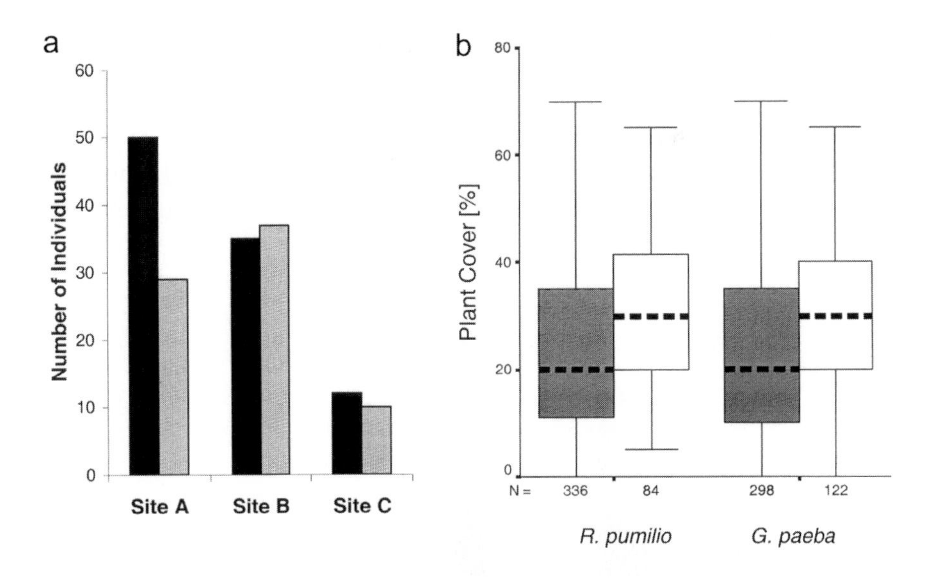

Figure 2. (a) Number of individuals trapped at the three study sites (*G. paeba*: black bars; *R. pumilio*: grey bars) and (b) mean plant cover (%) around traps with no trapping success (grey boxplots) and around those where at least one mouse was trapped (white boxplots). N= number of trap sites.

3.2 Parasite load

The faeces of *R. pumilio* revealed eight nematodes that also occurred in the stool samples of *G. paeba*. Significant differences existed in the number of nematode-infected individuals per site (*G. paeba*: χ^2= 9.03, p= 0,01; *R. pumilio*: χ^2= 5.72, p = 0,057). No differences existed between sites A and B, whereas the number of infected individuals was lower on site C compared to the two other sites (*G. paeba*: site A-C: χ^2= 7.72, p= 0.0074; site B-C: χ^2= 7.60, p= 0.0078; *R. pumilio*: site A-C: χ^2= 5.65, p= 0.023; all Bonferroni significant) (Fig. 3).

3.3 Parasite load and MHC variability

In both species, the genetic diversity of MHC DRB gene exon 2 measured in the number of alleles and gene diversity (Nei, 1987) was high and did not show marked differences between sites (*G. paeba*: site A: 13 alleles / 0.92 ± 0.02, site B: 11 alleles / 0.73 ± 0.04, site C: 11 alleles / 0.94 ± 0.03, total: 34 alleles / 0.95 ± 0.01; *R. pumilio*: site A: 22 alleles / 0.95 ± 0.01, site B: 22 alleles / 0.96 ± 0.01, site C: 16 alleles / 0.95 ± 0.02, total: 20 alleles / 0.91 ± 0.01).

On the individual level, significant associations were found between specific alleles and the number of infected individuals. In *G. paeba*, allele *Gepa*-DRB*15 was significantly associated with uninfected individuals (χ^2= 5.60, p= 0.04, Bonferroni not significant) (Fig. 4a) whereas in *R. pumilio* allele *Rhpu*-DRB*1 occurred significantly more often in infected individuals (χ^2= 4.75, p= 0.03, Bonferroni not significant) (Fig. 4b). A significant odds-

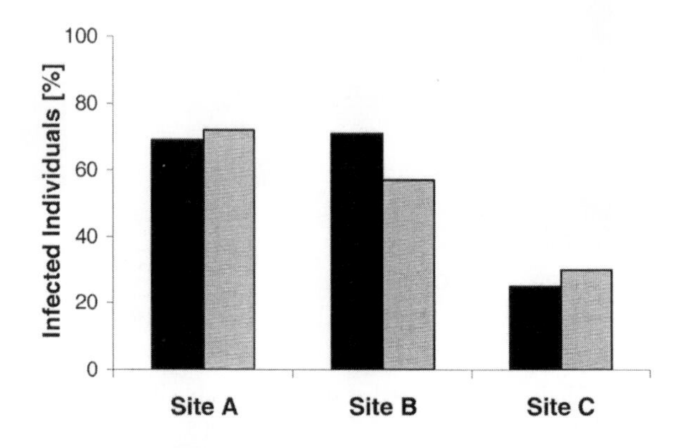

Figure 3. Percentage of infected individuals (*G. paeba:* black bars; *R. pumilio:* grey bars).

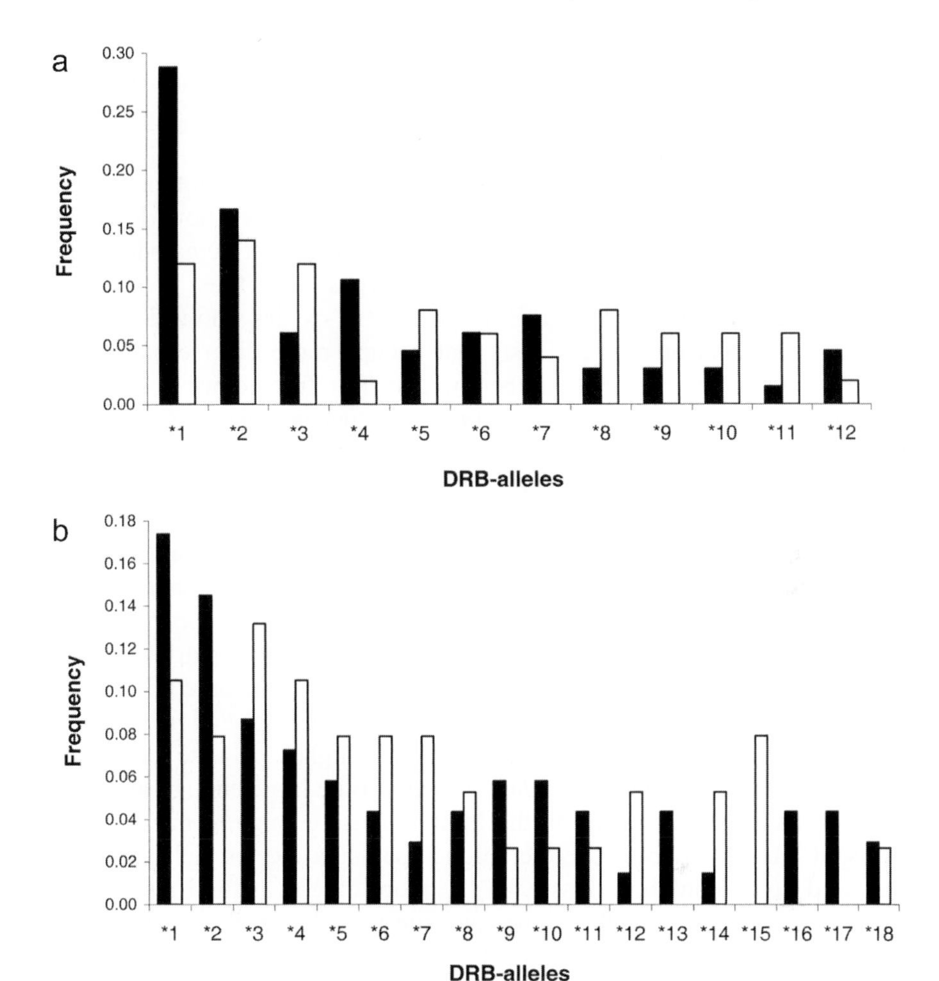

Figure 4. Frequency of alleles in infected (black bars) and not infected (white bars) *R. pumilio* (a) and *G. paeba* (b). Alleles with low prevalence (≤ 2 individuals) are not displayed.

ratio test ($p<0.05$) (Sachs, 1992) showed that individuals carrying allele *Rhpu*-DRB*1 had a 1.5-fold higher chance of belonging to the group of infected individuals than individuals without this allele.

4. DISCUSSION

In contrast to sites A and B on a private nature reserve, site C located on a commercially used farm showed indications of degradation. The reduced number of grass species and the high amount of shrubs are clear signs for bush encroachment due to overgrazing (Palmer and Van Rooyen, 1998). The mean plant cover on site C was significantly lower than on the two other sites and might be an explanation for the lower trapping success of *R. pumilio* and *G. paeba* because both rodent species were significantly associated with plant cover.

The parasite analyses revealed the same nematode morphotypes in both rodents suggesting that these parasite species do not show high host specificities. However, it might be possible that different nematode species might show similar shaped egg morphotypes and thus can not be distinguished. On sites A and B both rodent species showed a higher infection rate than on site C. Hosts that occupy more diverse habitats are likely to encounter a larger number of parasites from other host taxa or environments, leading to an increased parasite species richness (Nunn et al., 2003). Another explanation for the different parasite load might be different host population density which represents a key factor influencing the basic reproductive ratio, R_0, reflecting the ability of any directly transmitted parasite species to spread within a population (Anderson and May, 1991).

With a total number of 20 different MHC-DRB alleles in *R. pumilio* (located on a single gene locus) and 34 different alleles in *G. paeba* (located on two DRB loci) both rodent species show a high MHC variability. The MHC diversity did not differ between sites. Thus, the outcome of this study contradicts the hypothesis that populations in a more impacted habitat might have lower immune gene diversity and thus higher parasite prevalence. Obviously, the investigated populations seem to be large enough and/or the selection mechanisms acting on the MHC intense enough to maintain high levels of MHC variability. Either higher parasite species richness or the increased host population density in the more diverse natural reserve sites might be responsible for the higher parasite prevalence. So far, we cannot control for many unknown factors. Further studies on the effect of habitat degradation on parasite resistance and genetic variability will include more sites in undisturbed and degraded habitats.

The combined data analyses of genetic and parasitic data revealed significant associations between the common allele *Rhpu*-DRB*1 and infected individuals and between the rather rare allele *Gepa*-DRB*15 and not infected individuals. This gives evidence for actual association between certain alleles and resistance to nematode infections in both species. These results are in accord with the 'rare allele advantage hypothesis', one of the most debated MHC-hypotheses, which presumes a coevolutionary arms race between hosts and parasites (Takahata and Nei, 1990; Bernatchez and Landry, 2003). An allele, which is associated with resistance to parasites, will cause an advantage to the host and will spread through the population. This induces a corresponding shift in the genetic composition of the parasite population, the parasites will get adapted to this specific allele and it will become a disadvantage for the host. This might have happened with allele *Rhpu*-DRB*1. In contrary, the allele *Gepa*-DRB*15 might be an example of a new allele that is maintained due to its association with resistance to nematode infections. Our study adds to the small body of empirical evidence of the importance of immune gene variability in ecology and conservation. For further investigation of the selection hypotheses, allele frequencies and parasite burden need to be followed through time to see whether allele frequencies change in a cycling pattern.

ACKNOWLEDGMENTS

We are grateful to J. U. Ganzhorn, N. Jürgens, U. Schmiedel, M. Veste, T. Robinson and A. Rasa for stimulating discussions, logistic support and permission to carry out this project. We thank I. Tomaschewski for helping in the lab, Y. Meyer-Lucht for sharing the parasite analysing method, D. W. Büttner for helpful advice on coprological screening, and an anonymous referee for helpful comments on a previous version. The study has been conducted under accordance of the Northern Cape Nature Conservation Service, Department of Nature and Environmental Conservation (permit no 050/2002). The study was supported by the Hansische Universitätsstiftung and the German Science Foundation (DFG).

REFERENCES

Altizer, S., Harvell, D. and Friedle, E., 2003, Rapid evolutionary dynamics and disease threats to biodiversity, *TREE* **18:**589-596.

Anderson, R. M., and May, R. M., 1991, *Infectious diseases of humans: dynamics and control,* Oxford University Press, Oxford.

Barnard, P. 1998. *Biological diversity in Namibia: a country study*, Namibian National Biodiversity Task Force, Windhoek.

Bernatchez, L. and Landry, C., 2003, MHC studies in nonmodel vertebrates: what have we learned about natural selection in 15 years? *J. Evolution. Biol.* **16:**363-377.

Cassinello, J., Gomendio, M. and Roldan, E. R. S., 2001, Relationship between coefficient of inbreeding and parasite burden in endangered gazelles, *Conserv. Biol.* **15:**1171-1174.

Dudash, M. R. and Fenster, C. B., 2000, Inbreeding and outbreeding depression, in: *Genetics, Demography and Viability of Fragmented Populations*, A. G. Young and G. M. Clarke eds., Cambridge University Press, Cambridge.

Hedrick, P. W., 1999, Balancing selection and MHC, *Genetica* **104:**207-214.

Janeway, C. A. and Travers, P., 2002, *Immunologie*, Spektrum Akademischer Verlag GmbH, Heidelberg, Berlin, Oxford.

Jeffery, K. J. and Bangham, C. R., 2000, Do infectious diseases drive MHC diversity? *Microb Infect* **2:**1335-41.

Jeltsch, F., Milton, S., Dean, W. R. J. and Van Rooyen, N., 1997, Analysing shrub encroachment in the southern Kalahari: a grid-based modelling approach, *J. Appl. Ecol.* **34:**1497-1508.

Jeltsch, F., Weber, G., Paruelo, J., Dean, W. R. J., Milton, S. and Van Rooyen, N., 2000, Beweidung als Degradationsfaktor in ariden und semiariden Weidesystemen, *Brandenburg. Umwelt Berichte* **8:**89-90.

Keller, L. F. and Waller, D. M., 2002, Inbreeding effects in wild populations, *TREE* **17:**230-241.

Klein, J., 1986, *Natural History of the Major Histocompatibility Complex*, New York.

Lande, R., 1994, Risk of population extinction from fixation of new deleterious mutations, *Evolution* **48:**1460-1469.

Nei, M., 1987, *Molecular Evolutionary Genetics*, Columbia Univ. Press, New York.

Nunn, C. L., Altizer, S. and Jones, K. E. S., W. 2003, Comparative tests of parasite species richness in primates, *Amer. Nat.* **162:**597-614.

Palmer, A. R. and Van Rooyen, A. F., 1998, Detecting vegetation change in the southern Kalahari using Landsat TM data, *J. Arid Envir.* **39:**143-153.

Paterson, S., Wilson, K. and Pemberton, J. M., 1998, Major histocompatibility complex variation associated with juvenile survival and parasite resistance in a large unmanaged ungulate population (*Ovis aries* L.), *Evolution* **95:**3714-3719.

Penn, D. J., 2002, The Scent of Genetic Compatibility: Sexual Selection and the Major Histocompatibility Complex, *Ethology* **108:**1-21.

Rice, W. R., 1989, Analyzing tables of statistical tests, *Evolution* **43:**223-225.

Sachs, I., 1992, *Angewandte Statistik*, Springer Verlag, Berlin.

Schad, J., Sommer, S. and Ganzhorn, J. U., 2004, MHC variability of a small lemur in the littoral forest fragments of southeastern Madagascar, *Conserv. Genet.* **5:**299-309.

Schad, J., Ganzhorn, J. U. and Sommer, S., MHC constitution and parasite burden in the Malagasy mouse lemur, *Microcebus murinus.*, *Evolution*, in press.

Seeley, M. K., 1998, Environmental Change, in: *Biological Diversity in Namibia: a Country Study,* P. Barnard ed., Namibian National Biodiversity Task Force, Windhoek, pp. 67-72.

Sloss, M. W., Kemp, R. L. and Zajac, A., 1994, *Veterinary Clinical Parasitology*, 6th ed. Iowa State University Press, Ames.

Sommer, S., 2003, Effects of habitat fragmentation and changes of dispersal behaviour after a recent population decline on the genetic variability of non-coding and coding DNA of a monogamous rodent, *Mol. Ecol.* **12:**2845-2851.

Sommer, S., Schwab, D. and Ganzhorn, J. U., 2002, MHC diversity of endemic Malagasy rodents in relation to geographic range and social system, *Behav. Ecol. Sociobiol.* **51:**214-221.

Takahata, N. and Nei, M., 1990, Allelic genealogy under overdominant and frequency-dependent selection and polymorphism of major histocompatibility complex loci, *Genetics* **124:**967-978.

APPENDIX

Plant species in the growth form categories (grass, herb, shrub, tree) on the private nature reserve (sites A and B) and commercially used farm (site C). Trees: single-stemmed woody plants > 3m; shrubs: mostly multi-stemmed woody or semi-woody plants from 0.3 m – 3 m; herbs: herbaceous plants; grasses: plants with a tufted growth form and hollow stems.

Plant Species	Growth Form	Site A	Site B	Site C
Acacia ereoloba	tree	x	x	x
Acacia hebeclada	shrub	x	x	x
Acacia mellifera	shrub	x	x	
Acrotome inflata	herb	x	x	
Aptosimum marlothii	herb	x	x	x
Asparagus spec.	shrub	x	x	x
Barleria rigida	shrub		x	
Centropodia glauca	grass	x	x	
Chloris virgata	grass	x	x	
Chrysocoma obtusata	shrub		x	
Cleome spec.	herb		x	
Crotalaria spec.	shrub		x	
Cullen obtusifolia	herb	x		
Deverra denudata	herb			x
Dimorphotheca polyptera	herb		x	
Enneapogon scaber	grass	x	x	
Eragrostis lehmaniana	grass	x	x	
Geigeria ornativa	shrub	x		
Gisekia spec.	herb	x	x	
Gnidia polycephala	shrub	x		x
Gomphocarpus fruticosus	shrub	x		x
Hermannia tomentosa	herb	x	x	
Hermbstaedtia fleckii	herb	x	x	
Hirpicium spec.	herb		x	
Indigofera alternans	herb		x	
Kedrostis africana	shrub			x
Lessertia physodes	shrub		x	
Lycium cinereum	shrub	x	x	x
Nolletia arenosa	herb	x	x	
Plinthus spec.	shrub	x		
Radyera urens	shrub	x		
Requienia sphaerospermum	herb	x	x	x
Rhigozum trichotomum	shrub	x	x	x
Schmidtia kalihariensis	grass	x	x	
Sericorema remotiflora	shrub		x	
Solanum incanum	shrub		x	
Stipagrostis amabilis	grass	x	x	x
Stipagrostis ciliata	grass	x	x	
Stipagrostis obtusa	grass	x	x	

AN ASSESSMENT OF THE SYSTEMATICS OF THE GENUS *DESMOMYS* THOMAS, 1910 (RODENTIA: MURIDAE) USING MITOCHONDRIAL DNA SEQUENCES

Leonid A. Lavrenchenko[1] and Erik Verheyen[2]
[1] *Severtsov Institute of Ecology and Evolution RAS, Leninsky pr., 33, Moscow, 119071 Russia*
[2] *Royal Belgian Institute of Natural Sciences, Vautierstraat 29, 1000 Brussels, Belgium*

Abstract: We analyzed two mitochondrial gene fragments to assess genetic divergence within the genus *Desmomys* endemic to Ethiopia and its phylogenetic relationships with related genera. The phylogenetic analysis supported the monophyly of the Arvicanthini-Otomyini group and revealed that *Stochomys* is clearly a member of Arvicanthini. Our study demonstrated that *D. harringtoni* and *D. yaldeni* belong to remarkably different mitochondrial lineages, the estimated divergence time between them is 4.10-5.38 Myr. Such early splitting of specialized forest dweller, *D. yaldeni*, from its sole congener supposes a more ancient formation of some elements of Ethiopian forest rodent fauna than is assumed today.

Key words: Rodentia; Muridae; *Desmomys;* Ethiopia; phylogeny; evolution; mitochondrial DNA

1. INTRODUCTION

For a long time *Desmomys* has been treated as a subgenus of *Pelomys* (Ellerman, 1941; Corbet and Hill, 1991); presently, based on unique dental patterns, the generic status of the former taxon (including the single species, *D. harringtoni*, endemic to Ethiopia) was confirmed (Musser and Carleton, 1993). A recent phylogenetic study based on sequencing of mitochondrial DNA (Ducroz et al., 2001) revealed that *Desmomys* is a sister group to *Rhabdomys* and, hence, is a member of arvicanthines. Moreover, it was

B.A. Huber et al. (eds.), African Biodiversity, 363–369.

suggested that arvicanthine rodents (*Arvicanthis*, *Mylomys*, *Pelomys*, *Lemniscomys*, *Desmomys* and *Rhabdomys*) are part of tribe Arvicanthini that also comprises *Aethomys*, *Dasymys*, *Grammomys*, and *Hybomys*. The systematic position of other African genera considered by Misonne (1969) as members of the *Arvicanthis* "Division" still remains unclear. Recently, a second *Desmomys* species, *D. yaldeni*, was described from Ethiopia (Lavrenchenko, 2003). The aim of this paper is to enhance our knowledge on arvicanthine rodents by adding new data and to focus particularly on the phylogenetic relationships of the two *Desmomys* species.

2. MATERIALS AND METHODS

Specimens of *D. harringtoni* and *D. yaldeni* were collected in the course of the Joint Ethiopian-Russian Biological Expedition (for localities, field and museum numbers see Lavrenchenko, 2003). DNA was extracted from 96% alcohol preserved tissue samples by the standard phenol-chloroform method, PCR amplification and sequencing of segments of 16S rRNA (16S) and cytochrome-*b* (cyt-*b*) genes were carried out using primers and protocols described previously for *Lophuromys* by Lavrenchenko et al. (2004). We also used sequence data on relevant African murines from the database of the Royal Belgian Institute of Natural Sciences. Additional sequences for 16S and cyt-*b* genes were retrieved from GenBank: *Otomys irroratus* (AF141253; AF141222), *O. sloggetti* (AF141254; AF141223), *Aethomys namaquensis* (AF141246; AF141215), *Grammomys* sp. (AF141249; AF141218), *Hybomys univittatus* (AF141250; AF141219), *Dasymys rufulus* (AF141247; AF141216), *D. incomtus* (AF141248; AF141217), *Lemniscomys rosalia* (AF141238; AF141209), *L. zebra* (AF141235; AF141207), *L. macculus* (AF141237; AF141208), *L. bellieri* (AF141236; AF004586), *L. striatus* (AF141240; AF141211; AF141239; AF141210), *Arvicanthis* sp. (AF141230; AF004584; AF141231; AF141205; AF141232; AF004576), *A. abyssinicus* (AF141227; AF004566), *A. niloticus* (AF141228; AF004569), *Pelomys campanae* (AF141242; AF141213), *Mylomys dybowskii* (AF141241; AF141212), *Desmomys harringtoni* (AF141233; AF141206), and *Rhabdomys pumilio* (AF141244; AF141214). Phylogenetic relationships were analysed by maximum-parsimony (MP), neighbour-joining (NJ) and maximum-likelihood (ML) methods. The phylogenetic analyses were conducted using PAUP* v.4.0b4a (Swofford, 2000). ModelTest 3.06 (Posada and Crandall, 1998) was employed for the choice of the best model of sequence evolution for the NJ and ML analyses. This was the general time reversible model (GTR+G+I, alpha=0.695, pin=0.447). Simpler analyses were also carried out using the Tamura-Nei (TrN) and the Kimura's two-parameter (K2P) models. The relative stability of the NJ-MP and ML trees was assessed by bootstrapping with 10000 and 100 replicates, respectively. The estimates of divergence times were obtained on the basis of transversional changes in both 16S sequences and third codon positions of the cyt-*b* sequences. Traditionally accepted divergence time of 12 Myr for the dichotomy *Mus*/*Rattus* (Jaeger et al., 1986) was used as the calibration point. Relative-rate tests were conducted with Mega 2.1 program (Kumar et al., 2001).

3. RESULTS

Sequence data from 907 bp of combined 16S (535 bp) and cyt-*b* (372 bp) genes were used for the phylogenetic analyses. The combined sequences among ingroup haplotypes show 370 variable sites of which 269 are informative under parsimony. With a few exceptions, NJ, MP and ML results are quite consistent with each other. The results suggest the basal position of the lineage *Rattus-Mus-Nannomys* within the ingroup. Traditional association between *Mus* and *Nannomys* is supported with low bootstrap value in the MP and ML analyses (results not shown) although based on the NJ tree *Nannomys* is more closely related to *Rattus* (Fig. 1). All three reconstruction methods reveal that the "African lineage" (*sensu* Ducroz et al., 2001), including Arvicanthini and Otomyini, is the sister clade to the *Praomys* group *s. lat.* (represented here by *Praomys s. str.*, *Mastomys*, *Hylomyscus*, *Colomys* and *Malacomys*). With all these methods, the "African lineage" is well supported, however, the monophyly of the tribe Arvicanthini was not unequivocally confirmed. Moreover, in the NJ (GTR+G+I), MP and ML analyses, *Otomys* appears to be nested within the clade containing the Arvicanthini genera. Only NJ analyses using simpler models (K2P and TrN) indicated the basal position of the *Otomys* clade relative to all studied Arvicanthini genera (again without any statistical support) (Fig. 1). *Stochomys longicaudatus* always appears as a member of Arvicanthini, although its affiliations with other representatives of the tribe remain unresolved. All analyses indicate the monophyly of the arvicanthine genera (although only NJ results support this association with relatively weak bootstrap value) and suggest that *Desmomys* is the sister group to *Rhabdomys* (this association was constantly supported by higher bootstrap values). The relationships between other arvicanthine genera remain uncertain: only NJ analyses indicate the basal position of the *Desmomys-Rhabdomys* clade relative to the rest of the group with relatively weak bootstrap support.

Our study revealed that *D. harringtoni* and *D. yaldeni* form a monophyletic group, but also that they belong to deeply diverged

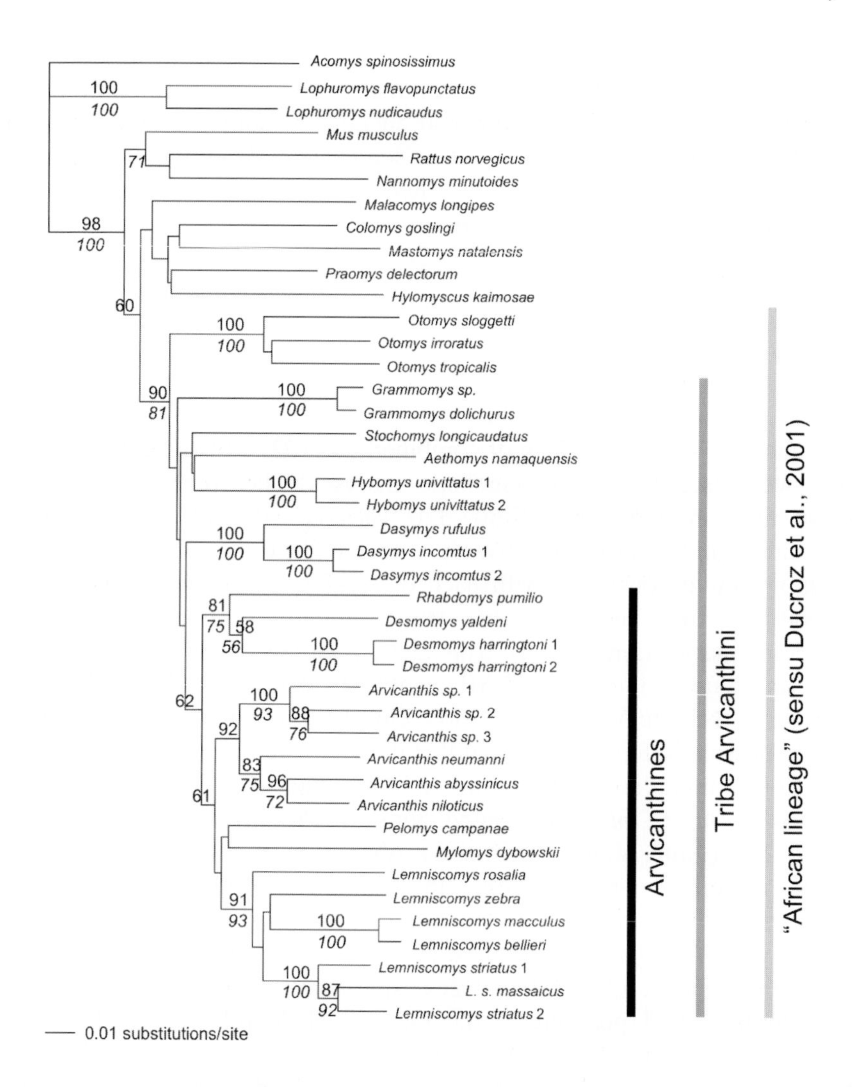

Figure 1. NJ tree constructed from TrN distances. Values above and below branches indicate percentage bootstrap support (>50%) for the NJ and MP trees, respectively (10,000 replicates). *Acomys spinosissimus*, *Lophuromys nudicaudus*, and *L. flavopunctatus* are used as outgroups.

mitochondrial lineages. The proportion of mismatches between these two haplotypes is 6.4% for 16S and 11.7% for cyt-*b*. These values lie well within the range typical for well-differentiated congeneric species or even related genera in Muroidea (e.g. Fadda et al., 2001; Lecompte et al., 2002). The results of the Tajima's relative-rate test demonstrate no significant variation of evolution rate. The estimated divergence times between the two *Desmomys* species are similar for both genes: 5.38 Myr (95% CI = 2.55-8.21 Myr) for 16S and 4.10 Myr (95% CI = 1.75-6.45 Myr) for cyt-*b*. At the same time, these results should be treated with caution because of the limited amount of data. Nevertheless, it seems reasonable to conclude that the diversification within the genus *Desmomys* might have started in the beginning of the Pliocene.

4. DISCUSSION

The results of our phylogenetic analysis are largely consistent with those in a previous study on arvicanthine rodents (Ducroz et al., 2001). The monophyly of the so-called "African lineage" (Arvicanthini + Otomyini) was confirmed. However, it is noteworthy that the position of tribe Otomyini still remains uncertain. Further study based on analyses of other mithohondrial and nuclear genes is necessary to find out whether this clade is the sister group to Arvicanthini or nests within the latter. Our analysis revealed that *Stochomys* is clearly a member of Arvicanthini. Additional studies should test the inclusion of other putative members (*Dephomys*, *Lamottemys*, *Oenomys*, *Thallomys*, and *Thamnomys*) in this tribe.

Clustering of *Desmomys* and *Rhabdomys* was constantly supported by significant bootstrap values in all analyses. Contrary to the previous phylogenetic analysis of arvicanthines (Ducroz et al., 2001) we have found no evidence for faster evolution in these two lineages. Given this result, it seems unlikely that the *Desmomys* – *Rhabdomys* grouping is an artifact of long branch attraction as supposed by Ducroz et al. (2001). The fact that *Desmomys* clusters with the predominantly South African taxon *Rhabdomys* suggests that there may have been an ancient association between faunas of the Ethiopian Plateau and South Africa.

Furthermore, the results of our study support the distinctness of the recently described *D. yaldeni*. As both known *Desmomys* species are endemic to Ethiopia, we can assume that the genus has evolved mostly, or even exclusively on the Ethiopian Plateau. Molecular clock estimates of divergence times suggest early diversification within *Desmomys* (4.10-5.38 Myr) relative to other Ethiopian endemic species assemblages studied so far: *Praomys albipes / Stenocephalemys* complex (1.34-2.24 Myr – Fadda et al., 2001; 1.9-2.4 Myr – Lecompte *et al.*, 2002) and Ethiopian *Lophuromys flavopunctatus* species complex (0.70-0.88 Myr – Lavrenchenko et al., 2004). *Desmomys yaldeni* inhabits humid afromontane forests and possesses some morphological characteristics that can be interpreted as adaptations to ecological specialization uncommon for arvicanthine rodents: a supposedly climbing life style and a diet that mainly consists of invertebrates and/or fruits and berries. Previously, we regarded *D. yaldeni* as an additional line of evidence supporting our hypothesis that the Ethiopian forest rodent fauna originated recently from native ancestors (Lavrenchenko, 2003). However, the inferred early split of this specialized forest dweller from its sole congener suggests that *D. yaldeni* may be older (presumably originated at the beginning of the Pliocene) than other elements of this forest fauna.

ACKNOWLEDGEMENTS

We wish to thank the Ethiopian Science and Technology Commission for support in the field work organisation. Dr. A.A. Darkov has coordinated field operations. Mr. A.A. Warshavsky has assisted in collecting specimens for this study. Two anonymous referees made helpful comments on the manuscript. The work of L.A.L. in the Royal Belgian Institute of Natural Sciences (Brussels) was supported by a Research Fellowship from the DWTC and a NATO Research Fellowship. This research was supported by the DWTC project 31.64 and the RFBR project 03-04-48924. Participation of the senior author at the Symposium was supported by the DFG.

REFERENCES

Corbet, G. B., and Hill, J. E., 1991, *A World List of Mammalian Species*, 3rd ed., Oxford University Press, Oxford.

Ducroz, J.F., Volobouev, V., and Granjon, L., 2001, An assessment of the systematics of Arvicanthine rodents using mitochondrial DNA sequences: evolutionary and biogeographical implications, *J. Mammal. Evol.* **8:**173-206.

Ellerman, J. R., 1941, *The families and genera of living rodents. Vol. II. Family Muridae,* British Museum (Natural History), London.

Fadda, C., Corti, M., and Verheyen, E., 2001, Molecular phylogeny of *Myomys/Stenocephalemys* complex and its relationships with related African genera, *Biochem. Syst. Ecol.* **29:**585-596.

Jaeger, J.-J., Tong, H., and Denys, C., 1986, Age de la divergence *Mus-Rattus*: comparaison des donnees paleontologique et moleculaires, *C. R. Acad. Sci. Paris, Ser. II* **302:**917-922.

Kumar, S., Tamura, K., Jakobsen, B., and Nei, M., 2001, *MEGA2: Molecular Evolutionary Genetics Analysis software,* Arizona State University, Tempe, Arizona, USA.

Lavrenchenko, L. A., 2003, A contribution to the systematics of *Desmomys* Thomas, 1910 (Rodentia, Muridae) with the description of a new species, *Bonn. zool. Beitr.* **50:**313-327.

Lavrenchenko, L. A., Verheyen, E., Potapov, S. G., Lebedev, V. S., Bulatova, N. Sh., Aniskin, V. M., Verheyen, W. N., and Ryskov, A. P., 2004, Divergent and reticulate processes in evolution of Ethiopian *Lophuromys flavopunctatus* species complex: evidence from mitochondrial and nuclear DNA differentiation patterns, *Biol. J. Linn. Soc.* **83**:301-316.

Lecompte, E., Granjon, L., Kerbis Peterhans, J., and Denys, C., 2002, Cytochrome b-based phylogeny of the *Praomys* group (Rodentia, Muridae): a new African radiation? *C. R. Biologies* **325:**827-840.

Misonne, X., 1969, African and Indo-Australian Muridae. Evolutionary trends, *Mus. Roy. l'Afrique Cent., Tervuren, Zool.* **172:**1-219.

Musser, G. G., and Carleton, M. D., 1993, Family Muridae, in: *Mammal species of the World - A taxonomic and geographic reference*, 2nd ed., D.E. Wilson and D.A.M. Reeder, eds., Washington: Smithsonian Institution Press, pp. 501-756.

Posada, D., and Crandall, K. A., 1998, Modeltest: testing the model of DNA substitution. *Bioinformatics* **14:** 817-818.

Swofford, D. L., 2000. *PAUP*. Phylogenetic Analysis Using Parsimony (*and Other Methods)*, 4th ed., Sunderland, M A: Sinauer.

INTEGRATIVE TAXONOMY AND PHYLOGENETIC SYSTEMATICS OF THE GENETS (CARNIVORA, VIVERRIDAE, *GENETTA*): A NEW CLASSIFICATION OF THE MOST SPECIOSE CARNIVORAN GENUS IN AFRICA

Philippe Gaubert[1]*, Peter J. Taylor[2] and Geraldine Veron[1]
[1] *Unité Origine, Structure et Evolution de la Biodiversité (CNRS FRE 2695), Département Systématique et Evolution, Muséum National d'Histoire Naturelle, CP 51, 57 rue Cuvier, 75231 Paris Cedex 05, France*
[2] *eThekwini Heritage Department, Natural Science Museum, P.O. Box 4085, Durban 4000, Republic of South Africa*
* *E-mail: gaubert@mnhn.fr*

Abstract: The taxonomy of the genus *Genetta* has been hotly debated. Following recent clarifications on the phylogeny of and species boundaries within the genets, we propose a new classification, including discriminant morphological diagnoses, which provides for the first time a synthetic description of the re-assessed species diversity within the genus (17 species). Prospects concerning further investigations on the systematics of the genets are discussed.

Key words: Carnivora; Viverridae; *Genetta;* phylogenetic systematics; integrative taxonomy; classification; diagnosis; discrete characters; distribution; habitat; Africa

1. INTRODUCTION

Members of the genus *Genetta* G. Cuvier, 1816 are small African carnivorans (Viverridae, Viverrinae) that diversified into a wide range of habitats (Kingdon, 1997; Gaubert, 2003a). The genets constitute by far the most speciose carnivoran genus in Africa, but their systematics has been hotly debated, especially within the large-spotted genet complex (Gray,

B.A. Huber et al. (eds.), African Biodiversity, 371–383.

1864; Matschie, 1902; Allen, 1939; Roberts, 1951; Wenzel and Haltenorth, 1972; Rosevear, 1974; Coetzee, 1977; Crawford-Cabral, 1981; Schlawe, 1981; Crawford-Cabral and Pacheco, 1992; Wozencraft, 1993; Kingdon, 1997; Crawford-Cabral and Fernandes, 2001).

Recent advances in the systematics of the genus *Genetta* have been made possible by the use of an integrative approach (i.e. combination of independent data sets). A molecular phylogenetic analysis (cytochrome *b*) allowed the redefinition of the boundaries of the genus, now including the aquatic genet (formerly *Osbornictis piscivora*; Gaubert et al., 2004a). A taxonomic revision of the large-spotted genet complex based on discrete morphological characters led to the characterization of three morphospecies: *G. poensis* Waterhouse, 1838, *G. "schoutedeni"* (Crawford-Cabral, 1970) and the new species *G. bourloni* Gaubert, 2003 (Gaubert, 2003b). In addition, a neotype was designated for the rusty-spotted genet *G. maculata* (Gray, 1830) (= *G. "rubiginosa" sensu* Crawford-Cabral and Pacheco, 1992) in order to stabilize the classification within the large-spotted genet complex (Gaubert et al., 2003a,b; but see Grubb (2004) for a different alternative). Species boundaries were further tested in a phylogenetic framework (cytochrome *b* and morphology). The results contradicted traditional taxonomy, the servaline and small-spotted genets not being monophyletic (Gaubert et al., 2004b). Branching patterns and genetic distances confirmed the species status of *G. cristata* and the three forest "forms" belonging to the large-spotted genet complex (*G. bourloni, G. poensis* and *G. "schoutedeni"*) and showed that the traditional subgenera *Paragenetta*, *Pseudogenetta* and *Genetta* had no phylogenetic value. In addition, a phylogeographic analysis using a combination of morphological and molecular approaches suggested the presence of a distinct species within *G. maculata,* namely a southeastern African population that we tentatively named *G. "letabae"* (Gaubert et al., in press a; in agreement with Crawford-Cabral & Fernandes, 2001). This last result was confirmed by the study of chromosomal morphology (Gaubert et al., in press b).

The aim of this paper is to establish the morphological diagnoses of the genets, together with geographic ranges and habitats, in order to provide for the first time a synthetic description of the re-assessed species diversity within the genus *Genetta* (17 species).

2. MATERIALS AND METHODS

In order to have an exhaustive representation of the morphological variability within genets, one of us (PG) defined diagnostic, discrete morphological characters from more than 5500 specimens of Viverrinae housed in 13 major museums worldwide (see Acknowledgments).

Information concerning distribution ranges (given by country) and habitats were taken directly from museum specimen labels and bibliographical sources when available. The locality names were checked using geographical atlas (*Atlas mondial Encarta,* 1998), gazetteers (*Gazetteer of collecting localities of African rodents*: Davis and Misonne, 1964; *Official Standard Names Gazetteer*, United States Board on Geographic Names; *Alexandria Digital Library Gazetteer Server*: http://fat-albert.alexandria.ucsb.edu:8827/gazetteer) and curators' resources. In order to maximize the discriminative power of the diagnoses, we focused on providing combinations of diagnostic characters allowing the distinction between sympatric species.

3. RESULTS

New classification of the genus *Genetta*, with distribution, habitat and discriminative diagnosis for each species.

Genus *Genetta* G. Cuvier, 1816

***Genetta abyssinica* (Rüppel, 1836)** [Abyssinian genet]

Synonyms: none.

Distribution: Djibouti (A. Laurent, Toulouse, pers. comm. 2001), Erytrea, Ethiopia, Somalia, Sudan.

Habitat: from high Ethiopian plateaux (montane moorland and grassland) to steppe and sub-desert areas on plains (Yalden et al., 1996; Diaz Behrens and Van Rompaey, 2002).

Sympatric species: *G. genetta, G. maculata.*

Identification: [body and coat] dark, continuous mid-dorsal line longitudinally crossed by a brighter (ground coloration) line – hair short, no dorsal crest – dorsal spots fused into several longitudinal stripes – feet bright (same as ground coloration) – central depression of forefeet hairless – alternation of dark and bright rings to end of tail – tip of tail dark; [skull] absence of premaxillary-frontal contact – caudal entotympanic bone not ventrally inflated – curve line of anterior part of caudal entotympanic bone (external side) continuous – maxillary-palatine suture anterior to main cusp of P^3.

***Genetta angolensis* Bocage, 1882** [miombo genet]

Synonyms: *hintoni* Schwarz, 1929; the type specimen of *mossambica* Matschie, 1902, is considered a hybrid between *G. maculata* and *G. angolensis* (Gaubert et al., in press a).

Distribution: Angola, Botswana, Democratic Republic of Congo, Malawi, Mozambique, Tanzania, Zambia, Zimbabwe.

Habitat: open miombo (*Brachystegia*) woodlands interspersed with savannah (Crawford-Cabral, in press), savannah-forest mosaic.

Sympatric species: *G. genetta, G. "letabae", G. maculata, G. "schoutedeni".*

Identification: [body and coat] full dark, continuous mid-dorsal line – hair long, presence of a dorsal crest – scapular region poorly spotted – posterior parts of feet dark, with hindfeet almost completely dark – width of bright rings relative to dark rings (middle of tail) = 50 to 75% – last bright ring of tail covered with dark – tip of tail dark – two pairs of nipples; [skull] *int1* = 1.00 ± 0.12 cm (*int1* = ratio between interorbital constriction and frontal width; see Gaubert et al., 2004b and in press a) – caudal entotympanic bone not ventrally inflated – curve line of anterior part of caudal entotympanic bone (external side) continuous.

***Genetta bourloni* Gaubert, 2003** [Bourlon's genet]

Synonyms: none.

Distribution: Ghana, Guinea, Ivory Coast, Liberia, Sierra Leone.

Habitat: rain forest.

Sympatric species: *G. johnstoni, G. pardina, G. poensis.*

Identification: [body and coat] full dark, continuous mid-dorsal line – dorsal spots partly fused at rump – feet dark – almost half of tail dark – tip of tail dark; [skull] *int1* = 1.00 ± 0.12 cm – posterior extension of frontal bones very large, almost completely overlapping dorsal region of interorbital constriction – curve line of anterior part of caudal entotympanic bone (external side) broken.

***Genetta cristata* (Hayman, 1940)** [crested servaline genet]

Synonyms: *bini* Rosevear, 1974.

Distribution: Cameroon, Nigeria; Gabon and Congo as potential zones of sympatry/intergradation with *G. servalina.*

Habitat: rain forest.

Sympatric species: *G. maculata, G. "schoutedeni"* (*G. servalina*: uncertain).

Identification: [body and coat] full dark, discontinuous mid-dorsal line – mid-dorsal line with hairs relatively long, giving a continuous aspect to the line – presence of a nuchal crest – feet dark – width of bright rings relative to dark rings (middle of tail) = 50 to 75% – alternation of dark and bright rings to end of tail – tip of tail bright; [skull] premaxillary-frontal contact present – *int1* = 1.00 ± 0.12 cm – caudal entotympanic bone ventrally inflated – curve line of anterior part of caudal entotympanic bone (external side) continuous.

***Genetta felina* (Thunberg, 1811)** [South African small-spotted genet]
Synonyms: *macrura* Jentink, 1892.
Distribution: Angola, Namibia, Orange Free State, South Africa, Zambia.
Habitat: woodland savannah, grassland, thickets, dry vlei areas, bordering of deserts.
Sympatric species: *G. genetta, G. "letabae", G. maculata, G. tigrina.*
Identification: [body and coat] full dark, continuous mid-dorsal line, always black – ground color whitish grey (brighter than in *G. genetta*) – hair long, presence of a dorsal crest – posterior parts of feet dark, with hind feet almost completely dark – confused annulation pattern at beginning of tail (contrary to *G. genetta*) – alternation of dark and bright rings to end of tail – width of bright rings relative to dark rings (middle of tail) = 200% – tip of tail bright (larger than in *G. genetta*); [skull] posterior extension of frontal bones moderated, overlapping ca. 50% of dorsal region of interorbital constriction – $int1 = 1.00 \pm 0.12$ cm – curve line of anterior part of caudal entotympanic bone (external side) broken.

***Genetta genetta* (Linnaeus, 1758)** [common small-spotted genet]
Synonyms: *afra* G. Cuvier, 1825; *albipes* Trouessart, 1904; *balearica* Thomas, 1902; *barbar* Wagner, 1841; *barbara* C. E. H. Smith, 1842; *bella* Matschie, 1902; *bonapartei* Loche, 1857; *communis* Burnett, 1830; *dongolana* Hemprich and Ehrenberg, 1832; *gallica* Oken, 1816; *grantii* Thomas, 1902; *guardafuensis* Neumann, 1902; *hararensis* Neumann, 1902; *hispanica* Oken, 1816; *isabelae* Delibes, 1977; *leptura* Reichenbach, 1836; *ludia* Thomas and Schwann, 1906; *lusitanica* Seabra, 1924; *melas* Graells, 1897; *neumanni* Matschie, 1902; *peninsulae* Cabrera, 1905; *pulchra* Matschie, 1902; *pyrenaica* Bourdelle and Dezillière, 1951; *rhodanica* Matschie, 1902; *senegalensis* J. B. Fischer, 1829; *tedescoi* de Beaux, 1924; *terraesanctae* Neumann, 1902; *vulgaris* Lesson, 1827.
Distribution: Algeria, Angola, Benin, Botswana, Burkina Faso, Chad, Djibouti (Künzel et al., 2000), Egypt, Eritrea, Ethiopia, Ghana, Ivory Coast, Kenya, Mali, Mauritania, Morocco, Mozambique, Namibia, Niger, Nigeria, Oman, Republic of Yemen, Saudi Arabia, Senegal, Somalia, South Africa, Sudan, Tanzania, Togo, Tunisia, Uganda, Zambia, Zimbabwe. Populations in France, Portugal and Spain resulting from introductions during historical times (Morales, 1994; Amigues, 1999). Individuals episodically found in Belgium, Germany, Holland, Italy and Libya.
Habitat: all habitats except densely forested areas (in Africa).
Sympatric species: *G. abyssinica, G. angolensis, G. felina, G. "letabae", G. maculata, G. pardina, G. "schoutedeni", G. thierryi.*
Identification: [body and coat] full dark, continuous mid-dorsal line, always black – hair long, presence of a dorsal crest – posterior parts of feet dark

– central depression of forefeet hairy – alternation of dark and bright rings to end of tail - width of bright rings relative to dark rings (middle of tail) = 100% – tip of tail bright; [skull] premaxillary-frontal contact absent – posterior extension of frontal bones moderated, overlapping ca. 50% of dorsal region of interorbital constriction – *int1* = 1.00 ± 0.12 cm – caudal entotympanic bone not ventrally inflated – curve line of anterior part of caudal entotympanic bone (external side) continuous – maxillary-palatine suture at same level than main cusp of P^3.

***Genetta johnstoni* Pocock, 1908** [Johnston's genet]
Synonyms: *lehmanni* Kuhn, 1960.
Distribution: Ghana, Guinea, Ivory Coast, Liberia.
Habitat: moist, mixed woodlands and savannah, rain forest (Gaubert et al., 2002).
Sympatric species: *G. bourloni, G. pardina, G. poensis.*
Identification: [body and coat] pattern of nuchal stripes confused – full dark, continuous mid-dorsal line – dorsal spots partly fused at rump – feet dark – alternation of dark and bright rings to end of tail – width of bright rings relative to dark rings (middle of tail) < 20% – tip of tail bright – one pair of nipples; [skull] flattened skull and mandible – *int1* = 1.00 ± 0.12 cm – posterior extension of frontal bones very large, almost completely overlapping dorsal region of interorbital constriction – curve line of anterior part of caudal entotympanic bone (external side) continuous – weak dentition (possibly adapted to piscivory).

***Genetta letabae* Thomas and Schwann, 1906** [provisional species name]
Synonyms: *zuluensis* Roberts, 1924.
Distribution: Lesotho, Mozambique, Namibia, South Africa, Swaziland.
Habitat: woodland savannah, savannah-forest mosaic.
Sympatric species: *G. genetta, G. "schoutedeni", G. tigrina* (*G. angolensis*: uncertain).
Identification: [body and coat] same as *G. maculata*; [skull] same as *G. maculata*, excepted: wider interorbital constriction relative to frontal width (characterized through morphometric geometric analysis; Gaubert et al., in press a).

***Genetta maculata* (Gray, 1830)** [rusty-spotted genet]
Synonyms: *aequatorialis* Heuglin, 1866; *albiventris* Roberts, 1932; *deorum* Funaioli and Simonetta, 1960; *erlangeri* Matschie, 1902; *fieldiana* Du Chaillu, 1860; *gleimi* Matschie, 1902; *insularis* Cabrera, 1921; *matschiei* Neumann, 1902; *pumila* Hollister, 1916; *schraderi* Matschie, 1902; *soror* Schwarz, 1929; *stuhlmanni* Matschie, 1902; *zambesiana* Matschie, 1902.
Distribution: Angola, Benin, Botswana, Burundi, Cameroon, Central African Republic, Chad, Congo, Democratic Republic of Congo, Equatorial

Guinea (Basilio, 1962), Eritrea, Ethiopia, Gabon, Ghana, Kenya, Malawi, Mozambique, Namibia, Nigeria, Rwanda, Somalia, South Africa, Sudan, Swaziland, Tanzania, Togo, Uganda, Zambia.

Habitat: rain forest, woodland savannah, savannah-forest mosaic, montane forest.

Sympatric species: *G. abyssinica, G. angolensis, G. cristata, G. genetta, G. poensis, G. "schoutedeni", G. servalina, G. thierryi, G. victoriae.*

Identification: [body and coat] full dark, continuous mid-dorsal line, of same color as dorsal spots – hair short, no dorsal crest – dorsal spots usually not fused at rump – feet bright (same as ground coloration) – central depression of forefeet hairy – last bright ring of tail covered with dark – width of bright rings relative to dark rings (middle of tail) = 50 to 75% – tip of tail dark – two pairs of nipples; [skull] *int1* < 1.00 – 0.07 cm – posterior extension of frontal bones very narrow – caudal entotympanic bone not ventrally inflated – curve line of anterior part of caudal entotympanic bone (external side) broken – maxillary-palatine suture at same level than main cusp of P^3.

***Genetta pardina* I. Geoffroy Saint-Hilaire, 1832** [pardine genet]

Synonyms: *amer* Gray, 1843; *dubia* Matschie, 1902; *pantherina* Hamilton-Smith, 1842. The series of specimens representing *genettoides* Temminck, 1853, are considered hybrids between *G. pardina* and *G. maculata* (Gaubert, 2003b).

Distribution: Gambia, Ghana, Guinea, Guinea Bissau (Crawford-Cabral, 1973), Ivory Coast, Liberia, Senegal, Sierra Leone.

Habitat: rain forest, woodland savannah, savannah-forest mosaic, dry savannah.

Sympatric species: *G. bourloni, G. genetta, G. johnstoni, G. poensis, G. thierryi.*

Identification: [body and coat] full dark, continuous mid-dorsal line – hair short, no dorsal crest – dorsal spots usually not fused at rump – feet dark – central depression of forefeet hairy – last bright ring of tail covered with dark – width of bright rings relative to dark rings (middle of the tail) < 20% – tip of tail dark – two pairs of nipples; [skull] premaxillary-frontal contact absent – *int1* < 1.00 – 0.12 cm – posterior extension of frontal bones moderated, overlapping ca. 50% of dorsal region of interorbital constriction – caudal entotympanic bone ventrally inflated – curve line of anterior part of caudal entotympanic bone (external side) broken – maxillary-palatine suture at same level than main cusp of P^3.

***Genetta piscivora* (Allen, 1919)** [aquatic genet]

Synonyms: none.

Distribution: Democratic Republic of Congo.

Habitat: rain forest.

Sympatric species: *G. maculata, G. "schoutedeni", G. servalina, G. victoriae.*

Identification: [body and coat] uniformly rufous brown coat pattern – uniformly dark tail – plantar pads hairless; [skull] premaxillary-frontal contact absent – dorsal region of frontal bone concave – *intl* < 1.00 – 0.12 cm – posterior extension of frontal bones very narrow – caudal entotympanic bone ventrally inflated – curve line of anterior part of caudal entotympanic bone (external side) continuous – premaxillary-maxillary suture at same level than P^1 – maxillary-palatine suture at same level than main cusp of P^3 – M_2 reduced – trenchant dentition (adapted to piscivory).

***Genetta poensis* Waterhouse, 1838** [king genet]

Synonyms: none.

Distribution: Congo, Equatorial Guinea (Bioko Island), Ghana, Liberia.

Habitat: rain forest.

Sympatric species: *G. bourloni, G. johnstoni, G. maculata, G. pardina, G. "schoutedeni", G. servalina.*

Identification: [body and coat] full dark, continuous mid-dorsal line – dorsal spots greatly fused in various regions of body – feet dark – almost half of tail dark; [skull] premaxillary-frontal contact present – *intl* < 1.00 – 0.12 cm – posterior extension of frontal bones very narrow – curve line of anterior part of caudal entotympanic bone (external side) broken.

***Genetta schoutedeni* (Crawford-Cabral, 1970)** [Schouteden's genet] [provisional species name]

Synonyms: [provisionally] *suahelica* Matschie, 1902.

Distribution: Angola, Burundi, Cameroon, Central African Republic, Congo, Democratic Republic of Congo, Ethiopia, Ghana, Kenya, Mozambique, Nigeria, Rwanda, Sudan, Tanzania, Togo, Uganda.

Habitat: rain forest, woodland savannah, savannah-forest mosaic, montane forest.

Sympatric species: *G. abyssinica, G. angolensis, G. genetta, G. "letabae", G. maculata, G. poensis, G. servalina, G. victoriae.*

Identification (provisional): [body and coat] same as *G. maculata*; [skull] similar to *G. maculata,* excepted: *intl* < 1.00 – 0.12 cm (mean value = 0.62 ± 0.16; inferior to *G. maculata*) – caudal entotympanic bone extremely inflated (Gaubert, 2003b).

***Genetta servalina* Pucheran, 1855** [servaline genet]

Synonyms: *aubryana* Pucheran, 1855; *archeri* Van Rompaey and Colyn, 1998; *bettoni* Thomas, 1902; *intensa* Lönnberg, 1917; *lowei* Kingdon, 1977; *schwarzi* Crawford-Cabral, 1970.

Distribution: Cameroon, Central African Republic, Congo, Democratic Republic of Congo, Equatorial Guinea, Gabon, Kenya, Tanzania, Uganda.

Habitat: rain forest, woodland savannah, savannah-forest mosaic, montane forest.

Sympatric species: *G. maculata, G. piscivora, G. poensis, G. "schoutedeni", G. victoriae* (*G. cristata*: uncertain).

Identification: [body and coat] full dark, discontinuous mid-dorsal line – hair short, no dorsal crest – feet dark – width of bright rings relative to dark rings (middle of tail) < 20 % – alternation of dark and bright rings to end of tail – tip of tail bright – one pair of nipples; [skull] premaxillary-frontal contact present – *intl* > 1.00 + 0.12 cm – caudal entotympanic bone ventrally inflated – curve line of anterior part of caudal entotympanic bone (external side) continuous.

***Genetta thierryi* Matschie, 1902** [Hausa genet]

Synonyms: *rubiginosa* Pucheran, 1855; *villiersi* Dekeyser, 1949.

Distribution: Benin, Burkina Faso, Cameroon, Gambia, Ghana, Guinea Bissau (Crawford-Cabral, 1973), Ivory Coast, Mali, Nigeria, Niger, Sierra Leone, Senegal, Togo.

Habitat: brush-grass savannah, Guinean savannah, moist woodlands.

Sympatric species: *G. genetta, G. maculata, G. pardina.*

Identification: [body and coat] brown-rufous, continuous mid-dorsal line, often longitudinally crossed by a brighter (ground coloration) line – hair short, no dorsal crest – feet bright (same as ground coloration) – central depression of forefeet hairless – width of bright rings relative to dark rings (middle of tail) = 100% – last bright ring of tail covered with dark – tip of tail dark – one pair of nipples; [skull] absence of premaxillary-frontal contact – *intl* = 1.00 ± 0.12 cm – caudal entotympanic bone ventrally inflated – curve line of the anterior part of the caudal entotympanic bone (external side) continuous – maxillary-palatine suture anterior to main cusp of P^3.

***Genetta tigrina* (Schreber, 1776)** [Cape genet]

Synonyms: *methi* Roberts, 1948.

Distribution: South Africa.

Habitat: fynbos, grassland, coastal forest.

Sympatric species: *G. felina, G. "letabae"* (narrow hybrid zone in KwaZulu-Natal; Gaubert et al., in press a).

Identification: [body and coat] full dark, continuous mid-dorsal line, always black – hair of intermediate length between *G. maculata* and *G. genetta*, marking a short dorsal crest – posterior parts of feet dark, with hind feet almost completely dark – width of bright rings relative to dark rings (middle of tail) = 50 to 75% – last bright ring of tail covered with dark –

tip of tail dark; [skull] posterior extension of frontal bones moderated, overlapping ca. 50 % of dorsal region of interorbital constriction – $intl \geq 1.00 \pm 0.05$ cm – curve line of anterior part of caudal entotympanic bone (external side) continuous – maxillary-palatine suture anterior to main cusp of P^3.

***Genetta victoriae* Thomas, 1901** [giant genet]

Synonyms: none.

Distribution: Democratic Republic of Congo, Uganda (Kingdon, 1977). The specimen from Cameroon illustrated in Depierre and Vivien (1992) is likely to correspond to *G. cristata.*

Habitat: savannah-forest mosaic, rain forest.

Sympatric species: *G. maculata, G. piscivora, G. "schoutedeni", G. servalina.*

Identification: [body and coat] largest size (head + body reaching 60 cm) - full dark, discontinuous mid-dorsal line – mid-dorsal line with hair relatively long, giving a continuous aspect to the line – presence of a nuchal crest – presence of a pair of wide nuchal stripes – small dorsal spots confusedly distributed – feet dark – tail short (ratio head + body / tail lengthes = 1.3 to 1.4) – width of bright rings relative to dark rings (middle of tail) < 20% – alternation of dark and bright rings to end of tail – tip of tail dark – one pair of nipples; [skull] premaxillary-frontal contact present – $intl < 1.00 - 0.12$ cm – caudal entotympanic bone ventrally inflated – curve line of anterior part of caudal entotympanic bone (external side) continuous – maxillary-palatine suture just behind main cusp of P^3.

4. DISCUSSION AND PROSPECTS

Genets constitute an ideal model for tackling a wide panel of nested evolutionary issues (see Gaubert et al., in press a). Their new classification, which represents a provisional number of 17 species, requires, however, additional investigations. A particular focus should concern the morphological and geographical delimitations of *G. cristata* (contact zone with *G. servalina*), *G. felina* (sympatry with *G. genetta*), *G. "schoutedeni"* and *G. "letabae"*, the latter two having -in addition- problematic species name attributions. Moreover, extended molecular studies will have to be undertaken in order to clarify the taxonomic boundaries within the large-spotted genet complex (see Crawford-Cabral and Fernandes, 2001) and the widely distributed common small-spotted genet *G. genetta.*

Genets are also a group of high conservation concern. Three of the recently identified species are restricted to rain forests and their biology and status in the wild are totally unknown (*G. bourloni, G. cristata, G. poensis*). More dramatically, the king genet *G. poensis*, represented by 10 known specimens in collections, may no longer exist *in naturae* (last specimen collected in 1946; Gaubert, 2003b). The clarification of the taxonomy and evolutionary history of the genets is a necessary step towards appropriate measures of conservation. The delimitation of new taxa, the identification of recent clades suggesting on-going speciation processes (e.g., large-spotted genets) and the detection of hybrid zones should promote a dynamic conservation of the genus throughout Africa.

ACKNOWLEDGMENTS

We are grateful to the following people and institutions for having given access to their collections: Paula Jenkins and Daphne Hills (Natural History Museum, London, UK), Chris Smeenk (National Museum of Natural History, Leiden, Holland), Wim Van Neer and Wim Wendelen (Musée Royal d'Afrique Centrale, Tervuren, Belgium), Georges Lenglet and Georges Coulon (Institut Royal des Sciences Naturelles de Belgique, Brussels, Belgium), Manfred Ade and Irene Thomas (Museum für Naturkunde, Humboldt University, Berlin, Germany), Rainer Hutterer (Zoologisches Forschungsinstitut und Museum Alexander Koenig, Bonn, Germany), M. Trainer, Jacques Cuisin and Alain Bens (Muséum National d'Histoire Naturelle, Paris, France), the Durban Natural Science Museum, South Africa (via PJT), Nico Avenant and Johan Eksteen (National Museum, Bloemfontein, South Africa), Judith Masters and Jason Londt (the Natal Museum, Pietermaritzburg, South Africa), D. Drinkrow (South African Museum, Cape Town, South Africa), Lawrence R. Heaney and Bill Stanley (Field Museum of Natural History, Chicago, USA) and Jean Spence (American Museum of Natural History, New York, USA). The work of PG in London, Chicago and New York was supported by a Sys-Resource grant (IHP Programme), a Travel Grant and a Collection Study Grant, respectively. We thank one anonymous reviewer for improving the quality of the early draft of the manuscript. We are grateful to the organizing committee of the Proceedings of the 5th International Symposium on Tropical Biology (Museum A. Koenig, Bonn, Germany) for their coordination and editorial efficiency.

REFERENCES

Allen, G. M., 1939, A checklist of African mammals, *Bull. Mus. Comp. Zool.* **83:**1-763.
Amigues, S., 1999, Les belettes de Tartessos, *Anthropozool.* **29:**55-64.

Atlas mondial Encarta, 1998, *[CD-ROM],* Microsoft Corporation.

Coetzee, C. G., 1977, Order Carnivora. Part 8, in: *1971-1977. The Mammals of Africa: an Identification Manual*, J. Meester and H. W. Setzer, eds., Smithsonian Inst. Press, Washington D.C.

Crawford-Cabral, J., 1970, As genetas da Africa Central, *Bol. Inst. Invest. cient. Ang.* **7:**3-23.

Crawford-Cabral, J., 1973, As genetas da Guiné portuguesa e de Moçambique, in: *Livro de Homenagem ao Prof. Fernando Frade*, Junta de Investigações do Ultramar, Lisboa, pp. 133-155.

Crawford-Cabral, J., 1981, The classification of the genets (Carnivora, Viverridae, genus *Genetta*), *Bolm. Soc. port. Ciênc. nat.* **20:**97-114.

Crawford-Cabral, J., in press, *Genetta angolensis*, in: *The Mammals of Africa. Vol. 4. Carnivora, Pinnipedia, Pholidota, Tubulidentata, Hyracoidea, Proboscidea, Sirenia, Perissodactyla,* J. Kingdon and T. Butynski, eds., Academic Press, London.

Crawford-Cabral, J., and Fernandes, C. A., 2001, The Rusty-spotted genets as a group with three species in Southern Africa (Carnivora: Viverridae), in: *African Small Mammals*, L. Granjon, A. Poulet and C. Denys, eds., I.R.D., Paris, pp. 65-80.

Crawford-Cabral, J., and Pacheco, A. P., 1992, Are the Large-spotted and the Rusty-spotted genets separate species (Carnivora, Viverridae, genus *Genetta*), *Garcia de Orta, Sér. Zool.* **16:**7-17.

Davis, D. H. S., and Misonne, X., 1964, Gazetteer of collecting localities of African rodents, *Doc. Zool.* **7:**1-100.

Depierre, D., and Vivien, J., 1992, *Mammifères sauvages du Cameroun*, O.N.C. / Ministère de la Coopération et du Développement, Fontainebleau.

Diaz Behrens, G., and Van Rompaey, H., 2002, The Ethiopian genet, *Genetta abyssinica* (Rüpell 1836) (Carnivora, Viverridae): Ecology and phenotypic aspects, *Small Carnivore Conservation* **27:**23-28.

Gaubert, P., Veron, G., Colyn, M., Dunham, A., Shultz, S., and Tranier, M., 2002, A reassessment of the distributional range of the rare *Genetta johnstoni* (Viverridae, Carnivora), with some newly discovered specimens, *Mamm. Rev.* **32:**132-144.

Gaubert, P., 2003a, *Systématique et phylogénie du genre* Genetta *et des énigmatiques "genet-like taxa"* Prionodon, Poiana *et* Osbornictis *(Carnivora, Viverridae): caractérisation de la sous-famille des Viverrinae et étude des patrons de diversification au sein du continent africain*, PhD Dissertation, Muséum National d'Histoire Naturelle, Paris.

Gaubert, P., 2003b, Description of a new species of genet (Carnivora; Viverridae; genus *Genetta*) and taxonomic revision of forest forms related to the Large-spotted Genet complex, *Mammalia* **67:**85-108.

Gaubert, P., Tranier, M., Veron, G., Kock, D., Dunham, A. E., Taylor, P. J., Stuart, C., Stuart, T., and Wozencraft, W. C., 2003a, Case 3204. *Viverra maculata* Gray, 1830 (currently *Genetta maculata*; Mammalia, Carnivora): proposed conservation of the specific name, *Bull. Zool. Nomencl.* **60:**45-47.

Gaubert, P., Tranier, M., Veron, G., Kock, D., Dunham, A. E., Taylor, P. J., Stuart, C., Stuart, T., and Wozencraft, W. C., 2003b, Nomenclatural comments on the Rusty-spotted Genet (Carnivora, Viverridae) and designation of a neotype, *Zootaxa* **160:**1-14.

Gaubert, P., Tranier, M., Delmas, A-S., Colyn, M., and Veron, G., 2004a, First molecular evidence for reassessing phylogenetic affinities between genets (*Genetta*) and the enigmatic genet-like taxa *Osbornictis, Poiana* and *Prionodon* (Carnivora, Viverridae), *Zool. Scr.* **33:**117-129.

Gaubert, P., Fernandes, C. A., Bruford, M. W., and Veron, G., 2004b, Genets (Carnivora, Viverridae) in Africa: an evolutionary synthesis based on cytochrome *b* sequences and morphological characters, *Biol. J. Linn. Soc.* **81:**589-610.

Gaubert, P., Taylor, P. J., Fernandes, C. A., Bruford, M. W., and Veron, G., in press a, Patterns of cryptic hybridisation revealed using an integrative approach: a case study on genets (Carnivora, Viverridae, *Genetta* spp.) from the southern African subregion, *Biol. J. Linn. Soc.*

Gaubert, P., Aniskin, V. M., Dunham, A. E., Crémière, C., and Volobouev, V. T., in press b, Karyotype of the rare Johnston's genet *Genetta johnstoni* (Viverridae) and a reassessment of chromosomal characterization among congeneric species, *Act. Ther.*

Gray, J. E., 1864, A revision of the genera and species of Viverrine animals (Viverridae) founded on the collection in the British Museum, *Proc. Zool. Soc. Lond.* **1864**:502-579.

Grubb, P., 2004, Comment on the proposed conservation of *Viverra maculata* Gray, 1830 (currently *Genetta maculata*; Mammalia, Carnivora), *Bull. Zool. Nomencl.* **61**:120-122.

Kingdon, J., 1977, *East African Mammals: an Atlas of Evolution in Africa*, Academic Press, London.

Kingdon, J., 1997, *The Kingdon Field Guide to African Mammals,* Academic Press, San Diego.

Künzel, T., Rayaleh, H. A., and Künzel, S., 2000, *Status assessment survey on wildlife in Djibouti*, Zoological Society for the Conservation of Species and Populations / Office National du Tourisme et de l'Artisanat, Djibouti.

Matschie, P., 1902, Ueber die individuellen und geographischen Abänderungen der Ginsterkatzen, in: *Verhandlungen des V. Internationalen Zoologen-Congresses*, Berlin, pp. 1128-1144.

Morales, A., 1994, Earliest genets in Europe, *Nature* **370:**512-513.

Roberts, A., 1951, *The Mammals of South Africa*, Trustees of "The mammals of South Africa" Book Fund, Johannesburg.

Rosevear, D. R., 1974, *The Carnivores of West Africa*, Trustees of the British Museum (N. H.), London.

Schlawe, L., 1981, Material, Fundorte, Text- und Bildquellen als Grundlagen für eine Artenliste zur Revision der Gattung *Genetta* G. Cuvier, 1816, *Zool. Abhandl. staatl. Mus. Tierk. Dresden* **37:**85-182.

Wenzel, E., and Haltenorth, T., 1972, System der Schleichkatzen (Viverridae), *Säugetierk. Mitt.* **20**:110-127.

Wozencraft, W. C., 1993, Order Carnivora, in: *Mammal species of the world - a taxonomic and geographic reference*, D. E. Wilson and D. M. Reeder, eds., Smithsonian Institution Press, Washington & London, pp. 279-348.

Yalden, D. W., Largen, M. J., Kock, D., and Hillman, J. C., 1996, Catalogue of the mammals of Ethiopia and Eritrea. 7. Revised checklist, zoogeography and conservation, *Trop. Zool.* **9:**73-164.

HOW TO DIGITIZE LARGE INSECT COLLECTIONS – PRELIMINARY RESULTS OF THE DIG PROJECT

Karl-Heinz Lampe[1] and Dirk Striebing[2]

[1]*Zoologisches Forschungsinstitut und Museum Alexander Koenig (ZFMK), Adenauerallee 160, 53113 Bonn, Germany, E-mail: k.lampe.zfmk@uni-bonn.de*

[2]*Museum für Naturkunde der Humboldt-Universität zu Berlin, Institut für systematische Zoologie (ZMB), Invalidenstraße 43, 10115 Berlin, Germany, E-mail: dirk.striebing @museum.hu-berlin.de*

Abstract: In practise the overall efficiency of data-basing the inventory of traditional entomological collections depends on two factors: suitable software and management measures to ensure the highest possible data quality already in the input process. Lessons learned from the development of the specimen-based database BIODAT and preliminary results of the DIG-(Digitization of key Insect groups at ZFMK) project, which is especially designed to develop a 'good practise', recommend: (1) a lockstep programme for data-basing, (2) data entry of collection units and split record function, (3) visualisation of geo-referenced locations/sites during data entry, (4) semi-automatic/automatic data transformation from original format into additional alternative ones, (5) semiautomatic data transfer of taxa- and geo-referenced information units. Current activities deal with the introduction of semantic feedback mechanisms into the practise of data-basing entomological collections.

Key words: specimen based databases; digitization; entomological collections; 'good practise'

1. INTRODUCTION

The aim of DIG (Digitization of Key Insect Groups at ZFMK; www.dig-gbif.de) is to computerize museum collections by digitizing specimens of representative insect orders (Orthoptera, Heteroptera, Homoptera s.l., Hymenoptera and some Diptera) and entering the associated label data into a

B.A. Huber et al. (eds.), African Biodiversity, 385–393.

database that can be used on-site and on the World Wide Web. The specimen based information units will be captured with and stored in BIODAT (www.biodat.de) – a database developed at the ZFMK and already connected both to the BIOCASE - (Biological Collection Access Service for Europe; ww2.biocase.org/abcdSimple/index.html) and to the GBIF- (Global Biodiversity Information Facility; www.gbif.net/portal/index.jsp) portals. Since digitization of mass collections – such as insects – at the specimen level is normally not integrated in the daily work of museum staff worldwide (Berendsohn et al., 1999) a 'good practise' has to be developed a key element of the DIG project.

2. LOCKSTEP PROGRAM FOR DATABASING

Dr. Hausmann from the Munich butterfly and moth collection (ZSM) estimated that 9 min. is needed for data capture for a single specimen in the traditional way (pers. comm.). That does not sound like very much but for 1 million specimens 85 man years would be needed (by working 8 h/day and 220 days/year). Additional time required for a validity check of a geo-referencing process of gathering sites is not considered here. This enormous expenditure of time might be one reason that most curators concentrate their efforts to the digitization of type specimens.

In our home institution we introduced a lockstep programme for data-basing entomological collections (Lampe, 2001) (Table 1). A primary data capture concerning taxonomic information can be followed by a secondary data capture completing all the information which is normally hidden in the collections. A practical example should elucidate the procedure. The situation is as follows: There are some insect collections which are deposited in old traditional cork-bottomed insect drawers. In order to cope with the demands of taxonomic changes and new accessions, pinned specimens eventually need to be transferred into a modern flexible container system, which consists of different sized boxes or unit trays.

Table 1. Lockstep program for databasing the inventory of traditional entomological collections at ZFMK.

1.	primary data capture of taxonomic information
2.	validity check –systematics-
3.	set up of a collection-based catalogue of taxa
4.	secondary data capture of sampling information
5.	validity check –geography- (incl. geo-referencing gathering sites)
6.	set up of a collection-based catalogue of named areas & completion of lists
7.	final data entry of existing specimens into the database

(1) While transferring these insects a primary data capture can be made simultaneously by listing the systematic taxa with their new storage place. (2) The next step is to check the validity of the systematic information against a current catalogue. (3) Next, the updated systematic information including synonyms, hierarchy of taxa, authors, year, etc. can be entered into the database establishing a taxonomic catalogue. At this point the various specimen collections can be merged into one main collection. (4) While the material is at hand again simultaneously a secondary data capture can take place by listing locations (gathering sites), collectors, determinators, etc. (5) The list of locations can be completed in a similar manner to the taxonomic catalogue by adding geographical attributes such as latitude, longitude and in a hierarchy such as province/area, state, country, continent/ocean and the link to the zoogeographical region. (6) The resulting geographical catalogue can be entered into the database like the previous lists. (7) The final data entry can be made now using all the previously formed catalogues and lists just by mouse clicking the information items that are needed. In this step databasing of collection units instead of single specimens speed up the digitizing process (Fig. 1). A 'collection unit' (or data set) in this context harbours all those specimens (males, females, larval stages, etc.), which carry the same identification and collecting dates.
In this example about 6000 data sets harbour the complete or full information of 18800 specimens. A split record function allows an easy update in case that one or more specimens of a 'collection unit' are identified differently from the previous identification or become separated in another way.

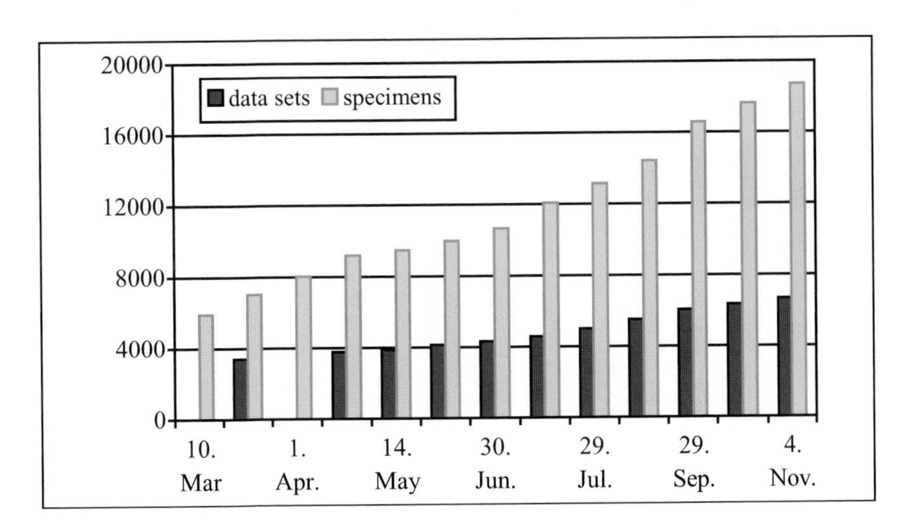

Figure 1. Progress of digitization of ZFMK-Orthoptera in 2003: collection units and specimens.

But how long does it take? For the primary data capture experienced staff require on average 5 min/taxon, for the secondary data capture again an average of 5 min/place name or gathering site and for the final data entry 4 min/collection unit (or about 1.5 min/specimen).

Each part of these proceedings is clearly separated from the next one. As a consequence the progress in the level of accessibility and or availability of the collections is visible for everyone. By profiling the collections in this way, which is similar to the 'Smithsonian Collection Standards and Profiling System' (McGinley, 1989) but more focused to the details and progress of digital accessibility, the whole procedure becomes transparent (Fig. 2).

Everyone involved in this work can clearly see his/her own part for which they are responsible and can follow the progress of their work. And last but not least a nice side effect is the allocation of the various jobs involved where they are most welcome. Someone who is interested in working with catalogues by looking for further geographic or systematic information can work easily together with someone who is more interested in an accurate data capture. That means a single person is no longer forced to complete the various tasks alone. Yet another advantage is that any of these procedures can be stopped or interrupted and even taken over by a third party with very little extra effort. The total numbers of a selection of various specimen based information units captured with and stored in BIODAT are listed in Table 2.

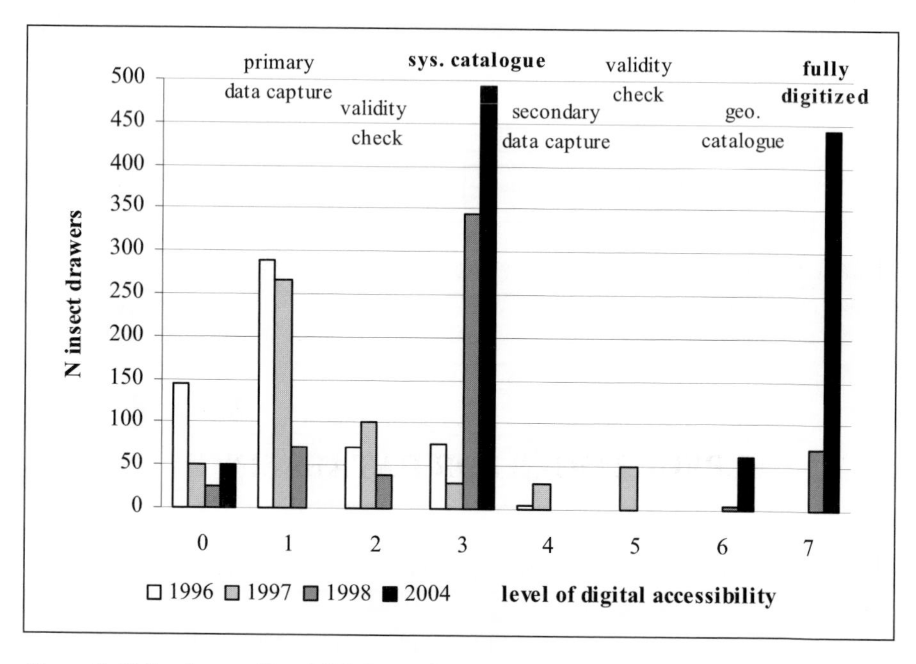

Figure 2. Collection profile of ZFMK section Hymenoptera in terms of digital accessibility (2004).

Table 2. Digitized information units of ZFMK Insect collections 2004 (Diptera curated by B. J. Sinclair, Odonata by B. Misof and remaining groups by K.-H. Lampe).

Taxonomic Catalogue (Level 3)	**Diptera**	**Heteroptera**	**'Homoptera'**	**Hymenoptera**	**Odonata**	**'Orthopteroidea'**	**Phthiraptera**	**Siphonaptera**
Taxa	4524	1808	1811	8520	1213	1905	201	29
Synonyms	275	457	381	2886	-	432	30	2
Genera	727	343	401	1181	219	436	41	4
Species	3517	1164	1168	6641	896	1160	144	9
Geographic Catalogue (Level 6)								
named areas	1403	4835	4835	4835	1	4835	4835	4835
Collection Catalogue (Level 7)								
specimens	**2124**	**4**	**8242**	**5460**	**-**	**20288**	**453**	**28**
localities	461	3	416	442	-	1147	136	9
primary types	194	4	86	51	-	72	6	2
secondary types	1962	-	408	756		154	437	29
images (types)	-	20	1001	603	-	1309	1068	62

It is noteworthy that the lockstep programme of data-basing entomological collections is clearly separated from scientific determination analysis of collection material. Non determined material for example does no longer exist in our data-based collections, because we distinguish different levels of determination. Even the identification of an insect as an insect is a precise determination, of course on a high taxonomic level. One advantage of this point of view is an easier handling of the problem of taxonomic impediment. The higher systematic classification helps us to record and identify where our lack of knowledge lays (Lampe and Riede, 2002).

3. HELPFUL TOOLS FOR DATABASING

Besides the already mentioned (1) lockstep program and the introduction of both (2) collection units and the split record function, Table 3 summarizes helpful tools for efficient databasing of entomological collections. (3) The visualisation of geo-referenced locations (named areas or sites) during data entry allows an immediate identification of outliers if, for example, longitude data are erroneously related to the East instead of the West (Fig.3).

Table 3. Tools to ensure data quality and to accelerate data input

1.	Lockstep program
2.	Collection units/split record function
3.	Visualisation of geo-referenced places/sites during data entry
4.	Automatic data transformation from original format into alternative ones (considering time- and spatial spans)
5.	Semiautomatic data transfer of taxa- and geo-referenced information units (ITIS, ADL)
6.	semantic feedback mechanisms (→ Perspectives)

A differentiation between named areas (listed in the geographic catalogue [e.g. Bonn] and linked to the respective specimens) and specific data on the gathering sites (directly attached to the respective specimen [e.g. a special birch tree in Bonn]) avoids a 'hypertrophic' geographic name index and improve the precision of the stored geographic information. (4) Other tools allow an automatic data transformation from original format into additional alternative ones. In this context the 'logidate' function transforms for example original label information consisting of 'VII 1913' to a time span ranging from July 1-31, 1913. In a similar manner geographical coordinates captured in degree, minutes and seconds can be transformed into decimal degrees and vice versa. Thereby the min/max values can be saved as line or alternatively as a bounded box. Quite similar the altitudinal information is captured in the original format (feet/meter/inch/fathom) with min/max values and additionally converted into alternative units. (5) The next point (still under construction at ZFMK) is a semiautomatic data transfer of taxa- and geo-referenced information units from remote or distinct authority files

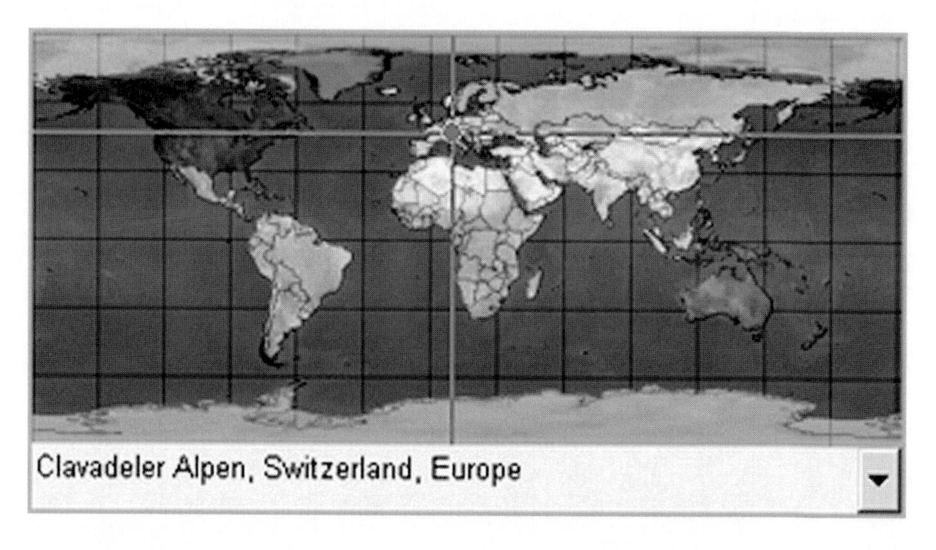

Figure 3. BIODAT's feedback during geographic data entry (screenshot).

MONITORING LANDCOVER CHANGES OF THE NIGER INLAND DELTA ECOSYSTEM (MALI) BY MEANS OF ENVISAT-MERIS DATA

Ralf Seiler[1], Elmar Csaplovics[2], and Elisabeth Vollmer
Department of Geosciences, University of Dresden, Helmholtzstrasse 10 13, D-01062 Dresden, Germany, E-mail: [1]rseiler@rcs.urz.tu-dresden.de [2]csaplovi@rcs.urz.tu-dresden.de

Abstract: The Niger Inland Delta is one of the most fragile ecosystems in the Sahelian zone of Africa. Annual flooding of vast plains during September to December depends on water levels of the main rivers Niger and Bani. The interaction among pre-flood, flood and post-flood conditions strongly affects land cover patterns in and around the delta. Assessing and mapping land cover dynamics is rendered possible by systematic application of remote sensing and image analysis. Thus support a better understanding of state and changes in the vulnerable ecosystem. Operational missions of earth observation contribute to the establishment of multi-seasonal and multi-annual monitoring schemes. This paper aims to show the potential of spectral high-resolution imaging spectrometry (MERIS) for monitoring spatio-temporal patterns of land use and land cover change in the Niger Inland Delta. 5 MERIS scenes, covering one vegetation-cycle from August 2002 until June 2003, have been used therefore.

Key words: multispectral remote sensing data; land cover classification; land cover changes

1. INTRODUCTION

Amount and spatial distribution of rainfall in Sahelian Africa vary extremely due to dynamics of northward moving ITCZ during the summer and due to the local and short-term character of the rainfalls. Niger Inland Deltas water regime is affected by annual variations in rainfall at the watersheds of upper courses of the Niger and Bani rivers in the highlands of *Futa Djallon*. Rising water levels arrive at the western to southern fringes of

B.A. Huber et al. (eds.), African Biodiversity, 395–404.

the delta in early to mid-October and fade away in January along the northern boundary of the delta. In consequence, large parts of the extremely flat region are flooded during early dry season (ORSTOM, 1986). Inundation waters the numerous flat pans and channels (*mare, mayo*) and induces remarkable growth of vegetation of high grazing value (*bourgoutière – vetiveria nigritana*). From this water regime result two seasonal variations (rainy-dry and flooding-drainage phase). In addition to these short-time periodical changes, a third variation has to kept in mind, which counts several years (between dry years and years with sufficient precipitation). This latter cycle is dominantly affected by the two seasonal cycles, but high spatial variability of precipitation does not permit a causal linkage. In particular, low amount of rainfall in the delta may profit from extended rainfall in the headwaters, thus inducing a reasonable extent of flooding.

Availability of water represents the main restricting factor for vegetation growth in the Sahel. Vegetation follows the above-described water cycles with a temporal delay, which may vary from a few days (germination of grasses) up to several months (death of trees caused by lack of water). Development of (annual) grasslands with sparsely distributed patches of shrubby vegetation dominantly composed of Combretaceae species is characteristic for the Sahelian landscape (Breman and DeRidder, 1991; Csaplovics, 1992). According to Csaplovics (1998) and LeHouérou (1989) vegetation can be categorised into three layers: (1) grass layer with annual grasses and herbs (height 40-80 cm); (2) shrub layer (height 50-300 cm); (3) tree layer, sparsely distributed single trees (height 3 m to 6 m).

Ligneous layers of shrubs and trees cover only small parts (up to 25%), while grassy layers extend over up to 80% of the surface (Kußerow, 1995). Annual grasses are withering during the dry season. Thus grassy layers are affected and/or destroyed by bush fires and strong winds. Areas of bare soil appear and extend during mid- and late dry season.

Besides semi-natural vegetation cover, large areas of the delta are used for rainfed (sorghum) and irrigated (rice) agriculture. Annual and perennial grasses prevail within the banks of the *mare* and along the *mayo*. These areas serve as pasture for nomadic or transhumant cattle breeders in late dry season. They are especially valuable, as *mare*s outside the delta are parched during mid-dry season at the latest. See Figure 1 for a landscape profile of the Inland Delta.

2. MERIS DATA CHARACTERISTICS

Medium Resolution Imaging Spectrometer (MERIS) as a payload of European environmental Satellite ENVISAT was launched into space in

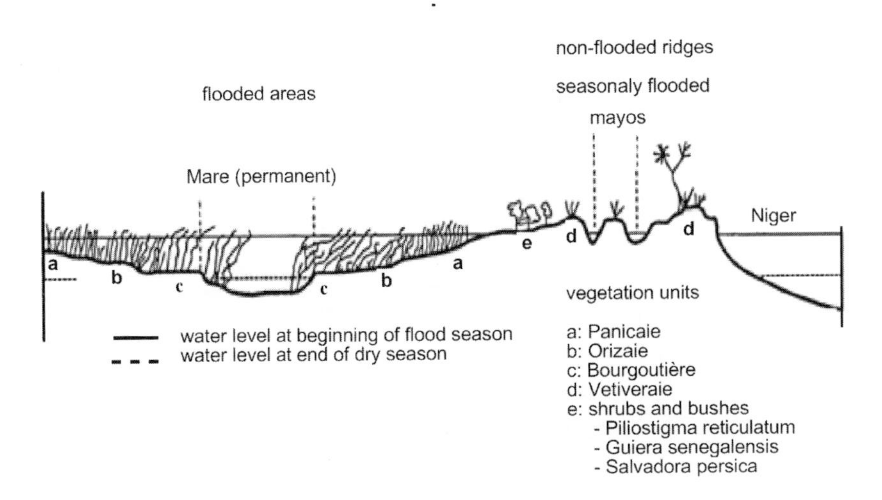

Figure 1. Landscape profile of a subsection of the Niger Inland Delta. Adapted from Diallo (2000).

March 2002. The optical sensor collects data at full mode with a spatial resolution of 300 m or at reduced mode with 1200 m resp., thus allowing for landcover analysis at regional to global scale. The instrument measures reflected *T*op *O*f *A*tmosphere radiances in 15 bands in the spectral range 408 - 905 nm. Coverage of the entire earth surface within an interval of 3 days is possible due to a swath width of 1150 km.

MERIS offers great potential for detection of absorption characteristics related to seasonal and annual variations in the photosynthetic activity of leaf pigments (chlorophyll a and b, antocyanine or carotenoids), thanks to its narrow bandwidths of mostly 10 nm. Ratios of near-infrared and red bands allow for computation of vegetation indices. The analysis of water colour provides information on the condition of open water surfaces, thus allowing for the detection of variations in sediment concentration indicating occurrence and extent of current. As water reflectance or absorption features occur in very small bandwidths of the electromagnetic spectrum (EMS), MERIS will allow for new applications in this field. The position of MERIS spectral bands within the EMS with regard to spectral signatures of certain surface features such as soil, vegetation or water is shown in Figure 2.

A limiting factor is the geometric resolution of MERIS data. As each MERIS pixel covers 90,000 m², monitoring at regional to global scale is possible. Investigation at the local level is restricted to areas with a homogenous landcover situation. While this restriction may sound rather deplorable, one has to keep in mind that MERIS offers improved spatial resolution anyway compared to long living NOAA-AVHRR and even in relation to MODIS.

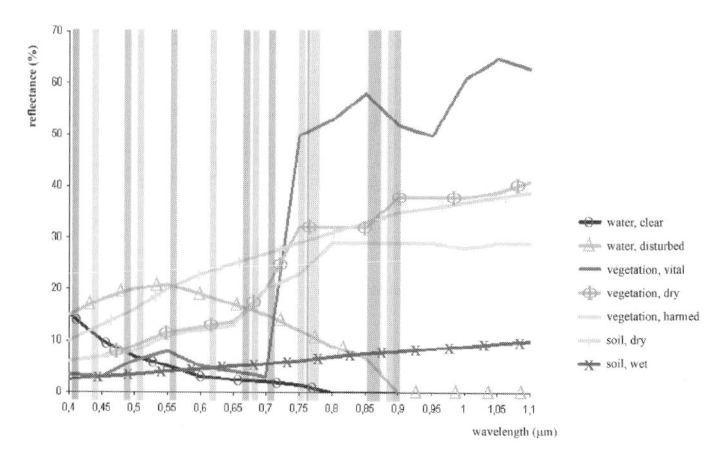

Figure 2. Position of MERIS spectral bands and spectral signatures of main surface features.

Research focuses on the central Inland Delta itself and the adjacent *Plateau du Bandiagara*, which is strongly related to the delta both in terms of socio-economic as well as socio-ecological aspects. The region of interest lies in between 15°37'N, 4°40'W / 13°38'N, 3°26'W. Seasonal changes were analysed with a time series of 5 datasets from 2002/03. August displays the situation at the end of the rainy season. October coincides with the beginning of the flood period. Imagery of December provides information about the flood extension. April data documents the situation at mid-dry season and June imagery shows late dry season.

3. STATISTICAL ANALYSES OF MERIS DATA

MERIS data cover the visible and near infrared part of the EMS. Therefore a high correlation between individual bands could be expected. Consequently we found correlation coefficients 0.8 - 1.0 for bands 1 to 5; 6 - 10 and 12 - 15 as shown in Table 1. Even band 11, which is strongly affected by oxygen absorption, shows high correlation to neighboured bands. One can easily recognise three groups of bands in Table 1. While the first five bands (408 to 565 nm) fall into the field of optical active water substances (e.g. Gelbstoff) the second group – band 6 - 10 - (615 to 756 nm) is domi-nantly affected by chlorophyll absorption/fluorescence features. The remaining bands (771 to 905 nm) lie at the near infrared part of the EMS and therefore serve in the field of land applications as indicator for vivid green vegetation and biomass.

Table1. Mean correlation coefficients for 5 datasets. MERIS bands 1 – 15

Band	1	2	3	4	5	6	7	8	9	10	11	12	13	14	15
1	1	0.99	0.96	0.95	0.87	0.69	0.64	0.63	0.59	0.47	0.55	0.46	0.46	0.46	0.44
2		1	0.99	0.98	0.92	0.75	0.71	0.70	0.66	0.54	0.61	0.53	0.54	0.53	0.51
3			1	1.00	0.95	0.80	0.76	0.75	0.72	0.59	0.65	0.59	0.60	0.60	0.57
4				1	0.97	0.82	0.79	0.78	0.75	0.62	0.68	0.62	0.63	0.63	0.60
5					1	0.92	0.89	0.88	0.87	0.77	0.79	0.76	0.74	0.74	0.71
6						1	1.00	0.99	0.98	0.84	0.81	0.83	0.78	0.77	0.75
7							1	1.00	0.98	0.84	0.80	0.83	0.78	0.77	0.76
8								1	0.98	0.84	0.80	0.83	0.78	0.77	0.76
9									1	0.92	0.88	0.92	0.87	0.86	0.85
10										1	0.95	1.00	0.97	0.96	0.95
11											1	0.95	0.94	0.94	0.92
12												1	0.98	0.97	0.96
13													1	1.00	0.99
14														1	0.99
15															1

High correlation values between the MERIS bands indicate that much of the information is contained more than once within the MERIS data. Thus a decorrelation of bands promises reasonable reduction in the amount of data by keeping all the information that is within the MERIS bands. Such a decorrelation can be calculated with a principal component analysis. PC analysis results in independent new bands, the principal components, where the information content fall within the first few bands and the signal-noise in the last few. It came out that the first six PCAs contain almost 99.95 % of the total information of all 15 MERIS bands. Table 2 gives the mean information content for the principal components of all five MERIS scenes. (Note that the norm of an eigenvalue represents the relative content of information).

4. CLASSIFICATION OF OPEN WATER AREAS

Several large lakes as well as the river Niger are situated within the region of interest. While the 250 to 400 m broad Niger is displayed at MERIS data by a line of only 1 pixel, larger lakes cover areas up to 4 by 8 km (14 by 27 pixels). Even smaller *mare* in the central part of the Inland Delta are detectable by visual interpretation of flood period data. MERIS Level 1B records contain a multitude of auxiliary data along with the image information itself. A land/ocean flag is among them. Unfortunately this flag seems not to work properly for inland lakes, as some lakes are masked, some not and others in a wrong manner.

Table 2. Mean eigenvalues of PCs from 5 MERIS datasets

PC	Norm of eigenvalue	% of total information	Cumulative
1	0.04334	89.809%	89.809%
2	0,00345	7,411%	97,220%
3	0,00095	2,274%	99,494%
4	0,00014	0,297%	99,791%
5	5,3E-05	0,119%	99,910%
6	2,3E-05	0,050%	99,960%
7	7,2E-06	0,016%	99,976%
8	3,8E-06	0,009%	99,986%
9	1,6E-06	0,004%	99,990%
10	1,4E-06	0,003%	99,993%
11	1,0E-06	0,003%	99,995%
12	7,9E-07	0,002%	99,998%
13	5,5E-07	0,001%	99,999%
14	3,2E-07	0,001%	100,000%
15	2,0^{E}-08	0,000%	100,000%

To avoid this uncertainty, we decided to establish an empirical approach. Water surfaces reflect in the red and near-infrared part of the EMS to a lesser extend compared with land surfaces. Thus we set up a threshold for reflectance in band 13. Figure 3 shows classification results for the central part of the Niger Inland Delta. It remained to conclude that August 2002 data is affected by some kind of artefacts. It seems that in this case the threshold misclassified some invalid pixel as open water areas. But as the water surfaces reflected nearly no radiation in this data record a separation remains difficult.

5. VEGETATION COVER ANALYSES

A second land cover feature we analysed so far is the photosynthetic active vegetation. For measuring biomass figures we calculated the widely introduced NDVI for every data as ratio of MERIS band 8 (680 nm) and band 13 (865 nm). Considering the fact that natural vegetation is rather sparsely distributed in the Sahel, we compared NDVI with SAVI (*S*oil *A*djusted *V*egetation *I*ndex). The latter Index allows for the soil-induced part of the measured signal by a correction term L ($0 \leq L \leq 1$), see also eq. 1. If one can assume that the remote sensing signal is not influenced by the soil (e.g. in case of dense vegetation), correction L = 0 yields to NDVI as a

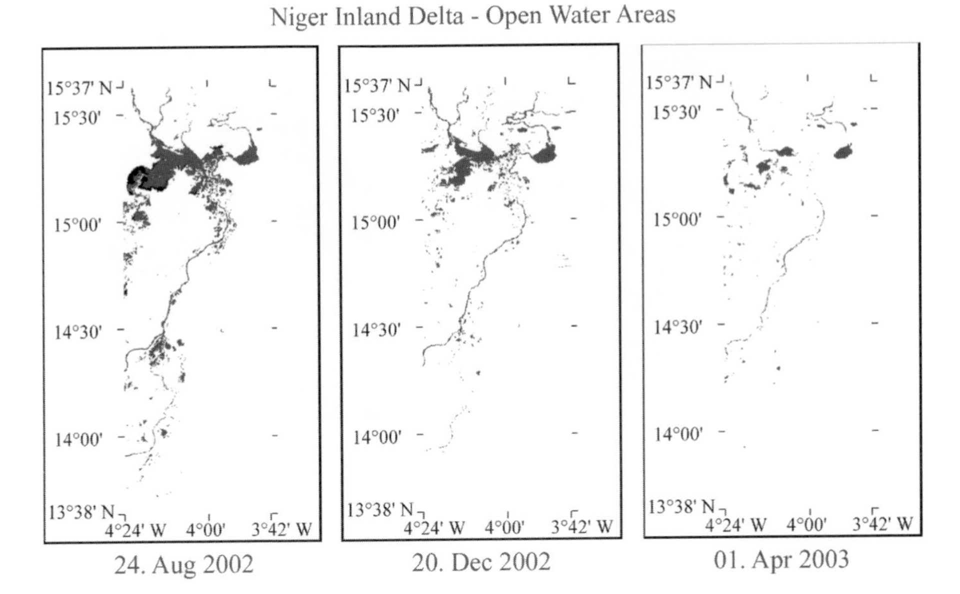

Figure 3. Open Water Areas at central Inland Delta for August 24, 2002, December 20, 2002, and April 01, 2003.

special case of SAVI. The stronger the influence of the soil component the higher values should be used for L. As the vegetation cover is not dense at all, we decided to work with L = 1.

$$\text{(eq. 1)} \qquad \mathbf{SAVI} = \frac{(\mathbf{NIR} - \mathbf{RED})}{(\mathbf{NIR} + \mathbf{RED} + \mathbf{L})} * (\mathbf{1} + \mathbf{L})$$

As an example for the situation during inundation serves Figure 4. This displays SAVI for Dec. 20, 2002. A comparison of NDVI with SAVI in Figure 5 indicates that the correction term L amplifies the vegetation signal. In case of low vegetation signal – VI produce small numbers – SAVI values are below NDVI values. For high VI numbers SAVI is above the NDVI. Although it remains to conclude that a unique correction value for the entire research area doesn't model the situation at any time. This becomes most evident during flood period when the Inland Delta is covered by a comparatively close grass layer. To overcome this problem, further investigations at a smaller scale will be necessary.

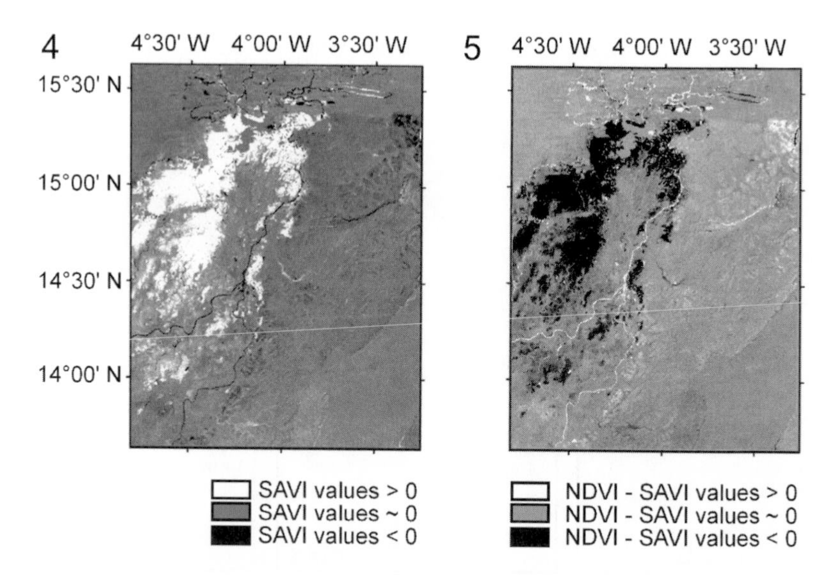

Figures 4 and 5. **4**. SAVI for 20. December 2002. L = 1. **5**. Difference Image NDVI – SAVI for 20. Dec. 2002. L = 1 (SAVI > NDVI indicated by dark areas; SAVI < NDVI indicated by light areas).

6. MONITORING CHANGES – SEASONAL DYNAMICS

As a first step to monitor annual vegetation cycles, we calculated difference images from the 5 SAVI images. Figure 6 shows one difference exemplarily for the time period from end of the rainy season to high tide of inundation. As one might expect, the resulting difference image indicates for the semi-arid neighbourhood of the Inland Delta a shrunken biomass. But the Delta shows half a year after the end of rainy season still a higher amount of biomass than during rainy season itself. This is strong evidence for the dominant influence which inundation has upon the vegetation cover at the Inland Delta.

7. OUTLOOK

This paper presents early results of investigations that have been undertaken to validate the potential of MERIS data for multitemporal monitoring of land use / land cover changes of the Niger Inland Delta. It has been determined that MERIS Level 1B full resolution data are suitable for

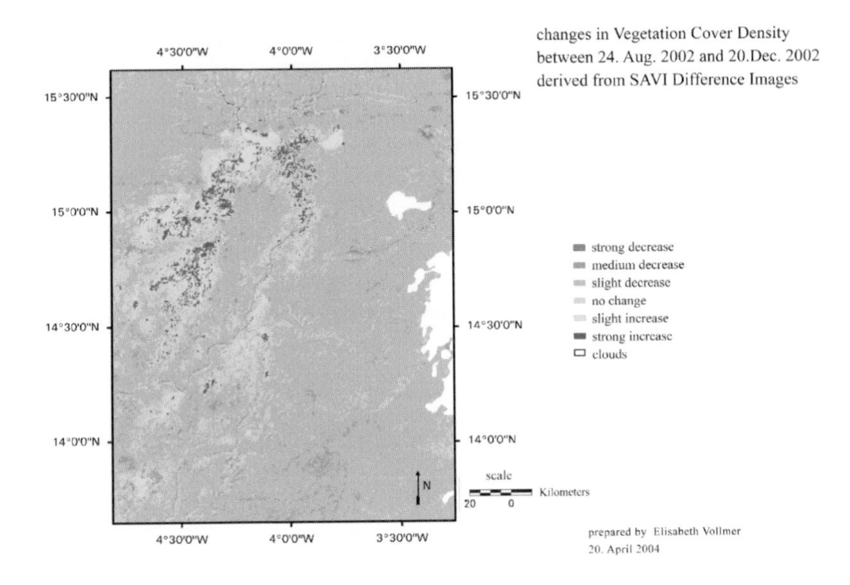

Figure 6. Difference Image. SAVI Dec. 20, 2002 minus SAVI Aug. 24, 2002.

detecting annual variations in vegetation (cp. biomass) cycle for wetlands in semi-arid regions. The low spatial resolution limits classification prospects mainly for open water surfaces. To reduce the huge amount of data within the 15 highly correlated MERIS bands a principal component analysis might be worth the effort, that is necessary to resolve the meaning of principal components. It is intended to extend the data record over more than one year, allowing investigation of longer periodical aspects such as soil degradation processes.

ACKNOWLEDGEMENTS

This study is partly funded by K. Adenauer Stiftung and ESA –AO 776. Elisabeth Vollmer has prepared the vegetation analysis. We'd like to thank also the referees for their comments on this paper.

REFERENCES

Breman, H., and DeRidder, N., 1991, *Manuel sur les pâturages des pays sahéliens*, Karthala, Paris.

Csaplovics, E., 1992, *Methoden der regionalen Fernerkundung – Anwendungen im Sahel Afrikas*, Springer, Berlin New York.

Csaplovics, E., 1998, Integrative Methoden der Dokumentation und Analyse lokaler Desertifikationsprozesse im Sahel Afrikas - zur Bedeutung von Fernerkundung und geographischem Informationssystem, *Die Erde* **129:**195-210.

Diallo, O. A., 2000, Contribution à l'étude de la dynamique des écosystèmes des mares dans le Delta Central du Niger au Mali, Thèse, Université Paris I.

Kußerow, H., 1995, Einsatz von Fernerkundungsdaten zur Vegetationsklassifizierung im Südsahel Malis, Verlag Dr. Köster, Berlin.

LeHouérou, H.N., 1989, *The grazing land ecosystem of the African Sahel*, Springer, Berlin-New York.

MERIS handbook. http://envisat.esa.int

ORSTOM, 1986, (ed.), *Monographie hydrologique du fleuve Niger, tomes I, II*, ORSTOM, Paris.

IS GEODIVERSITY CORRELATED TO BIO-DIVERSITY? A CASE STUDY OF THE RELATIONSHIP BETWEEN SPATIAL HETEROGENEITY OF SOIL RESOURCES AND TREE DIVERSITY IN A WESTERN KENYAN RAINFOREST

Winfred Musila[1, 2], Henning Todt[1], Dana Uster[1] and Helmut Dalitz[1]
[1] *University of Hohenheim, Institute of Botany (210), 70593 Stuttgart, Germany*
[2] *National Museums of Kenya, Nairobi, Kenya*

Abstract: Interactions of biotic and abiotic factors are critical in determining population dynamics and maintenance of high diversity in tropical rainforests. An investigation was conducted to test the hypothesis that tree species composition and dispersion influence spatial pattern of soil properties on a small scale of meters. Soil variables (pH, EC and total and water extractable (available) K, Ca, Mg and Mn) were highly heterogeneous in both vertical and horizontal dimensions. Available Mn was the most variable (coefficient of variation CV = 76%) parameter and pH the least (CV = 4%). There was an inverse relationship between abundant levels of available cations and heterogeneity. Plots with high heterogeneity in available cations had low tree diversity but there was no evident pattern between total concentration of cations and tree diversity. Tree diversity was negatively correlated to variability of available Mn ($r = -0.7$). These findings indicate that tree diversity influences soil heterogeneity hence contributing to maintenance of high diversity in tropical rainforests. The results show that the observed soil patterns are caused by geogenic processes as well as processes influenced by vegetation. The raising effect of the vegetation on the elements is expected in the order from Mn, Ca, Mg to K. This vegetation effect indicates the significance of maintaining a closed canopy in tropical rain forests. Geodiversity is thus correlated to biodiversity and needs to be more recognized in ecological research.

Key words: biodiversity; geodiversity; Kakamega rainforest; spatial heterogeneity; tree diversity

B.A. Huber et al. (eds.), African Biodiversity, 405–414.

1. INTRODUCTION

Multiple processes such as competition, predation, disturbance, resource heterogeneity have been proposed to explain the high diversity of tree species in tropical rainforests. A major biotic force is envisioned that is complemented by one or more other forces. However, abiotic factors are least appreciated among the mechanisms that interact with biotic factors in determining population dynamics and distributions. According to Barthlott et al. (1999), total diversity of an area is an outcome of interdependencies of the biotic and abiotic components, or biodiversity and geodiversity respect-tively. Thus, diversity not only depends on species diversity but also on the diversity of abiotic factors such as climate, soil, etc.

With the current unprecedented rate of deforestation of tropical rain-forests, it is crucial to understand the integrative role of biodiversity and geodiversity in order to conserve and manage tropical rainforests sustainably. Whilst biodiversity is studied by identifying and counting an indicator group such as plants, geodiversity is investigated by searching for spatial heterogeneity of an ecologically relevant parameter. Spatial hetero-geneity refers to variability of a parameter in space and time in terms of the range of values and also of the different kinds. Spatial heterogeneity is fundamental to the structure and dynamics of ecosystems (Levin, 1992).

The relationship between spatial pattern in soils and plants is less understood. Spatial heterogeneity in soil properties may underlie a part of the differences in tree species diversity. An investigation was conducted to establish whether spatial heterogeneity of soil properties is reflected on tree diversity in a tropical rainforest. Our hypothesis was that high spatial soil heterogeneity leads to high tree diversity. This hypothesis is part of a proposed positive feedback mechanism (Dalitz, 2004) which describes the influence of tree species diversity and dispersion on leaching processes, light conditions in the understorey and differences of water and nutrient input to soil. These differences may cause different conditions of various soil properties in space, which determine the conditions under which seedlings may establish. Changes in seedling establishment may then lead to changes in tree species composition. But, trees may also influence soil properties because of their different demands for nutrients and water. These processes range in small scales like meters or less. The objectives therefore were to characterize the small-scale spatial pattern of soil chemical properties and examine their relationship with tree diversity. The soil properties investigated were pH, Electrical conductivity (EC), total and water extractable (available) fractions of nutrient cations, i.e. K, Ca, Mg and Mn. Soil pH is a useful attribute, which affects directly and indirectly the ability of plants to utilize nutrients. Soil EC is a measure of ion concentration and is an indicator of those soil properties affecting it such as nutrient, texture and soil moisture. The nutrient cations indicate different pattern and processes regulating the nutrient cycling process. These soil attributes were selected to describe the soil diversity because they are easily measurable since a large amount of soil samples is required to investigate soil variability on a small scale. It is also possible to detect subtle changes among these parameters.

2. MATERIALS AND METHODS

2.1 Study site

The research was conducted in Kakamega rainforest, located in Shinyalu division of Kakamega District in the Western Province of Kenya (Kokwaro, 1988). It lies between latitudes $0^{o}10'$N and 0° $21'$N and longitudes 34^{0} $47'$E and 34^{0} $58'$E at an altitude of 1500m and 1600m above the sea level (Ojany and Ogendo, 1987). The mean annual rainfall recorded at Kakamega Forest Station is 2215 ($\pm$ 26mm) (Cords, 1987). Kakamega forest is comprised of a mixture of degraded forest sites, shrublands and grasslands (glades). Within the forested sites a high number of tree species occur with some more frequent species such as *Funtumia africana, Prunus africana*, *Diospyros* sp. *Olea capensis* and *Celtis africana*. The soils are heavily weathered, deep to very deep, clayey, acidic and of low fertility (Musila et al, 2004).

2.2 Methods

The investigation was done within triplicate plots of 20 m by 20 m in four sites. The distance within a triplet is around 200 to 300 m and the distance between the triplets is between 1000 and 1500 m. All tree species with more than 5 cm diameter at breast height (DBH) were identified, marked and mapped within each plot. Soil was sampled more intensively than normal in 25 points at 5 m intervals in 7 horizons up to 2 m depth. All field data were collected in 2001. The soil samples were air-dried, ground and sieved to pass through 2 mm sieve. Electrical Conductivity (EC) and pH were determined potentiometrically in 1:2.5 soil/deionised water suspension. Water extractable (available) and total concentrations of K, Ca, Mg and Mn were determined by Atomic Absorption Spectrometer (Perkin Elmer 5100 model). Available cations were extracted with millipore water while the total concentrations with concentrated HNO_3 under pressure (Heinrichs et al., 1986).

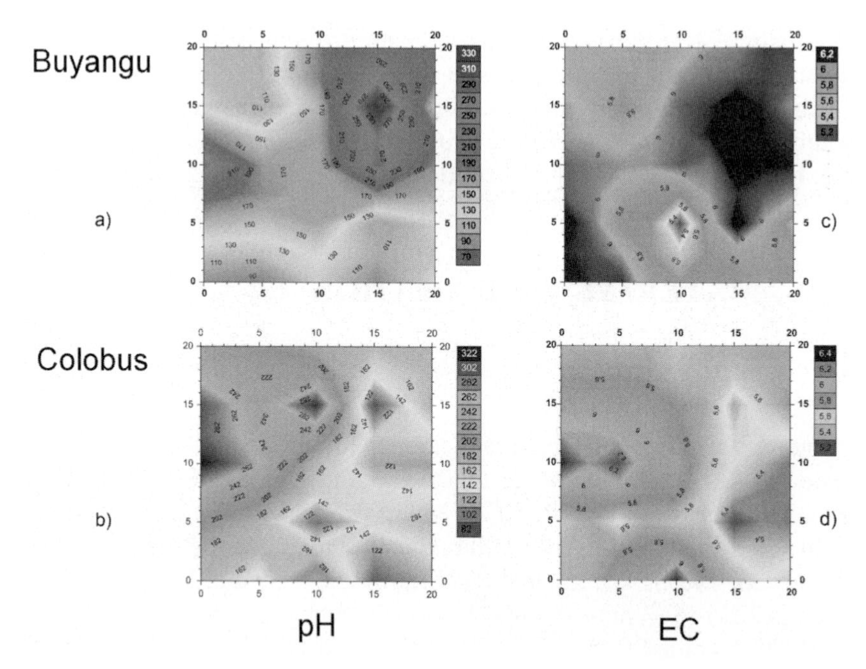

Figure 1. Spatial heterogeneity of pH (a and b) and EC (c and d) in 20 m by 20 m plots for the soil horizon 0 to 10 cm. Part a and b compares pH between Buyangu and Colobus, whereas part c and d show spatial differentiation of EC. Both sites are at Kakamega forest in the northern part.

2.3 Calculations and statistical analysis

Shannon-Wiener diversity index was used to calculate the tree diversity per plot. The variability of each soil variable was determined by calculating the coefficient of variation (CV) in each horizon within the 12 plots in 4 sites. A Two-Way cluster analysis of the CVs of the elements was done to show how combinations of the elements vary from one plot to the others. Pearson correlation coefficients were calculated to determine if there were relationships between spatial soil heterogeneity and tree diversity. Statistical significance was set at $p<0.05$. Statistical analyses were performed with STATISTICA ('99 Edition; Version 5.5 StatSoft Inc.; Tulsa; USA), and XACT (Version 7.21; SciLab GmbH; Germany).

3. RESULTS

3.1 Spatial heterogeneity of soil properties

All soil variables (pH, EC and total and available K, Ca, Mg and Mn) were spatially structured with zones of aggregation with high values and zones of depletion with lower values. This is true for both vertical and horizontal dimensions. Examples for the spatial differentiation between plots in horizontal dimensions are given in Figure 1 showing the different pattern

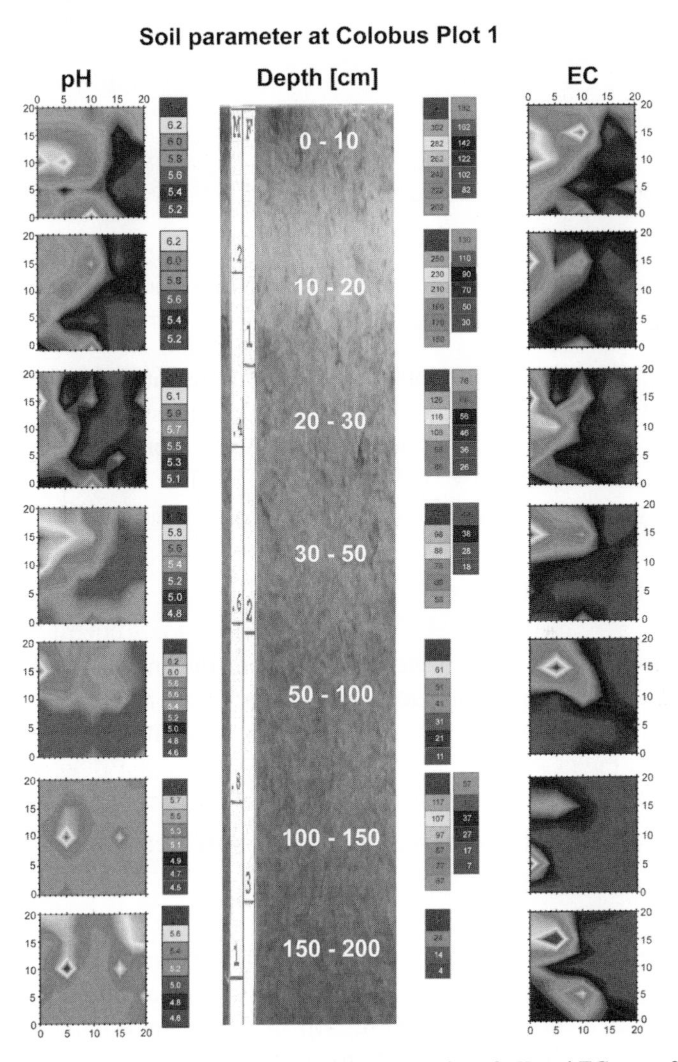

Figure 2. The graph shows vertical and spatial heterogeneity of pH and EC up to 2m depth in one plot. EC values are given in µS • cm^{-1}. For each horizon of the soil the spatial differentiation is shown as interpolation graphs.

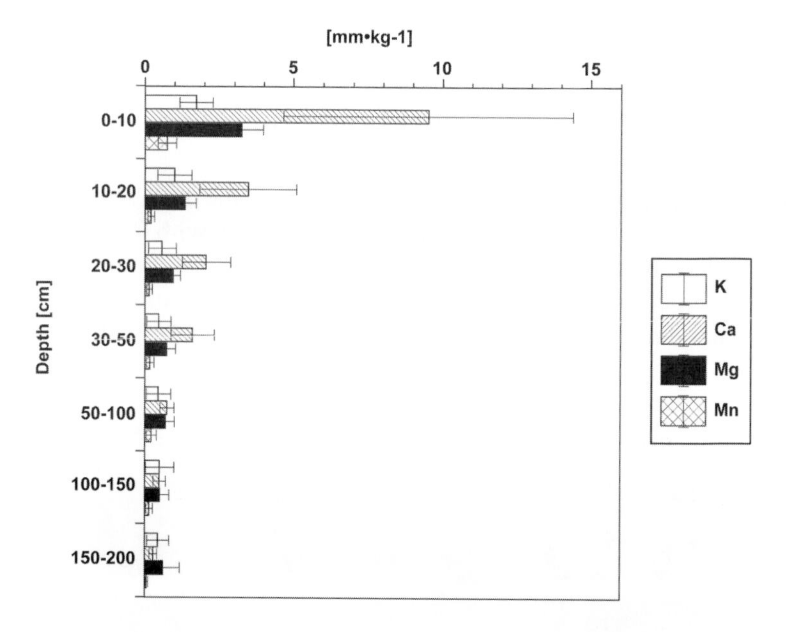

Figure 3. Vertical distribution of available cations in Colobus plot 1 for the elements K, Ca, Mg and Mn.

of EC and pH in two of the plots. Figure 2 demonstrates the changes in the pattern of EC and pH in one plot in the vertical dimension indicating that the spatial pattern is more pronounced in the upper horizons. The available cations, pH and EC decrease with depth (Fig. 3). The most abundant cation was Ca and the least abundant and was Mn.

3.2 Tree diversity

A total of 99 tree species were recorded in the investigated plots. The most frequent species species were *Antiaris toxicaria*, *Heinsenia diervillioides* and *Funtumia africana*. The tree diversity ranged from 3.5 to 4.2.

3.3 Relationship between spatial soil heterogeneity and tree diversity

In Figures 4 and 5 the Shannon-Wiener diversity index (Fig. 4a and Fig. 5a) are compared with a two way cluster analysis of available cations (Fig. 4a) and total content of cations (Fig. 5b) in the 12 investigated plots. The cluster analysis of the available cations in the 12 plots showed that Mn and K were the most heterogeneous elements (Fig. 4b). Neighboring plots within the same triplet like S1, S2 and S3 showed differences in the diversity

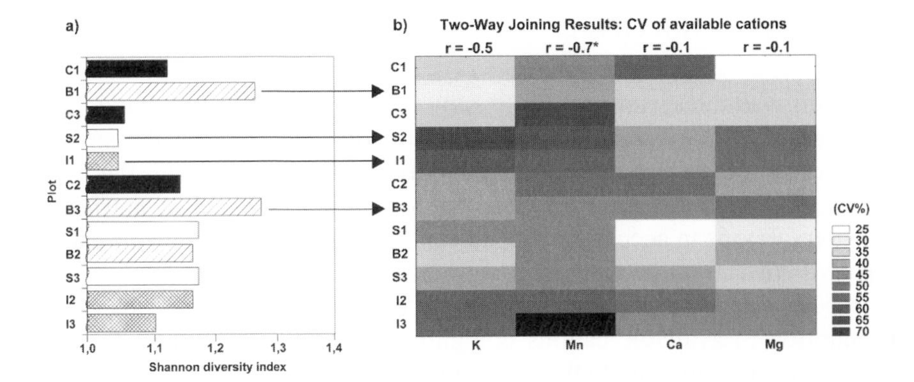

Figure 4. Part a shows the tree diversity within 12 plots in Kakamega forest and part b shows a two-way indicator cluster analysis of CV of available cations within the 12 plots. The correlation between the spatial heterogeneity of each element and the tree diversity is shown on top of each column.

index and, in the heterogeneity of the elements. The plots with high available cation heterogeneity (S2, I1) were found to have low tree diversity, whereas plots with high Shannon-Wiener values showed either a low cation heterogeneity (B1) or an intermediate (B3). There was a significant negative correlation between the variability of available Mn and tree diversity (r = -0.7). The analysis of the heterogeneity of total contents of the cations showed that Ca and Mg were the most heterogeneous elements (Fig. 5b). There was no clear pattern in the variability of the total contents between the plots and the tree diversity (Fig. 5a).

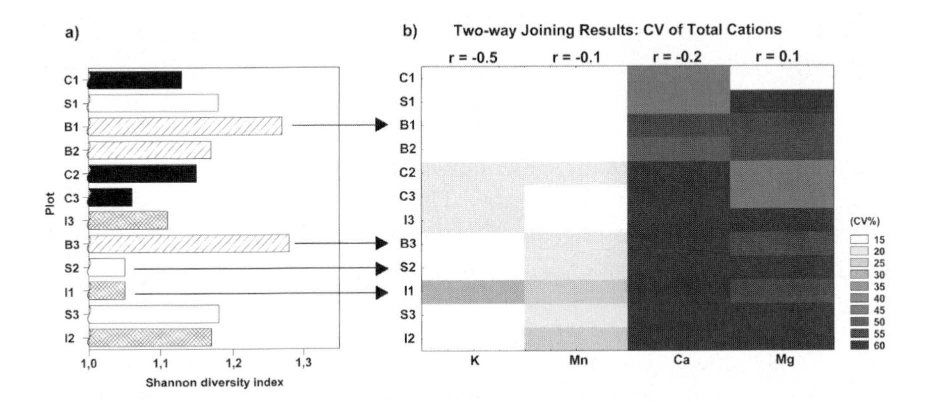

Figure 5. Part a shows the tree diversity within 12 plots in Kakamega forest (different order but same data like in 4a) and part b shows a two-way indicator cluster analysis of CV of total cations within the 12 plots. The correlation between the spatial heterogeneity of each element and the tree diversity is shown on top of each column.

4. DISCUSSION

All soil variables were spatially structured in vertical and horizontal dimensions. We found a general negative relationship between the spatial heterogeneity of the water extractable (available) cations and the tree diversity but there was no correlation with the total contents of nutrients. This observation is in accordance to Huston's (1980) findings, who observed a negative correlation between soil nutrient (K, Ca, & Na) availability and tree species richness but not with Mg and Mn in Costa Rica. The concentration of available cations is highly fluctuating and highly affected by vegetation as compared to the total contents, which are depended more on the parent material of a site. This suggests that soil nutrient heterogeneity of the water extractable fraction is influenced by vegetation. Trees create heterogeneity in nutrient distribution, initially by accumulating nutrients in tissues and creating zones of depletion around roots. The zones of depletion may extend for a few millimeters or more depending on the nutrient and the demands of the trees. Several studies have demonstrated the existence of tree effects on soil properties (Zinke, 1962; Hobbie, 1992; Van Breemen, 1993). Therefore the measurement of water extractable nutrients may give a good indication for the influence of vegetation on soils. This is also true for the high mobile ions of potassium because the pattern is similar in the vertical dimension even when it is extenuating with depth.

The pattern of each soil attribute indicates the dominant process affecting it. The pH values are increasing from deeper layers, indicating the effect of the organic layer by addition of more bases in the topsoil. Although most of the cations are leached down the profile, Ca and Mg are added through the organic layer in large amounts and retained in the soil. This may explain why the total contents of Ca and Mg are not reflected in the tree diversity. Potassium is a highly mobile and easily leached element and this may explain why it shows a remarkable correlation with vegetation, though not significant. Available Mn was the least abundant element but the most variable (CV = 76%). This reflects an inverse relationship between resource levels and heterogeneity. Manganese shows highly changing availabilities in soil due to its different oxidation status. The availability of Mn is controlled by the soil but in this case vegetation seems to influence the availability of Mn through retention of water. The observed soil patterns are hence caused by geogenic processes as well as processes influenced by vegetation. Our results show that a raising effect of the vegetation on the elements is expected in the order from Mn, Ca, Mg to K. This vegetation effect indicates the significance of maintaining a closed canopy in tropical rain forests. This is especially true when leaching processes from the canopy to soil were taken into account, which indicate that K is most leachable element followed by Mn. The spatially differenciated canopy leads to huge spatial heterogeneities in nutrient support from the canopy (Dalitz et al., 2004).

The choice for investigating the water extractable fraction of nutrient elements in the soils reflects the idea that short-term effects are easier to detect than long-term effects. This corresponds to results from Oesker et al (2004a, b) who found that different species of trees have different losses of nutrients out of leaves. This effect of species-specific nutrient leaching processes results in spatially differentiated nutrient inputs onto the soil, which can be utilized by the same or by other tree individuals.

In most studies, a single measurement or mean values of soil properties are used to describe the soil characteristics of a site and soil nutrient variability is ignored. Our results show that this may give a misleading picture since spatial heterogeneity of soil attributes exists at finer scale of less than 5m. The study also shows the importance of using a sampling grid to assess for the variation of soil attributes with respect to the spatial location of the sampling points and to detect systematic effects of different tree stands on soil properties.

In conclusion, soil heterogeneity, especially of the water extractable fraction is enormous and it is possible that it permits various species to exploit distinct soil horizons leading to coexistence. There is a possibility that tree diversity influences soil heterogeneity contributing to maintenance of high diversity in tropical rainforests. Geodiversity is hence correlated to biodiversity and needs to be more recognized in ecological research.

ACKNOWLEDGEMENTS

We are grateful to Patrick Kilei, Lena Sagita and Jared Sagita for their help in the field, Luis Martinez, Robert Gliniars, and Gunther Klippel for assistance in the lab. World Agroforestry Centre (ICRAF) for facilitating postage of soil samples to Germany and BMBF for financial support.

REFERENCES

Barthlott, W., Biedinger, N., Feig, F., Kier, G., Mutke, J., and Braun, G., 1999, Terminological and methodological aspects of the mapping and analysis of the global biodiversity, *Acta Bot. Fennica* **162:**103-110.

Cords, M., 1987, Mixed species association of *Cercopithecus* monkeys in Kakamega forest, Kenya. *Univ. Calif. Publ. Zool.* **117:**1-109.

Dalitz, H., 2004, Spatial heterogeneity of parameters likely influencing tree species diversity and regeneration, in: *Results of worldwide ecological studies. Proceedings of the 2nd Symposium of the A.F.W. Schimper Foundation,* S. W. Breckle, B. Schweiter, and A. Fangmeier, eds., Günther Heimbach, Stuttgart, pp. 387-388.

Dalitz, H., Homeier, J., Salazar, H. R., and Wolters, A., 2004, Spatial heterogeneity Generating plant diversity?. in: *Results of worldwide ecological studies. Proceedings of the 2nd Symposium of the A.F.W. Schimper Foundation*, S. W. Breckle, B. Schweizer, and A. Fangmeier, eds., Günther Heimbach, Stuttgart, pp. 199-213.

Heinrichs, H., König, N., Schultz, R., 1985, Atom-Absorptions und –emissions – spectroskopische Bestimmungmethoden für Haupt und Spurelement in Probelösungen aus Waldökosystem-Untersuchungen, *Ber. Forschungsz. Waldökosyteme/ Waldsterben*, **8**:1-92.

Hobbie, S. E., 1992, Effects of plant species on nutrient cycling, *Trends Ecol. Evol.* **7**:336-339.

Huston, M., 1980, Soil nutrients and tree species richness in Costa Rican forests, *J. Biogeogr.* **7**:147-157.

Kokwaro, J.O., 1988, Conservation status of the Kakamega forest in Kenya: the easternmost relic of the equatorial rainforest of Africa, *Monogr. Syst. Bot. Mo Bot. Gard.* **25**:471-489.

Levin, S.A., 1992, The problem of pattern and scale in ecology, *Ecology* **73**:1943-1967.

Musila, W., Kamoni, P. T., Albrecht, A., Todt, H., Uster, D., and Dalitz, H., 2004, Soil characteristics of Kakamega Forest, Kenya, in: BIOTA report no. 2, H. Dalitz, ed., *Bielefelder Ökol. Beitr.* **21** (in press).

Oesker, M., Homeier, J., and Dalitz, H., 2004a, Spatial Heterogeneity of Canopy Throughfall Quantity and Quality in a Tropical Montane Forest in South Ecuador. Conference proceedings of the International conference TMCF "Mountains in the mist", Hawaii 2004, (submitted)

Oesker, M., Dalitz, H., Homeier, J., and Breckle, S.-W., 2004b, Spatial and Temporal Heterogeneity of Throughfall Water and the Influence of Canopy Structure in three different Forest Types of a Tropical Montane Forest in south Ecuador, *J. Trop. Ecol.* (submitted)

Ojany, F.F., and Ogendo, R.B., 1987, *Kenya: A study in physical and human geography,* Longman, Kenya.

Van Breemen, N., 1993, Soils as biotic constructs favouring net primary productivity, *Geoderma* **57**:183-211.

Zinke, P. J., 1962, The pattern of influence of individual forest trees on soil properties, *Ecology* **43**:130-133.

DECLINE OF WOODY SPECIES IN THE SAHEL

Alexander Wezel
International Nature Conservation, University of Greifswald. Corresponding address: Wilhams 25, 87547 Missen, Germany, E-mail: alexanderwezel@tiscali.de

Abstract: Vegetation changes in Sahelian West Africa were investigated by analysing and summarising findings from different case studies conducted in Burkina Faso, Niger and Senegal, which were based on the local knowledge of rural people. At all locations analysed in the different countries, the local population perceives a decrease or even the disappearance of woody species. For a single location 4 to 59 different woody species were mentioned. In total, 88 species were named as having disappeared or as decreasing, whereas only 12 species were reported as increasing or as new. The high numbers of declining or disappeared species undoubtedly indicates the need to improve local and regional resource management by generally promoting regeneration and protecting young trees as well as improving reforestation of suitable species. In addition, on local scales, prioritisation of species for resource management should be further worked out with different user groups to improve local acceptance and implementation for species conservation.

Key words: local knowledge; vegetation change; Sahelian West Africa; semi-arid

1. INTRODUCTION

Natural long-term vegetation changes in Sahelian West Africa have been documented for the last centuries (Lézine, 1989; LeHouérou; 1997), but also short-term changes in particular associated with the droughts of the 1970s and 1980s. Tucker et al. (1991) showed that these short-term changes are strongly linked to inter-annual rainfall variability altering vegetation cover enormously from year to year. Although vegetation changes always existed, human activity played a major role in the last decades because of a strong increase in population (LeHouérou, 1997). Tree cover decreased due to firewood cutting and overgrazing (Breman and Kessler, 1995, Gonzalez,

B.A. Huber et al. (eds.), African Biodiversity, 415–421.

2001). Many former tree savannas degraded to shrub/grass savannas or to bare land.

Apart from the changes of climate, vegetation cover or species composition, the knowledge of local populations about vegetation changes was increasingly investigated in the last years. In different countries of Sahelian West Africa, a decrease or even the disappearance of various species was perceived by the local population (Hahn-Hadjali and Thiombiano, 2000; Lykke, 2000; Sadio et al., 2000; Wezel and Haigis, 2000; Müller and Wittig, 2002). In most cases, valuable woody species were mentioned that are used for different purposes, and which have often a socio-economic importance for the individual households. Although all investigations report vegetation changes, they only reflect findings from case studies under specific local conditions. In this paper, an attempt is made to summarise the data of the different case studies by mainly analysing two questions. First, is there a widespread decline or disappearance of certain species in the Sahel? Secondly, what implications for local resource management can be derived from the different case studies?

2. MATERIALS AND METHODS

Local knowledge of rural people investigated in five case studies from three West African countries were used to analyse vegetation changes in the Sahel in this paper. Three investigations have been carried out in Burkina Faso, one in Senegal and one in Niger (see Table 1). Number of villages studied, number of local people interviewed and climatic parameters such as average precipitation of study villages are presented in Table 1. Although interviews were carried out either with single persons or in group discussions, and the decision which species were selected and finally listed in the different case studies changed from study to study, the studies are comparable on the basis of general trends of species changes. All studies present and discuss the perception and knowledge of the majority of the people interviewed about the changes of woody species.

The scientific names of the plant species follow Lebrun and Stork (1991-1997).

3. RESULTS AND DISCUSSION

In total, 110 different woody species were named by the local population in Burkina Faso, Niger and Senegal to have changed in numbers compared

Table 1. Different parameters of the case studies on the local knowledge about vegetation change

	Lykke et al. (2004)	Müller and Wittig (2002)	Hahn-Hadjali and Thiombiano (2000)	Sadio et al. (2000)	Wezel and Haigis (2000)
Country	Burkina Faso	Burkina Faso	Burkina Faso	Senegal	Niger
Average precipitation	400	400	570	400-500	330-580
Number of villages analysed	5	2	1	4	7
Total number of interviews	100	group discussions	15	group discussions	253
Asked for species that are/have	- increasing, - declining	- increasing, - decreasing, - disappeared	- endangered (decreasing), - disappeared	- still existing, - becoming rare, - disappeared	- new (introduced), - increasing, - decreasing, - disappeared
Selection criteria of species for enrolment on species list	first 59 species were selected in 21 group discussions; then 100 individual interviews were conducted and species that were attributed by at least 20% of the interviewees per village to the different categories were enrolled[1]	first a species list was created in several group discussions; then all species on the list were classified to the different categories in the group discussions	species that were mentioned by at least 30% of interviewees per village for the respective classification were enrolled	species change categories were first attributed to four different time periods (<1960, 1960-1972, 1972-1990, 1999); then the changes over the four time periods were used for final categorisation	species that were mentioned by at least 20% of interviewees per village were enrolled

[1]In Lykke et al. (2004), there was no distinction between the five villages. Nevertheless, the 20% threshold could be applied for each of the five different villages because the raw data was kindly provided by the authors.

to the past. In general the past means 30-50 years ago, depending on the age of the informants. Most species (88) were classified as decreasing or as having disappeared (Fig. 1). Only 12 species have increased compared to the past or were newly introduced, and the remaining 10 species were differently classified at different locations in the case studies.

The most prominent species that were mentioned at 30 % of the analysed 19 locations as having decreased or disappeared are: *Acacia ataxacantha*, *A. senegal*, *A. seyal*, *A. sieberiana*, *Adansonia digitata*, *Boscia senegalensis*, *B.*

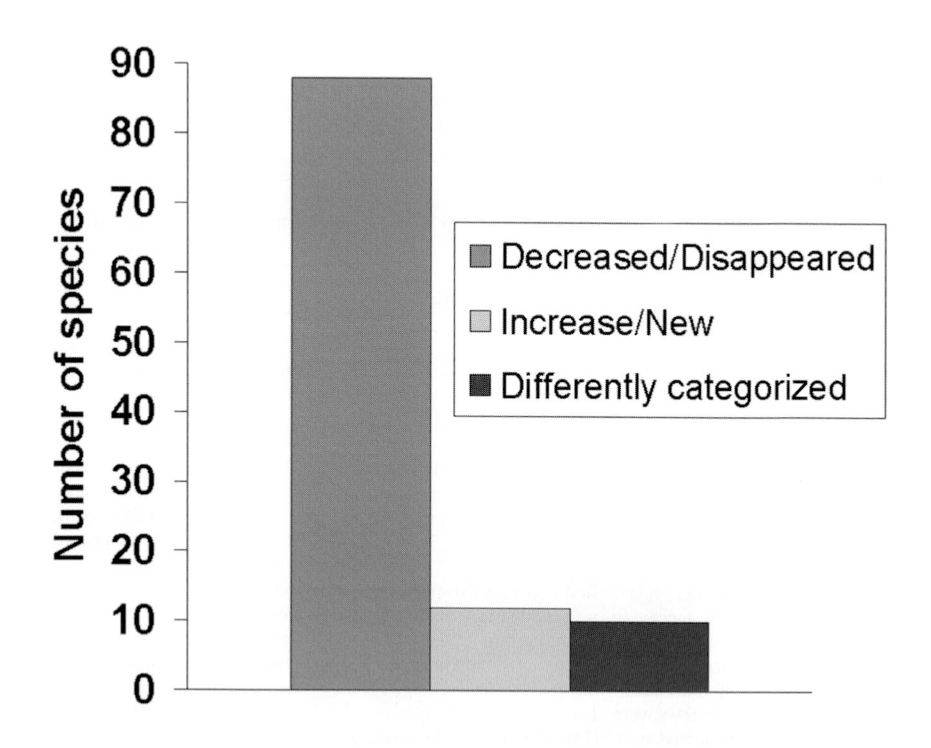

Figure 1. Number of different woody species mentioned by rural people in the Sahel as having changed.

angustifolia, *Cadaba farinosa*, *Ceiba pentandra*, *Combretum glutinosum*, *C. micranthum*, *Commiphora africana*, *Dalbergia melanoxylon*, *Diospyros mespiliformis*, *Ficus gnaphalocarpa*, *Gardenia ternifolia*, *Grewia bicolor*, *Hyphaene thebaica*, *Khaya senegalensis*, *Maerua crassifolia*, *Mitragyna inermis*, *Parkia biglogosa*, *Prosopis africana*, *Sclerocary birrea*, *Tamarindus indica*, *Ximenia americana* and *Vitex doniana* (for details see Wezel and Lykke, submitted). Almost all new or increasing species are exotic, often introduced by different developing projects, e.g. *Eucalyptus camadulensis*. Some species such as *Acacia albida*, *A. tortilis*, *A. nilotica* or *Parinari macrophylla* indicate significant difference between case studies, e.g. the latter species disappeared in Senegal, but increased in Niger.

In average, 22 species were mentioned at each village as decreasing/disappeared and four species as increasing/new species (Fig. 2). There exists a wide range of minimum and maximum values for the different villages.

The high number of decreasing or disappeared species throughout the Sahel reveals the need to discuss different issues related to local and regional

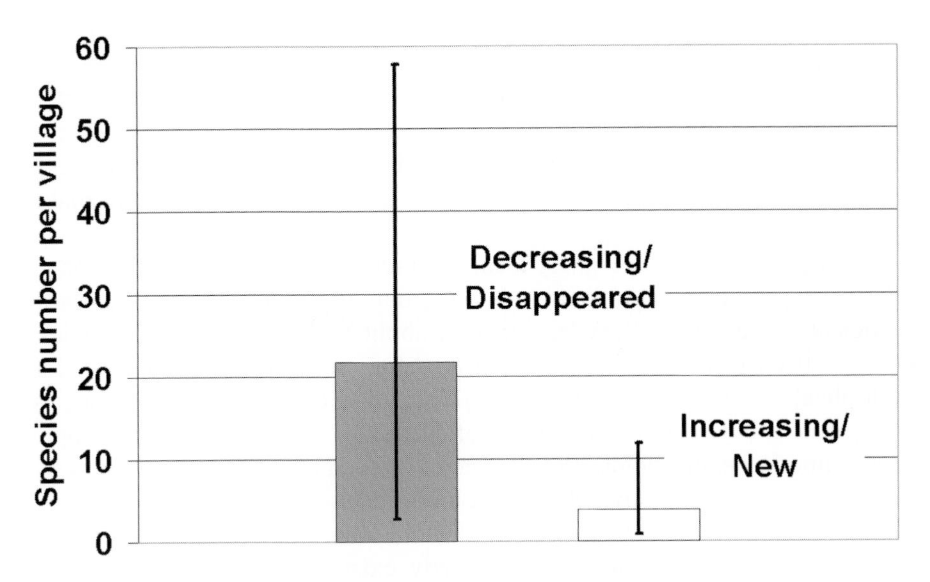

Figure 2. Average, minimum and maximum species numbers per village.

resource management. In the first place the causes for species change must be examined. The most important factors stated in various publications in the last years are human impact and climate change. In general, many authors report a persistent decline in precipitation in the Sahel since the 1970s (e.g. Hulme, 2001; Hôte et al., 2002). Other researchers see indications of an increase in precipitation since 1990, e.g. in Senegal and Niger (Ozer et al., 2003), and in other areas of the Sahel after the 1980s drought (Eklundh and Olsson, 2003). An increasing aridity can cause the decline of at least some species, e.g. *Acacia senegal* or *Guiera senegalensis* which had a mortality of about 50% at a protected site, caused by the drought in 1973 (Poupon and Bille, 1974). Nevertheless, Poupon and Bille (1974) reported no species to have disappeared, instead, in years after drought many woody plants increased again. In general, a decline of certain species can be related to years of low rainfall, but they do not disappear totally. This is only indicated for species occurring at the northern limit of their geographic range, e.g. *Sclerocarya birrea*, *Bombax costatum* and *Vitellaria paradoxa*. Thus, it seems that human impact is the most important factor for the many species reported to have disappeared at different locations in the Sahel. Therefore, local and regional management of woody vegetation should play a more important role in the future. As there are quite a lot of species, which disappeared or decreased, there should be a focus on them. These species were already listed in the results section. The common characteristic among them is that they are declining or have disappeared in many areas of the West African Sahel. Species endangered and confined to very few or single locations should of course not be left out, but the fact that most of the key species identified here are multi-purpose species used in many ways by the local population (for details see Hahn-Hadjali and Thiombiano, 2000; Sadio et al., 2000; Wezel and Haigis, 2000; Lykke et al., 2004), underlines that management of natural vegetation plays a crucial role. In most of the different studies, the local people mentioned using the species for firewood, for construction, for medicinal purposes, as fodder, for nutrition or to make tools.

A further discussion point for local resource management is that although interviews provide a better database, few group discussions might be already sufficient to get satisfactory information about locally endangered species. Group discussions are also effective for selection of species for multiplication or regeneration. For multiplication of species, exotic species are almost exclusively used. This is often due to the fact that rural people were trained to multiply introduced species or even received the seedlings by the projects, whereas knowledge about multiplication of local species is very limited. Thus an intensive collection of more local knowledge is necessary, as well as a better distribution of already existing information about the multiplication of local species. Further, natural regeneration of many species is low because young seedlings are often grazed by livestock or cut by annual clearing of fields. To improve regeneration, certain species need to be specifically protected by selective clearing of fields or from being grazed. This seems to be a rather feasible option as already successfully applied by projects (Taylor and Rand, 1992; Joet et al., 1998). When propagating natural regeneration, the different land and tree tenure systems in the Sahelian countries have to be considered, otherwise no sustainable impact can be expected. Lastly and surely not a new, but still actual point, is that improved stoves or fireplaces are needed for more people to reduce consumption rates of firewood.

ACKNOWLEDGEMENTS

I gratefully acknowledge the ideas and comments of Anne Mette Lykke, Elke Mannigel and Jonas Müller on an early version of this manuscript.

REFERENCES

Breman, H., and Kessler, J.-J., 1995, *Woody plants in agro-ecosystems of semi-arid regions - with an emphasis on the Sahelian countries*, Advanced Series in Agricultural Science 23, Springer, Berlin, Heidelberg, 340 pp.

Eklundh, L., and Olsson, L., 2003, Vegetation index trends for the African Sahel 1982-1999, *Geophys. Res. Letters* **30(8):**13-1 - 13-4.

Gonzalez, P., 2001, Desertification and a shift of forest species in the West African Sahel, *Clim. Res.* **17:**217-228.

Hahn-Hadjali, K., and Thiombiano, A., 2000, Perception des espèces en voie de disparition en milieu gourmantché: causes et solutions, *Ber. Sonderforsch. 268* **14:**285-297.

Hôte, Y., Mahé, G., Somé, B., and Triboulet, J. P., 2002, Analysis of a Sahelian annual rainfall index from 1896 to 2000; the drought continues, *Hydro. Sci. J.* **47(4):**563-572.

Hulme, M., 2001, Climatic perspectives on Sahelian desiccation: 1973-1998, *Glob. Environ. Change* **11:**19-29.

Joet, A., Jouve, P., and Banoin, M., 1998, Le défrichement amélioré au Sahel. Une pratique agroforestière adoptée par les paysans, *Bois For. Trop.* **255(1):**31-44.

LeHouérou, H. N., 1997, Climate, flora and fauna changes in the Sahara over the past 500 million years, *J. Arid Environ.* **37:**619-647.

Lebrun, J.-P., and Stork, A. L., 1991-1997, *Enumération des plantes à fleurs d'Afrique tropicale*, Vol. I-IV, Editions des Conservatoire et Jardin botaniques, Genève, Suisse.

Lezine, A. M., 1989. Le Sahel: 20 000 ans d'histoire de la végétation - reliques végétales, zonation de la végétation, paléoclimats, *Bull. Soc. Géol. Fr.* **5(1):**35-42.

Lykke, A. M., Kristensen, M. K., and Ganaba, S., 2004, Valuation of local use and dynamics of 56 woody species in the Sahel, *Biodivers. Conserv.* **13:**1961-1990.

Müller, J., and Wittig, R., 2002, L'état actuel du peuplement ligneux et la perception de sa dynamique par la population dans le Sahel burkinabé - présenté à l'exemple de Tintaboora et de Kollangal Alyaakum, *Etud. flore vég. Burkina Faso pays avoisin.* **6:**19-30.

Ozer, P. Erpicum, M., Demarée, G., and Vandiepenbeeck, M., 2003, The Sahelain drought may have ended during the 1990s, *Hydrol. Sci. J.* **48(3):**489-492.

Poupon, H., Bille, J.C., 1974, Recherches écologiques sur une savane sahélienne du Ferlo septentrional, Sénégal: influence de la sécheresse de l'année 1972-1973 sur la strate ligneuse, *La Terre et la Vie* **28(11):**49-75.

Sadio, S., Dione, M., and Ngom, M. S., 2000, Région de Diourbel: Gestion des ressources forestière et de l'arbre, Drylands Research Working Paper 17, Drylands Research, Somerset, Great Britain, 34 pp.

Taylor, G. F. II, and Rands, B. C., 1992, Trees and forests in the management of rural areas in the West African Sahel: Farmer managed natural regeneration, *Desert. Control Bull.* **21:**49-51.

Tucker C. J., Dregne H. E., and Newcomb W. W., 1991, Expansion and contraction of the Sahara Desert from 1980 to 1990, *Science* **253:**299-301.

Wezel, A., and Haigis, J., 2000, Farmers' perception of vegetation changes in semi-arid Niger, *Land Degrad. Develop.* **11:**523-534.

Wezel, A., and Lykke, A. M., (submitted), Woody vegetation change in Sahelian West Africa: evidence from local knowledge.

DIVERSITY OVER TIME AND SPACE IN ENSET LANDRACES (*ENSETE VENTRICOSUM*) IN ETHIOPIA

Karin Zippel
Humboldt-Universität zu Berlin, Fachgebiet Obstbau, Albrecht-Thaer-Weg 3, 14195 Berlin, Germany, E-mail: kzippel@gmx.net

Abstract: Enset (*Ensete ventricosum*: Musaceae) is a staple crop in mixed subsistence farming systems in the most densely populated regions in southern and south-western Ethiopia. Mainly grown for its starch-containing leaf sheaths and underground corm, it also serves purposes like animal feed, uses in households, and in agriculture. A large number of landraces are cultivated for manifold purposes, as well as for different requirements of site and climate. To retain the specific characteristics of each landrace, enset is propagated vegetatively by sprouts. In each region differently named landraces are grown. However, changing weather and soil causes phenotypic variability, and make identification difficult. Both increasing as well as decreasing number of landraces have been observed over the years in the different regions due to changing food preferences, climate, pests and diseases, cultivation systems, and infrastructure. Farmers tend to increase the number of cultivated landraces to broaden its use, and to respond to annual climate fluctuations.

Key words: enset; *Ensete ventricosum*; landraces; Ethiopia

1. INTRODUCTION

Enset (*Ensete ventricosum*) belongs to the order Zingiberales, which includes several aromatic and ornamental plants (Musaceae, Heliconiaceae, Strelitziaceae, Zingiberaceae, Costaceae, Maranthaceae, Cannaceae), and fruit and fibre crops (*Musa* spp.). *Ensete* is one out of three genera in the Musaceae family, and consists of six good species: *E. glaucum* and *E. superbum* in Asia, *E. gilletii* and *E. homblei* in mainland Africa, *E. perrieri*

B.A. Huber et al. (eds.), African Biodiversity, 423–438.

in Madagascar and *E. ventricosum* in Africa and Latin America. *Ensete ventricosum* grows wild from Eritrea, Ethiopia and Sudan in the north, to Angola, South Africa and Mozambique in the South (Fig. 1). Its altitudinal distribution ranges between 800 m (Angola) and 2000 m (Eastern Africa) with an average rainfall between 1000 and 1500 mm (Cheesman, 1947; Simmonds, 1962).

Main cultivation areas are restricted to the densely populated regions in southern and south-western Ethiopia at altitudes from 1300 to 3300 meters (Fig. 1). Here, the climate is influenced by the tropical interconvergence zone resulting in unpredictable weather. Enset cultivation systems integrate animal husbandry (cattle, sheep, poultry, bees), and crop production (cereals, pulses, vegetables, root and tuber crops, fruit trees, cash crops; Westphal, 1975).

Enset is cultivated for food products made from the starch-containing leaf sheaths and corm. Leaf sheaths are scraped, and the corm is cut into pieces and pounded. Both provide a starchy porridge. Fermentation with lactic acid bacteria allows storage for more than one year (Ayele, 1997, 1998). Farmers in Gurage, Sidamo and Arsi report storage of the fermented product for up to seven years. The corm is also eaten boiled fresh. Fine fibres remain after extracting the starch from the leaf sheaths, and are twisted into ropes. Furthermore, enset is used as animal fodder. All plant parts are used for various purposes in the household and agriculture, i.e. to make ropes, mats, towels, seats, fuel, manure, shade, umbrella and wrapping material for food products. Dried leaf sheaths are used for construction, and all plant parts for traditional medicinal treatments (Huffnagel, 1961; Kefale and Sandford, 1991; Westphal, 1975).

Enset is grown in plantations consisting of plants of varying age. This guarantees a permanent food supply. Farmers harvest enset shortly before or after the emergence of the inflorescence, when most assimilates are accumulated in the leaf sheaths and the corm. Length of pre-flowering phase depends on climatic conditions, soil, and farm management, and lasts three to 12 years. Under unfavourable conditions it takes longer, or plants do not flower at all (Kefale and Sandford, 1991; Zippel and Kefale, 1995). Yield per plant increases with altitude up to 2700 m. Enset can withstand drought periods up to 6 months. Limitations in enset cultivation are pests and diseases, water logging, drought (rainfall below 800 mm/a), wind and frost (Huffnagel, 1961; Kefale and Sandford, 1991; Westphal, 1975). Especially young plants are affected, and need special protection.

Farmers distinguish a large number of landraces by use, colour and shape (Fig. 2), and demands of site and climate. Furthermore, landraces differ in their ability to resist or tolerate pests and diseases. Vegetative propagation

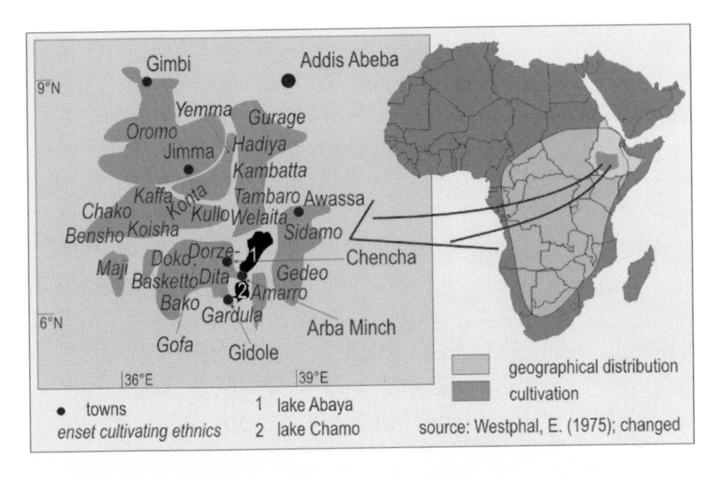

Figure 1. Enset cultivation regions in southern and south-western Ethiopia.

provides homogenous plant material with known characteristics (Asnaketch et al., 1993; Bizuayehu, 2002; IAR, 1991/92; Jensen, 1959; Kefale and Sandford, 1991; Olmstead, 1974; Shigeta, 1990; Zippel and Kefale, 1995). Despite their heterogeneity, seeds are used sometimes for propagation. Promising seedlings are propagated vegetatively, and regarded as a new landrace, if not resembling an existing ones (Kefale and Sandford, 1996; Shigeta, 1990; Zippel and Kefale, 1995). Over the years, and in the regions, the number of landraces varies; and so do the cultivated landraces. Genetical correspondence of plant material in the regions is assumed for at least a few of the landraces by striking similarity in morphological characteristics (Zippel and Kefale, 1995) as well as by phylogenetical analyses (Almaz et al., 2002). A completely purple coloured landrace with a cylindrical pseudostem was identified in Arsi, Borana (south of Gedeo), Dorze, Gardula, Gedeo, Gurage, Hadiya, Kambatta, Sidamo and Welaita ('Lochingii'; Zippel and Kefale, 1995). 'Agiiniia' with whitish midribs and leaf sheaths, and 'Suiitiia' containing a pink sap, are both cultivated in Welaita, and were also identified in Sidamo and Gedeo, and Dorze and Gurage, respectively (Zippel and Kefale, 1995). Straube (1963) describes the 'Agiiniia' type for Ammarro, and Shigeta (1990) for Bako.

2. MATERIALS AND METHODS

This study investigates the occurrence of different landraces in ten regions in 1994 and 1998/99 (Figs. 1, 3), and compares results with data

Figure 2. Enset landraces after three years.

from field experiments in 1998/99 investigating vegetative parameters, and leaf anatomy. At least ten randomly selected farms were visited in each region. Data have been collected on the number of landraces grown at each visited farm, their properties, and their importance. Further studies in the 1990s complemented additional data (Asnaketch et al., 1993; Bizuayehu, 2002; IAR, 1991/92; Shigeta, 1990). For Bako, Gurage, Sura, and Sidamo, results from the 1990s were compared with previous literature on studies from the same localities (Huffnagel, 1961; Jensen, 1959; Olmstead, 1974; Smeds, 1955).

Results have been compared with data from field experiments. Ten landraces bought at four neighbouring farms in Soddo/Welaita, namely 'Agiiniia' (also called 'Agena'), 'Allagena', 'Arrkiia', 'Gefetenna', 'Lemmbo', 'Lochingii', 'Mazziia', 'Shelekuummiia', 'Suiitiia' and 'Wannidiia', and one widely distributed type from Addis Ababa bought at Selam nursery in Addis Ababa ('AddisAbaba') were established at two altitudes (1850 m and 2350 m; for a short description of the landraces see Appendix 1). They were investigated for vegetative parameters and leaf anatomy. Plants have been propagated in August 1998. Average temperature and rainfall patterns varied throughout the seasons: annual means over ten years are 17.1°C, 2805 hours of sunshine, and 858.4 mm rainfall in 96.6 days at 1850 m; and 16.1°C, 2576 hours, and 1127.4 mm in 169.9 days at 2350 m.

The term 'landrace' is defined according to Melaku (1991, p. 104-105):

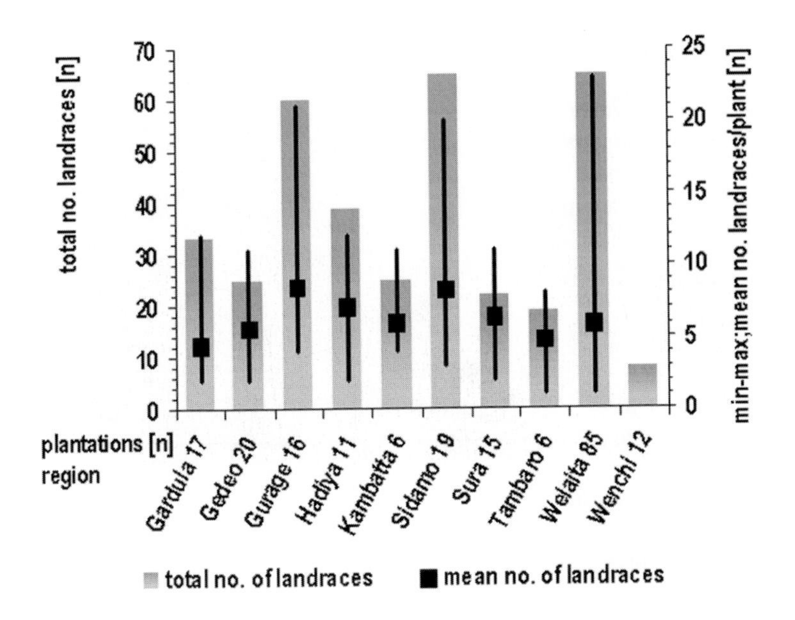

Figure 3. Total number of landraces counted in each region, and minimum, maximum and mean number of landraces per plantation in each region.

"Landraces are crop populations that have not been bred as varieties by scientists but which farmers have adapted to local conditions through years of natural and artificial selection".

3. RESULTS

The number of landraces grown is closely associated with the importance of enset for a certain ethnic group (Fig. 3). Farmers tend to increase the number of grown landraces to broaden the use of the crop, and to respond to the annual weather fluctuations. Highly esteemed landraces are often cultivated despite unsuitable conditions requiring special cultivation measurements, i.e. frost protection or irrigation. Furthermore, farmers always grow some less favoured landraces that tolerate extreme weather conditions, i.e. drought or frost, as well as diseases and pests to ensure permanent food supply.

The field experiments indicate that landraces differ in their site requirements. Overall, sprout numbers were much lower at 1850 m than at 2350 m (Fig. 4) connected with drought from November 1998 to February

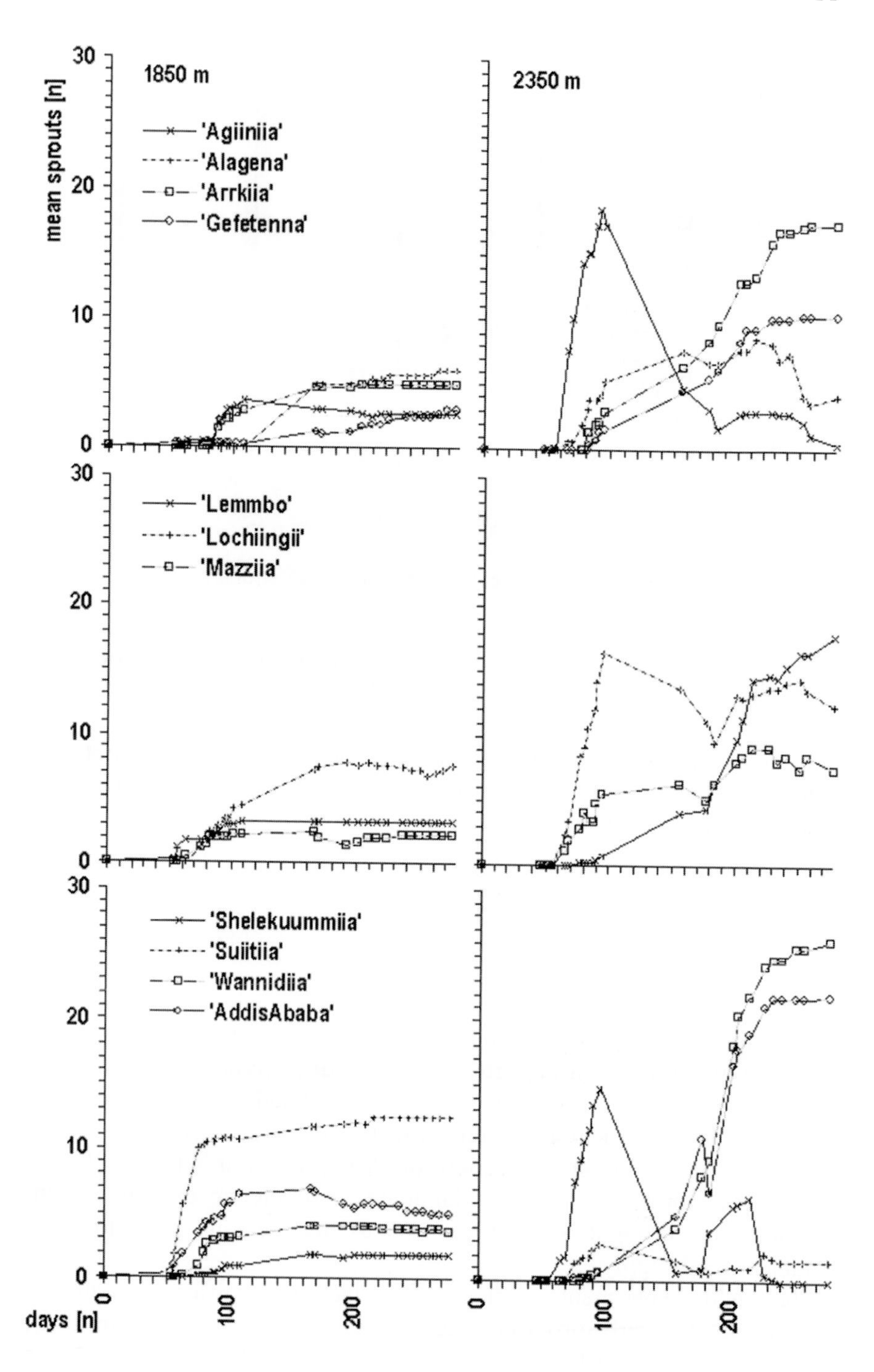

Figure 4. Sprout development of 11 enset landraces at 1850 and 2350 m.

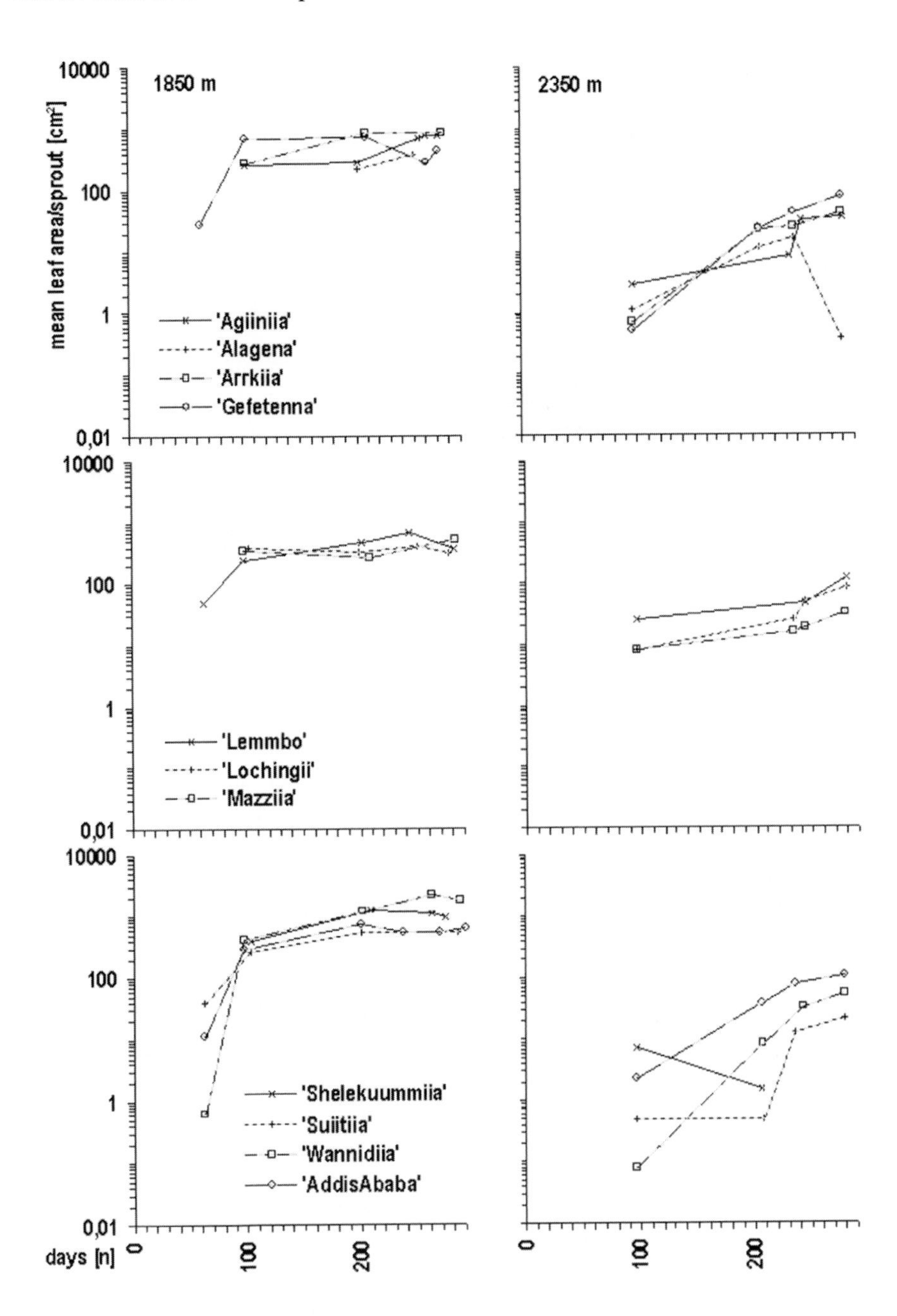

Figure 5. Leaf area per plant of 11 enset landraces at 1850 and 2350 m.

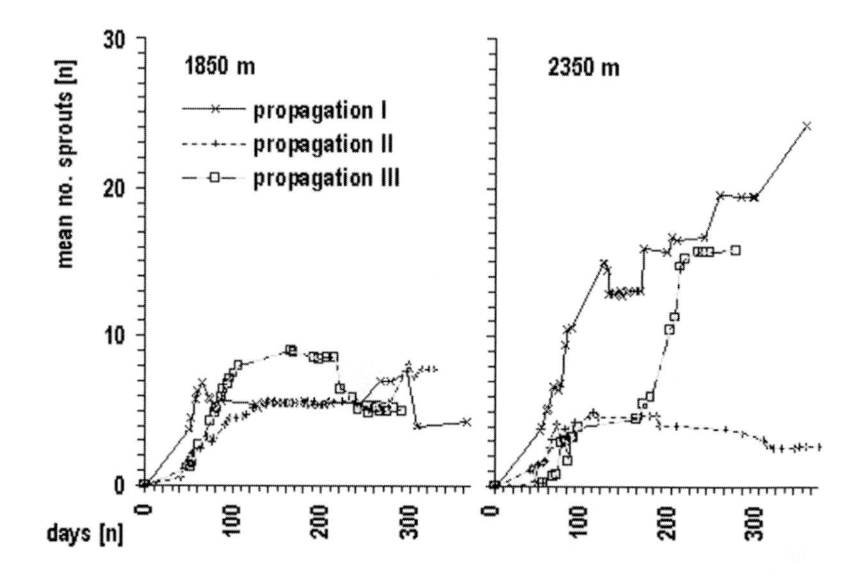

Figure 6. Sprout development at three propagation dates (February, April and August 1998) and two altitudes (1850 and 2350 m).

1999, and maximum temperatures up to 32°C. Meanwhile, total height of sprouts, leaf number and leaf area was much higher at 1850 m (Fig. 5). Some landraces failed completely at 2350 m during a period with ground frost starting after 120 days and lasting for three months (Fig. 4). Experiments with one landrace ('Addis Ababa') indicate differences between the altitudes in optimum propagation time (Fig. 6). At 1850 m, plants propagated in January showed additional hypodermis layers during the dry period.

Comparing previous investigations in Sidamo (Smeds, 1955), Sura (Straube, 1963) and Gurage (Shack, 1966) with results from the 1990s for the same localities indicated considerable changes. An increasing number of landraces was observed in Sidamo (Bizuayehu, 2002; Smeds, 1955; Zippel and Kefale, 1995), while other regions showed a slight decline (Asnaketch et al., 1993; Huffnagel, 1961; IAR, 1991/92; Jensen, 1959; Kefale and Sandford, 1996; Olmstead, 1974; Shigeta, 1990; Zippel and Kefale, 1995; Fig. 7). Investigations that provided precise full names of the landraces, allowed a more detailed comparison of the cultivated landraces (Bizuayehu, 2002; Jensen, 1959; Kefale and Sandford, 1996; Olmstead, 1974; Shigeta, 1990; Zippel and Kefale, 1995; Fig. 8). It became obvious that some of the landraces had disappeared, while new ones had emerged. Some landraces used for very long periods were reported to have especially high value for

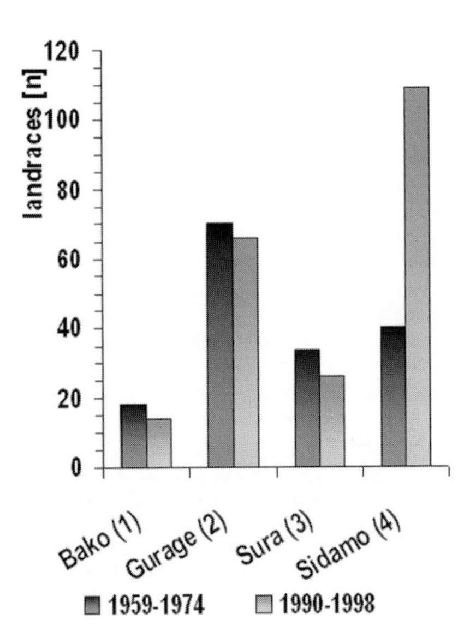

Figure 7. Landraces counted at two periods at four regions; (1) Jensen (1959), Shigeta (1990); (2) Asnaketch et. al. (1990), Huffnagel (1961), IAR (1991/92), Zippel and Kefale (1995); (3) Kefale and Sandford (1996), Olmstead (1974), Zippel and Kefale (1995); (4) Bizuayehu (2002), Smeds (1955), Zippel and Kefale (1995).

human consumption (Bizuayehu, 2002; Kefale and Sandford, 1996; Shigeta, 1990; Zippel and Kefale, 1995). According to the farmers, new landraces with favoured characteristics replace older ones. These characteristics include better taste, adaptability to fluctuating weather conditions, and low susceptibility to pests and diseases. Furthermore, improved infrastructure allows transport of sprouts for long distances, and therefore enables exchange of plant material between the different regions.

New landraces are obtained either by exchange and trade, or from seedlings and mutants. In Sidamo, traditional trading of enset sprouts and a comparatively well-developed infrastructure may explain the immense increase in number of presently grown landraces (Bizuayehu, 2002; Smeds, 1955; Zippel and Kefale, 1995). Some landrace names are found in almost all enset growing regions (Bizuayehu, 2002; Jensen, 1959; Kefale and Sandford, 1996; Olmstead, 1974; Shigeta, 1990; van Treuren, 2004; Zippel and Kefale, 1995). This indicates that such landraces are traded, i.e. for the landrace ‘Astara’. It seems to originate in Gurage (Semitic ethnic group), but is also found in Sidamo (Cushitic ethnic group). In Hadiya, Kambatta, Sidamo and Welaita, traded landraces are often named according to their origin. In Kambatta, Hadiya and Welaita mutants are used to breed new landraces.

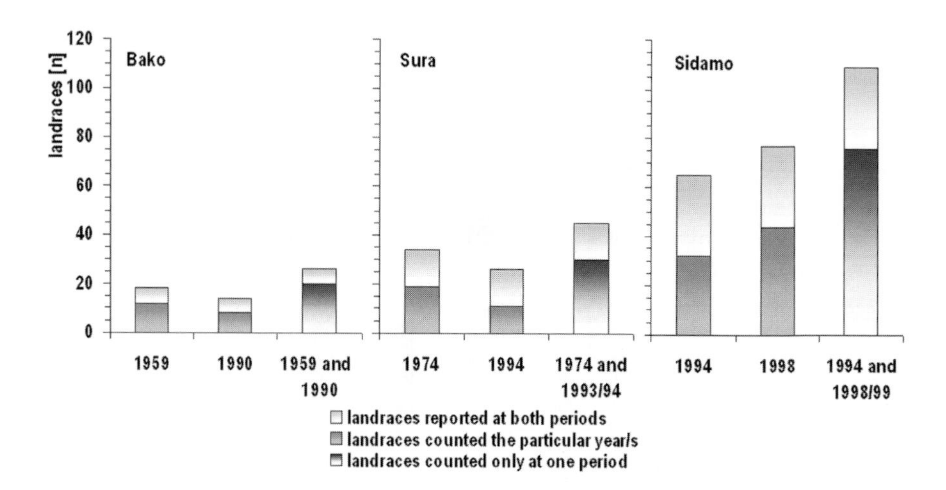

Figure 8. Landraces reported at two periods in Bako (1600-2400 m), Sura (2200-3000 m), and Sidamo (1700-2800 m); Bako: Jensen (1959), Shigeta (1990); Sura: Kefale and Sandford (1996), Olmstead (1974), Zippel and Kefale (1995); Sidamo: Bizuayehu (2002), Smeds (1955), Zippel and Kefale (1995).

Mutants differ in colour and shape from non-mutated sprouts of the same mother plant. Mutants with promising features are kept for future cultivation. In these regions neighbours and relatives exchange landraces, too (Zippel and Kefale, 1995). In Bako and Gardula seedlings are a source for new landraces (Shigeta, 1990; Zippel and Kefale, 1995). Despite a decline in some regions, enset cultivation recently has been introduced to new regions in Western Wellega and northern Shewa (Fig. 1), where farmers have started growing enset for its high yield and drought tolerance.

4. DISCUSSION AND CONCLUSIONS

Enset cultivation in subsistence agriculture depends on a high diversity of plant material to meet the farmers' requirements. Decline in the number of landraces is caused by drought, pests and diseases, but also by changes in the cultivation system. Farmers have always insisted on the advantage of growing many landraces. There is a tendency to grow a higher number of landraces, and to enlarge plantations. Spreading of enset cultivation to new regions has occurred due to the high yield and storability of enset food products, and its drought tolerance. The droughts in 1984/85 and 1994 seem to have been most important for the introduction of new landraces from distant regions, and have led to an increase in number of the cultivated landraces. However, in past centuries enset cultivation was reported in these regions (Stiehler, 1948).

Enlarging of enset cultivation may lead to failure caused by reduced soil fertility, pests and diseases. Moreover, rising human population causes increased competition for space used for enset cultivation, for other crops or as grazing land.

Changes in consumption preferences, and in the cultivation system may explain the differing numbers of landraces. It is not known yet, whether change of climate affects the cultivation of distinct landraces. Some highly esteemed landraces show sensitivity to drought, frost, pests and diseases, while bitter landraces are cultivated for their tolerance. In the future, a warmer climate may lead to an increase of the highly favoured sensitive landraces, and a simultaneous decrease of the bitter landraces. For example, the landrace 'Agena' is highly frost sensitive and failed in the field experiment at 2350 m. In 1994, Sidama farmers reported recent introduction of this landrace to altitudes above 2400 m, but according to Bizuayehu (2002), it was fully established in 1999.

Enset cultivation declines with increased cultivation of cash crops and pasture for animal husbandry. Demand for cash needed to pay taxes and purchase goods, forces more emphasis on the cultivation of cash crops or cereals. In Gedeo (Fig. 1), with ecological conditions similar to Sidamo, cultivation of cash crops in the lowlands has replaced enset cultivation, and only few landraces are grown. At higher altitudes, the climate does not allow cultivation of coffee and chat (*Catha edulis*; a stimulant). Consequently, enset plantations are larger, and number of landraces higher.

A rising population forces an earlier harvest; therefore, plants are harvested at a younger stage and thus provide less yield. This may promote fast maturing landraces. In Welaita, one landrace is highly regarded for its short maturing period, requiring only three years from propagation to harvest.

The changes in the cultivated landraces reflect the farmers' need to adapt to changes in climate, economy and social conditions. Landraces need to be more closely investigated on their site requirements and their chemical contents, i.e. phenols. Combined with intensified phylogenetic analyses this should provide additional information on the diversity of enset, and may enable farmers to improve and optimize cultivation.

ACKNOWLEDGEMENTS

I would like to thank my supervisor, Professor Peter Lüdders from the Fachgebiet Obstbau, Landwirtschaftlich-Gärtnerische Fakultät of the

Humboldt-Universität zu Berlin. In Ethiopia, research was supported by Dr. Stephen Sandford and Kefale Alemu from FARM-Africa in Sodo/Welaita, Mrs. Sue Edwards and Professor Sebsebe Demissew from the National Herbarium Ethiopia in Addis Ababa, and Dr. Jean Hansen and Abate Tedla from ILRI Addis Ababa and Debre Zeit. To all of them I would like to express my appreciation and gratitude for their immense help in various ways. The Heinrich-Boell Stiftung I thank for providing a PhD scholarship. Furthermore, I thank anonymous referees for proof-reading and improving the paper.

REFERENCES

Almaz Negash, Admasu Tsegaye, van Treuren, R., and Visser, B., 2002, AFLP analysis of enset clonal diversity in south and south-western Ethiopia for conservation, *Crop Science* **42:**1105-1111.

Asnaketch Wolde Tensaye, Lindén, B., Ohlander, L., and Petrini, F., 1993, personal communication.

Ayele Nigatu, and Berhanu Abegaz Gashe, 1998, Effect of heat treatment on the antimicrobial properties of *tef* dough, *injera, kocho* and *aradisame* and the fate of selected pathogens, *World J. Microbiol. Biotechnol.* **14:**63-69.

Ayele Nigatu, Berhanu Abegaz Gashe, and Terekegn Ayele, 1997, Bacillus spp. from fermented tef dough and kocho: Identity and role in the two Ethiopian fermented foods, *Sinet, Ethiopian J. Sci.* **20**(1):101-114.

Bizuayehu Tesfaye Asres, 2002, *Studies on Landrace Diversity, in vivo and in vitro Regeneration of Enset (Ensete ventricosum Welw.,* PhD thesis, Landwirtschaftlich-Gärtnerische Fakultät, Humboldt-Universität zu Berlin. vii+12 pp., Köster Berlin.

Cheesman, E. E., 1947, Classification of the bananas. 1. The genus *Ensete* Horan, *Kew Bulletin* **2:**97-106.

Huffnagel, H. P. (ed.), 1961, *A Report of the Agriculture in Ethiopia*, FAO Rome, pp. 248-251, 282-293.

IAR, 1991/92, Chapter 3. *Progress Report 1990/91,* IAR Addis Ababa, 18 pp.

Jensen, A. E. (ed.), 1959, *Völker Süd-Äthiopiens: 1. Altvölker Süd-Äthiopiens*, Kohlhammer Stuttgart, pp. 2-6, 16-17, 90-95.

Kefale Alemu, and Sandford, S., 1991, *Enset in North Omo Regio,* FRP Technical Pamphlet No. 1, 49 pp., FARM-Africa Addis Ababa.

Kefale Alemu, and Sandford, S., 1996, Enset landraces, in: *Enset-based sustainable agriculture in Ethiopia.* Proceedings from the International Workshop on Enset Addis Ababa, 13-20 December 1993. Tsedeke Abate; C. Hiebsch, S. A. Brandt; Seifu Gebremariam eds., IAR Addis Ababa, pp. 138-148.

Melaku Worede, 1991, Crop genetic resources conservation and utilization: an Ethiopian perspective, in: *Science in Africa - Achievements and Prospects,* A Symposium at the 1991 AAAS Annual Meeting, AAAS Washington, D.C, pp. 104-123.

Olmstead, J., 1974, The versatile ensete plant: Its use in the Gamu highland, *J. Ethiopian Studies* **12**(2):147-158.

Shack, W.A., 1966, *The Gurage - a people of the Ensete Culture,* Oxford University Press London, New York, Nairobi.

Shigeta, M., 1990, Folk in-situ conservation of *Ensete* (*Ensete ventricosum* (Welw.) E. E. Cheesman): Towards the interpretation of indigenous agricultural science of the Ari, south-western Ethiopia, *African Study Monographs* **10**(3):93-107.

Simmonds, N.W., 1962, *The Evolution of the Bananas*, Longman London, xii+170 pp.

Smeds, H., 1955, The enset planting culture of eastern Sidamo, Ethiopia, *Acta Geographica* **13**(4):1-40.

Stiehler, W., 1948, Studien zur Landwirtschafts- und Siedlungsgeschichte Äthiopiens, *Erdkunde* **2**:257-282, map 1.

Straube, H., 1963, *Völker Süd-Äthiopiens. BD. III. Westkuschitische Völker Südäthiopiens,* W. Kohlhammer Stuttgart, pp. 30-32, 88-90, 154-155.

van Treuren, 2004, personal communication.

Westphal, E., 1975, *Agricultural Systems in Ethiopia,* Agricultural Research Reports 826, Center for Agricultural Publishing and Documentation Wageningen, viii+278 pp., 10 maps.

Zippel, K., and Kefale Alemu, 1995, *A Field Guide to Enset Landraces of North Omo, Ethiopia,* FARM-Africa Technical Pamphlet No. 9, FARM-Africa Addis Ababa, iv+208 pp.

APPENDIX

Appendix 1. Description of landraces investigated (Zippel and Kefale, 1995)

Name of landrace	Zones	Altitude [m]	Size of plant [m]	Shape of pseudostem	Leaf sheathes
Agiiniia/ Agena	Welaita/ North Omo, Sidamo, Gedeo	1900-2500	3-6	almost cylindrical	short, yellowish to whitish, sometimes red patches
Arrkiia	Welaita/ North Omo	1800-2300	6-8	moderately bulbous	long; yellowish-green, sometimes with a few reddish-brown stripes; base sometimes turns bluish green; becomes yellowish towards the top
Gefetenna	Welaita/ North Omo	1900-2300	5.8-8	moderately bulbous	short to medium; yellowish-green, becoming dark purple at the top where it becomes glaucous
Lemmbo	Welaita/ North Omo	1900-2200	6.4-7.8	moderately bulbous	long; yellowish-green with reddish stripes and black pigments
Lochingii	Sura and Welaita/ North Omo, Kambatta, Hadiya, Gurage, Gardula, Sidamo, Gedeo, Borana	1800-2600	6-8	cylindrical, slim	medium length; dark red/reddish purple; few greenish pigments may occur, esp. on young plants
Mazziia	Sura and Welaita/ North Omo	1900-2950	5.5-9.2	moderately bulbous	long; greenish with reddish pigments or stripes, and sometimes black stripes, often increasing near the bottom
Sheelekuum-miia	Welaita/ North Omo; Gurage	1800-2300	7.2-8.3	markedly bulbous	long; yellowish-green, with red, purple or black spots or patches, yellowish on the top; glaucous
Suiitiia	Sura and Welaita/ North Omo; Gurage	1800-2600	3-7	slightly bulbous	medium; yellowish-green with a reddish to pinkish pigmentation
Wannidiia	Welaita/ North Omo	1825-2300	5-10	moderately bulbous	long; yellowish-green to yellowish, reddish to reddish-brown stripes or pigments dominating; to the top dark red and glaucous
AddisAbaba	Addis Ababa	2000-2700	7-10	moderately bulbous	medium, greenish to yellowish greenish at the top, with red, purple or reddish-brown patches, glaucous

Appendix 1. continued.

Name of landrace	Midrib	Lamina	Sap	Flower appearance [after n years]	Gender*
Agiiniia/Agena	whitish petiole, midrib turns light green near the top, sometimes red pigmentation appears	wide; green	transparent, brownish (dry)	3-6	female
Arrkiia	yellowish-green, sometimes with black or reddish brown stripes and spots on the petiole	narrow to medium; green	transparent, brownish (dry)	3-6	female
Gefetenna	dark purple, fading where it joins the petiole; colour of old midribs fades, leaving reddish and black spots on the midrib	narrow to medium; green	transparent, brownish (dry)	3-4	ambiguous (male/ female)
Lemmbo	yellowish-green with black spots or stripes on the petiole	narrow; green	transparent, brownish (dry)	3-5	female
Lochingii	dark red/reddish purple; on the petiole greenish pigments may occur, esp. on young plants	narrow; dark red/reddish purple; greenish pigments may occur, esp. on young plants	milky, sticky	3-8	male; rarely female
Mazziia	petiole greenish, with black stripes or patches; midrib reddish which becomes darker near the top	medium; green	transparent, brownish (dry)	4-7 (Welaita); 12-14 (Sura)	male
Sheelekuummiia	yellowish-green with black stripes or spots on the petiole	medium; green	transparent, brownish (dry)	3-6 years (Welaita); 7-14 (Gurage)	female
Suiitiia	green; petiole often pinkish; sometimes reddish pigments occur from the base of the leavers onwards	medium; green	reddish	3-7 (Welaita); 10-13 (Sura); 7-13 (Gurage)	female
Wannidiia	green petiole, turns reddish at the leaf bease or at the middle; old leaves yellowish	narrow to medium; green	transparent, brownish (dry)	5-9	male; sometimes female
AddisAbaba	turns reddish or brownish-reddish at the petiole onwards to the top	medium; green	transparent, brownish (dry)	3-7	not used in Addis Ababa

* Gender describes the plants' characteristics which farmers relate to sex, i.e. short/tall size, low/high fibre content, fragile/strong fibre, 'sweet'/bitter taste.

Appendix 1. continued.

Name of landrace	Human consumption	Household and farming	Traditional medicine (humans)	Traditional medicine (animals)
Agiiniia	ammicho preferred; in Gedeo only qocho		bulla is eaten after birth to strengthen the mother	
Arrkiia	ammicho, qocho and bulla of minor importance			
Gefetenna	good bulla	livestock feed: all parts during food shortages	bulla for broken bones and during pregnancy to improve growth of the foetus, after childbirth; ammicho to cure back injuries together with milk, butter and spices	leaves to strengthen weak animals and cows after delivery of calves; broken bones are covered with the chopped and boiled corm
Lemmbo	ammicho; qocho and bulla from the leaf sheathes and the corm	livestock feed; fibre		
Lochingii	ammicho; qocho and bulla from the leaf sheathes and the corm	whole plant as livestock feed; ornamental, protection against 'evil eye'	boiled fresh corm: contraception, abortions, to fasten the birth, hepatitis, malaria and AIDS to induce sweating; sap from the midrib: cold and coughing	whole plant to fasten the birth and to deliver the afterbirth
Mazziia	qocho and bulla from the leaf sheathes; qocho is due to its dark colour disliked, the corm has a bitter taste	long and strong fibres which are preferred to make ropes; dried leaf sheathes are storing and used for constructing houses etc.		
Sheele-kuummiia	ammicho; qocho and bulla of minor importance	mattresses from the dried leaf sheathe		
Suiitiia	ammicho preferred		diarrhoea: the chopped corm is roasted and eaten plain; broken bones: the corm is roasted in fire and eaten with milk; medical treatment was mentioned only in Welaita	
Wannidiia	qocho and bulla from the leaf sheathes and the corm	long and strong fibre, preferred		
Addis-Ababa		for special festivities a bread is wrapped with the leaves and baked		

Index